I0787953

China Semiconductor Technology International Conference 2011 (CSTIC 2011)

Editors:

H. Wu
Semiconductor Manufacturing International
Corporation
Shanghai, China

C. Claeys
imec
Leuven, Belgium

Y. Kuo
Texas A&M University
College Station, Texas, USA

K. Lai
IBM, Thomas J. Watson Research Center
Yorktown Heights, New York, USA

A. Philipossian
The University of Arizona
Tucson, Arizona, USA

T. Jiang
Maxim Integrated Products Inc.
Sunnydale, California, USA

S. Xiaoping
imec
Leuven, Belgium

Q. Lin
IBM, Thomas J. Watson Research Center
Yorktown Heights, New York, USA

D. Huang
Praxair
Danbury, Connecticut, USA

R. Huang
Peking University
Beijing, China

Y. Zhang
Yorktown Heights, New York, USA

R. Liu
Fudan University
Shanghia, China

P. Song
IBM, Thomas J. Watson Research Center
Yorktown Heights, New York, USA

Published by
The Electrochemical Society

65 South Main Street, Building D
Pennington, NJ 08534-2839, USA
tel 609 737 1902
fax 609 737 2743
www.electrochem.org

ECStransactions ™

Vol. 34 No. 1

Copyright 2011 by The Electrochemical Society.
All rights reserved.

This book has been registered with Copyright Clearance Center.
For further information, please contact the Copyright Clearance Center,
Salem, Massachusetts.

Published by:

The Electrochemical Society
65 South Main Street
Pennington, New Jersey 08534-2839, USA

Telephone 609.737.1902
Fax 609.737.2743
e-mail: ecs@electrochem.org
Web: www.electrochem.org

ISSN 1938-6737 (online)
ISSN 1938-5862 (print)

ISBN 978-1-56677-884-1 (CD-ROM)
ISBN 978-1-60768-234-9 (PDF)
ISBN 978-1-60768-235-6 (Soft-cover)

Printed in the United States of America.

Preface

On behalf of the conference committee and organizers, we would like to express our sincere thanks to all attendees for their active participation and significant contribution to one of the largest semiconductor conferences in China, CSTIC-2011 (China International Semiconductor Technology Conference). The 2011 conference will be the tenth international conference since ISTC was first initiated in 2001 and it has been constantly well organized, jointly, by The Electrochemical Society and SEMI, co-organized by CHTEC, and sponsored by IEEE, CEMIA, MRS and CSE.

Our mission is to provide a forum for world experts to discuss technologies, address the growing needs associated with silicon technology, and exchange their discoveries and solutions for current issues of high interest. We encourage collaboration, open discussion, and critical reviews at this conference. Furthermore, we hope that this conference will also provide collaborative opportunities for those who are interested in the semiconductor industry in Asia, particularly in China.

The conference committee consists of more than 100 world-class experts from high profile companies including, but not limited to Intel, IBM, SMIC, IMEC, Maxim, Applied Materials, TEL, Praxair, MEMC, Cabot Microelectronics, Dow, and also local companies and universities such as Anji, Gritek, Peking University, Fudan University and Zhejiang University. Guest speakers came from companies and institutions like those mentioned above, as well as renowned universities such as UC Berkeley, Stanford, Yale, Peking University, Fudan University, and other top universities around the world. We are proud of this prestigious volunteer team and feel honored by the distinguished speakers. With their contributions, we believe that we have prepared a world class international conference.

The conference offers more than 390 high quality presentations and posters in ten symposia to cover most of the aspects of semiconductor and solar technology. It is well known that the silicon technology development trend is primarily dependent of innovative materials application. In this volume, there are more than 50% presentations are new material and process related technologies, while the rest are related to device and process integration. As a part of low-carbon society, we have enhanced symposium X "Silicon Materials for Electronic and Photovoltaic Applications". The conference started with a plenary session before all the symposia, covering a critical review and high-level summary of the technology.

We have achieved a record number of paper submissions this year from all around the world. We've seen strong participation from the USA, Europe, areas in Asia, and other countries, and especially growing paper submission from China. The following table will provide you with a brief summary of the paper distribution in each symposium:

iii

Table 1 – Conference Symposia and Papers

Symposium	Total Presentation
Plenary Session	4
I - Design and Device Engineering	55
II - Lithography and Patterning	35
III - Dry &Wet Etch and Cleaning	38
IV - Thin Film Technology	42
V - CMP and Post-CMP Cleaning	40
VI - Materials and Process Integration for Device and Interconnection	32
VII - Packaging and Assembly	33
VIII - Metrology, Reliability and Testing	42
IX - Emerging Semiconductor Technologies	35
X - Silicon Materials for Electronic and Photovoltaic Applications	39
Total	395

There is little doubt that our conference would have been as successful without active participation from the audience and its readers. We appreciate your support and trust that this issue of *ECS Transactions* will be a worthwhile addition and reference to your collection.

Han-Ming Wu (SMIC), Conference Chairman
Allen Lu, President of SEMI China
Roque J. Calvo, Executive Director of ECS
CSTIC Committee
March 13, 2011, Shanghai, China

ECS Transactions, **Volume 34, Issue 1**
China Semiconductor Technology International Conference 2011 (CSTIC 2011)

Table of Contents

Preface   *iii*

Chapter 1
Design and Device Engineering

Embedded Non-Volatile Memory Technologies   3
    D. Shum (Infineon Technologies Taiwan Co. Ltd.)

A Novel High Programming Efficiency and Highly Scalable Flash Memory   9
Cell Based on Tunneling FET (TFET)
    S. Qin, P. Tang, Y. Cai, Q. Huang, Y. Tang, and R. Huang
    (Peking University)

Design of a 1-T Image Sensor by Simulation   17
    X. Liu, S. Zang, X. Lin, C. Cao, P. Wang, and D. Zhang
    (Fudan University)

Deposition of ZnO Films by Sputtering and its Resistive Switching Properties   25
    F. Wang, K. Zhang, B. Yang, L. Wang, and K. Song
    (Tianjin University of Technology)

Leakage Engineering Enabling PDSOI Ring Oscillators Operating in   31
Sub-100pA/μm I_{off} Regime
    Z. Ren (IBM Semiconductor Research & Development Center), J. Cai
    (IBM Research Division), R. R. Robison, B. Jagannathan
    (IBM Semiconductor Research & Development Center), D. Park, and
    T. H. Ning (IBM Research Division)

Ultra-Thin Body and BOX (UTBB) Device for Aggressive Scaling of
CMOS Technology
Q. Liu (STMicroelectronics), A. Yagishita (Toshiba), A. Kumar (IBM),
N. Loubet (STMicroelectronics), T. Yamamoto (RENESAS Electronics),
P. Kulkarni (IBM Research), F. Monsieur (STMicroelectronics),
A. Khakifirooz, S. Ponoth, K. Cheng, B. Haran (IBM), M. Vinet
(CEA-LETI), J. Cai (IBM Research Division), P. Khare, S. Monfray,
F. Boeuf (STMicroelectronics), S. Mehta, J. Kuss (IBM), E. Leobandung
(IBM Research), M. Hane (RENESAS Electronics), H. Bu
(IBM Research), K. Ishimaru (Toshiba), T. Skotnicki, W. Kleemeier
(STMicroelectronics), M. Takayanagi (Toshiba), T. Hook (IBM); M. Khare
(IBM Research), S. Luning (GLOBALFOUNDRIES), B. Doris
(IBM Research), and R. Sampson (STMicroelectronics)

37

Simulations of FDSOI CMOS with Sharing Contact between Source/Drain
and Back Gate
M. Xu, Q. Liang, H. Zhu, H. Yin, Z. Luo, D. Chen, and T. Ye
(Chinese Academy of Sciences)

43

Scaling MOSFETs with Self-aligned Super-Steep-Retrograded Halo (3SRH)
B. Wu, W. Xiao, H. Zhu, Q. Liang, H. Wu, H. Yin, Z. Luo
(Chinese Academy of Sciences), H. Yu (Nanyang Technological
University), D. Chen, and T. Ye (Chinese Academy of Sciences)

49

Electrostatic Discharge (ESD) Protection Challenges of Gate-All-Around
Nanowire Field-Effect Transistors
W. Liu, J. Liou (University of Central Florida), N. Singh, G. Lo
(Agency for Science, Technology and Research), J. Chung, and Y. Jeong
(Pohang University of Science and Technology)

55

Characterization of Random Telegraph Signal Effects for 0.18um Technology
Y. Ji, S. Dai, M. Wei, X. Lu, S. Zhang, and D. Xu
(Grace Semiconductor Manufacturing Corporation)

61

Effect of AlGaN Barrier Thickness on the Noise of AlGaN/GaN High
Electron Mobility Transistors
R. Yahyazadeh and Z. Hashempour
(Islamic Azad University of Khoy Branch)

67

Extraction and Analysis of Substrate Parameters in On-Chip Spiral Inductor
Model
X. Li (East China Normal University), Z. Ren
(Shanghai Integrated Circuits Research & Development Center), D. Chen,
and Y. Shi (East China Normal University)

75

vi

Opportunities and Challenges of FinFET as a Device Structure Candidate for 14nm Node CMOS Technology 81
T. Yamashita, V. Basker, T. Standaert, C. Yeh, J. Faltermeier (IBM Research), T. Yamamoto (RENESAS Electronics), C. Lin, A. Bryant (IBM Research), K. Maitra (GLOBALFOUNDRIES), P. Kulkarni, S. Kanakasabapathy (IBM Research), H. Sunamura (RENESAS Electronics), J. Wang, H. Jagannathan (IBM Research), A. Inada (RENESAS Electronics), J. Cho, R. Miller (GLOBALFOUNDRIES), B. Doris, V. Paruchuri, H. Bu, M. Khare, J. O'Neill, and E. Leobandung (IBM Research)

Structural effects of Channel Cross-Section on the Gate Capacitance of Silicon Nanowire Field-effect Transistors 87
S. Sato, K. Kakushima, P. Ahmet (Tokyo Institute of Technology), K. Ohmori (University of Tsukuba), K. Natori (Tokyo Institute of Technology), K. Yamada (University of Tsukuba), and H. Iwai (Tokyo Institute of Technology)

Process Impact and Design Optimization on the Soft Yield of 25nm FinFET SRAM Cells 93
M. Li, Q. Liang, H. Zhu, H. Zhong, D. Chen, and T. Ye (Chinese Academy of Sciences)

TiN/W/La$_2$O$_3$/Si High-k Gate Stack for EOT below 0.5nm 99
P. Ahmet, D. Kitayama, T. Kaneda, T. Suzuki, T. Koyanagi, M. Kouda, M. Mamatrishat, T. Kawanago, K. Kakushima, K. Tsutsui, A. Nishiyama, N. Sugii, K. Natori, T. Hattori, and H. Iwai (Tokyo Institute of Technology)

PMOS Source/Drain Extension Dopant Species effect on Device and SRAM Performance 103
J. Liu, J. Zhou, W. Wang, R. Guo, L. Zhang, Z. Shen, B. Wang, A. Zhou, H. Hao, J. Cui, and J. Ning (Semiconductor Manufacturing International Corporation)

A Novel Tunnel Oxide Based Tunnel FET 107
H. Wang, Z. Luo, H. Yin, H. Zhu, J. Liu, and Z. Zhu (Chinese Academy of Sciences)

STI CMP: Exploration of a Colloidal Silica Based Slurry System 113
P. Song, D. Yaoying, and J. Daw Sun (Anji Microelectronics (Shanghai) Co., Ltd.)

Linearity Improvement on MIM Capacitors 119
T. Chu, P. Yang, E. S. Kho, Y. Ang, and S. Tia (X-FAB Sarawak Sdn. Bdn.)

vii

Modeling of Electron Transport in III- Nitride Compound Semiconductors for   125
Low Field and Low Temperature Applications
  S. Chakrabarti, S. Gupta Chatterjee, D. Chattopadhyay, and S. Chatterjee
  (Techno India)

Fast Flexible Electronics Based on Printable Thin Mono-Crystalline Silicon   137
  Z. Ma, K. Zhang, J. Seo, H. Zhou, L. Sun, H. Yuan, G. Qin, H. Pang
  (University of Wisconsin Madison), and W. Zhou
  (University of Texas Arlington)

HHNEC 0.18um BCD Technology for High Density Power Integration   143
  Z. Shuai, Q. Wensheng, and D. Ke
  (Shanghai Hua Hong NEC Electronics Company, Limited)

Performance Improvement of Si-NC Memory Device by Using a Novel   149
Junction Assisted Programming Scheme
  D. Jiang, Z. Huo, M. Zhang, Q. Wang, J. Liu, Z. Yu, X. Yang,
  Y. Wang (Chinese Academy of Sciences), B. Zhang
  (Grace Semiconductor Manufacturing Corporation), J. Chen
  (Anhui University), and M. Liu (Chinese Academy of Sciences)

Dual Floating Gate Flash Cell Using Single Poly Processes   155
  X. Lin (Fudan University), L. Liu (Oriental Semiconductor Co. Ltd.),
  X. Liu, S. Zanga, C. Cao, P. Wang, and D. Zhang (Fudan University)

Anomalous Behaviors of Cubic GaInN Ternary Alloys   161
  N. Tit (United Arab Emirates University)

0.18um Scalable 7~45V pLDMOS for Smart Power Application   167
  Z. Liu, S. Tang, J. Shen, and C. Shao
  (Grace Semiconductor Manufacturing Corporation)

0.18 Micron BiCMOS Process with Novel Structure SiGeC HBT   173
  D. Liu, W. Qian, X. Chen, F. Chen, J. Hu, S. Xiao, Y. Wang, and T. Chiu
  (HHNEC)

Temperature Insensitive Clock Buffer and Its Application on Clock Tree   183
  M. Tie and X. Li (IBM Systems & Technology Group)

Low-Power Design of Double Edge-Triggered Static SOI D Flip-Flop   189
  W. Xing, J. Song, and D. Gang (Peking University)

Effects of Oxygen Flow Ratios and Annealing on TiO_x Deposited by   195
Reactive Magnetron Sputtering
  L. Wang, K. Zhang, Q. Wang, F. Wang, and X. Wei
  (Tianjin University of Technology)

Chapter 2
Lithography and Patterning

Robustness Enhancement in Optical Lithography: From Pixelated Mask
Optimization to Pixelated Source-Mask Optimization   203
N. Jia and E. Y. Lam (The University of Hong Kong)

Mask Synthesis for Aerial Image Fidelity in Optical Lithography Using a
Coarse-Grid-Approximation Level-Set Approach   209
Y. Peng, J. Zhang, Y. Wang, and Z. Yu (Tsinghua University)

A Fast OPC Algorithm for IC Layout Based on 1-D Cells after Optimization
of Gap Distribution   215
B. Lin, C. Xie, and Z. Shi (Zhejiang University)

Extension Use of Immersion Lithography for the 22nm Half-Pitch and
Beyond   223
R. Kanaya (Nikon Corporation)

Cymer LPP EUV Source System Development Status   231
*B. Lin (Cymer Southeast Asia Ltd.), B. La Fontaine, D. Brandt, and
N. Farrar (Cymer Inc.)*

Advanced Packaging Stepper for 300mm Wafer Process   237
Z. Chang and H. Ling (Shanghai Micro Electronics Equipment Co., Ltd.)

Foundry Efficiency Gains Through Common Photolithography Themes   243
*J. E. Lamb III, C. Chris, Z. Zhu, D. Drain, and D. Sullivan
(Brewer Science, Inc.)*

Use of DBARCs Beyond Implant   249
C. Washburn, J. A. Lowes, and A. Guerrero (Brewer Science, Inc.)

Development of Under Layer Material for EUV Lithography   257
*R. Sakamoto, B. Ho, N. Fujitani, T. Endo, and R. Ohnishi
(Nissan Chemical Industries, Ltd.)*

Evaluation of 193 nm Photoresist Material at Advanced Immersion Nodes   263
*J. Hao, Y. Xu, and C. Liu
(Semiconductor Manufacturing International Corporation)*

Limit of Line End Shortening Correction under Single Exposure in 193 nm
Immersion Lithography   269
*Q. Wu, Y. Xu, J. Hao, C. Liu, X. Shi, and Y. Gu
(Semiconductor Manufacturing International Corporation)*

248nm Process Is Capable for sub 0.09 um Groundrules 277
L. Wang, X. Guo, Y. Tong, H. Meng, B. Su, and S. Xiao
(Shanghai HuaHong NEC Electronic Company Ltd.)

Study to Transfer 0.11µm DRAM ArF Process to KrF Process in Litho 285
J. Liu, E. Yao, E. Fan, K. Chang, T. Lv, J. Zhang, L. Liang, J. Hong, and
M. Li (Semiconductor Manufacturing International Corporation)

Studying Photoresist Type for Sub-32nm Node Dense SRAM 2nd GT Layer 303
Y. Xu, J. Hao, C. Liu, X. Shi, Q. Wu, and Y. Gu
(Semiconductor Manufacturing International Corporation)

Chapter 3
Dry and Wet Etch and Cleaning

Selective Removal of High-k Dielectrics 311
D. Shamiryan and V. Paraschiv (imec)

Active Area Width and Topography Effects on Sub 45nm Poly Gate CD 319
M. Shen, X. Meng, Y. Huang, H. Zhang, S. Chang, K. Lee, and Y. Gu
(Semiconductor Manufacturing International Corporation)

Reverse Phase Solution for Mesa Chamber Uniformity Improvement 325
Q. Ge, Y. Huang, and X. Tang (Semiconductor Technology Group Applied
Materials China Globe Account)

Plasma Etch Challenges for Porous Low k Materials for 32nm and Beyond 329
C. Labelle (GLOBALFOUNDRIES), R. Srivastava
(GLOBALFOUNDRIES Singapore), J. C. Arnold, Y. Yin (IBM Research),
M. Ishikawa (Toshiba America Electronic Components, Inc.), Y. Mignot
(STMicroelectronics), H. Yusuff (IBM Microelectronics), J. Linville
(GLOBALFOUNDRIES), D. Horak, N. Fuller (IBM Research), R. Patz,
A. Darlak, K. Zhou, Y. Zhou, and J. Pender (Applied Materials)

Dry Etch Process Effects on Cu/low-k Dielectric Reliability for Advanced 335
CMOS Technologies
J. Zhou, W. Sun, H. Zhang, M. Hu, F. Li, X. Song, S. Chang, and K. Lee
(Semiconductor Manufacturing International Corporation)

New Al Post-Etch Residue Remover with Al Surface Passivation Function 343
J. C. Wei and M. Huang (DuPont Electronics and Communications)

WAT and VBD Distribution Improvement on Low-K Trench All-in-one Process — 349
J. Hendrianto, H. Zhijie (Lam Research Corporation), and A. Liu (Semiconductor Manufacturing International Corporation)

Clean Mode Al Etch Process Development for Defect Reduction — 355
F. Qiang, C. Huang, J. Hendrianto (Lam Research Corporation), J. Song, M. Lv, K. Wang, and C. Shi (Semiconductor Manufacturing International Corporation)

Dummy Poly Silicon Gate Removal by Wet Chemical Etching — 361
T. Young, H. Yin, Q. Xu, C. Zhao, J. Li, and D. Chen (Chinese Academy of Sciences)

Theoretical And Experimental Development Of Advanced Dopant-Sensitive Systems — 365
P. Zhang (Qingdao Feiyang Vocational & Technical College), L. Zhang (China Electronics Technology Group Corporation), Y. Ye (Nanjing University), and Y. Yang (China Electronics Technology Group Corporation)

Ultrapure Water-Related Problems and Waterless Cleaning Challenges — 371
T. Hattori (n/a)

Dry Etch Fin Patterning of a Sub-22nm Node SRAM Cell: EUV Lithography New Dry Etch Challenges — 377
E. Altamirano-Sanchez (imec), Y. Yamaguchi, J. Lindain (Lam Reseach), N. Horiguchi, M. Ercken, M. Demand, and W. Boullart (imec)

Effect of O_2/Ar Ratio on Etching of Diamond Films by MPCVD — 383
S. Wang, K. Zhang, Z. Taofeng, and J. Ren (Tianjin University of Technology)

Porous SiOCH Integration: Etch Challenges with a Trench First Metal Hard Mask Approach — 389
N. Possémé, T. David (CEA-LETI), T. Chevolleau, M. Darnon (CNRS-LTM), P. Brun, M. Guillermet (CEA-LETI), J. Oddou (ST Microelectronics), S. Barnola (CEA-LETI), F. Bailly, R. Bouyssou (ST Microelectronics), J. Ducote (CNRS-LTM), R. Hurand (CEA-LETI), C. Vérove (ST Microelectronics), and O. Joubert (CNRS-LTM)

Plasma Etching Parameters Impact To Low-k Damage — 395
J. Zhang, H. Pei, and L. Cheng (Lam Research Corporation)

Prevention of AlCu Line Galvanic Corrosion after Fluoride Containing
Stripper Cleaning: A Case Study   399
  V. Luo, J. Chang, K. Shi
  (Semiconductor Manufacturing International Corporation),
  B. Liu, L. Peng, A. Wang, and J. Sun
  (Anji Microelectronics (Shanghai) Co., Ltd.)

Highly Selective Etching Solutions for Advanced Logic Technologies   405
  X. Wang, H. Zhang, S. Chang, and K. Lee
  (Semiconductor Manufacturing International Corporation)

Discovering Practical Use of Sensor Wafers in CCP Reactors   409
  A. Milenin, M. Demand, W. Boullart (imec), and P. Arleo (KLA-Tencor)

Study on Silicon Sieve Holes Array for Future Lithography Application   415
  W. Si, M. Yin (Tsinghua University), J. Qin (Hunan University), and Z. Liu
  (Tsinghua University)

Controlled Etching of III-V Materials with Optical Emission Interferometry   421
(OEI)
  C. Johnson, D. Johnson, R. Westerman, D. Geerpuram, L. Martinez, and
  J. Plumhoff (Plasma-Therm LLC)

Low Silicon and SiGe Loss in High Dose Implant Resist Strip   427
  X. Meng, M. Shen, Y. Huang, H. Zhang, S. Chang, and K. Lee
  (Semiconductor Manufacturing International Corporation)

Effluent Management for Non-Oxidizing Plasma Strip Processes   433
  S. Luo, C. Waldfried, O. Escorcia, I. Berry, P. Geissbühler, A. Srivastava,
  and D. Roh (Axcelis Technologies, Inc.)

Wafer Backside Particle Reduction By Optimizing AC3 Coating for Poly   439
Etch Chamber
  B. Ma, W. Liu (Lam Research Co., Ltd), F. Niu, J. Xia
  (Semiconductor Manufacturing International Corporation), L. Cheng
  (Lam Research Co., Ltd), and K. Liang
  (Semiconductor Manufacturing International Corporation)

The Study of Dry Etching Process on Plasma Induced Damage in Cu   445
Interconnects Technology
  J. Zhou, H. Zhang, W. Sun, X. Wang, M. Hu, F. Li, L. Fu, S. Chang, and
  K. Lee (Semiconductor Manufacturing International Corporation)

xii

Chapter 4
Thin Film Technology

Selective Epitaxial Growth: Trends in a Modern Transistor Device Fabrication   455
A. Y. Hikavyy, W. Vanherle, J. Dekoster, L. Witters, T. Hoffmann, and R. Loo (imec)

Electron States at Interfaces of Semiconductors and Metals with Insulating Films   467
V. V. Afanas'ev, M. Houssa, and A. Stesmans (University of Leuven)

Atomic-Layer Deposition of Lutetium Aluminate Thin Films for Non-Volatile Memory Applications   473
C. Adelmann, J. Swerts, T. Conard, B. Brijs, A. Franquet (imec), A. Hardy (Hasselt University), H. Tielens, K. Opsomer, A. Moussa (imec), M. K. Van Bael (Hasselt University), M. Jurczak, J. A. Kittl, and S. Van Elshocht (imec)

Evolution of STI Gap Fill Technology   479
J. C. Chen, Y. Chen, R. Gao, C. Cheng, X. Li, G. Zhao (Applied Materials China), D. Chan, and T. Lee (Applied Materials)

Annealing Effect on the Electrical Properties of La_2O_3/InGaAs MOS Capacitors   483
T. Kanda, D. Zade (Tokyo Institute of Technology), Y. C. Lin (National Chiao-Tung University), K. Kakushima, P. Ahmet, K. Tsutsui, A. Nishiyama, N. Sugii (Tokyo Institute of Technology), E. Y. Chang (National Chiao-Tung University), K. Natori, T. Hattori, and H. Iwai (Tokyo Institute of Technology)

Metal Inserted Poly-Si Stacks with La_2O_3 Gate Dielectrics for Scaled EOT and V_{FB} Control by Oxygen Incorporation   489
T. Kawanago, K. Kakushima, P. Ahmet, K. Tsutsui, A. Nishiyama, N. Sugii, K. Natori, T. Hattori, and H. Iwai (Tokyo Institute of Technology)

Characteristics of HfSiAlON Gate Dielectric Prepared by Physical Vapor Deposition   495
G. Xu and Q. Xu (Chinese Academy of Sciences)

Deposition of VO_X Films by Reactive Sputtering and its Properties   503
X. Wei, K. Zhang, W. Fang, L. Wang, Y. Zhang, and K. Song (Tianjin University of Technology)

xiii

ALD Ru and its Application in DRAM MIM-Capacitors and Interconnect 509
   M. Schaekers, J. Swerts, L. Altimime, and Z. Tőkei (imec)

Evaluation of Metallization Options for Advanced Cu Interconnects 515
Application
   N. Jourdan, L. Carbonell, N. Heylen, J. Swerts, S. Armini, A. Caro,
   S. Demuynck, K. Croes, G. Beyer, Z. Tökei, S. Elshocht, and E. Vancoille
   (imec)

Fine Pitch Micro-Bump Interconnections for Advanced 3D Chip Stacking 523
   W. Zhang, P. Limaye, A. La Manna, E. Beyne, and P. Soussan (imec)

Temperature and Stress effects on IMC Behavior of Thin Film Cu-Al System 529
in Wire Bond
   X. Ming (CETC 58th Research Institute) and K. Fan
   (ASM Pacific Technology)

Review of Silicon Nanowire Oxidation 535
   X. Shi, R. Kurstjens, I. Vos, J. Everaert, and M. Schaekers (imec)

Effect of Film Thickness on Resistance Switching Characteristics for 541
Cu/NiO/Pt Structure
   Y. Zhang, K. Zhang, W. Fang, X. Wei, and J. Zhao
   (Tianjin University of Technology)

Optical Constants of ZnO Films 547
   B. Huang and H. Yang (Jinan University)

Influence of the Pressure on ZnO:Al Film Deposited by DC Magnetron 551
Reactive Sputtering
   S. Yu, H. Yang, B. Huang, J. Shi, and L. Zeng (Jinan University)

Study of the Electrical and Optical Properties of the Silicon Carbide Thin 557
Film
   R. Luo, H. Yang, B. Huang, and B. Y. Xu (Jinan University)

Influence of Vacuum-Annealing Temperature on the Properties of Direct 563
Current (DC) Magnetron Sputtered ZAO Thin Films
   J. Shi, H. Yang, B. Huang, B. Xu, and S. Yu (Jinan University)

Study of Phosphorus Out-Diffusion from High Density Plasma CVD 567
Phosphosilicate Glass Process
   L. Min, Z. Ying, and Q. Xu
   (Semiconductor Manufacturing International Corporation)

Effects of Substrate Temperature on Resistive Switching of TiO_X Thin Film   571
L. Wang, K. Zhang, W. Fang, and K. Song
(Tianjin University of Technology)

Electrical and Optical Properties of Zinc Oxide Thin Films Deposited by   577
Magnetron Sputtering
X. Ding and Y. Lai (Fuzhou University)

A Highly Conductive Bimodal Isotropic Conductive Adhesive and Its   583
Reliability
D. Li, H. Cui, S. Chen, Q. Fan, Z. Yuan (Shanghai University), L. Ye
(SHT Smart High Tech AB), and J. Liu (Shanghai University)

Chapter 5
CMP and Post-CMP Cleaning

Challenges and Mechanisms of CMP Slurries for 32nm and Beyond   591
H. Morinaga and K. Tamai (Fujimi Incorporated)

A Study on Optimized Conditioner for Soft Pad in Cu Barrier Removal   597
S. Yoon and J. H. Lee (Ehwa Diamond Ind. Co. Ltd.)

Overcome Challenges in TSV CMP via Slurry Formulation   603
K. Luo, C. Wang, J. Jing, and S. Xu
(Anji Microelectronics (Shanghai) Co., Ltd.)

Investigation on the Correlationship between Process Performances and   609
Composition of CMP Slurry Designed for GST Alloy Polishing
K. Pang (Anji Microelectronics (Shanghai) Co., Ltd.), F. Chen, L. Jiang,
M. Li, and M. Zhong
(Semiconductor Manufacturing International Corporation)

Fundamental Characterization Studies of Condensed Chemical Mechanical   615
Polishing Waste Slurry
Y. Yamada, M. Kawakubo, S. Watanabe, and T. Sugaya (Hitachi, Ltd.)

Correlation of Pad Topography, Friction Force and Removal Rate during   621
Tungsten Chemical Mechanical Planarization
Y. Sampurno, A. Rice, Y. Zhuang, and A. Philipossian
(The University of Arizona)

Tribological and Kinetical Analysis of Barrier Metal Polishing for Next 627
Generation Copper Interconnects
 R. Duyos Mateo, X. Gu, T. Nemoto, S. Sugawa (Tohoku University),
 Y. Zhuang (Araca Incorporated), Y. Sampurno, A. Philipossian
 (The University of Arizona), and T. Ohmi (Tohoku University)

Finite Element Analysis (FEA) of Pad Deformation Due to Diamond Disc 633
Conditioning in Chemical Mechanical Polishing (CMP)
 E. Baisie (North Carolina Agricultural & Technical State University),
 B. Lin (Tianjin University), X. Zhang (Seagate Technology), and Z. Li
 (North Carolina Agricultural & Technical State University)

Data Driven CMP Manufacturing Modeling for Process and Design 639
Optimization
 L. J. Song and V. Mehrotra (Ascertin LLC)

Ge- and III/V-CMP for Integration of High Mobility Channel Materials 647
 P. Ong, L. Witters, N. Waldron, and L. Leunissen (imec)

Advanced Direct-Polish Process on Organic Non-Porous Ultra Low-k 653
Fluorocarbon Dielectric on Cu Interconnects
 X. Gu, T. Nemoto, Y. Tomita, R. Duyos Mateo, A. Teramoto, S. Kuroki,
 S. Sugawa, and T. Ohmi (Tohoku University)

Effect of Slurry Application/Injection Methods and Polishing Conditions on 659
Bow Wave Characteristics
 X. Liao, Y. Sampurno, Y. Zhuang (The University of Arizona), F. Sudargho
 (Araca, Inc.), A. Rice, and A. Philipossian (The University of Arizona)

Evolution of Post CMP Cleaning Technology 665
 G. Banerjee (Air Products)

Cleaning Aspects of Novel Materials after CMP 671
 R. Vos (imec), M. Wada (Dainippon Screen Mfg. Co.), S. Arnauts,
 H. Takahashi, D. Cuypers, H. Struyf, and P. Mertens (imec)

Study on the Ring Type Crater Defect Reduction in Cu CMP Process 677
 J. Xu, P. Lin, C. Xing, P. Li, and Z. Ma
 (Semiconductor Manufacturing International Corporation)

New Application of Optical Endpoint System: In Situ Cu Residue Detection 683
 W. Zhang, X. Wang, C. Tan, S. Wang (Applied Materials China), W. Shen
 (Applied Materials USA), and G. Ge
 (Semiconductor Manufacturing International Corporation)

xvi

The Mechanism of Organic Base and Surfactant in Silicon Wafer CMP
Process
    L. Weiwei (Hebei University of Technology)

691

Modeling Copper Chemical Mechanical Polishing Processes Using Linear
System Method
    L. Wu and C. Yan (Lanzhou University of Technology)

699

Effect of pH on CMP of VOx Thin Films for RRAM
    Y. Liguo, K. Zhang, W. Fang, X. Wei, and Z. Taofeng
    (Tianjin University of Technology)

705

Study of Inhibition effects on Copper CMP Slurry Performance
    J. Jing (Anji Microelectronics (Shanghai) Co., Ltd.), Z. Ma, P. Li, C. Lu,
    P. Lin (Semiconductor Manufacturing International Corporation),
    J. Zhang, and X. Cai (Anji Microelectronics (Shanghai) Co., Ltd.)

711

Chapter 6
Materials and Process Integration for Device and Interconnection

SiON Gate Dielectric Optimization for NBTI Improvement
    Y. Chen, Y. He, W. Wang, R. Guo, Z. Tang, J. Liu, J. Wu, and J. Ju
    (Semiconductor Manufacturing International Corporation)

719

Analysis of the Temperature Dependence of Trap-Assisted-Tunneling in
Ge pFET Junctions
    M. Bargallo Gonzalez, G. Eneman, G. Wang, B. De Jaeger, E. Simoen,
    and C. Claeys (imec)

725

eSiGe Global and Micro Loading Effect Study in High Performance 45nm
CMOS Technology
    Y. He, H. Tu, J. Lin, H. Song, J. Wang, G. Ma, W. Xu, B. Ye, T. Yu, and
    J. Wu (Semiconductor Manufacturing International Corporation)

731

Investigation of Laser Spike Anneal Dwell Time and It's Compatibility with
Embedded-SiGe
    Y. He, Y. Chen, J. Lu, J. Wu, C. Xu, T. Yu
    (Semiconductor Manufacturing International Corporation),
    D. M. Owen, Y. Zhang, and S. Shetty (Ultratech)

737

A Robust Shallow Trench Isolation High Density Plasma Chemical Vapor
Deposition Void Free Process for 0.13μm CMOS Technology
    G. Ning, P. Lin, C. Xing, A. Bian, H. Zhao, and Y. Cao
    (Semiconductor Manufacturing International Corporation)

743

xvii

CMP-Less Planarization Technology with SOG/LTO Etchback for Low Cost 70nm Gate-Last Process
H. Yin, L. Men, T. Yang, G. Xu, Q. Xu, C. Zhao, and D. Chen (Chinese Academy of Sciences)
749

Etch and Wet Clean Challenges and Joint Optimization
B. Yen, J. Lin, C. Lee, M. Hegarty, and P. Loewenhardt (Lam Research Corporation)
755

Growth and Processing Defects in CMOS Homo- and Hetero-Epitaxy
E. Simoen, M. Bargallo Gonzalez, G. Eneman, E. Rosseel, A. Y. Hikavyy, D. Kobayashi, R. Loo, M. Caymax, and C. Claeys (imec)
761

Precise Control of Spike Anneal Process for Advanced CMOS
Z. Zhao, J. Tang, and G. Zhao (Applied Materials China)
769

Improving Copper Interconnect Reliability via Ta/Ti Based Barrier
X. Hu, P. Lin, J. Ma, J. Jiang, and P. He (Semiconductor Manufacturing International Corporation)
775

Glue Layer Study of Inter Via between Cu and Al Metal Lines
J. Chen, C. Qiao, L. Yang, and K. Chang (Semiconductor Manufacturing International Corporation)
781

The Influence of The SIN Cap Process on The Voltage Breakdown and Electromigration Performance of Dual Damascene Cu Interconnects
Y. Cao, C. Xing, N. Xu, H. Zhou, A. Bian, and P. Lin (Semiconductor Manufacturing International Corporation)
787

Effect of RF Power on Carbon Nanotubes Synthesized at Low Temperature by RF PECVD
X. Lin, K. Zhang, K. Hu, X. Qiang, and S. Wang (Tianjin University of Technology)
793

Improving Yield with High-Performance Cables
P. Warren (W. L. Gore & Associates)
799

Study on the Reliability of Fast Curing Isotropic Conductive Adhesive
W. Du, H. Cui, S. Chen, Z. Yuan (Shanghai University), L. Ye (SHT Smart High Tech AB), and J. Liu (Shanghai University)
805

The Effect of Functionalized Silver on Rheological and Electrical Properties of Conductive Adhesives
Q. Fan, H. Cui, C. Fu, D. Li, X. Tang, Z. Yuan (Shanghai University), L. Ye (SHT Smart High Tech AB), and J. Liu (Shanghai University)
811

Chapter 7
Packaging and Assembly

Microstructural Evolution of Sn3.0Ag0.5Cu3.0Bi0.05Cr/Cu Solder Joints
During Thermal Aging and Its Effects on Mechanical Properties 819
F. Lin, W. Bi, G. Ju, and X. Wei (Shanghai University)

Study of EMC for Cu Bonding Wire Application 825
T. Takeda, H. Seki, S. Itoh, and S. Zenbutsu (Sumitomo Bakelite Co., Ltd.)

Corrosion of Gold and Copper Ball Bonds 831
*C. D. Breach (ProMat Consultants), H. Ng, T. Lee (ITE College Central),
and R. Holliday (World Gold Council)*

Cost-effective Use of Gold Wire in Semiconductor Packaging 843
*C. J. Vath III (ComSol Consulting Pte. Ltd.) and R. Holliday
(World Gold Council)*

Copper Wire Bonding in High Volume Manufacturing 857
B. K. Appelt, A. Tseng, Y. Lai, and C. Chen (ASE Group)

MUF Technology Development for SiP Module 865
*Y. Kweon, J. Ha, K. Kim, M. Jang, J. Doh, C. Lee, and D. Yoo
(Samsung Electro-Mechanics Co.)*

Multi Beam Grooving and Full Cut Laser Dicing of IC Wafers 873
*J. V. Borkulo and R. Hendriks
(Advanced Laser Separation International NV)*

Advanced Bump Structure for Improving the Board Level Characteristics of 879
WLCSP
*C. Lee, J. Choi, J. Kim, S. Choi, D. Yoo, S. Park, and Y. Kweon
(Samsung Electro-Mechanics Co.)*

Plasma Cleaning Effect on Automotive Devices 887
Y. P. Chew and T. Aw (Infineon Technologies Sdn Bhd)

Packaging Issues for High-Voltage Power Electronic Modules 893
*S. S. Ang, T. Evans, J. Zhou, K. Schirmer, H. Zhang, B. Rowden, J. Balda,
and A. Mantooth (University of Arkansas)*

xix

Chapter 8
Metrology, Reliability and Testing

IDDQ Test Practice in Nanotechnologies   901
    S. X. Ye, C. Shen, Z. Liu, and Q. Liyun (Availink, Inc.)

Cost-Effective and Accurate Solution for Jitter Performance Test in   907
High-Speed Serial Links
    M. Lu (Verigy)

Plasma Etching for Failure Analysis of Integrated Circuit Packages   913
    J. Tang, J. Schelen, and C. Beenakker (Delft University of Technology)

Process Optimization of Contact Module in NOR Flash Using High   919
Resolution e-Beam Inspection
    H. C. Liao, C. L. Hung, T. Luoh, L. Yang, T. Yang, K. Chen, and C. Lu
    (Macronix International Co., Ltd.)

Verification of Systematic Defects Using e-Beam Defect Review System   925
    T. Luoh, L. Yang, T. Yang, K. Chen, and C. Lu
    (Macronix International Co., Ltd.)

Determining Coherence length of X-ray Beam Utilizing Line Grating   931
Structures
    H. Lee, C. L. Soles, and W. Wu
    (National Institute of Standards and Technology)

TSV/3DIC Profile Metrology Based on Infrared Microscope Image   937
    J. Tang (Southern Taiwan University), Y. Lay, L. Chen, and L. Lin
    (National Cheng Kung University)

Endpoint Detection in Plasma Etching Using Principal Component Analysis   943
and Expanded Hidden Markov Model
    M. Kim, S. Kim, S. Zhao, S. Hong, and S. Han (Myongji University)

Improvement of In-line SCD Metrology on BEOL Copper CMP Erosion   949
Layers for 65nm Technology Node Logic Production Application
    C. Rong, Z. Wang, Z. Yin
    (Semiconductor Manufacturing International Corporation),
    Z. Tan (KLA-Tencor FaST Division), and L. Zhao
    (KLA-Tencor China)

Spectral Sensitivity Analysis of OCD Based on Muller Matrix Formulism   955
    S. Yaoming, Z. Zhensheng, L. Guoxiang, L. Zhijun, and X. Yiping
    (Raintree Scientific Instruments (Shanghai) Corp.)

A Method to Determine Process Capability Cpk and Corresponding 961
Percentage of Non-Conforming for Non-Normally Distributed and Limited
Production Data
    S. F. Yang (Semiconductor Manufacturing International Corporation)

High Voltage Device Negative Bias Temperature Instability Improvement 967
with Different Process Conditions
    P. Sim, S. Koo, and D. Pal (X-FAB Sarawak Sdn. Bhd.)

Study The Mixed-Mode Delamination of The Epoxy/Cu Interface 973
    Y. Liu and J. Wang (Fudan University)

LDMOS Thermal SOA Investigation of a Novel 800V Multiple RESURF 979
with Linear P-top Rings
    A. P. Herlambang, G. Sheu (Asia University), Y. Guo
    (Nanjing University of Posts and Telecommunications), and H. Wasisto
    (Asia University)

Investigation of Lateral Die Crack Failure at Reliability Test 985
    Y. Soh, C. Tan, X. Chen (Infineon Technologies (Wuxi) Co. Ltd.),
    K. Chua (Infineon Technologies (Malaysia) Sdn. Bhd.), R. Du, Y. Xi
    (Infineon Technologies (Wuxi) Co. Ltd.), and T. Lim
    (Infineon Technologies (Malaysia) Sdn. Bhd.)

Study on the Reliability of Nano-Structured Polymer-Metal Composite for 991
Thermal Interface Material
    L. Zhang, X. Luo, X. Lu, and J. Liu (Shanghai University)

Failure Mechanism and Testing of PCB Pad Cratering 997
    D. Xie (Flextronics International USA), M. Cai, B. Wu
    (Flextronics Manufacturing Zhuhai), D. Geiger, D. Shangguan
    (Flextronics International USA), and I. Martin (Flextronics)

Chapter 9
Emerging Semiconductor Technologies

FPGA Design with Double-Gate Carbon Nanotube Transistors 1005
    M. Ben Jamaa, P. Gaillardon (Commissariat a l'Energie Atomique),
    S. Frégonèse (Université Bordeaux), M. De Marchi, G. De Micheli
    (Ecole Polytechnique Federale de Lausanne), T. Zimmer
    (Université Bordeaux), I. O'Connor
    (Institut des Nanotechnologies de Lyon), and F. Clermidy
    (Commissariat a l'Energie Atomique)

xxi

Three-Dimensional (3D) Integration Technology 1011
  T. Ohba (The University of Tokyo)

Electrical Quality of III-V/Oxide Interfaces: Good Enough for MOSFET 1017
Devices
  G. Brammertz, A. Alian, H. Lin, L. Nyns, S. Sioncke, C. Merckling,
  W. Wang, M. Caymax, and T. Hoffmann (imec)

Low Temperature Bonding with Thin Wafers for 3D Integration 1023
  T. Matthias (EV Group), B. Kim (EV Group Inc.), P. Kettner,
  M. Wimplinger, and P. Lindner (EV Group)

Vertical LED with Diamond-Like Carbon Interface for High-Power 1029
Illumination
  J. C. Sung, K. Kan, and M. Sung (SinoDiamond LED)

Alumina Abrasives for Sapphire Substrate Polishing 1035
  D. Merricks (Ferro Electronic Materials)

Experimental and Modeling on Atomic Layer Deposition Al_2O_3/n-InAs 1041
Metal-Oxide-Semiconductor Capacitors with Various Surface Treatments
  H. Trinh, E. Chang (National Chiao Tung University), G. Brammertz
  (imec), C. Lu, H. Nguyen, and B. Tran (National Chiao Tung University)

Effects of Surface Pretreatments on p-GaN/GZO Contact by rf Magnetron 1047
Sputter
  W. Wang, X. Li, J. Zhang, and J. Zhang (Shanghai University)

A Phase Change Memory Device Fabrication Technology Using $Si_2Sb_2Te_6$ 1053
for Low Power Consumption Application
  Y. Li (Chinese Academy of Sciences), X. Wan
  (Semiconductor Manufacturing International Corporation),
  Z. Song (Chinese Academy of Sciences), J. Xie
  (Semiconductor Manufacturing International Corporation),
  B. Chenc (Silicon Storage Technology, Incorporated),
  B. Liu (Chinese Academy of Sciences), G. Wu, N. Zhu
  (Semiconductor Manufacturing International Corporation),
  M. Zhong (Chinese Academy of Sciences), J. Xu
  (Semiconductor Manufacturing International Corporation), and
  Y. Chen (Chinese Academy of Sciences)

Smart Systems 1059
  T. Gessner, M. Vogel, T. Otto, S. Schulz, and R. Baumann
  (Fraunhofer Institute for Electronic Nano Systems)

xxii

Electrical Characterization of the MOS (Metal-Oxide-Semiconductor) System: High Mobility Substrates 1065
D. Lin, G. Brammertz, S. Sioncke, L. Nyns, A. Alian, W. Wang, M. Heyns, M. Caymax, and T. Hoffmann (imec)

Characterization and Optical Properties of CdS Thin Films Grown by Chemical Bath Deposition 1071
W. Zhang and S. Cheng (Fuzhou University)

Electroluminescence of End-Capped Poly[9,9-di-(2'-ethylhexyl)fluorenyl-2,7-diyl] Blended with F8BT 1077
Q. Zhang (University of Electronic Science and Technology of China) and S. Zhang (Xihua University)

Enhancement of Luminance via Blending F8BT with Tetraphenyldiaminobiphenyl-Containing Hole Transport Polymer 1087
Q. Zhang (University of Electronic Science and Technology of China)

Chapter 10
Silicon Materials for Electronic and Photovoltaic Applications

Improvements on the Uniformity of a-Si Solar Thin Films by Using Auxiliary Magnetic Field 1097
L. C. Hu, Y. P. Chen, J. Chang, J. J. Lee, I. Chen, and T. T. Li (National Central University)

Hydrogenated Silicon Thin Film and Solar Cell Prepared by Electron Cyclotron Resonance Chemical Vapor Deposition Method 1103
C. Lee, J. Chang, Y. Chu, C. Lien, I. Chen, and T. Li (National Central University)

Properties of Multicrystalline Silicon Wafers Based on UMG Material 1109
T. Jiang, X. Yu, X. Li, X. Gu, P. Wang, and D. Yang (Zhejiang University)

Defect Evaluation by Photoluminescence for Uniaxially Strained Si-On-Insulator 1117
D. Wang, K. Yamamoto, H. Gao, H. Yang, and H. Nakashima (Kyushu University)

Effects of Transverse Magnetic Field on Thermal Fluctuations in the Melt of a Cz-Si Crystal Growth 1123
X. Liu, L. Liu, and Y. Wang (Xi'an Jiaotong University)

xxiii

Light Trapping for High Efficiency Heterojunction Crystalline Si Solar Cells   1129
   Q. Wang, Y. Xu, E. Iwaniczko, and M. Page
   (National Renewable Energy Laboratory)

Fabrication and Quantum Confinement Investigation of Ge Multiple   1135
Quantum Wells with Si_3N_4 Barriers
   J. Chen, S. Lee, and S. Huang (University of New South Wales)

Structural and Optical Properties of Porous SiGe/Si Multilayer Films   1145
   B. Zhou, X. Li (Minjiang University), S. W. Pan, S. Y. Chen, and C. Li
   (Xiamen University)

On the Impact of Heavy Doping on Grown-In Defects in Czochralski-Grown   1151
Silicon
   X. Zhang, W. Xu, J. Chen, X. Ma, D. Yang (Zhejiang University), L. Gong,
   D. Tian (QL Electronics), and J. Vanhellemont (Ghent University)

The Influence Of Silicon Orientation On Surface Blistering Behaviors for   1159
Molecular Hydrogen Ion Implantation
   Y. Hsiao, J. Liang (National Tsing Hua University), and C. Lin
   (National Hsinchu University of Education)

Very High Deposition Rate of a-Si:H Thin Films by ECRCVD   1165
   H. F. Chiu, Y. S. Chang (National Tsing Hua University), J. Y. Wu,
   Y. S. Li, J. Chang, C. C. Lee, I. Chen (National Central University),
   C. C. Su (Chung-Shan Institute of Science & Technology), and T. T. Li
   (National Central University)

Author Index

xxiv

Facts about ECS

The Electrochemical Society (ECS) is an international, nonprofit, scientific, educational organization founded for the advancement of the theory and practice of electrochemistry, electrothermics, electronics, and allied subjects. The Society was founded in Philadelphia in 1902 and incorporated in 1930. There are currently over 7,000 scientists and engineers from more than 70 countries who hold individual membership; the Society is also supported by more than 100 corporations through Corporate Memberships.

The technical activities of the Society are carried on by Divisions. Sections of the Society have been organized in a number of cities and regions. Major international meetings of the Society are held in the spring and fall of each year. At these meetings, the Divisions and Groups hold general sessions and sponsor symposia on specialized subjects.

The Society has an active publications program that includes the following.

Journal of The Electrochemical Society — JES is the peer-reviewed leader in the field of electrochemical and solid-state science and technology. Articles are posted online as soon as they become available for publication. This archival journal is also available in a paper edition, published monthly following electronic publication.

Electrochemical and Solid-State Letters — ESL is the first and only rapid-publication electronic journal covering the same technical areas as JES. Articles are posted online as soon as they become available for publication. This peer-reviewed, archival journal is also available in a paper edition, published monthly following electronic publication. It is a joint publication of ECS and the IEEE Electron Devices Society.

Interface — *Interface* is ECS's quarterly news magazine. It provides a forum for the lively exchange of ideas and news among members of ECS and the international scientific community at large. Published online (with free access to all) and in paper, issues highlight special features on the state of electrochemical and solid-state science and technology. The paper edition is automatically sent to all ECS members.

Meeting Abstracts (formerly Extended Abstracts) — Abstracts of the technical papers presented at the spring and fall meetings of the Society are published on CD-ROM.

ECS Transactions — This online database provides access to full-text articles presented at ECS and ECS-sponsored meetings. Content is available through individual articles, or as collections of articles representing entire symposia.

Monograph Volumes — The Society sponsors the publication of hardbound monograph volumes, which provide authoritative accounts of specific topics in electrochemistry, solid-state science, and related disciplines.

For more information on these and other Society activities, visit the ECS website:

www.electrochem.org

A Highly Conductive Bimodal Isotropic Conductive Adhesive and Its Reliability

Dongsheng Li[a], Huiwang Cui[a], Si Chen[a,c], Qiong Fan[a], Zhichao Yuan[b], Lilei Ye[d] and Johan Liu[a, c]

a) Key Laboratory of Advanced Display and System Applications, Ministry of Education and SMIT Center, School of Mechatronics Engineering and Automation, Shanghai University, Shanghai 200072, China
b) Shang Da Rui Hu Microsystem Integration Technology Co.Ltd, Room 101, Science &Technology Building, ,Shanghai University, No.149 Yanchang Road, Shanghai 200072,China
c) SMIT Center, Department of Microtechnology and Nanoscience, Chalmers University of Technology SE-412 96 Gothenburg, Sweden
d) SHT Smart High Tech AB, Fysikgränd 3, 41296 Gothenburg, SWEDEN

In this paper, micro silver flakes and micro spherical particles were incorporated into the matrix resin of isotropic conductive adhesives (ICAs). Their electrical properties were investigated. The total weight ratio of silver fillers was kept at 75 wt% for all samples. When the content of micro spherical particles was 8 wt%, the bulk resistivity of the bimodal ICAs reduced dramatically to as low as 1.26×10^{-4} $\Omega\cdot$cm and its viscosity was 24,289 cP under 5rpm at 25^0C. Scanning electronic microscopy (SEM) images of the bimodal ICAs showed silver fillers well distributed in the matrix resin. In addition, the lap shear strength of different metal surfaces, and the bulk resistivity shifts during aging time under 85^0C/85% RH for more than 500 hours were also measured. The results showed that bulk resistivity shifts of bimodal ICAs remained stable after further cured and the bond strength on the copper surface was the greatest among the three metal surfaces tested.

Introduction

Isotropic conductive adhesives (ICAs) have been commonly used as interconnect material and a lead-free alternative in advanced packaging systems, owing to their environmental and technical benefits over traditional tin/lead (Sn/Pb) solders [1-2]. However, compared with Sn/Pb solder, two main factors that limit the development of conductive adhesive are high resistivity and poor mechanical behavior [3-5].

Most ICAs are filled with silver fillers which have high electrical and thermal conductivity at room temperature. More importantly, silver can be conductive even after oxidation. In order to meet the requirements of high electrical conductivity, high loading of silver fillers needs to be added in the formulation of ICAs. However, high filler loading will lead to the increased viscosity, increased cost and a decrease in shear strength of ICAs, while decreased filler loading often results in poor electrical properties.

In this paper, we developed some high conductive ICAs that incorporated both micron-sized silver flakes and micron-sized silver spherical particles into the matrix resin of ICAs. The appropriate loading of silver spherical particles was beneficial to the interconnection between silver flakes. The lap shear strength on different metal surfaces, and the bulk resistivity shifts of bimodal ICAs with the lowest bulk resistivity under 85 ^{0}C/85% RH were also investigated and discussed.

Experiment

Materials

A matrix resin named SHT6 (Supplied by SHT Smart High-Tech AB), containing epoxy resin, silane coupling agent, curing agent and some reactive diluents to reduce viscosity, was selected as the binder matrix for the bimodal ICAs. In this paper, we chose two kinds of micron-sized silver fillers to introduce into SHT6, one was 6-8 μm and the other was 1.5-2 μm. They were both supplied by Guangzhou Litop Non-ferrous Metals Co., Ltd. The matrix and silver fillers were strong-stirred in a planetary mixer (MIX500D SLOPE) for 30 min to scatter the silver fillers uniformly through the resin, and then cured in a pre-heated oven at 150^{0}C for 1 hour. The weight ratio of all samples is shown in table I.

TABLE I. Weight Ratio of all Samples.

Sample ID	SHT6	Ag flakes (6-8μm)	Ag spherical particles (1.5-2μm)
SM	25wt%	75wt%	0wt%
BM-1	25wt%	70wt%	5wt%
BM-2	25wt%	67wt%	8wt%
BM-3	25wt%	63wt%	12wt%
BM-4	25wt%	55wt%	20wt%
BM-5	25wt%	37.5wt%	37.5wt%

* Single modal and bimodal ICAs will be abbreviated as SM and BM, respectively.

Bulk Resistivity Measurements

Bulk resistivity test samples were prepared based on the GJB548A-5011 standard. The procedures were as follows: two parallel strips of adhesive tape were applied onto a pre-cleaned standard 25.4 mm by 76.2 mm glass slide. The ICA was then squeezed into the space between the two strips of tape. The tape was removed and the glass slide put in a pre-heated oven at 150^{0}C for 1 h. After curing, the samples were taken out from the oven and stored at room temperature for 1 hour before the bulk resistivity test. The in-plane bulk resistivity was measured by means of standard In-Line four-probe method

(Figure 1). A constant current of 0.2 A was applied to the samples by DC Power Supply. The voltage was measured by a multi-meter (True RMS Multimeter187). The width and thickness of the samples were measured by vernier caliper and micrometer, respectively.

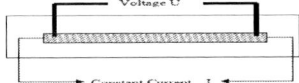

Fig. 1 Schematic of 4-probe method

The bulk resistivity was obtained by the following equation:

$$\text{Bulk resistivity}(\Omega \bullet cm) = \frac{U}{I} \times \frac{wt}{L}$$ [1]

Where U and I are voltage and constant current, L=2.54cm, w and t are the width and thickness of the sample. Five specimens for each ICA were tested. Average bulk resistivity and standard deviation for each sample were calculated and reported in figure 2. From this figure we can see, BM-2 ICA had the minimum bulk resistivity 1.26×10^{-4} $\Omega \cdot$cm. Next we carried out some reliability tests on BM-2 ICA.

Bulk resistivity shifts during aging. Bulk resistivity shifts of BM-2 ICA during 85^{0}C /85%RH aging were studied. Five specimens were tested for each sample. The bulk resistivity of each specimen was recorded periodically and the results are shown in Figure 3. From this figure we can see that the bulk resistivity shifts of BM-2 ICA decreased rapidly at the beginning of the aging time and then remained stable in the late stage. The decline of bulk resistivity should be due to further cure of the ICAs [6].

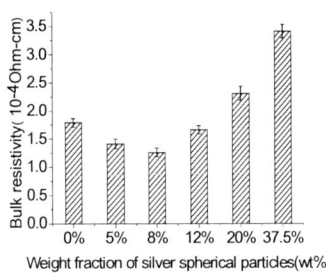
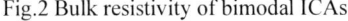
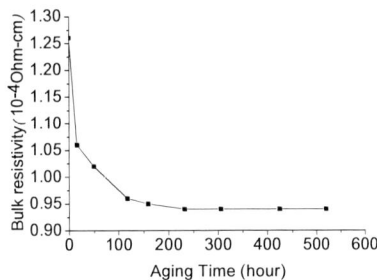

Fig.2 Bulk resistivity of bimodal ICAs   Fig.3 Bulk Resistance Shifts

Measurement of lap shear strength

Compared with traditional Sn/Pb solders, another weakness of conductive adhesives is low shear strength. Shear strength is affected by many factors such as types of epoxy matrix, metallization of pads, loading of fillers, and adhesive thickness [7~8]. In this paper, a lap shear strength of BM-2 ICA was processed on an electronic universal testing machine (XLD-1000D) at room temperature. A shear rate of 5 mm/min was used for the shear tests. The thickness of conductive adhesives was guaranteed at 1 mm by self-made mold. The shear strength can be calculated by the following equation:

$$\tau = \frac{P}{LB} \tag{2}$$

Where L=8 mm and B=8 mm are the length and width of bonding part respectively. P (N) is the max-force applied to break off the bonding parts. Five specimens were tested for each mental type on FR-4 PCB. Average shear strength for each sample was calculated and summarized in Figure 4. The average lap shear strength on the copper surface was 12.2 MPa, the largest among Cu, Au/Ni/Cu and Sn/Cu surfaces.

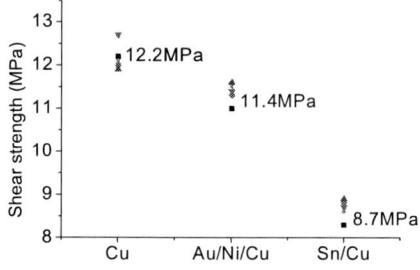

Fig.4 Lap shear strength of BM-2 ICA with different mental types on PCB

Rheology studies

Viscosity is one of the important process performances of conductive adhesives. The viscosity of the BM-2 ICA, according to GB7124-86, was tested by HBDV-II+Pro Viscometer under the spindle CPE-52 with conductive adhesive volume was 0.5 ml. All measurements were carried out at 25^0C. The results based on the lowest value of three tests are given in Table II. BM-2 ICA had medium viscosity and high thixotropic index, which was suitable for printing process.

TABLE II.   The Rotational Viscosity of BM-2 ICA.

Viscosity cP @0.5RPM	Viscosity cP @5RPM	Thixotropic index
125,000	24,289	5.15

Morphology studies

To further understand the bulk resistivity results obtained in the present work, morphology studies of three samples were observed by SEM, magnified 4000 times. Figure 5 well illustrates the conductive mechanism of bimodal ICAs filled with micron-sized silver flakes and micron- sized silver spherical particles. The BM-2 ICA showed a very good uniform distribution in the matrix.

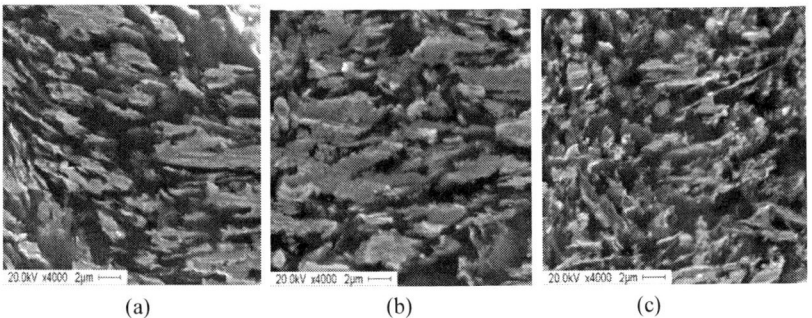

(a)   (b)   (c)

Fig.5 SEM cross-sectional view of bimodal ICAs with
(a) only silver flakes (b) silver flakes and 8 wt% silver spheres (c) silver flakes and 37.5 wt% silver spheres

Results and Discussion

The total weight ratio of silver filler was kept at 75 wt% and when the content of micron silver spherical particles was 8 wt%, the bulk resistivity of the bimodal ICAs was the lowest, at 1.26×10^{-4} $\Omega \cdot$cm, and its viscosity was 24,289 cP under 5rpm at 25^{0}C. From SEM cross-sectional views of the bimodal ICAs, we see that an appropriate loading of micron-sized silver spherical particles filled the space and increased the contact area between silver flakes. However, adding more silver spherical particles caused them to occupy the position of silver flakes and decrease the contact area of silver flakes and then increase the bulk resistivity. Under 85^{0}C/85% RH for more than 500 h, the bulk resistivity shifts of bimodal ICAs remained stable after being further cured. The lap shear strength of ICAs on the copper surface was the largest among the three metal surfaces tested. More reliability tests, such as contact resistance shifts and moisture absorption under $85^0C/85\%$ RH aging will be studied in the future.

Acknowledgments

This work was supported by the STC Torch program, contract no: 0903H195300, EU programs "Thema-CNT", "Mercure" and Nanopack. This work was also carried out within the Sustainable Production Initiative and the Production Area of Advance at Chalmers. This support is gratefully acknowledged. The authors are also grateful for the support of the Swedish National Science Foundation under the project "Nanointerconnect" (621-2007-4660) and the Vinnova Program on "Designade Material" through the contract No. 2009-03230.

References

1. Zwolinski, M., *IEEE Trans-CPMT Part C, Vol.* 19, No. 4 (1996), pp. 241-250.
2. Detert, M. & Herzog, Th., *Proc 49th Electronic Components and Technology Conf*, San Diego, CA,(1999).
3. Kotthaus S, Günther BH, Hang R, Schafer H., Study of Isotropically Conductive Bondings Filled with Aggregates of Nano-Sized Ag-Particles. *IEEE Trans Component Packag Technol 1997*; 20(1):15–20.
4. Yang R, Lu D, Wong CP, A study of impact performance of conductive adhesives. *Int. J. Adhes and Adhes 2004*; 24:449–453.
5. Ye L, Lai Z, Johan L, Tholen, A., Effect of Ag particle size on electrical conductivity of isotropically conductive adhesives. *IEEE Trans Electron Pack Manuf 1999*; 22(4):299–302.
6. Liu J, Lai Z, Kristiansen H and Khoo C, Overview of Conductive Adhesive Joining Technology in Electronics Packaging Applications. *Adhesive Joining and Coating Technology in Electronics Manufacturing, 1998*. PP.1-18(1998)
7. S.G. Prolongo and A. Urena, Effect of surface pre-treatment on the adhesive strength of epoxy–aluminium joints. *J. Adhes and Adhes 2009*; 29(1):23-31
8. R.Kahraman, Mehmet Sunar BekirYilbas, Influence of adhesive thickness and filler content on the mechanical performance of aluminum single-lap joints bonded with aluminum powder filled epoxy adhesive, *J. Mater. Process. Tech.205(2008)*:183-189

CHAPTER 5

CMP AND POST-CMP CLEANING

Challenges and Mechanisms of CMP Slurries for 32nm and Beyond

Hitoshi Morinaga and Kazusei Tamai

Fujimi Incorporated, 1-8, Techno Plaza, Kakamigahara, Gifu Pref. 509-0109, Japan

CMP slurries for 32nm and beyond are required to effectively planarize surfaces with fine and delicate device patterns without causing any nanodefects, without contributing nanocontaminants, and without generating atomic-level-roughness. Key solutions to these challenges are material removal using homogeneous and efficient mechanical action, surface protection using appropriate additives, and thorough contamination control (from large particles to dissolved gasses and light). The important parameter for mechanical action with abrasives is friction energy. Key factors to increase friction energy are: i) the number of active particles, ii) the contact area of each particle, and iii) the friction coefficient of each particle. These factors can be controlled with size, shape, morphology or surface charge of abrasive particles. Controlling abrasive shape and surface charge are especially important to achieve higher removal rate.

Introduction

Chemical Mechanical Polishing (CMP) processes are required to achieve i) planarized surfaces that are ii) defect free and , iii) surface roughness free and , iv) contamination free, and with v) higher productivity (Figure 1). These requirements by the industry are becoming more severe along with further integration and cost-reduction of semiconductor devices. CMP technology, therefore, must innovate to meet these ever-increasing requirements. Technology innovation is facilitated by a correct understanding of underlying mechanisms. This paper will review the challenges and mechanisms of CMP slurries for 32nm and beyond.

Challenges of CMP Technology for 32nm and Beyond

CMP slurries, use abrasive particles such as colloidal silica, ceria or alumina particles with a particle size of about 30nm (range of 10-50nm). This typical particle size has not changed much since the 350nm device generation even though the device pattern size has been reduced to one-tenth of the size. The 32-22nm generation is a major turning point in terms of nanodefects or nanocontamination; also the the size of abrasive particles becomes critical. Various new materials have been introduced since the 90nm generation. Notable in CMP processes in the 32nm generation and beyond are: i) An Al damascene gate was adopted in FEOL (Front End Of Line), ii) ULK (Ultra Low-k) or new barrier materials including CuMn were adopted in BEOL (Back End Of Line), and iii) Coexisting Cu and Si on the surfaces are being introduced in TSV (Through Silicon Via) processes. Al damascene gate process is especially critical because this process must

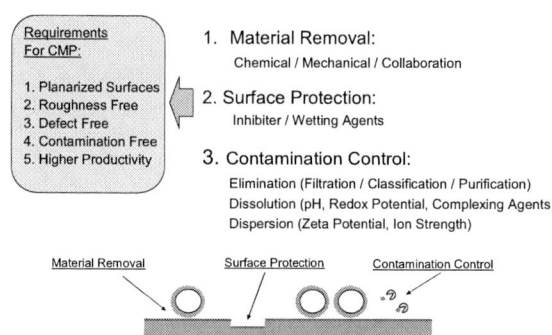

Figure 1. Requirements and Essential Functions for CMP.

utilize CMP on the surface of a chemically/mechanically unstable metal (Al). This gate process is most sensitive to contamination. On the other hand, improvements in productivity are more important in terms of device cost-reduction. Efficient CMP technology which can achieve higher polishing rates with lower material cost is essential.

Mechanism of CMP and the Solutions to Key Challenges

The essential functions required for CMP are: i) material removal, ii) surface protection and iii) contamination control (Figure 1). Authors will discuss the mechanisms of each function and the solution to the challenges of 32nm and beyond in detail below.

Material Removal

In order to remove the target surface layer, chemical action such as etching with chemistry and mechanical action using abrasive particles, and their collaboration such as particle adsorption/desorption control with additive chemistry are utilized.

Parameters which are important for chemical etching include pH, redox (reduction-oxidation) potential, and the use/omission of complexing agents. In particular, pH and redox potential constitute core parameters. While chemical etching is effective in increasing the material removal rate, the excess etching can cause corrosion or topography degradation of metal interconnects. In order to improve the material removal rate while also suppressing this chemical damage and the topography degradation, it is important to enhance the mechanical action.

The important parameter for mechanical action with abrasives is friction energy. The mechanical energy given by the polishing machine is converted to friction energy through the abrasive particles. This friction causes the mechanical stripping of surface materials. Additionally, a portion of the friction energy is converted to friction heat which can accelerate chemical etching. Friction energy can be calculated from the electric energy of the rotating platen of the polishing machine. Figure 2 indicates that there is a good correlation between polishing rate and mechanical energy (friction energy - heat loss) when SiO_2 is polished using various abrasives with different particle size, shape and materials.

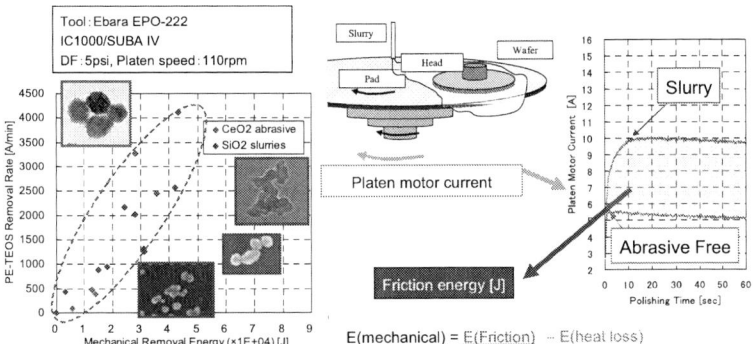

$$E(mechanical) = E(Friction) - E(heat loss)$$

Figure 2. Removal rate of SiO_2 (PE-TEOS) as a function of mechanical energy (friction energy - heat loss) (1).

The key factors to increase friction energy are: i) the number of active particles, ii) the contact area of each particle, and iii) the friction coefficient of each particle which is related to the shape, surface morphology and type of material. It is effective to increase the number of active particles not only by increasing the particle concentration but also by helping the particles to adhere onto substrate surface. The particles in CMP slurry distributed to the platen are constantly adsorbing and desorbing onto the substrate surface. In order to increase the number of adhered particles, it is important to induce the attractive force between particle and substrate by controlling the zeta potential of each material. The force acting between particle and substrate in a liquid can be calculated as the sum of electrostatic force due to electrical double layers and Van der Waals force, according to the DLVO theory. Figure 3 indicates that the attractive force between particle and substrate (induced by the opposite zeta potential) causes an increase in the number of active (adhered) particles which leads to a friction increase. As a result, the polishing rate is effectively improved (2).

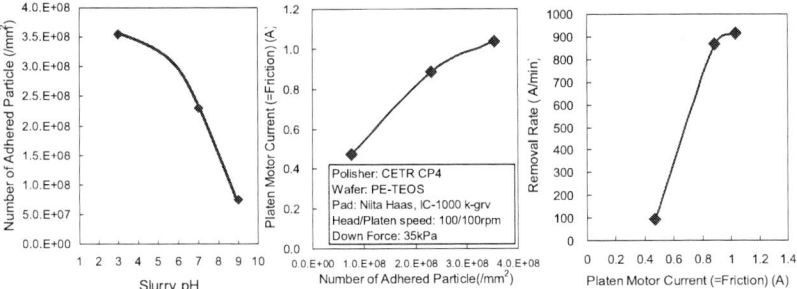

Figure 3. Relationship between Si_3N_4 removal rate with colloidal silica slurry, particle adhesion number, and platen motor current (friction). Colloidal silica and Si_3N_4 substrate attract each other as they have opposite polarity of zeta potential in pH 4 (Si_3N_4: +10mV, SiO_2: -5mV,) while they have same polarity in pH 9 (Si_3N_4: -40mV, SiO_2: -75mV,).

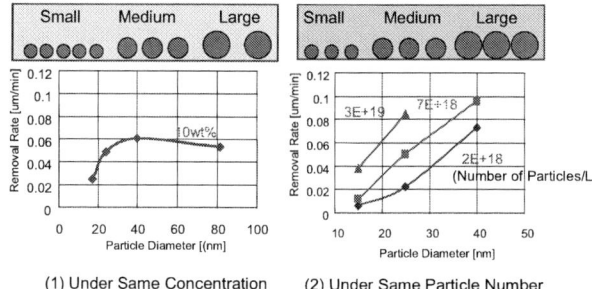

(1) Under Same Concentration   (2) Under Same Particle Number

Figure 4. Effect of particle size on SiO$_2$ polishing rate (Abrasives: colloidal silica, pH10, Substrate: glass hard disk, Soft pad).

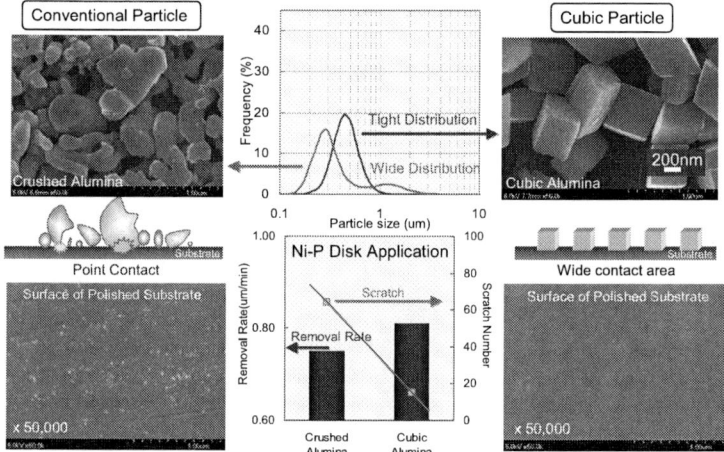

Figure 5. Effectiveness of cubic abrasives on removal rate, scratches and residual contaminants (Abrasives: cubic or crushed alumina, Substrate: Ni-P hard disk, Soft pad).

The contact area of each particle depends on the size and the shape of particles. Figure 4 indicates that larger particles can achieve higher removal rate when the number of particle in the slurry is fixed. Under the same particle number, removal rate of larger particle is higher as its contact area is larger than that of smaller particle. The friction coefficient of each particle heavily depends on the shape and surface morphology. Figure 5 tries to apply the cubic particles which have a wide contact area and higher friction coefficient to improve the polishing rate. Cubic alumina features higher polishing rates than conventional crushed alumina. Moreover, it was found that polishing with cubic alumina can reduce the residual particle count remaining on the polished surface because the fine particle count was reduced. Also, cubic alumina can reduce the scratch count because the larger particle count was reduced (3).

Surface Protection

Technology using corrosion inhibitors or surfactants is used for planarizing device surfaces. These additives serve as a protective film on the surface which prevents the corrosive materials in the system from excess etching. This additive technology is effective to prevent dishing and corrosion of metallic materials. It is necessary, however, to pay sufficient attention to the selection and the cleanliness of such additives because they may remain on the substrate surface due to their high adsorbability. Moreover, local heterogeneity of the protection film must be avoided as they can cause local heterogeneity of the surface morphology.

Contamination Control

Planarization by CMP is achieved with material removal and surface protection. If these functions act homogeneously at the atomic level, then polishing induced defects such as scratches, pits, dishing and roughness should not occur. In the process, however, there are various factors (contamination) which affect the surface uniformity. This contamination can include intervention by large particles or foreign matters, agglomeration of abrasive particles, pad clogging by polishing byproducts, and ambient factors such as dissolved gasses and light. These contaminants affect not only surface cleanliness but also defects and roughness of the surface. As for the surface roughness, it is also necessary to pay attention to cleaning and rinsing process following after CMP process as they may roughen the surface (4, 5). Figure 6 indicates that silicon surface is roughened just by UPW (ultrapure water) rinsing process when atomically flat silicon (argon-annealed) surface is used as the starting surface material. The surface roughness induced in the rinsing process increases more at the presence of dissolved oxygen and light.

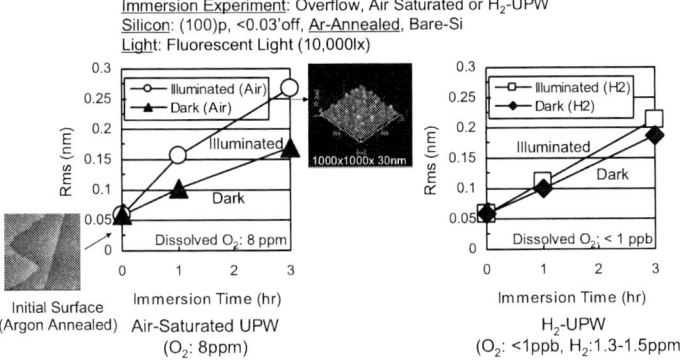

Figure 6: Surface microroughness of Ar-annealed silicon surfaces after UPW rinsing under illumination vs darkness. Air-saturated UPW (O_2: 8ppm) or H_2-added (O_2: < 1 ppb) UPW were used. Light and oxygen affect the surface microroughness of silicon. Photoinduced roughness can be suppressed by suppressing dissolved oxygen concentration in UPW.

Elimination, dissolution and dispersion are three important functions for contamination control in CMP process. Large particles and foreign matter must be eliminated by advanced filtration technology, particle classification and purification. Agglomeration or pad clogging can be prevented by utilizing dispersion technology of abrasive particles and the dissolution technology of byproducts. Technology using chelating agents is effective to prevent metal-cross-contamination during advanced CMP processes such as Cu-TSV (Figure 7) or metal damascene gate process.

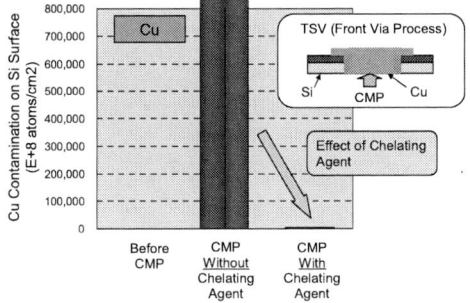

Figure 7: The Effectiveness of adding chelating agent in preventing Cu cross-contamination onto silicon surface during TSV-CMP process.

Conclusion

In CMP processes for 32nm devices and beyond, fine and delicate device patterns must be effectively planarized without causing any nanodefects, nanocontaminants, and atomic-level-roughness. Material removal with homogeneous and efficient mechanical action, surface protection with appropriate additives, and thorough contamination control are all key to effective CMP polishing.

Acknowledgments

The authors wish to thank Mr. T. Ohtsu, Mr. K. Ohashi, Mr. M. Tahara, Mr. T. Matsunami, Dr. A. Yasui, Dr. M. Serikawa, Mr. T. Shinoda and Mr. N. Yasufuku of Fujimi Incorporated, Mr. P. Kreiter of Fujimi Corporation, Mr. K. Shimaoka of Tokuyama Corporation, Prof. T. Ohmi of Tohoku University, Prof. T. K. Doi and Prof. S. Kurokawa of Kyusyu University.

References

1. K. Tamai, H. Morinaga, T. Doi, and S. Kurokawa, Proceedings of ICPT 2008, pp.22-28, Hsinchu (2008).
2. K. Tamai, A. Yasui, M. Serikawa, H. Morinaga, T. Doi and S. Kurokawa, Proceednigs of ICPT 2010, Arizona, (2010).
3. M. Tahara, Y. Matsunami, F. Saeki, K. Tamai, and H. Morinaga, Proceedings of ICPT 2009, pp.71-75, Fukuoka (2009).
4. H. Morinaga, K. Shimaoka, and T. Ohmi, J. Electrochem. Soc., 153(7), pp.G626-G631, (2006).
5. H. Morinaga, K. Shimaoka, and T. Ohmi, Solid State Phenomena, 134, pp.45-48, (2008)

A Study on Optimized Conditioner for Soft Pad in Cu Barrier Removal

S.Y. Yoon , J.H. Lee

Semiconductor Biz. Div. of Ehwa Diamond Ind. Co., Ltd., Osan, Korea

In removing the Cu barrier in Cu CMP process, high- porosity soft pads have been commonly used. However, due to its high surface roughness of its initial state, the high-porosity pads have caused the continuous increase in wafer removal rate until being conditioned by diamond conditioners for a certain period of time which results in process instability and the excessive waste of consumables including slurries and dummy wafers. In this paper, the mechanism that soft pads causes the increase in wafer removal rate at the early stage of Cu barrier polishing and the methodology to stabilize the polishing process by minimizing the period of increase in wafer removal rate are studied. In order to do so, the pad morphology, pad wear rate, wafer removal rate, wafer non-uniformity of soft pads were evaluated as a function of the configurations of diamond conditioners, diamond protrusion, diamond density, grit size, diamond distributions.

1. INTRODUCTION

Chemical mechanical planarization(CMP), a surface planarization process, has enabled the production of advanced semiconductor devices by producing a globally planar wafer surface. The CMP process requires a poly-urethane polishing pad; a slurry composed of abrasive, chemicals, and water; and pad conditioner etc [1-3]. In the CMP process, chemical reaction occurs between chemicals (slurry) and the wafer. Consequently, the reaction product would gradually accumulate onto the holes and grooves of the pad surface leading to the so-called glazing of the pad [4]. Therefore, CMP pad conditioner used to regenerate a new pad surface and recover its role in the process is necessary.

In general, Ni plating or brazing technology has been utilized for bonding the diamond grits on the substrates. Up to the present time, a number of tests and studies have been conducted to find the optimum diamond disc for slurries and pads for Oxide/Metal polishing along with the studies on various factors that affect the disc's performance and life time including diamond size, grade, and separation, diamond protrusion, design and flatness of disc. In the wake of abundant studies, we have made continuous progresses on development of CMP process [5-6]. However, it is evident that the studies on the soft pad have not been actively performed so far. Especially, on Cu barrier polishing process where the soft pads are widely used, shifting of wafer removal rate for a length of polishing time until the stage of wafer removal rate become stable has become a critical issue resulting in the process instability and lapse of process time. In order to remedy this issue, the development of conditioner for soft pad is highly needed. In this study, we have conducted tests to get correlations between conditioner configurations (diamond size, concentration, protrusion) and pad morphology, pad wear rate and wafer removal rate.

2. EXPERIMENTAL

Among several soft pads for Cu barrier removal, our tests have been solely conducted to find the optimum conditioners for Fujibo pad. First, we did run screen tests evaluating pad wear rate and pad morphology to select discs for marathon test. The selected discs information is shown in table 1. (Disc-A: sintering type disc, Disc-B: electroplating, Disc-C & D: brazing type discs.)

TABLE 1. Conditioner types

Mark	Method	Grit size (US mesh)	Density (ea/cm²)	Protrusion (μm)
Disc-A	Sintering	170 mesh	4000	30~60
Disc-B	Electroplating	80 mesh	1000	25~65
Disc-C	Brazing	170 mesh	2000	30~60
Disc-D	Brazing	170 mesh	2000	25~35

3. RESULTS AND DISCUSSION

Conditioning test result on the 10 inch pad

Due to its softness, Fujibo pad is hard to measure the pad wear rate. In this study, profiler and SEM has been utilized to measure the pad thickness difference before and after polishing. Pad thickness change on Disc-C is shown in Figure. 1. In Table 2 and Figure 2, pad wear rate and pad morphology after polishing are shown.

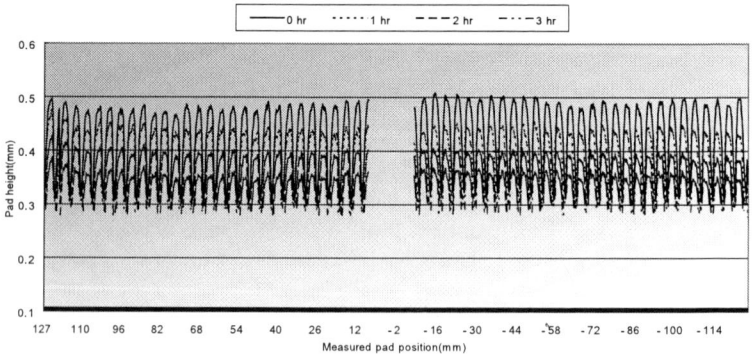

Figure. 1. Pad thickness change according to conditioning time by pad profiler. (Disc-C)

TABLE 2. Pad wear rate(10" conditioning test) according to conditioning time and pad surface morphology after 3 hours conditioning

Mark	Pad wear(μm)			
	1 hour	2 hours	3 hours	Average
Disc-A	57	48	51	52.0
Disc-B	13	11	10	11.6
Disc-C	34	28	27	29.6
Disc-D	14	12	11	12.3

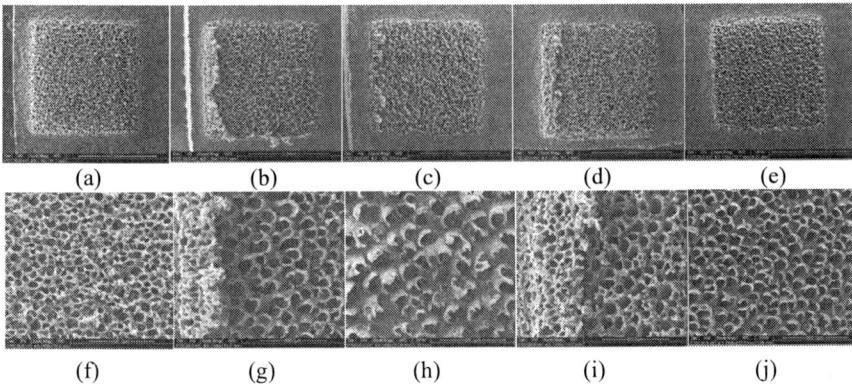

Figure. 2. Pad morphology after 3 hours conditioning by SEM.
(a)~(e) : x 30 magnification, (f)~(j) : x 100 magnification.
(a), (f) : Initial pad, (b), (g) : Disc-A, (c), (h) : Disc-B
(d), (i) : Disc-C, (e), (j) : Disc-D

Burrs were found on the surface of the pad polished by Disc-A and Disc-C having a higher value of pad wear rate. Although Disc-B showed the low pad wear, pore on the pad were pushed out and tilted. This phenomenon is believed to be brought out by a poor pad conditioning. For Disc-B, the side of diamond contacted the pad by design to lower the pad wear. For Disc-D, pad wear was moderate and the small pores on pad were desirably open. It is believed that the narrow diamond protrusion distributions and fine sized diamond grits of the conditioner generates the uniform pad conditioning.

From this test, we found that for soft pad, a disc with coarse diamond size and wide diamond protrusion distribution demonstrates high pad wear resulting in a short conditioner life time. In order to realize a low pad wear and uniform pad conditioning, fine diamond size and narrow diamond protrusion distribution are recommended.

Conditioning test result on the 30 inch pad

When removing Cu barrier with Fujibo pad without pad conditioning, it is found that wafer removal rate is continually going up for a certain period of time. In order to find the root cause of this issue, we measured the pad morphology and wafer removal rate over wafer polishing time. In Figure 3 and 4, pad morphology changes over time with pad conditioning with Disc-D and without conditioning are shown.

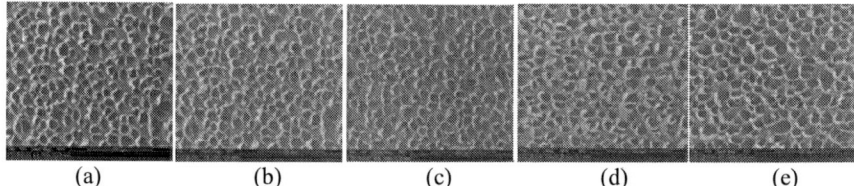

(a)   (b)   (c)   (d)   (e)

Figure. 3. Pad morphology change according to polishing time (without conditioning). (a) Initial, (b) after 1 hour, (c) after 2hours, (d) after 3 hours, (e) after 4 hours

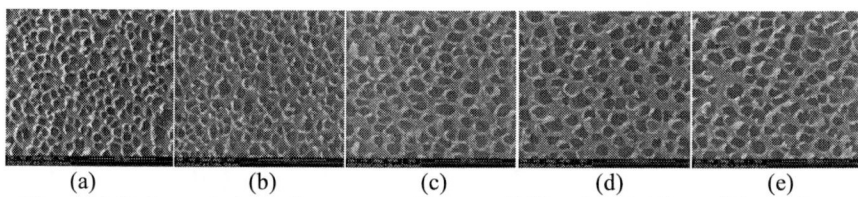

(a)   (b)   (c)   (d)   (e)

Figure. 4. Pad morphology change according to polishing time (with conditioning). (a) Initial, (b) after 1 hour, (c) after 2hours, (d) after 3 hours, (e) after 4 hours

On wafer polishing without pad conditioning, some sludge was found on pad pores and pore sizes after 4 hour polishing are bigger than initial pore size. In contrast, when wafer polishing was concurring with pad polishing, pore size became bigger after 1 hour polishing process. The increase of pad pore size means that pores neighboring are merged so that pad contact areas are increased. After 2 hours polishing, pore size seems similar. However, pad morphology became stable more rapidly. Further tests to identify the and pore size and contact area change over long period of polishing time without pad polishing could provide us better understanding of this issue. In Figure. 5, pad morphology after 1500 wafers processed is shown.

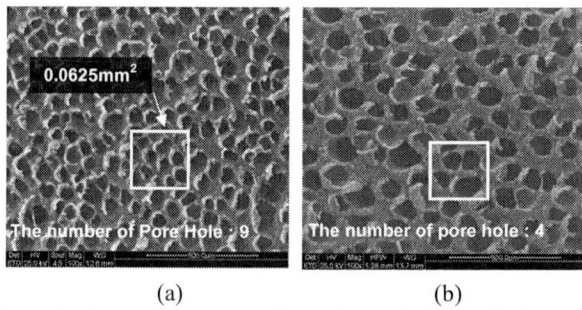

(a)   (b)

Figure. 5. Pad morphology comparison before using and after polishing. (a) initial pad, (b) after 1500 piece wafers polishing

In Figure. 6, wafer removal rate over polishing time is shown. It is confirmed that as pad pore size become bigger, wafer removal rate had increased and became stable as shown in Figure 4~5. From this test, we found that wafer polishing with pad conditioning demonstrated relatively higher wafer removal rate and shorter time period for the wafer removal rate to become stable.

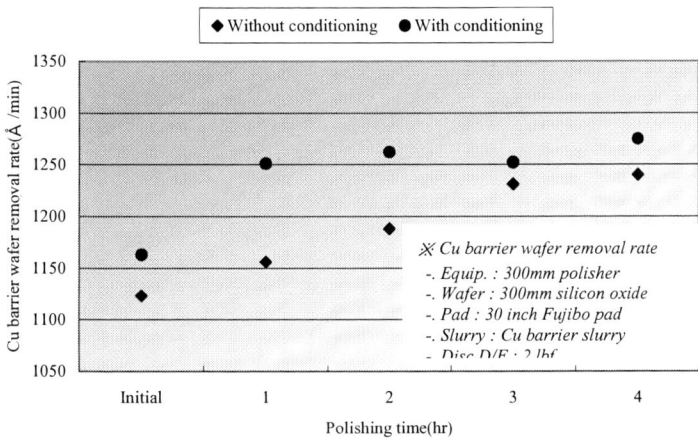

Figure. 6. Wafer removal rate over polishing time.

4. CONCLUSION

In our study, we found that a conditioner with finer grit size and narrow diamond protrusion distribution demonstrate substantial advantages in pad wear and pad conditioning ability for the soft pads. Also for Fujibo pad, pad pore size had increased over the wafer polishing time which triggered the increase of pad contact area and wafer removal rate as a result. In removing Cu barrier, wafer removal rate can be stable in a short period time using pad conditioning. There would be multiple reasons that wafer removal rate is continuously increasing for a certain of time after pad change and we could have more understanding on this issue after further tests on pad pore size, friction force changes and slurry dispersion over the polishing time.

5. REFERENCES

1. T. Dyer and J. Schlueter, Micro, Vol. 20, p. 47-54 (2002)
2. S. Nag and A. Chatterjee, Solid State Technology, p. 129 (1997)
3. K. Smekalin, Solid State Technology, p.187 (1997)
4. S. Sivaram, H. Bath, R. Leggett, A. Maury, K. Monnig and R. Tolles, Solid State Technology, Vol. 35, p.87 (1992)
5. J. Zimmer and A. Stubbmann, Proc. NCCAVS CMP'98 Symposium, pp. 88 (1998)
6. C.C. Garretson, S.T. Mear, J.P. Rudd, G. Prabhu, T. Osterneld and D. Flynn, Proc. CMP-MIC, p. 1 (2000)

Overcome Challenges in TSV CMP via Slurry Formulation

K. Luo[a], C. Wang[a], J. Jing[a], and S. Xu[a]

[a] Anji Microelectronics (Shanghai) Co., Ltd., Shanghai 201203, China

Along with the advancement of IC 3D technology, TSV (through silicon via) has been paid close attention to practically enable the 3D integration and packaging processes with addressing various technical and economical concerns in most recent years. Due to its intrinsic nature of the via size and the amount of via filling materials, the technical requirements in TSV CMP are much more different than those in traditional IC CMP technology. These technical difficulties form new challenges for CMP users to enable TSV CMP process in 3D technology in IC industry. In this work, we are going specifically present some slurry formulation work to demonstrate how to address some of these challenges in two TSV CMP applications.

Introduction

In the CMP process of TSV wafers, new challenges exist in both wafer front-side and back-side polishing. Although via sizes in TSV wafers vary for different applications/designs, they typically range in 1-30 um in diameter and 50-200 um in depth. As an interconnecting metal, Cu is mostly used to fill TSV vias. To avoid voids, Cu plating is commonly applied with a Cu overburden layer left on the surface. As illustrated in Figure 1(A), the thickness of Cu overburden layer is often 2-10 um thick, which depends on the integration and design. Apparently, there is a significant difference between the TSV integration scheme and traditional IC Cu CMP process with a few thousand angstroms of Cu overburden. In TSV applications, Cu RR is required to be 2-5 um/min for a Cu TSV slurry. Moreover, the dishing, surface roughness, and defectivity requirements still remain stringent. All these scenarios together challenge slurry designers to reform current slurry formulations and break into new slurry design regimes.

As we know, inhibitors often play key roles in Cu slurry formulation. In this paper, we will focus on the inhibitors effect on removal rates, surface quality, dishing, etc. to illustrate how to select and use right inhibitors to overcome the challenges for TSV Cu slurries.

In the mean time, wafer back-side thinning/polishing is another technical difficulty for TSV CMP technology. During the process, TSV wafers are bonded to wafers carriers to do back-side thinning and polishing. The common method of wafer thinning is grinding, but the surface quality certainly does not meet TSV integration standards. Usually, about 10 um Si or more on top of Cu vias is left after wafer grinding as shown in Figure 1(B), and then CMP is applied to finish the wafer thinning and achieve high surface quality required for further processes. In most cases except that the vias are opened after wafer bonding in a via last approach, a CMP process is required to remove

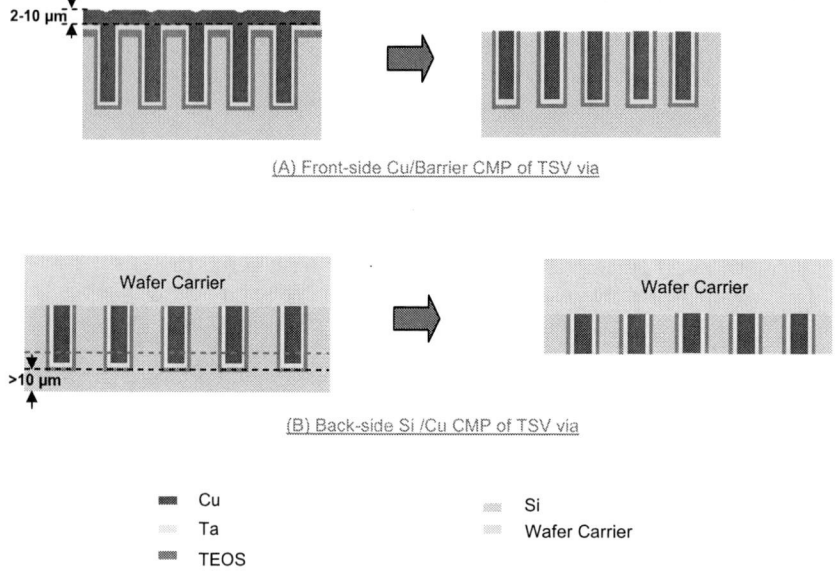

Figure 1. (A) TSV front-side Cu/Barrier CMP process flow; (B) TSV back-side Si/Cu CMP process flow.

the Si overburden on the wafer back-side as well as the TEOS and Ta barriers to open the Cu vias, as the illustrated in the right picture of Figure 1(B). The throughput requirement appears to set the target of Si RR at 1-5 um/min, and similar RR requirement for Cu applies. Additionally, relatively high removal rates of TEOS and Ta are required to open the Cu vias from the wafer back-side.

To provide solutions for the issues above, we introduce a new designed Si slurry system. The performance of these slurries shows high removal rates for all back-side materials, including Si, Cu, TEOS, and Ta. The effects of rate promoters, oxidizers, and abrasive loadings are discussed to understand how to achieve the RR targets from a formulation aspect.

Experimental

The work of front-side Cu polishing was performed on a 8 inch Applied Materials Mirra tool. IC1010 pads were applied. The head down force, platen/head speed, and slurry flow are 3 psi, 93/87 rpm, and 200 ml/min, respectively. The Cu film thickness was measured based on film resistivity. The structure of 5 um wide Cu line with a density of 50% was used for dishing performance study at the point of zero overpolishing.

The data of back-side polishing for all Cu, Si, TEOS, and Ta were collected on coupon wafers in a lab scale. Politex pads were used for back-side Cu, Si, TEOS and Ta polishing. The head down force is 3 psi and the platen speed is 70 rpm. The Cu and Ta thickness were measured based on film resistivity. The poly Si film thickness is measured by nanospec 6100, and the TEOS thickness measured by a thermowave opti-probe.

Results and Discussions

For Cu CMP, a plot of Cu RR as a function of down force is often utilized to evaluate the performance of a slurry in terms of removal rates and planarization effect. With all the new performance requirements for TSV Cu slurries, the key is how to balance Cu complexing and surface inhibiting effects [1,2].

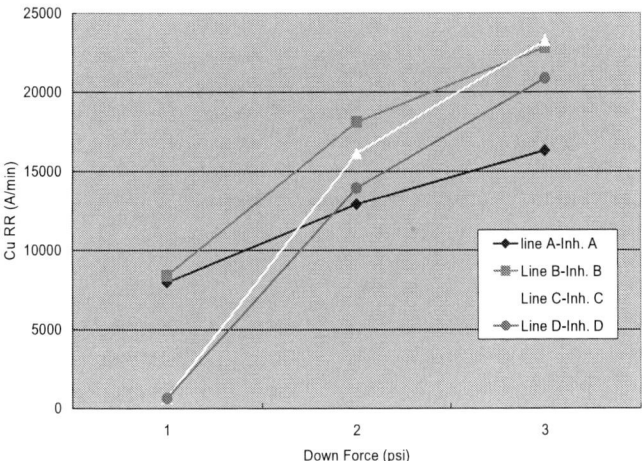

Figure 2. Cu RR as a function of down force in different inhibitor systems.

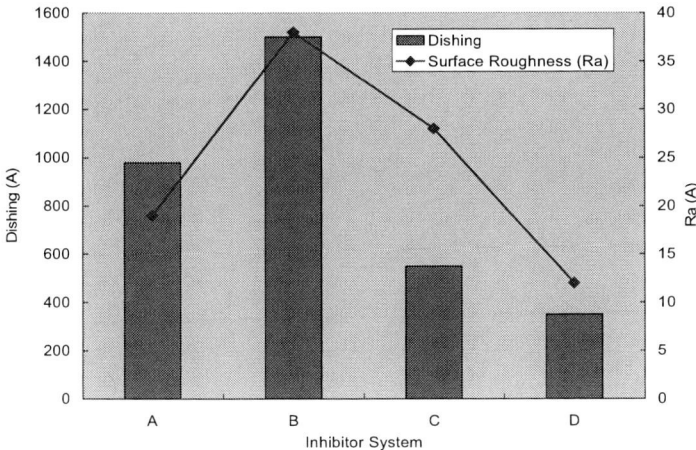

Figure 3. Dishing performance (854 mask, 5 um wide Cu line with a density of 50%) and surface roughness (Ra) for different inhibitor systems.

In our study, the focus is to leverage inhibiting effects of different inhibitor systems with a fixed complexing system to achieve the performance targets. In Figure 2, four lines were drawn to show the performance of Cu RR as a function of down force for four inhibitor systems (A, B, C and D) in a TSV Cu slurry platform. As shown, the inhibitors significantly alter Cu RRs at low/high down forces. For example, line A represents a slurry with a near prestonian behavior-meaning Cu RR increases with the increase of down force proportionally, might not be optimal for a TSV Cu slurry. As shown, inhibitor B system is effective to lift the Cu RR at high down force ranges. However, a high Cu RR (~8000 A/min) at low down force (1 psi) for slurries with inhibitor system A and B may not yield the best final topographic performance, particularly dishing. This was evidenced in Figure 2, where the dishing values of 5 um wide Cu lines with a density of 50% for both inhibitor systems A and B are 1000 A and 1500 A, respectively, which are quite high. With more inhibitor screening, we found that inhibitor C system similarly has a high Cu RR (23,000 A/min) at 3psi, but exhibits a very low removal rate (700 A/min) at 1 psi. This is more preferred for a Cu slurry to achieve a high throughput at a high down force at a stage of P1 polishing and a good planarization effect at a low down force at a stage of P2 polishing. As a result, the dishing value drops from 1500 A to 500 A at the same structure from inhibitor B system to inhibitor C system. In the case of Inhibitor C system, a denser film might be formed at a low down force to prevent extensive Cu removal, but a high down force works effectively to break the surface protection mechanically and results in a high Cu RR. To achieve a smoother Cu surface, inhibitor system D was found to further decrease surface roughness (Ra), while maintaining a similar curve of RR vs. DF to system C. With inhibitor D system in Figure 3, Ra drops 10 A, much lower than 30 A with inhibitor C system. In the case above, we are demonstrating a method how to use different inhibitors in formulation to balance Cu RR, planarization, and surface roughness to meet the challenges for TSV Cu slurries. The mechanistic studies on this platform are on-going, and can be available in future. Note that the Cu RR non-uniformity (NU) can be well controlled with a process tuning in our slurry platform, and a within wafer NU was demonstrated to be less than <3% for our experimental slurries in commercial 300 mm wafer tools.

In our previous work [3], we had invented a Si slurry with a high Si RR up to 1 um/min. As presented in Figure 4, a chemical additive A was put into a colloidal silica abrasive system. The square connected blue line shows that the Si RR increases along with the increase of the normalized concentration of additive A. When the concentration of A is at a normalized concentration of one, Si RR reaches about twice as much as the RR of the slurry without any additive inside, Note that the data shown here reflects a situation on coupon wafer (2 inch in diameter) polishing, and the removal rates are 2-3 times higher in large-scale commercial CMP tools with our correlation studies. However, the Si RR with additive A alone is not high enough to meet TSV Si RR requirements. With continuous formulation work, additive B was identified to similarly promote Si RR in addition to additive A. As the triangle connected pink line represents in Figure 4, the Si RR also increases with increasing the concentration of additive B by following a similar RR trend to that with Additive A. Moreover, with the concentration of additive B at a normalized value of one, the Si RR is plotted in a circle connected red line as a function of normalized concentration of additive A. For the A+B system, it is found that the rate promoting effect of A+B does not simply behave as a summation of two rates. Instead, a synergetic effect of additive A and B to boost Si RRs was clearly observed, especially when the concentration level of additive A approaches to the high end in

Figure 4. As a result, Si RR was almost tripled, compared of the RR with A or B only. Apparently, additive A and B chemically work together to speed up the Si removal rates. The exact mechanism is to be further studied.

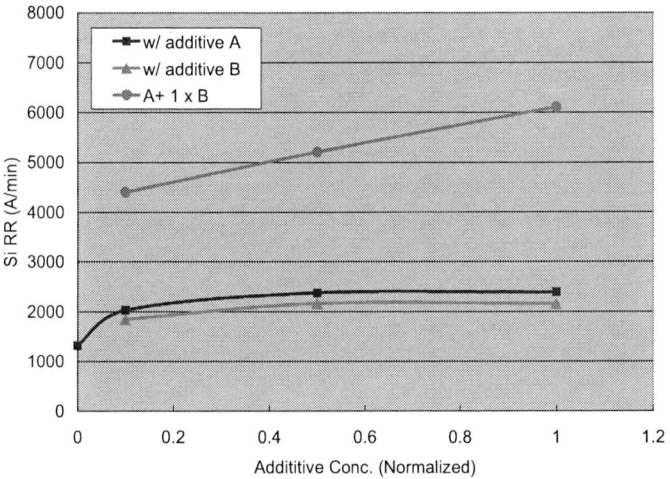

Figure 4. Si RR as a function of the concentration of additive A, B, and A+B.

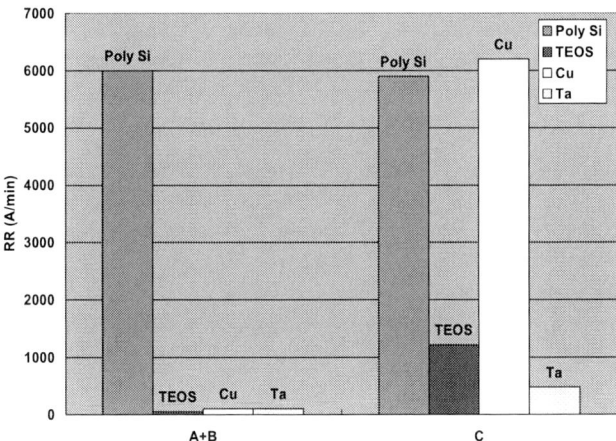

Figure 5. Poly Si, TEOS, Cu and Ta RR for Slurry A+B vs. Slurry C.

Moreover, when back-side Si is removed to get close to the bottom TSV vias as shown in Figure 1(B), a slurry capable of removing Si, Cu, Ta and TEOS simultaneously becomes necessary. In general, Si is almost inert in an acidic slurry, although a high Cu RR is achievable. On contrary, a basic slurry can be highly efficient for Si removal. In the mean time, to achieve a high Cu RR, an effective oxidizer, like H_2O_2, is required. The problem is that Si RR diminishes rapidly with the addition of H_2O_2, which results in a conflict. In our studies here, a brand new oxidizer other than H_2O_2 was found to be capable of not only oxidizing Cu to achieve high Cu removal rates, but also maintaining high Si RRs in a basic ambient. Although the fundamental mechanism is not fully understood, we propose that this oxidizer may not be strong enough to fully oxidize Si to SiO_2 and passivate the surface. Instead, it enables high Cu RR without affecting much on Si removal rates.

As illustrated in Figure 5, the Si RR of slurry A+B, which was discussed previously in Figure 4, exhibits a superior Si RR, but lack of efficient removal for TEOS, Cu, and Ta. With the new oxidizer in slurry C, both Si and Cu are efficiently removed at high rates. At an optimized concentration in Figure 5, the removal rate selectivity of Cu to Si can be tuned to be ~1:1. Meanwhile, to promote TEOS RR, an increase on abrasive loadings is effective here. With the increase of abrasive loadings and introduction of the new oxidizer, Ta RR of slurry C was also observed to be significantly higher than that of slurry A+B.

As an example here, through working on different aspects in formulation, including rate promoters, oxidizer, and abrasive loading, we successfully demonstrate a path to overcome removal rate challenges for four different materials in TSV back-side CMP process. Certainly, more work is necessary to meet the commercial requirements.

Summary

In short, through extensive slurry formulation work on inhibitors, rate promoters, oxidizers, abrasive loadings, etc., we have evidenced the solutions to overcome certain technical challenges in TSV CMP processes, which shields some light in slurry development field.

References

1. K. Cheemalapati, J. Keleher, and Y. Li, in *Microelectronic Applications of Chemical Mechanical Planarization,* Z. Li, Editor, p 214, Wiley Interscience, New Jersey, 2008; S. K. Govindaswamy, in *Microelectronic Applications of Chemical Mechanical Planarization,* Z. Li, Editor, p 249, Wiley Interscience, New Jersey, 2008.
2. K. W. Chen, Y. L. Wang, C. P. Liu, L. Chang, F. Y. Li, Thin Solid Films **498**, 50(2006).
3. C. Wang, Y. Cao, C. Yang, H. Zhou, J. Jing, Z. Xia, J. Tang, J. Lin, C. Chiu, S. Wang, and C. Yu, ECS Transaction **18**(1), 511 (2009).

ECS Transactions, 34 (1) 609-613 (2011)
10.1149/1.3567646 ©The Electrochemical Society

Investigation on the Correlationship between Process Performances and Composition of CMP Slurry Designed for GST Alloy Polishing

Keliang Pang[a], Feng Chen[b], Li Jiang[b], Mingqi Li[b] and Min Zhong[b]

[a] Anji Microelectronics (Shanghai) Co., Ltd., Shanghai 201203, China
[b] Semiconductor Manufacturing International Corporation, Shanghai 201203, China

> Newly developed technology utilizes phase change materials in integrated circuit to produce a new type of memory device, namely PRAM. The delicate design of such device poses new challenges for CMP (Chemical Mechanical Polishing), which is one important process to construct memory cells in PRAM. In this paper, recent efforts in developing slurries for GST CMP are described. The preliminary investigation focuses on some of the most concerned issues in GST CMP, such as GST removal rate and removal rate selectivity, post-CMP surface roughness, galvanic effect between different materials and composition consistency of GST alloy. Performance of the testing slurries is analyzed on a laboratory level using physical analytical instruments, such as SEM, EDX, AFM, as well as electrochemical analytical tools.

Introduction

First explored in the late 1960s, chalcogenide-based phase change materials (PCM) have been successfully used in some commercialized memory devices, such as CD and DVD (1). Recent developed techniques enable the utilization of phase-change materials in integrated circuit to produce a new type of nonvolatile memory device, which is usually called PRAM (or PCRAM, Phase Change Random Access Memory). Although controversies are heard from time to time, PRAM technology has been considered to be a promising technology to shine in the next generation of nonvolatile memory (2, 3).

Among phase change materials that have been explored, GST (Ge-Sb-Te) alloy is one of the most promising materials that suitable for PRAM manufacture. Early design of GST memory device applies a GST layer in a diode (Figure 1a), which switches the physical states of the GST alloy between amorphous state and crystalline state by joule heating and recognizes such physical states by measuring electrical resistivity of the GST alloy. Although such layer structure is simple and relatively easy to process, it has some intrinsic problems, such as low heating efficiency and tension-induced device failure caused by volume change during phase switching. To address these problems, newly developed technology utilizes confined cell structure in each memory bit, as illustrated in Figure 1b. In a confined cell, heating efficiency is increased by increasing current density, while the phase change tension is reduced by introducing a barrier consisting of proper materials. Since confined cell structure is more complicated compared to single layer structure, a fine-tuned CMP process is required to provide a delicate planarized surface after depositing GST onto the device, as illustrated in Figure 2. The challenge of such GST CMP process is obvious, not only because GST alloy itself is composed of more

609

than one element, but also because there are more than one material exposed to the CMP slurry during the final CMP stage (4, 5).

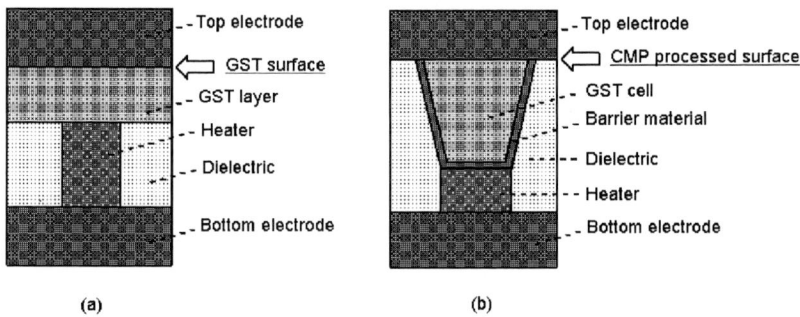

(a)   (b)

Figure 1. An illustrative scheme of a PCM memory diode. Each diode represents one memory bit in a memory device. (a) A GST layer structure. (b) A GST confined cell structure.

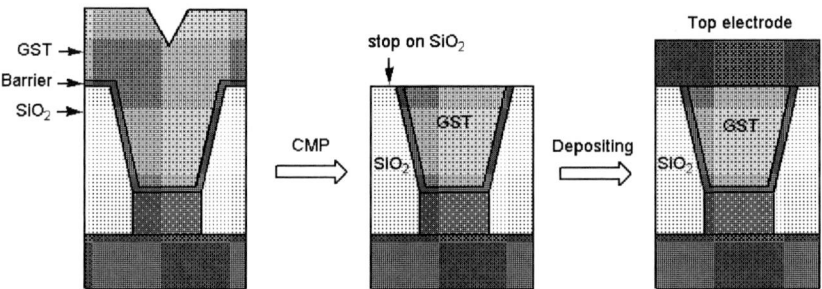

Figure 2. Schematic processes to construct a PCM memory diode. Only the last couple steps are shown here. From (a) to (b) is a GST CMP process.

Reported in this paper are some recent efforts that have been paid to develop a CMP slurry for GST polishing. Our preliminary investigation focuses on the following aspects: 1) GST removal rate and removal rate selectivity; 2) Post-CMP surface roughness; 3) Galvanic effect between different materials; 4) Composition consistency of GST alloy.

Experimental

Polishing experiments are performed on a Logitech 1PM52 polisher equipped with a Politex pad. 4 cm × 4 cm coupon wafers are used with the platen and polishing head rolling speeds range from 50 to 100 rpm, while the down force ranges from 2 to 4 psi. The amorphous GST films on GST wafers are deposited by PVD. The thickness of the original GST film is measured with SEM. The polishing time is monitored till the entire

GST film is removed from the wafer. The GST removal rate is then calculated by dividing the original thickness of the GST film by the polishing time. The silicon oxide film is a TEOS silicon oxide film and its thickness is measured optically.

The electrochemical experiments are performed on a CHI600B Echem-station manufactured by Chenhua (Shanghai) Company. The working electrode is made of amorphous GST, which is fabricated into a cylinder shape with a diameter of 6.0mm. Saturated calomel electrode (SCE) is used as reference electrode. The counter electrode is a Pt wire. Data generated by the Echem-station are collected and analyzed using a PC equipped with a CHI program (version 5.1).

Results and Discussion

Coupon Wafer Polishing Results

Taking technological maturity and SHE (Safety, Health and Environment) concerns into account, Anji developed a series of testing slurries based on the H_2O_2/SiO_2 aqueous system. Other than H_2O_2/SiO_2 and the aqueous carrier, the formulation further consists of removal rate promoters, corrosion inhibitors and surfactants. Listed in Table I is a performance summary of a selected testing slurry and a control sample, which is composed of H_2O_2/SiO_2 and water. GST removal rate of 300~350 nm/min is achieved by adjusting slurry composition. The GST/TEOS selectivity is tunable by adjusting both of abrasive loading and pH value. Despite of the softness of the GST alloy, a satisfactory post-CMP surface with an RMS value at 0.228 nm is obtained. Interestingly, GST removal rate is significantly affected when different type of SiO_2 abrasive is used. Table II shows that GST removal rate changes from 187 to 350 nm/min when different commercial brands of SiO_2 are used. The similarity in the mean sizes of these abrasives suggests that the removal rate difference is probably more chemically induced.

TABLE I. Performance summary of one testing slurry designed for GST CMP, with a control sample for comparison.

Slurry	GST Removal Rate (nm/min)	TEOS Removal Rate (nm/min)	GST/TEOS Selectivity	Pre-CMP GST Surface Roughness, R_a (nm)	Post-CMP GST Surface Roughness, R_a (nm)
A	350	3.2	109	0.280	0.228
Control*	<100	-	-	-	-

* Sample contains H_2O_2/SiO_2 and water only.

TABLE II. Comparison of GST removal rates for three slurries using different abrasives. Solid loading and other components are the same in all three slurries.

Type of SiO_2	Commercial Brand A	Commercial Brand B	Commercial Brand C
Abrasive Mean Size (nm)	ca. 80	ca. 80	ca. 100
GST Removal Rate (nm/min)	350	233	187

One challenge of GST CMP is to maintain composition consistency of the GST alloy. Thus, the mole percentages of three elements in the alloy should remain largely unchanged. An EDX analysis shows that the mole percentage of Germanium in the GST alloy drops from X to 0.963X after polishing. Our next primary goal is to find a corrosion inhibitor suitable for protecting Germanium from over-removal.

Electrochemistry Analysis Results

Since electrochemical reactions are involved in most corrosion processes occurred in an aqueous system, electrochemistry analysis can provide critical information for addressing corrosion problems in CMP processes. In the other hand, well established electrochemical experiments can also provide valuable forecasts on slurry behaviors in on-line CMP process, thus reducing research frustration and cost. Therefore, electrochemical experiments are conducted in our early research stage to investigate possible corrosion effects on GST alloy and to chemically minimize these effects.

In figure 3, a Tafel analysis shows that addition of a corrosion inhibitor to the GST CMP slurry substantially lowers the GST corrosion current (I_{corr}), which is reduced from 0.15mA/cm^2 (curve a) to 0.017mA/cm^2 (curve b). Noteworthy is that in the entire GST anodic region shown in the Tafel plot, the current intensity in the slurry with inhibitor (curve b) is lower than that of the slurry without inhibitor (curve a), indicating an immediate formation of passivating layer on GST alloy when it becomes anodic.

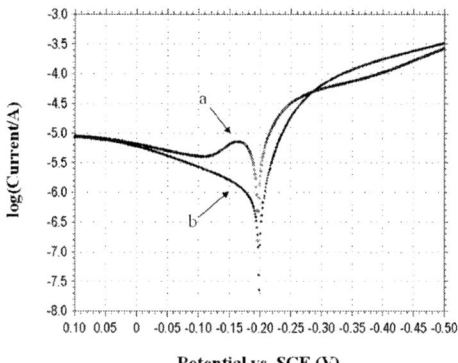

Potential vs. SCE (V)

Figure 3. Tafel analysis results of a testing slurry containing corrosion inhibitor (curve b), with comparison to a controlled sample without inhibitor (curve a). Scanning direction: from right to left.

While more than one material is exposed to the CMP slurry at the final stage of GST polishing (Figure 2b), galvanic corrosion may occur between GST and other conductive material. Thus, a galvanic analysis of GST alloy versus another metal (metal A) is performed and the results are summarized in Table III. The magnitude of ΔE indicates the difference of electronegativity of the two materials, while a positive ΔE indicates that GST alloy is more electronegative. A lower ΔE is preferred in order to reduce the potential galvanic corrosion occurred on the two materials.

As shown in Table IV, the magnitudes of ΔE obtained from OCP method are substantially lower than those from Tafel method. Such difference arises from the fact that in Tafel method, fresh polished surface on each of the materials is exposed to the testing slurry, while stable passivating layers on both materials are allowed to form in OCP method. A smaller ΔE in OCP method indicates that the formation of the passivating layers on both of the materials reduces their electronegativity difference. However, if the formations of these passivating layers are evenly disturbed during a CMP process, the electronegativity difference of the two materials in a dynamic CMP process should be in between of ΔE_{Tafel} and ΔE_{OCP}. Yet the situation becomes more complicated when the formations of the two passivating layers are unevenly disturbed. Nevertheless, if ΔE values are low in both methods, then a relatively milder galvanic corrosion is expected.

TABLE III. Electrochemical analysis results for investigation of galvanic effect between GST alloy and metal A.

Sample Number	Tafel plot method					OCP method
	GST		Metal A		ΔE_{Tafel} (mV)	ΔE_{OCP} (mV)
	I_{corr} (mA/cm^2)	V_{corr} (V)	I_{corr} (mA/cm^2)	V_{corr} (V)		
1	0.086	-0.314	0.939	-0.497	183	75
2	0.079	+0.272	0.290	+0.065	207	37
3	0.061	-0.210	0.682	-0.370	160	6
4	0.105	-0.149	0.543	-0.300	151	4
5	0.038	-0.160	0.277	-0.287	127	2
6	0.125	-0.184	0.612	-0.336	152	-1

Also noteworthy is that although the ΔE values are positive in most of the case, it does change into a negative value in specific condition, indicating GST becomes a corroding anode in such condition.

Conclusion

Laboratory level slurries designed for GST CMP have been developed. By adjusting slurry compositions, the GST removal rate is tunable at the range of 300 to 400 nm/min, while the GST/SiO$_2$ selectivity reaches over 100. Post-CMP surface roughness on the GST alloy is as low as 0.228 nm. However, EDX analysis indicates that the post-CMP alloy suffers from a Germanium mole percentage dropping, which would be the major issue to be solved in the future formulating work. In addition, electrochemical analysis is performed to address potential corrosion issues in GST CMP. It is believed that such corrosions can be relieved by introducing corrosion inhibitors into the slurry.

References

1. For a recent review on phase change materials, see: S. Raoux, W. Welnic, and D. Ielmini, *Chem. Rev.*, **110**, 240-267, (2010).
2. H. Mehling: *Phase Change Memory: The Next Big Thing in Data Storage?* http://www.enterprisestorageforum.com, February 3, 2010. (Retrieved on Nov. 12, 2010)
3. C. D. Wright, M. Armand and M. M. Aziz, *IEEE Transactions on Nanotechnology*, **5**(1), 50-61, (2006).
4. M. Zhong, Z. Song, B. Liu, S. Feng, F. Zhang and Y. Xiang, *ECS Transactions*, **22**(1), 161-166, (2009).
5. F. Q. Liu, C. Ge, K. Xu, M. Ye, Y. Wang, Y. Chen, S. Xia, A. Rosenbusch, A. Duboust, W. Tu and L. Karuppiah, *ECS Transactions*, **19**(7), 73-19, (2009).

Fundamental Characterization Studies of Condensed Chemical Mechanical Polishing Waste Slurry

Yohei Yamada, Masanori Kawakubo, Shusuke Watanabe and Takahiro Sugaya

Micro Device Division, Hitachi, Ltd., Ome-shi, Tokyo 198-8512, Japan

We investigated the possibility of efficiently reclaiming fumed silica slurry in an oxide chemical-mechanical polishing (CMP) process in order to reduce the cost of consumables. We demonstrated the feasibility of reclaiming condensed waste slurry for reuse. The reclaimed slurry had a ten-fold increase in number of particles larger than 0.56 μm over the original slurry. However, there was significant decrease in the number of particles larger than 1.01μm in the reclaimed slurry compared with that in the original one because depth filters used during reclamation effectively remove agglomerated particles and foreign materials from the condensed waste slurry. The removal rate and uniformity of the reclaimed slurry was comparable with original one. Even though there was a slight increase in particles on the wafers, no changes in level of micro-scratches on wafers was observed. Therefore, the CMP performance of the reclaimed slurry is comparable to that of the original one.

Introduction

Chemical-mechanical polishing (CMP) plays a key role in the fabrication of ultra-large scale integrated (ULSI) devices, as electronic devices become increasingly fast with downsizing dimensions and integrated circuit design becomes more complex. There are two major applications of CMP in ULSI manufacturing: to create smooth surface topography of inter-level dielectrics and to remove excess material to produce an inlaid metal structure or isolation trenches.

For certain processes, CMP is relatively high cost process in semiconductor industry. The main consumables in CMP include the polishing pad, polishing slurry, and pad conditioner. Slurry cost in particular represents almost half the total cost of consumables (COC) of the entire CMP operation. However, slurry utilization in oxide CMP applications is known to be inefficient; a large percentage of the abrasives in the slurry is lost during CMP process (1). Thus, the high cost of CMP consumables warrants novel solutions to reduce the slurry consumption. The most cost-effective approach to minimizing the consumption is recycling of CMP slurries for reuse (2). The deionized water used in the CMP process changes the concentration and the pH of the spent slurry, resulting in a large volume of diluted waste CMP slurry. Reusing the diluted waste slurry is difficult because the polishing rate is greatly reduced and aggregation of the slurry particles causes micro-scratches to wafer surfaces. Therefore, effective slurry recycling requires separation of the deionized water from waste slurry by use of filtration techniques (3). In this recycling, the waste slurry is condensed at a predetermined concentration. Despite its importance, there is relatively little published work on the chemical and physical properties of condensed oxide CMP waste slurry. In this work, we investigated the possibility of efficiently reclaiming fumed silica slurry in order to reduce the COC. The characteristics of the reclaimed slurry were determined according to pH value, specific gravity, conductivity, particle size and trace-metal levels. Furthermore, we determined whether the reclaimed slurry could be functionally comparable to the original one.

Materials and Methods

Reclaimed slurry preparation

Ammonia (NH_3)-based fumed oxide silica slurry with the pH level at about 10.4~10.5 (hereinafter referred to as the original slurry) was used for the polishing plasma-enhanced tetraethylorthosilicate (PE-TEOS) wafers. The silica abrasive concentration was 12.5 wt%. A slurry condensation system was designed and installed in a CMP apparatus to separate the deionized water from waste slurry, as shown in Fig. 1. The diluted waste slurry passes through slurry circulating passages to ultra filters. Then, the ultra filters separate excess water from the waste slurry. The waste slurry is circulated with ultra filtration until the concentration reaches about 12 wt% solids (hereinafter referred to as the condensed waste slurry). A combination of pH adjustment and microfiltration is used to reclaim the condensed waste slurry in a slurry reclamation system (hereinafter referred to as the reclaimed slurry). Ammonia solution is added to the condensed waste slurry as the pH adjuster. The native slurry abrasives are passed through depth filters with multiple ratings to trap the oversized agglomerates and foreign materials.

Characteristics of reclaimed slurry

In the case of high-pH slurry, the silica abrasives can be held in a stable liquid suspension because they possess highly negative surface charge and repel adjacent abrasives (4). The dispersibility of suspensions can be determined from important chemical slurry characteristics such as pH value, conductivity and zeta potential. The pH of the slurry was measured using a pH meter (D-54, Horiba, Ltd., Japan). The conductivity of the slurry was measured by a conductivity meter (ES-54, Horiba, Ltd., Japan). The specific gravity of the slurry was measured to be in the 1.068-1.080 range using a specific gravity meter (MMM7050K, Tokyo Keiso Co., Ltd, Japan).The distribution of slurry particles less than 0.5 μm in size was measured through acoustic attenuation spectroscopy (APS-100, Matec Applied Sciences, U.S.A.). Particles larger than 0.5 μm were analyzed using a laser diffraction particle sizer (AccuSizer 780, Particle Sizing Systems, U.S.A.). The trace-metal level in permeate water produced in the slurry condensation system was detected by an inductively coupled plasma atomic emission spectrometry (ICP-AES).

CMP conditions

A 200-mm wafer CMP tool (Mirra, Applied Materials Inc., USA) was used. The polishing pad was IC1520 (Nitta Haas Inc., Japan), and its surface had concentric circle-like grooves. We prepared blanket wafers with a 800-nm-thick PE-TEOS film.

We evaluated the basic characteristics of CMP processes: material removal rate, polish uniformity, defect level, and planarity of pattered wafer surfaces. The oxide film thickness was measured using a film thickness measurement system (Opti-Probe 3260 DUV, Therma-Wave Inc., USA) to calculate removal rate. In oxide CMP, the generation of micro-scratches on oxide wafer surfaces is the most serious defect. Micro-scratch can lead to severe circuit failure. Defects such as particles and scratches were detected using a laser scanning wafer surface inspection system (LS6500, Hitachi High-Technologies

Corp., Japan). A particle defect size larger than 0.25 μm was detected. The experimental conditions are shown in Table. I

Results and Discussions

Characteristics of condensed waste slurry
A comparison of the large particle distributions of the condensed waste slurry and the original one is shown in Fig. 2. The experimental results showed that condensation affected the large particle count (LPC) of particles larger than 0.56 μm. The condensed waste slurry had a ten-fold increase in LPC over original slurry. This could be because fractured pieces of silica were generated due to polishing. High hydrodynamic stresses during polishing break the large agglomerates of the fumed silica into finer particles.
Characteristics of reclaimed slurry
The change in specific gravity and the conductivity of the reclaimed slurry as a function of pH of the slurry is shown in Fig. 3. With the increase of the specific gravity, the pH values decreased. This was because ammonia concentration was decreased to fix conductivity to 1.5 mS/cm. The pH value and conductivity are proportional. A comparison of the large particle distributions of the reclaimed slurry and the original one is shown in Fig. 4. Reclamation affected the LPC of particles larger than 0.56 μm. The reclaimed slurry had a ten-fold increase in LPC over original slurry. However, there was significant decrease in the count of particles larger than 1.01 μm in the reclaimed slurry compared with that in the original one because the depth filters in the slurry reclamation system effectively remove agglomerated particles and foreign materials from the condensed waste slurry. The characteristics of both the reclaimed slurry and the original one are shown in Table II. The mean particle size of these slurries was about 129-144 nm. The mean particle size and the zeta potential of the reclaimed slurry were respectively smaller and slightly higher than that of the original one. The presence of metal contamination in silica-based alkaline slurries can cause adjacent silica abrasives to aggregate through dehydration reactions between the hydroxyl groups on the surfaces of the silica abrasives and metal hydroxides (5). The ICP-AES data for trace metals in the permeate water is shown in Table II. The slurry reclamation process had little effect on trace-metal levels.
Polishing performance
The changes in removal rate of the reclaimed slurry and removal rate uniformity as a function of the specific gravity and conductivity of the slurry are shown in Figs. 5 and 6 respectively. For each removal experiment, at least ten wafers were polished. The material removal rate increased when either the conductivity or the specific gravity of the slurry increased. We believe that the removal rate increased because the number of the silica abrasives increases due to the increase of the specific gravity of the slurry. The increase of the conductivity raises the pH value shown in Fig. 3. Silica has a higher dissolution rate at higher alkaline pH range (6). Thus, we believe that the removal rate increases due to both the chemical effect of soft glass easily forming on the surface of the PE-TEOS and the mechanical effect of the dispersibility of the slurry abrasives being increased. The removal rate of the reclaimed slurry was 10% higher than that of the original one. This increase is likely a result of the increases in pH values shown in Fig. 3. The oxide removal rate uniformity degraded when either the conductivity or the specific gravity of the slurry increased. The removal rate particularly in the area within 5 mm of the silicon wafer edge increased. However, the uniformity of the reclaimed slurry was comparable with original one at the same specific gravity or conductivity of the slurry.

The changes in particle count and the number of micro-scratches on a PE-TEOS wafer as a function of specific gravity and conductivity of the reclaimed slurry are shown in Figs. 7 and 8 respectively. There was a slight increase in slurry residues on the wafer in the reclaimed slurry compared with that in the original slurry. The levels of the micro-scratch on the wafers did not differ according to whether reclaimed or original slurries were used. Figure 9 shows cross-sectional scanning electron microscope (SEM) images of isolated active areas and dense active areas of the wafers after shallow trench isolation (STI) CMP. No difference of the recess at the STI and the silicon nitride (SiN) thickness on the active area after CMP was observed between wafers from the reclaimed slurry and from the original one.

These results indicate the feasibility of reclaiming condensed waste slurry for reuse. The particle size distribution of the condensed waste slurry and that of the reclaimed slurry were clearly correlated. A proper choice of depth filtration enables oversized agglomerates to be trapped when the native slurry abrasives are passed through the filter. These experimental results show that the CMP performance of the reclaimed slurry is comparable with that of the original one when the pH, the specific gravity and the conductivity are adjusted. Consequently, use of reclaimed slurry could decrease the high cost of slurry consumption.

Conclusions

This study focused on determining the correlation between polishing performance and the characteristics of reclaimed slurry. We found that the reclaimed slurry resulted in a ten-fold increase in LPCs over original slurry. However, there was significant decrease in the number of particles larger than 1.01 μm in the reclaimed slurry compared with the original one due to the removal of agglomerated particles from the condensed waste slurry through effective use of depth filters. We also found that the material removal rate increased as either the conductivity or the specific gravity of the slurry increased. The removal rate and uniformity of the reclaimed slurry was comparable with those of the original one. No change in level of micro-scratches on wafers was observed. We thus demonstrated the feasibility of reclaiming condensed waste slurry for reuse. Use of the reclaimed slurry could help decrease the high cost of slurry consumption.

Acknowledgements

The authors express their gratitude to Kiefer Tech Co., Ltd. for their technical contributions and experimental assistances. The authors are also grateful to Mr. T. Matsuo, Mr. T. Mizuta and Mr. M. Tokunaga of Kiefer Tech Co., Ltd. for stimulating discussions

References

1. A. Philipossian and E. Mitchell, *Jpn. J. Appl. Phys.*, **42** Part1, 7259 (2003).
2. H-J Kim, D-H Eom, and J-G Park, *Jpn. Appl. Phys.*, **40** Part1, 1236 (2001).
3. J. Hsu, Y. J. Wann, and J. R. Pai, in Proc *IEEE Int. Symp. Semicond. Manuf. Conf.*, p. 200, Tokyo, Japan (2010).
4. Y. Hayashi, M. Sakurai, T. Nakajima, K. Hayashi, S. Sasaki, S. Chikaki, and T. Kunio, *Jpn. J. Appl. Phys.*, **34** Part1, 1037 (1995).
5. A. Philipossian and E. Mitchell, *Micro*, **20**, 85 (2002).
6. R. K. Iler, *The Chemistry of Silica*, p. 47, Wiley, New York (1979).

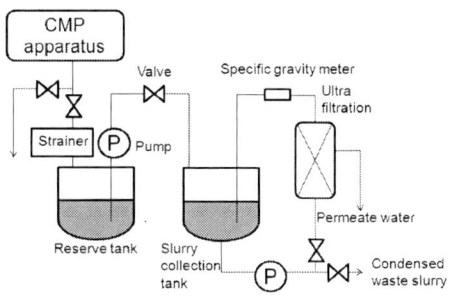

Fig. 1 Schematic diagram of slurry condensation system

Table I. Experimental conditions

	Parameters	Unit	Polishing	Pad conditioning
1	Down pressure	(kPa)	27.6	-
2	Retaining ring down pressure	(kPa)	37.1	-
3	Down force	(kgf)	-	2.3
4	Platen rotational speed	(min^{-1})	93	93
5	Head rotational speed	(min^{-1})	87	65
6	Pad conditioning		-	In situ
7	Sweep period	(sec)	-	12
8	Slurry		Ammonia-based fumed silica slurry	
9	Slurry flow rate	(ml/min)	170	-
10	Polishing pad		Polyurethane pad	

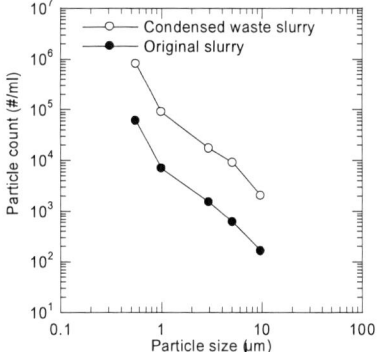

Fig. 2 Comparison of large particle counts between condensed waste slurry and original slurry

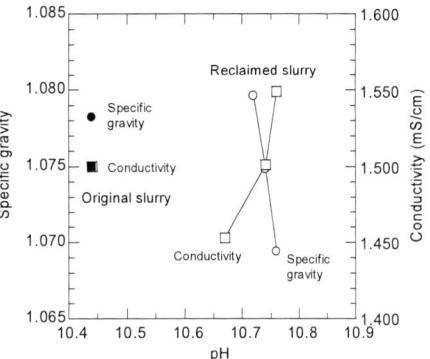

Fig. 3 Specific gravity and conductivity as function of pH value of slurries

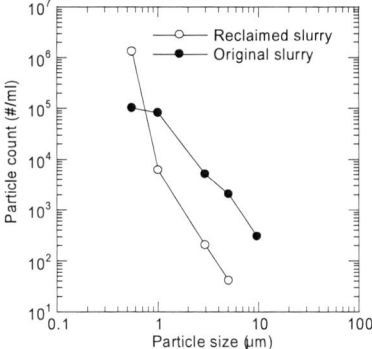

Fig. 4 Comparison of large particle counts between reclaimed slurry and original slurry

Table II. Characteristics of slurries

	Item	Unit	Reclaimed slurry	Original slurry
1	Abrasive size (mean)	(nm)	129	144
2	Abrasive size (median)	(nm)	120	127
3	Abrasive size (maximum)	(nm)	240	250
4	pH	-	10.49	10.44
5	Specific gravity	-	1.077	1.078
6	Conductivity	(mS/cm)	1.49	1.5
7	Zeta potential	(mV)	-38	-35
8	Trace metals (12 elements)	(ppm)	< 0.1 (Na < 1.0)	< 0.1

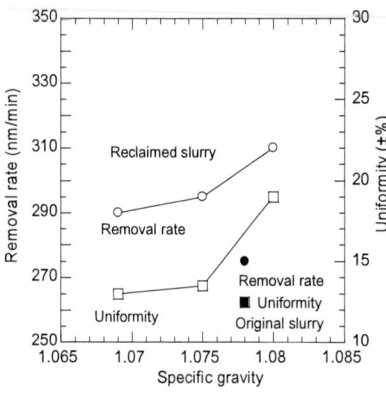

Fig. 5 Removal rate and uniformity as function of specific gravity of slurries.

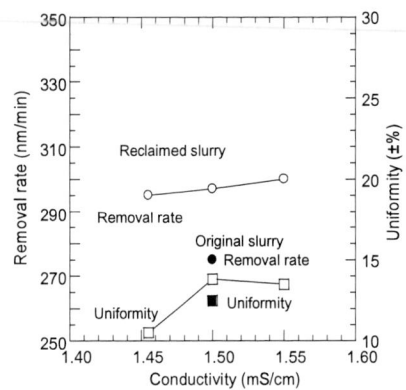

Fig. 6 Removal rate and uniformity as function of conductivity of slurries.

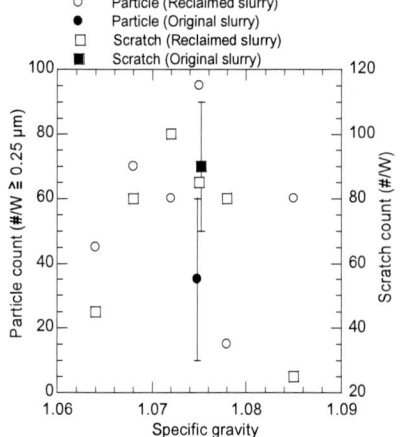

Fig. 7 Defect counts on PE-TEOS wafer as function of specific gravity of slurries.

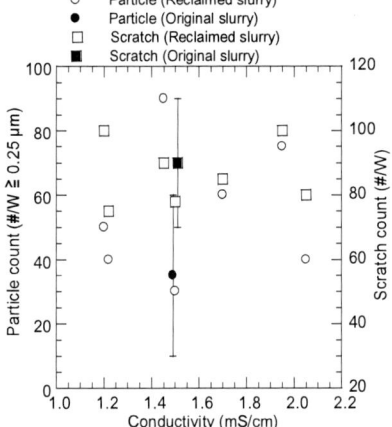

Fig. 8 Defect counts on PE-TEOS wafer as function of conductivity of slurries.

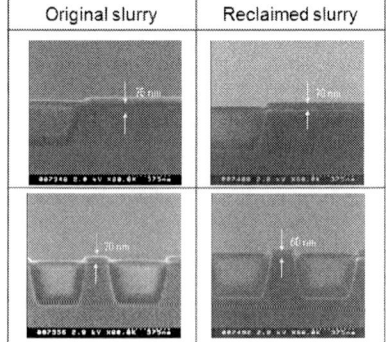

Fig. 9 Cross-sectional SEM images of STI region after CMP (×80K).

Correlation of Pad Topography, Friction Force and Removal Rate during Tungsten Chemical Mechanical Planarization

Yasa Sampurno [1,2], Adam Rice [2], Yun Zhuang [1,2], and Ara Philipossian [1,2]

[1] Araca, Inc., 2550 East River Road, Suite 12204, Tucson, Arizona 85718 USA
[2] University of Arizona, 1133 James E. Rogers Way, Tucson, Arizona 85721 USA

The evolution of coefficient of friction and removal rate during 8.5 hours of tungsten chemical mechanical planarization is correlated to pad surface topography via a novel pad surface descriptor termed 'abruptness'. Interferometric analysis results indicate that during the first 2.5 hours of polishing, pad abruptness remains stable. After 5.5 hours, pad abruptness decreases (i.e. surface becomes smoother). Results from polishing show similar trends whereby removal rate and coefficient of friction are stable during the first 2.5 hours period and decrease significantly thereafter. The coefficient of correlation between pad abruptness and coefficient of friction as well as pad abruptness and removal rate are 0.98 and 0.77, respectively.

Introduction

Pad surface properties play a critical role in establishing consistent chemical mechanical planarization (CMP) processes. As such, it is widely known that 'soft' polyurethane pads (typically having Shore D Hardness values of less than 15) are the preferred type of pads for tungsten CMP.1 While there is an abundance of published work highlighting how diamond disc conditioning is used to ensure consistent pad surface properties for 'hard' polyurethane pads (typically having Shore D hardness values of over 65), little is known in the form of published work for 'soft' pads and their longevity during tungsten CMP.

For a 'hard' polyurethane pad, diamond disc acts to regenerate the pad surface by opening the pores and refreshing the surface asperities. Without conditioning, the ability of the pad to hold and transport slurry, and to promote the polishing process, decreases. Stavreva et al. 2 showed that this can adversely affect wafer uniformity and the removal rate of copper CMP, thus making in-situ conditioning the preferred method for CMP. Zhuang et al. 3 also showed that conditioning down force on a 'hard' pad has a significant impact on pad surface properties and, therefore, removal rate and coefficient of friction (COF). Meled et al. 4 also indicated diamond disc has a significant contribution in pad cut rate during conditioning.

In contrast, diamond disc is not commonly used in conjunction with 'soft' pads applications because of the pad's fibrous characteristics which cause the pad to rip apart and disintegrated when conventional diamond discs are used. Instead, pad brushing,5 applied at low down forces, is employed in an ex-situ manner after a relatively long polishing time to clean up the pad surface.

In this study, surface topography of a 'soft' pad was characterized over 8.5 hours of tungsten CMP. A previous study used scanning electron microscopy (SEM) to obtain images of the pad surface before and after tungsten polishing.6 SEM produces high-resolution images of a polishing pad surface from which general insight regarding changes in pad topography can be described. This study builds upon and improves on the previous study by utilizing optical interferometry, a non-contact method that allows extraction of surface height probability density functions (PDF) and the surface abruptness (λ) parameter (a novel pad topography descriptor discussed later in this paper). This paper also correlates surface abruptness to friction force and removal rate.

Experimental Apparatus and Procedure

All experiments were performed using a 100-mm tribometer and polisher having the unique ability to record shear force in real-time during polishing. The polisher and its associated accessories have been described in detail elsewhere.7 The Politex REG E II pad was conditioned with a Duraclean 3020-SST brush at a constant down force of 2 lbf. During polishing, slurry was injected at a constant flow rate of 43 ml/min. The slurry was made up of a mixture of Cabot Microelectronics Corporation SS W2000 (43.3 % by volume), ultra pure water (43.3 % by volume) and 30 % hydrogen peroxide (13.3 % by volume). Polishing was performed at a constant pressure and sliding velocity of 4 psi and 1.1 m/s. To carry out the 8.5 hour polishing test, both 100-mm tungsten wafers as well as a 100-mm tungsten metal disc were used. The 100-mm tungsten wafers consisted of 4,500 angstroms of chemical vapor deposited tungsten on top of 500 angstroms of sputtered titanium on top of the silicon substrate. The tungsten metal disc having a purity of 99.9 % was used mainly to age the pad in between polishes with 100-mm wafers. Removal rate was determined by measuring the thickness of the tungsten wafer with an automated four point probe prior to, and after, polishing.

Prior to polishing, the pad was broken in for 5 minutes with the slurry flowing at 43 ml/min and with the brush down force maintained at 2 lbf. Next, the tungsten disc was polished continuously until a stable coefficient of friction was achieved. Initially, tungsten wafers were polished to obtain friction force and removal rate data at the beginning of the 8.5 hour polishing process after which a small pad sample was excised for interferometry. To age the pad, tungsten metal disc was then polished for 30 minutes followed by 2 minutes of pad brushing to remove the slurry residues. Tungsten wafers were polished next to obtain COF and removal rate data after which 3 pad samples were excised for interferometry. The above-described cycles were then repeated for pad life values of 2.5, 5.5 and 8.5 hours. The excised pad samples were dehydrated prior to performing interferometry within a 1.5 × 1.5 mm area of the pad sample.

Results and Discussion

Fig. 1 shows examples of the interferometric images obtained in this study. The general features are: (1) asperities which contact the substrate surface, and (2) bulk which consists of the majority of the pad. The color bar ranged from blue to red color depicts the pad surface height. Pad asperities are depicted in yellow-red color. The asperities extend vertically up from the reference plane determined by the interferometer. The pad bulk or pad pores are depicted in blue tones. White areas in the image correspond to values less than - 200 microns. The interferometric images are then further processed for PDFs. It must be noted that the interferometer measures surface heights over a selected area relative to an arbitrary reference plane used by the tool. The range of heights is divided into equal bins and a histogram is made of the number of times that the surface height falls into each one of these bin. Since the total count in each bin depends on the size of the region that is measured, it is convenient to normalize the histogram by dividing it by the area under the curve. It should be noted that by doing so, the result does not depend on the sampling area. The histogram generated using this method is called a PDF. The area under the PDF in a given height range indicates the probability of finding a point on the pad surface within that range. Because the reference plane is arbitrary, the mean of the PDF is then shifted to zero to facilitate the comparison of PDFs from different pad surfaces or different parts of the same pad surface.

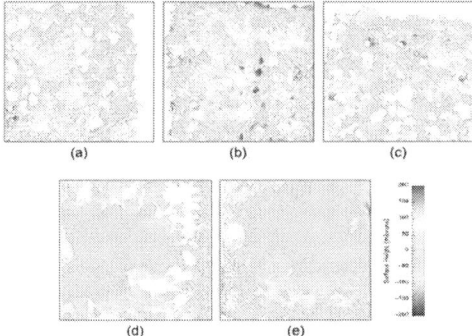

Figure 1. Interferometric images of pad surface taken at (a) the initial as well as after (b) 0.5,(c) 2.5, (d) 5.5, and (e) 8.5 hours of tungsten CMP

Heights to the left of the mean on the PDF correspond to surface points on the bulk side of the pad. On the other hand, heights to the right of the mean correspond to the pad asperities that contacts the wafer. When the asperity summits have exponentially distributed heights, then the right hand tail of the PDF will be linear on a log plot and can be characterized by a decay length λ. Therefore, λ is defined as the distance over which the right hand tail of the surface height probability density drops by a factor of e.8 Pad surface with a higher λ value indicates that its right hand tail of surface height probability density function decrease more slowly with surface height. This represents a rougher surface due to more variation in pad asperity heights.It must be noted that a simple pad surface roughness parameter, such as Ra, is not a proper metric to be studied in CMP simply due to the fact that the valley regions of the pad surface (which are captured by the parameter Ra) do not contact the wafer during polishing.

Fig. 1 shows interferometric images of pad surface taken at the initial as well as after 0.5, 2.5, 5.5, and 8.5 hours of tungsten CMP. At the early stages of polishing (i.e. up to 2.5 hours), a considerable number of pad pores and asperities are visible as shown in Fig. 1(a) to 1(c). On the other hand, Fig. 1(d) indicates that, after 5.5 hours, the pad surface has flattened significantly. The flattened surface is shown by the large green area on the interferometric image. Comparison of Figs. 1(c) and 1(d) indicates that the evolution of flattened pad surfaces increases significantly between 2.5 and 5.5 hours of polishing.

Flattened areas are notable deviations from the original pad topography and they expand as polish time approached 8.5 hours.

Fig. 2 shows surface height PDFs at 5 different polishing times. At a given polishing time, three surface height PDFs are constructed from 3 different land area of the extracted pad samples. Referring to Fig. 2, initially, topography of the land areas is uniform as indicated by their similar surface height PDFs shown in Fig. 2(a). However, it is interesting to note that the pad wear does not occur uniformly as polishing progresses. This is clearly seen after 5.5 hours of polishing when the surface height PDFs of the 3 different land areas becomes more varied compared to those at initial and after 2.5 hours. This observation confirms that the pad wear process does not occur uniformly. Later, the extent of pad wear saturates more uniformly after 8.5 hours of polishing as shown by Fig. 2(d).

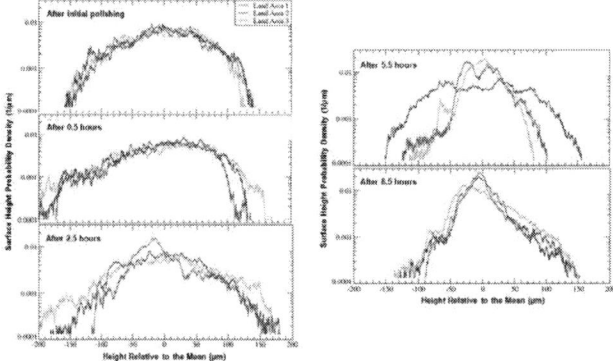

Figure 2. Surface height probability density functions of pad surface

Fig. 2 also indicates that, as the polishing time increases, the slope of right hand tail of PDF histogram increases indicating that surface abruptness, λ, is decreasing. The surface abruptness, λ, is inversely proportional to the slope.[8] The surface abruptness parameters are extracted from Fig. 2 and plotted in Fig. 3. The relative standard deviation of surface abruptness data is 30 percent. Fig. 3 clearly shows that the surface abruptness for the first 2.5 hours is relatively stable. The surface abruptness decreases significantly after 5.5 and up to 8.5 hours of polishing. Fig. 3 confirms that pad topography flattens significantly as polishing proceeds in between 2.5 and 5.5 hours which is due to rapid pad wear.

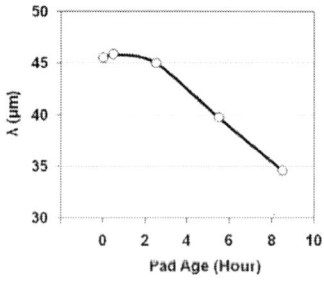

Figure 3. Pad surface abruptness as a function of polishing time

Fig. 4(a) shows the COF of tungsten wafer polishing as a function of polishing time. COF is calculated by dividing the average shear force over the applied down force. The relative standard deviation of COF data shown in Fig. 4(a) is 1.5 percent. During the first 2.5 hours, average coefficient of friction remains stable at 0.96. After 5.5 and up to 8.5 hours of polishing, average coefficient of friction decrease to 0.87 and 0.82, respectively. The most considerable decrease in COF is observed after 5.5 hours of polishing when a large flattened area is developed on the pad land area suggesting that pad flattening contributed to a lower COF. Figure 4(b) shows the removal rate data over the 8.5 hour period. The relative standard deviation of removal rate data shown in Fig. 4(b) is 2.0 percent. Similar to the surface abruptness and COF data, the removal rate is relatively stable during the first 2.5 hours at approx. 5,400 Angstroms/min. After 5.5 hours, removal rate drops to approx. 5,100 Angstroms/min and then levels off. By adopting the consumables and polishing conditions described in Experimental section, it is clear that the pad becomes significantly worn out thereby degrading the polishing performance sometime between 2.5 and 5.5 hours of operation. In high volume manufacturing, assuming a typical polish time of 2 minutes, such a result means that the Politex REG E II pad will, on average, last a mere 100 or so polishes.

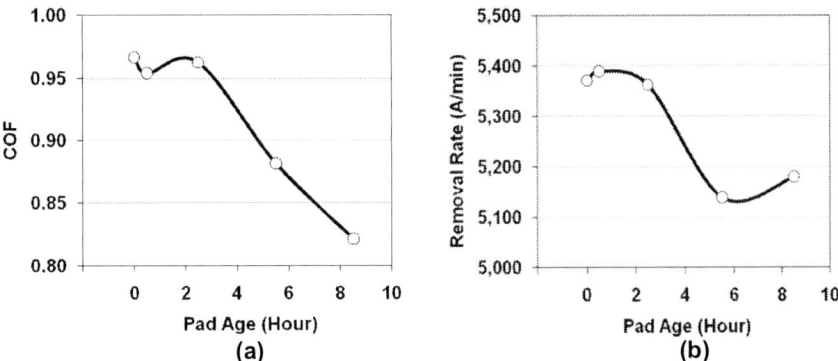

Figure 4. (a) Coefficient of friction and (b) removal rate as a function of polishing time

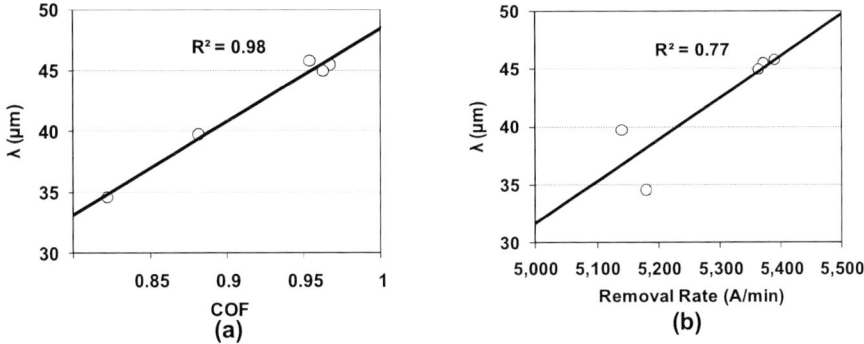

Figure 5. Correlation of pad surface abruptness with (a) COF and (b) removal rate

The plots of the pad surface roughness as a function of COF and removal rate, shown in Figs. 5(a) and 5(b), indicate a linear correlation with an R2 values of 0.98 and 0.77, respectively thus indicating that higher pad surface abruptness values result in higher values of COF and removal rate.

Conclusions

The evolution of pad surface topography, coefficient of friction and removal rate during 8.5 hours of tungsten chemical mechanical planarization is studied. Interferometry, a non-contact method, is used to measure a key parameter of pad surface known as surface abruptness, a metric that considers the variation of height of the pad asperities only. Results indicate that during the first 2.5 hours of polishing, the distribution of pad surface height probability density function, removal rate and coefficient of friction remain stable. After 5.5 hours of polishing, the distribution becomes narrower indicating that the pad surface becomes smoother and removal rate and coefficient of friction decreases accordingly. Friction force and removal rate results show qualitative correlations among surface abruptness, fiction force and removal rate whereby fiction force and removal rate decrease with decreasing surface abruptness.

References

1. C. Wang, E. Paul, T. Kobayashi, and Y. Li, in *Microelectronic Applications of Chemical Mechanical Planarization*, Y. Li, Editor, p. 125, Wiley Interscience, New York (2008).
2. Z. Stavreva, D. Zeidler, M. Plotner, and K. Drescher, *Applied Surface Science*, **108,** 39 (1997).
3. Y. Zhuang, L. Borucki, N. Rikita, F. Sudargho, Y. Sampurno, X. Wei, G. Steward and A. Philipossian, in *Proceedings of the CMP-MIC*, Institute for Microelectronics on Chip Interconnection, pp. 277–282 (2007).
4. A. Meled, Y. Zhuang, X. Wei, J. Cheng, Y. Sampurno, L. Borucki, M. Moinpour, D. Hooper, and A. Philipossian, J. Electrochem. Soc., 157(3) H250 (2010)
5. B. Neo, C. Lin, and L. Lau, Chemical Mechanical Polishing Process, United States Patent Application 20070111517.
6. A. Moy, J. Cecchi, D. Hetherington, and D. Stein, *Mat. Res. Soc. Symp. Proc.* **671** (2001), p. M1.7.1.
7. A. Philipossian and S. Olsen, *Jap. Soc. App. Phys.* **42** (2003) p. 6371.
8. T. Sun, L. Borucki, Y. Zhuang, and A. Philipossian: Microelectron. Eng. **87** (2010) 553.

ECS Transactions, 34 (1) 627-632 (2011)
10.1149/1.3567649 ©The Electrochemical Society

Tribological and Kinetical Analysis of Barrier Metal Polishing for Next Generation Copper Interconnects

R. Duyos-Mateo[a&b], X. Gu[a], T. Nemoto[a], S. Sugawa[a], Y. Zhuang[b&c],
Y. Sampurno[b&c], A. Philipossian[b&c] and T. Ohmi[a]

[a]Tohoku University, Aza-Aoba 6-6-10, Aramaki, Aoba-ku, Sendai, 980-8579, Japan
[b]Araca, Inc., 2550 East River Road, Suite 12204, Tucson, AZ 85718 USA
[c]University of Arizona, 1133 East James E. Rogers Way, Tucson, AZ 85721 USA

In this study, the tribological, thermal and kinetic attributes of Ti CMP process was investigated. Hitachi Chemical HS-T815 and HS-T605 slurries with different H_2O_2 concentrations were used to polish 200-mm blanket Ti wafers under different polishing conditions. Under the polishing pressure of 10.3 KPa, the measured shear force between the pad and wafer surface decreased significantly while the Ti removal rate increased significantly when small amount of H_2O_2 was added to the slurries. On the other hand, the shear force decreased and the removal rate increased slowly with further increase in the H_2O_2 concentration. A particle indentation model was used to explain the shear force behavior and Ti removal rate mechanism. The shear force decreases slightly when 0.06% H_2O_2 was added to the HS-T815 slurry and then remained stable with further increase in the H_2O_2 concentration.

Introduction

One of the challenges for chemical mechanical planarization (CMP) in the semiconductor industry is copper/ultra low-k dielectric integration. A barrier layer is used between copper interconnect and low-k dielectric materials to prevent copper diffusion into the low-k dielectrics. This layer also functions as an adhesive layer between copper interconnect and low-k dielectrics. The deposition and subsequent polishing of the barrier layer is one of the critical steps for ultra large-scale integrated (ULSI) circuits manufacturing. Currently, titanium (Ti) has been used as the barrier metal of choice for copper interconnects. However, Ti CMP process has not been studied extensively and Ti removal rate mechanism has not been fully understood. In this study, the tribological, thermal and kinetic attributes of Ti CMP process was investigated, and a particle indentation model was used to explain the observed shear force behavior and removal rate mechanism.

Experimental

All experiments were performed on an Araca APD-800 polisher and tribometer. A detailed description of the apparatus can be found elsewhere (1- 3). Blanket 200-mm Ti wafers were polished on an embossed Politex pad, which was conditioned ex-situ by a 3M PB32A brush under the conditioning force of 12.7 N. For the Ti wafers, Ti films were deposited on the silicon substrate through physical vapor deposition (PVD). Hitachi Chemical HS-T815 and HS-T605 slurries were used in this study. The main difference between these two slurries was that the HS-T815 slurry had a higher silica abrasive concentration. For each slurry, four different hydrogen peroxide concentrations (0, 0.06, 0.12 and 0.6 volume percent) were tested. The pH with different H_2O_2 concentrations was maintained at 2.5 due to the buffering agent in the slurries.

Results and discussions

Before wafer polishing, a blanket 200-mm Ti wafer was dipped in the Hitachi Chemical HS-T815 slurry with 0.06% H_2O_2 for 2 minutes. Electron Spectroscopy for Chemical Analysis (ESCA) was performed on the wafer before and after the dipping test. Figure 1 shows the ESCA results. There is a major peak at about 459 eV and two minor peaks at about 454 and 465 eV. Based on the binding energies for Ti (4), Ti peak appears at 454 eV and Ti oxide binding energy is between 458 and 466 eV depending of the oxide valence. Figure 1 indicates that there is Ti oxide layer on the Ti wafer surface before the wafer dip, and the oxide layer grows and becomes thicker after the wafer dip due to the reaction with H_2O_2 in the slurry.

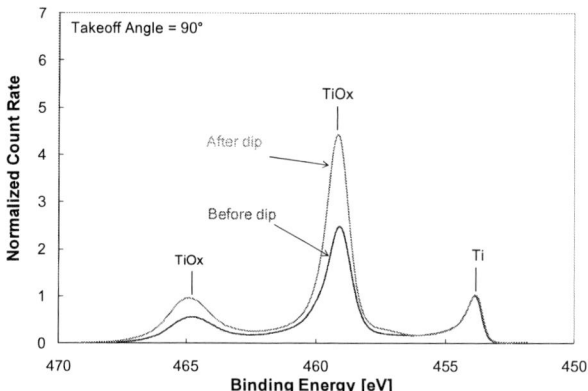

Figure 1: ESCA analysis of Ti wafer surface before and after wafer dip

Figure 2 shows the measured shear forces between the wafer and pad and Ti removal rates for the HS-T815 and HS-T605 slurries with different H_2O_2 concentrations. The polishing pressure and wafer/pad sliding velocity were maintained at 10.3 kPa and 1.2 m/s, respectively. In Fig. 2, there are two regions in which the shear force and removal rate behave differently. In Region 1, the shear force decreases dramatically while the Ti removal rate increases significantly with the increase of H_2O_2 concentration. In comparison, the shear force decreases and the removal rate increases slowly with further increase of H_2O_2 concentration in Region 2.

A removal rate mechanism based on particle indentation (5) is proposed in Fig. 3. When small amounts of H_2O_2 are added to the slurry, a thin TiO film is formed on the Ti wafer surface as shown in Fig. 3(a). During polishing, the abrasive particles penetrate the thin TiO film and generate indentations on the Ti surface under the applied pressure resulting in material removal. The shear force generated by the abrasives, F_s, is assumed to be proportional to the number of abrasives (N_a) and abrasive contact area with the Ti surface and TiO layer (δ_{Ti} and δ_{TiO}) as follows:

$$F_s = N_a \times f_s = N_a \times (f_{s,Ti} + f_{s,TiO}) = N_a(\mu_{Ti}A_{Ti} + \mu_{TiO}A_{TiO})p = N_a(\mu_{Ti}f(\delta_{Ti}) + \mu_{TiO}f(\delta_{TiO}))p \quad [1]$$

where f_s, is the average shear force generated by a single abrasive particle, μ is the coefficient of friction, δ is the indentation depth, and p is the polishing pressure. As the TiO layer is softer than Ti, it is assumed that μ_{Ti} is significantly larger than μ_{TiO}. When a very thin TiO film is formed on the Ti wafer surface for slurries with low H_2O_2 concentrations, the shear force and removal rate is dominated by the indentation depth of Ti (δ_{Ti}). As the thin TiO layer grows quickly with the increase of H_2O_2 concentration (6-8), δ_{Ti} decreases significantly leading to a dramatic decrease in the shear force for both slurries as shown in Region 1 of Fig. 2. On the other hand, the softer TiO layer is easily removed by mechanical abrasion, leading to a significant increase in the removal rate for both slurries. When a relatively thicker TiO film is formed on the Ti wafer surface for slurries with high H_2O_2 concentrations as shown in Fig. 3(b), the removal rate becomes dominated by δ_{TiO}. In Region 2 of Fig. 2, as the indentation depth of Ti (δ_{Ti}) continues to decrease and the TiO layer grows slowly, the shear force decreases slowly and the removal rate increases slowly with further H_2O_2 concentration increase for both slurries. As the HS-T815 slurry has a higher silica abrasive concentration, it generates consistently higher shear forces and removal rates than the HS-T605 slurry with different H_2O_2 concentrations as shown in Fig. 2.

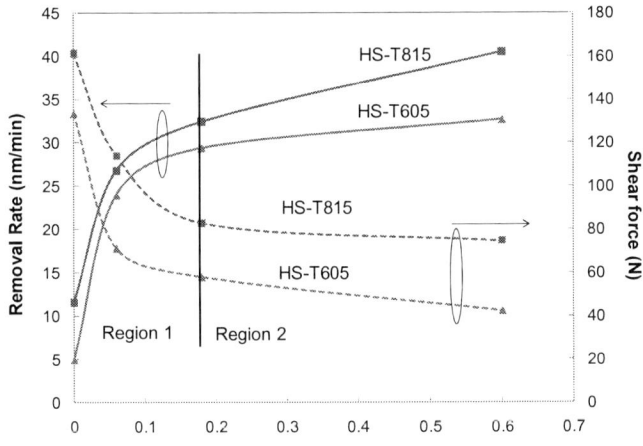

Figure 2: Ti removal rate and shear force vs. H_2O_2 concentration.

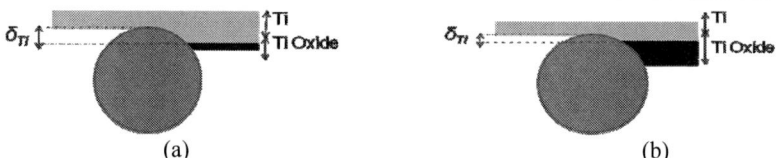

(a)   (b)

Figure 3: Removal rate mechanism by particle indentation.

Figure 4 shows the Ti removal rate and shear force comparison between the HS-T815 slurry without H_2O_2 and with 0.06% H_2O_2 under different polishing pressures. The wafer/pad sliding velocity was maintained at 1.2 m/s. The shear force of the HS-T815 slurry without H_2O_2 grows faster than the HS-T815 slurry with 0.06% H_2O_2 when the polishing pressure increases from 6.9 to 17.2 KPa. This is consistent with the shear force theory described by Eq. [1]. For the HS-T815 slurry without H_2O_2, the indentation depth of Ti (δ_{Ti}) increases significantly with the increase of polishing pressure, leading to a significant increase in the shear force. In comparison, for the HS-T815 slurry with 0.06% H_2O_2, the indentation depth of Ti (δ_{Ti}) increases only slightly with the increase of polishing pressure due to the formation of TiO layer on the Ti wafer surface, rendering a smaller increase in the shear force. The removal rate of the HS-T815 slurry without H_2O_2 is consistently lower than the HS-T815 slurry with 0.06% H_2O_2, confirming that the formation and indentation of the TiO layer enhances Ti removal.

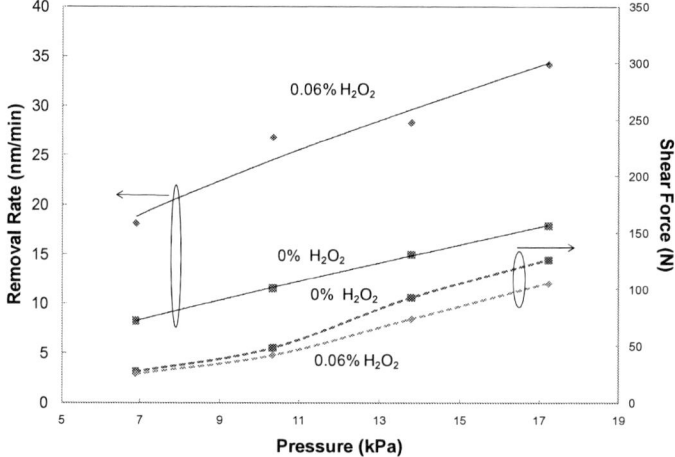

Figure 4: Ti removal rate and shear force under different polishing pressures.

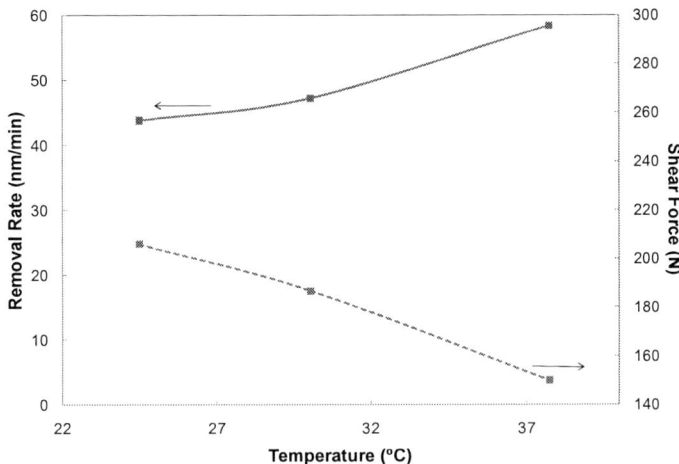

Figure 5: Ti removal rate and shear force under different platen temperatures.

Figure 5 shows the shear force and removal rate results for a separate set of experiments under three different platen temperatures (25, 30 and 38 °C). The platen temperature was maintained by an external heater during polishing. The HS-T815 slurry was used and the H_2O_2 concentration was 0.06%. The polishing pressure and wafer sliding velocity were maintained at 16.8 kPa and 1.2 m/s, respectively. As TiO layer formation is enhanced and indentation depth of Ti (δ_{Ti}) decreases at higher platen temperatures, the shear force decreases and the removal rate increases confirming the shear force theory described by Eq. [1] and removal rate mechanism shown in Fig. 3.

Conclusions

In this study, the tribological, thermal and kinetic attributes of Ti CMP process was investigated. Hitachi Chemical HS-T815 and HS-T605 slurries with different H_2O_2 concentrations were used to polish 200-mm blanket Ti wafers on an embossed Politex pad under different polishing conditions. Under the polishing pressure of 10.3 KPa, the measured shear force between the pad and wafer surface decreased significantly while the Ti removal rate increased significantly when small amount of H_2O_2 was added to the slurries. On the other hand, the shear force decreased and the removal rate increased slowly with further increase in the H_2O_2 concentration. A particle indentation model was used to explain the shear force behavior and removal rate mechanism. For slurries with very low H_2O_2 concentrations, a thin oxide film was formed on the Ti wafer surface. During polishing, silica particles penetrated the thin oxide film and made indentations on the underneath Ti layer. This resulted in high shear forces and low Ti removal rates. When the H_2O_2 concentration continued to increase in the slurry, a thicker TiO film was formed on the Ti wafer surface. During polishing, silica particles made indentations on the thick TiO film, rendering low shear force and high Ti removal rate. The shear force and Ti removal rate results under different polishing pressures and platen temperatures confirmed the particle indentation model with a bilayer mechanism.

Acknowledgments

The authors thank Hitachi Chemical Co., Ltd. for providing the slurries for this work.

References

1. X. Gu, T. Nemoto, Y. Sampurno, J. Cheng, S. Theng, A. Philipossian, Y. Zhuang, A. Teramoto, T. Ito, S. Sugawa and T. Ohmi, *Mat. Res. Soc. Symp.* Proc. Vol. 1157-E13-03 (2009).

2. http://www.aracainc.com/products/apd-800.

3. Y. Sampurno, X. Gu, T. Nemoto, Y. Zhuang, A. Teramoto, A. Philipossian, and T. Ohmi, Jap. *J. of Appl. Phys.* 45, 05FC01 (2010).

4. K. Hamrin, G. Johansson, A. Fahlman, C Nordling and L. Ramquist, *J. Phys. Chem. Solids*, 30, 1835 (1969).

5. D. Bozkaya and Sinan Muftu, *J. Electrochem. Soc.*, 156, H890 (2009).

6. S. Chiu, Y. Wang, C. Liu, J. Lan, C. Ay, M. Feng, M. Tsai and B. Dai, *Mater. Chem. Phys.*, 82, 444 (2003).

7. R. Duyos-Mateo, X. Gu, T. Nemoto, S. Sugawa, Z. Han, Y. Zhuang, Y. Sampurno, A. Philipossian and T. Ohmi, *International Conference on Planarization/CMP Technology*, pp.13-18, (2010).

8. V. S. Chathapuram, T. Du, K. B. Sundaram and V. Desai, *Microelectron. Eng.*, 65, 478 (2003).

Finite Element Analysis (FEA) of Pad Deformation Due to Diamond Disc Conditioning in Chemical Mechanical Polishing (CMP)

E. A. Baisie [a], B. Lin [b], X. H Zhang [c] and Z. C. Li [a]

[a] Department of Industrial & Systems Engineering, North Carolina Agricultural & Technical State University, Greensboro, North Carolina, USA 27411
[b] College of Mechanical Engineering, Tianjin University, Tianjin, China 300072
[c] Seagate Technology, Minneapolis, Minnesota, USA 55435

Chemical mechanical polishing (CMP) is widely used to planarize semiconductor wafers and smooth the wafer surface. In CMP, diamond disc conditioning is traditionally employed to restore pad planarity and surface asperity. Despite the advancement of studies on diamond disc pad conditioning, there are very few published reports about the effect of diamond disc conditioning on pad deformation. A two-dimensional (2-D) axisymmetric quasi-static finite element analysis (FEA) model is proposed to investigate the interaction between the diamond disc conditioner and the polishing pad for the first time. The FEA model is developed by using the ANSYS software to study the effects of process parameters (conditioning pressure, pad thickness and pad hardness) on pad deformation. The study leads to a better understanding of the relationship between the diamond disc conditioner and the polishing pad in CMP. It also provides a basis for three-dimensional and dynamic analysis in the future.

Introduction

Chemical Mechanical Polishing (or Planarization) (CMP) is a final major manufacturing step extensively used in semiconductor fabrication for planarizing semiconductor wafers or other substrates. As shown in Figure. 1, the CMP machine consists of a wafer carrier with a retaining ring and a rotating polishing pad mounted on a rotatable platen. First, corrosive slurry containing fine abrasive particles is released onto the porous pad and attacks the wafer chemically weakening it. This allows the mechanical action of the abrasive particles held by the pad to easily facilitate material removal. Here, the wafer is held in the carrier by which a down force is applied to press the wafer against the rotating polishing pad. During the polishing process, slurry and materials removed from the wafer glaze the surface of the polishing pad, making the polishing pad slick. In the absence of a pad regeneration process, this results in degradation of the pad surface planarity and roughness.

Conditioning is used to regenerate the pad surface by breaking up the glazed areas generally via an abrasive device. Today, diamond disc conditioning is the most widely used method of pad conditioning in wafer fabrication facilities. Conditioning plays a key role in maintaining removal rates , within-wafer non-uniformity (WIWNU %) and extending the life of the pad (1). Zhou (2) showed that material removal rate (MRR) can be maintained at the same level with proper pad conditioning and NU can also be improved. During the conditioning process, a conditioning assembly is attached to the CMP machine.The rotating conditioner is fed down to the pad from an overhead position and pressed against the pad surface under a predetermined down force. Then the conditioner shifts in a straight line from the pad centre to the pad periphery at desired intervals and lifts up to end the cycle. As this ucurrs, local compression is introduced on the pad suface resulting in deformation. Pad deformation is crucial to the fundamental understanding of the material removal mechanism of CMP since pad shape, stress and strain are realated to cut rate during conditioning, pad wear rate and wafer MRR during polishing (3, 4).

Despite the advancement of experimental and theoretical studies on the diamond disc pad conditioning, there are only few published reports about the effect of conditioning on the pad deformation. Horng (4) developed pad deformation equations for conditioning using Hertzian contact theorem and principle of elasticity. The theoretical results revealed that the deformation at the centre of the conditioner is significant. But, Horng's method is limited to certain conditioning configurations. Finite Element Analysis (FEA) may be used to perform more extensive studeis on conditioner-pad interaction. A set of CMP investigators have used the FEA approach to model the interactions between the pad, wafer and abrasive particles to predict wafer MRR (5-8), WIWNU (9, 10), wafer flatness (11), pad surface asperities (3) and pad wear (12-14). This powerful computational approach allows for 3-D geometries and more detailed representation of physical characteristics and mechanics of the process components in the model. However, none of the available FEA modeling reports considers pad conditioning and its effects on the process.

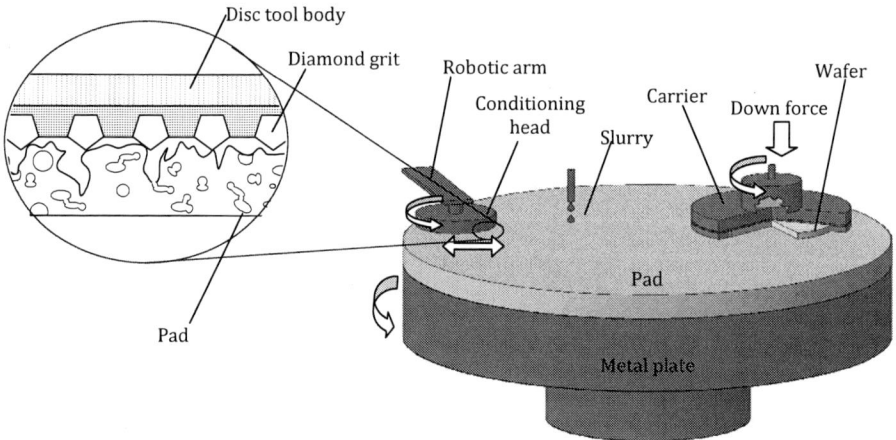

FIGURE. 1 SCHEMATIC OF CMP PROCESS WITH INSET SHOWING
CONDITIONER-PAD INTERACTION

In this paper, a two-dimensional (2-D) axisymmetric quasi-static FEA model is proposed to investigate the interaction between the diamond disc conditioner and the polishing pad for the first time. First assumptions are drawn and the model is developed. Then the model is utilized in the ANSYS software to study the effects of some process parameters on the pad deformation.

Model development

The inset in Figure. 1 illustrates the interaction between one type of diamond disc conditioner and the pad. Stresses on the pad surface arise mainly from two sources; (1) down pressure from the conditioner head and (2) shear forces due to the relative motion of the conditioner and pad. For simplicity, the contact surface between the conditioner and pad can be considered as a smooth plane. Then a uniformly distributed normal pressure acts on the conditioner head. The sliding contact causes a traction force which can be correlated to the relative velocity between the conditioner and pad. But this can be ignored if the pad and conditioner rotating speeds are similar. With the given configuration, static equilibrium is necessitated to solve the deformation equations. A quasi-static simulation is thus employed. Considering symmetry of pad shapfe and quasi-static conditions, loading can be assumed to be axisymetrically distributed. In this model, the pad and the conditioner are assumed to be free from internal forces prior to the application of loads. The pad is porous in nature and does not expand significantly in the lateral directions when compressed, therefore a Poisson ratio of 0.2 is assumed. For the conditioner, a Poisson ratio of 0.3 is assumed. Depite the anisotropic property of the pad , it is assumed to possess isotropic mechanical properties due to the complexity in obtaining anisotropic data.

The deformation problem is modeled using Surface-to-Surface contact approach in ANSYS. Since deformation of the conditioner is considerably less than that of the pad, a rigid-flexible contact is assumed. The bottom of the conditioner is modeled as the target surface while the top surface of the pad is modeled as the contact surface with a friction coefficient of 0.3. In this way, the target surface elements can impact the contact surface upon loading. Figure. 2 shows the boundary conditions for the 2-D FEA model. For a cylindrical shaped conditioner, the pad deformation is more critical at the areas near the edge of the conditioner. Thus the model is contrained to a neigbourhood of 2mm from the conditioner edge. The model components are discretized into 2560 elements and 2553 nodes. Due to the high aspect ratio of the pad, a much finer mesh was used for the pad in camparison to the conditioner. Meshing was performed such that the discplacements of adjacent nodes along the conditioner-pad line of contact are identical in all directions. Since the bottom surface of the pad is fixed, all nodes lying on the bottom plane are fixed in all degrees of freedom. Displacement of all nodes lying along the axis of symmetry are restricted to the z-axis.

Results and discussion

To investigate the effect of process parameters including pad thickness, pad hardness and conditioning pressure on the pad deformation, three levels of each parameter are simulated. Table I. provides a list of the parameter values used at each level. The chosen parameter values are considered from literature and standard manufacturer specifications (4, 13). Figure. 3 shows the variation of deformation of the pad surface and subsurface. It can be seen that upon deformation, the pad surface directly under the conditioner remains flat and experiences the most deformation. The surface out of contact with the conditioner curves toward the edge of the conditioner. Also, the intensity of deformation reduces away form the corner of the conditioner.

TABLE I. Parameters used for simulation

Parameter	Level		
	1	2	3
Pad Thickness t (mm)	0.762	1.397	2.032
Pad Elastic Modulus E (MPa)	90	110	130
Conditioning Pressure P (N)	10	12	14

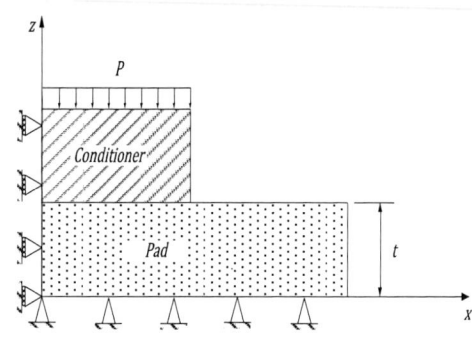

FIGURE.2 BOUNDARY CONDITIONS OF 2-D FEA MODEL

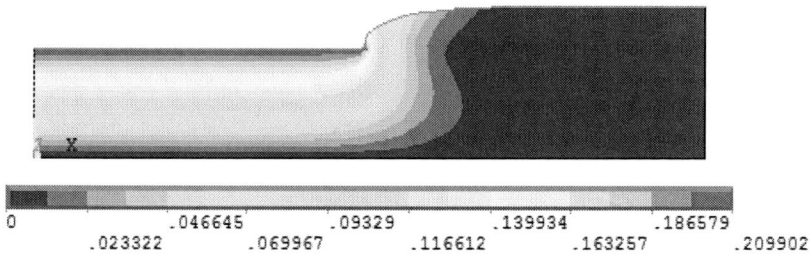

0		.046645		.09329		.139934		.186579	
	.023322		.069967		.116612		.163257		.209902

FIGURE. 3. VARIATION OF AVERAGE PAD DEFORMATION (t = 0.763 mm, E=130 Mpa, P =10 Psi, Length of pad = 4 mm)

Pad thickness is highly correlated to CMP process outputs. For three pads of different thickness t, simulation is conducted with the pad hardness and conditioning pressure held constant. From Figure. 4, the conditioner has less of a drawing effect on pad as the thickness increased. In terms of depth, it is observed that deformation increases with pad thickness. This can be attributed to reduced ability of the pad structure to resist deformation. Seok (15) reported that when the pad is modeled as a simple Wrinkle elastic foundation or "mattress", the effective pad stiffness modulus is inversely proportional to the thickness. This could mean that processes which neccessitate thicker pads may not require additional conditioning force to maintain the same level of conditioning. The effect of pad hardness is investigated by varying the Elastic Modulus E of the pad material. Results from Figure.5, show that deformation increases as expected when pad material becomes softer. However, unlike the resulting shapes for t, the deformed shapes are similar for all three levels of E. The deformed shapes exhibit a steep curve with a slight bulge above the top surface of the pad. In CMP, conditioning pressure is usually increased to intensify the regeneration of pad surface. In Figure. 6, the conditioner is subjected to three conditioning loads. Results show that deformation increases with increasing conditioning pressure. The evenly spaced curves suggest that conditioning prressure may have a linear relationship with the depth of deformation.

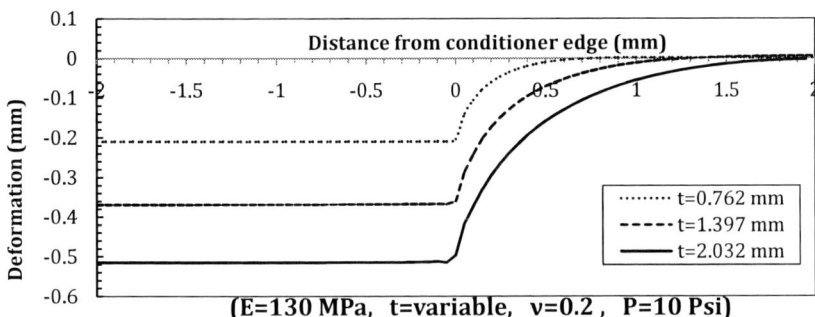

FIGURE 4. EFFECT OF PAD THICKNESS (t) ON DEFORMATION

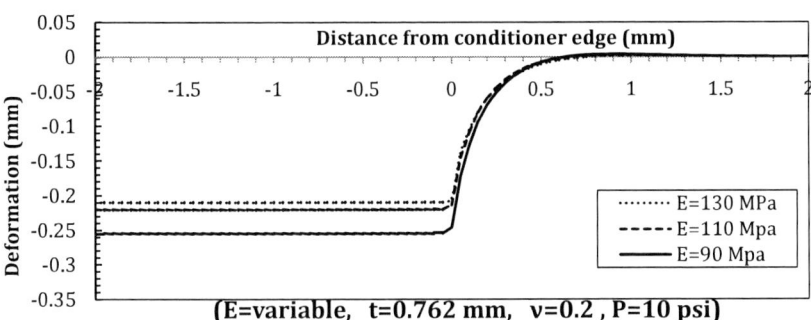

FIGURE. 5. EFFECT OF PAD HARDNESS (E) ON DEFORMATION

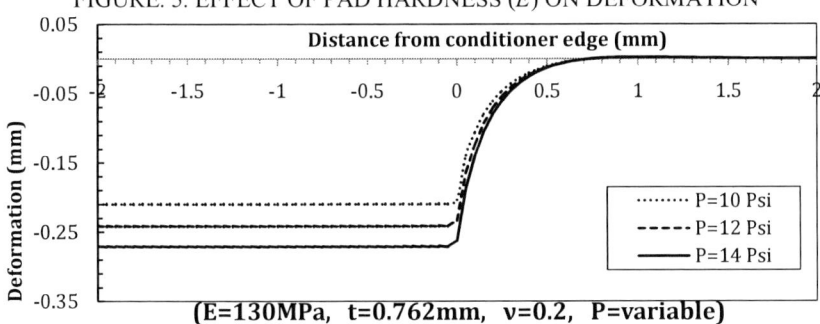

FIGURE. 6. EFFECT OF CONDITIONER PRESSURE (P) ON DEFORMATION

Conclusions

Pad deformation is crucial to the fundamental understanding of the material removal mechanism of CMP. Based on key assumptions about the configuration of typical pad conditioning, a 2-D axisymmetric quasi-static FEA model was developed utilizing the ANSYS software. Surface-to-Surface contact elements under rigid-flexible contact conditions were used. The model was used to investigate the effect of process parameters including pad thickness, pad hardness and conditioning pressure on the pad deformation. Three levels of each parameter were simulated. Results show that: (1) the conditioner slightly indents the pad with its top surface bending towards the edge of the conditioner, (2) deformation increases with pad thickness, (3) deformation increases as the pad's elastic modulus decreases, and (4) deformation increases with increasing conditioning pressure. The study has led to a better understanding of the relationship between the diamond disc conditioner and the polishing pad in CMP. Future work could include the investigation of the effect of the conditioner design and more accurate modeling of the pad's anisotropic nature. A three-dimensional dynamic analysis of the interaction between the conditioner and the polishing pad could also be considered in the future.

References

1. E. A. Baisie, Z. C. Li and X. H. Zhang, in *ASME International Manufacturing Science and Engineering Conference 2009, MSEC2009, October 4, 2009 - October 7, 2009*, p. 661, Proceedings of the ASME International Manufacturing Science and Engineering Conference 2009, MSEC2009, West Lafayette, IN, United states (2009).
2. Z. Zhou, J. Yuan, B. Lv and J. Zheng, in, p. 309, Key Engineering Materials, Laubisrutistr.24, Stafa-Zuerich, CH-8712, Switzerland (2008).
3. B. Jiang and G. P. Muldowney, in *2007 MRS Spring Meeting, April 10, 2007 - April 12, 2007*, p. 39, Materials Research Society Symposium Proceedings, San Francisco, CA, United states (2007).
4. T.-L. Horng, *Key Engineering Materials*, **238-239**, 241 (2003).
5. D. Bozkaya and S. Muftu, *Journal of the Electrochemical Society*, **156**, H890 (2009).
6. W. Che, Y. Guo, A. Bastawros and A. Chandra, in *2002 MRS Spring Meeting, April 1, 2002 - April 5, 2002*, p. 90, Materials Research Society Symposium Proceedings, San Francisco, CA, United states (2002).
7. B. Yan, X.-M. Zhang and X. Lu, *Gongcheng Lixue/Engineering Mechanics*, **21**, 126 (2004).
8. J. McGrath and C. Davis, in, p. 305, Proceedings - Electrochemical Society, Orlando, FL., United states (2003).
9. D. H. Lee, D. J. Kwon, Y. K. Hong and J. G. Park, *Key Engineering Materials*, **257-258**, 433 (2004).
10. Y.-Y. Lin, S.-P. Lo, S.-L. Lin and J.-T. Chiu, *Journal of materials processing technology*, **202**, 156 (2008).
11. X. H. Zhang, Z. J. Pei and G. K. Fisher, in, p. 893, American Society of Mechanical Engineers, Manufacturing Engineering Division, MED, New York, NY 10016-5990, United States (2005).
12. T. Pei-Lum and H. Rick, *International Journal of Advanced Manufacturing Technology*, **32**, 682 (2007).
13. M. Li, Y. Zhu, J. Li and K. Lin, in *10th International Conference on Machining and Advanced Manufacturing Technology, November 7, 2009 - November 9, 2009*, p. 318, Key Engineering Materials, Jinan, China (2010).
14. T. Nishioka, S. Iwami, T. Kawakami, Y. Tateyama, H. Ohtani and N. Miyashita, in *Chemical-Mechanical Polishing 2000 - Fundamentals and Materials Issues. Symposium, 26-27 April 2000*, p. 1, Warrendale, PA, USA (2001).
15. J. Seok, C. P. Sukam, A. T. Kim, J. A. Tichy and T. S. Cale, *Wear*, **254**, 307 (2003).

Data Driven CMP Manufacturing Modeling
for Process and Design Optimization

Li J. Song and Vikas Mehrotra

Ascertin LLC, Fremont, CA 94539, USA

A hybrid model for CMP process recipe and IC design optimization is introduced. It combines data mining techniques, CMP physical aspects, and uses existing CMP metrology data. The model has been validated to within 7% of the actual thickness measurements on a test chip, and the simulated topography profiles closely match those measured by an AFM. The hybrid model provides intermediate results for all the process steps, including ECD, bulk, touchdown and barrier removal and runs an order of magnitude faster than physical based model. The calibrated hybrid model has been used for wafer level CMP hotspots detection, CMP process recipe change, optimization, dummy fill generation, and rule file development. A novel CMP metal fill tool is also developed to incorporate a field solver for critical nets RC parasitic and timing impact estimation, and a mechanical model for ultra-low k dielectric mechanical strength consideration in addition to a hybrid model.

Introduction

In this paper we introduce a hybrid model for CMP process recipe and IC design optimization. It is well known that dishing and erosion in CMP process, topography and thickness variations [1,2] cause chip performance and yield issues[3-5]. Traditionally, look up tables have been used in the design to account for such variations, and dummy fills are used to reduce metal density variation. However, complex physical effects such as long range and multi-level effects of CMP process are ignored and the predictions have limited accuracy. On the other hand, a physics based model has also been used in the design to simulate the topography and thickness profiles based on calibrated model parameters and principles of the chemical and mechanical processes[6,7]. Although much more accurate, such a physics based model is not only difficult to establish and calibrate, but also takes up to several hours to predict a full chip design, limiting its application to final verification stage when design change implications are significant. Therefore a fast process variation simulation model is important in both process and design optimization.

Hybrid Modeling Approach

The hybrid model is a data driven model which uses existing CMP metrology data. No special test structures are required to calibrate the model and the model establishment is relatively easy. It combines data mining techniques and CMP physical principles as shown in Figure 1. As a result, long range and multilevel CMP effects are incorporated in the model, to achieve equivalent accuracy to physics based models. But the hybrid model does not solve physical equations like traditional physics based model does, hence the simulation time is much faster and the run time has been reduced to seconds or minutes from hours when compared to a physics based model. It is also capable of simulating wafer level designs, which is currently considered to be extremely computationally intensive by a physics based model.

The input of the model is similar to physics based model; it requires design information such as metal density and effective line width, process recipe and calibrated model library. The output of the model includes intermediate results for all the process steps, including ECD, bulk, touchdown and barrier removal in CMP process. This is a feature typically not found in the lookup table based approach.

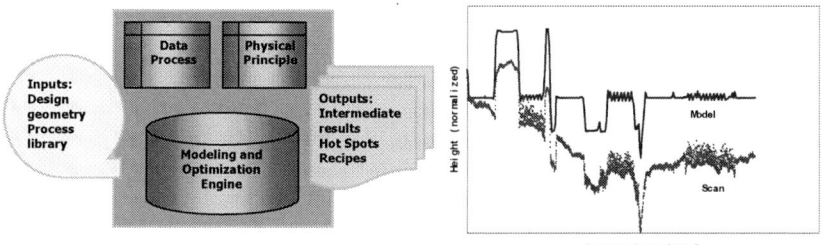

Figure 1. A Hybrid modeling approach and its inputs and outputs.

Figure 2. Modeled surface height and the comparison with AFM unleveled scans.

The model has been validated to within 10% of the actual thickness measurements on a test chip, as shown in Table 1. Most of the modeled metal thickness values are within 7% of the thickness values measured by SEM. The simulated topography profiles also closely match AFM scans as shown in Figure 2. In the figure, the AFM scan is not leveled, but it can be seen that the model not only predicts the surface variations from block to block well, but also provides excellent details for the structures within the block.

TABLE 1. Modeled metal thickness comparison with SEM measurement

Sites	1	2	3	4	5	6	7	8	9	10	11	12	13	14
Error%	10.7	0.7	0.3	-3.8	-7.1	-2.4	-0.1	-5.5	6.0	1.6	3.2	-3.4	5.4	8.9

Process Optimization With Wafer Level Simulation

The calibrated hybrid model has been used for wafer level CMP hotspot detection, and CMP process recipe change and optimization. Simulation results show that new hotspots are detected when wafer level variations are included. This demonstrates the importance of wafer level simulations as CMP effects are long range in nature. Using a hybrid model developed for various process conditions such as polish pressure and time at different process steps, the process recipe can be changed to reduce or eliminate hotspots and identify the origin step of the hotspot, saving much guess work, DOE and wafer split runs.

Given wafer level thickness measurement data, the hybrid model can simulate the thickness profile for the entire wafer. Figure 3 shows an example of the wafer level prediction capability. Full wafer level simulations of various parameters such as step height, dishing, erosion, and residual copper thickness can be generated for each step in the CMP process. In this example, a wafer with 39 chips size 6mm by 6mm with 2 metal levels can be simulated within about 8 minutes on a laptop with 2GHz processor and 4GB ram, at a resolution of 20 micron grid size.

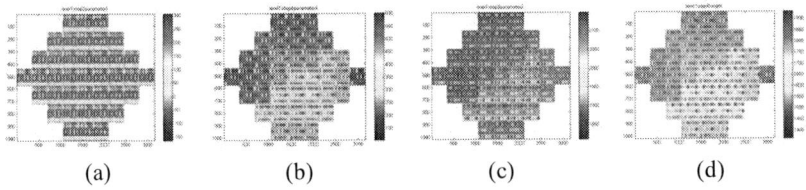

(a)   (b)   (c)   (d)

Figure 3. Wafer level thickness predictions for a logic chip: (a) bulk step height, (b) barrier step height, (c) touchdown surface height, and (d) barrier copper thickness.

Once the wafer level simulation is completed, it can be used to detect wafer level hotspots that are outside of manufacturing specifications, and correct them through process optimization. Since the simulation is performed at intermediate steps, the optimal process settings may not necessarily involve modifying the pressure or polish time in the final (eg. barrier) CMP step. Figure 4 shows an example where the optimal process setting (least number of hotspots) results by modifying the pressure at an intermediate (touchdown) step.

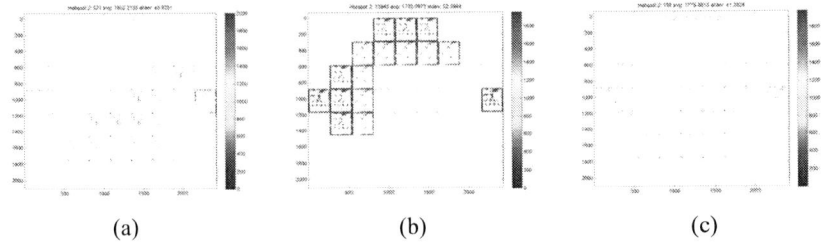

(a)   (b)   (c)

Figure 4. (a) The number of hotspots with the original process is 521. (b) Increasing the polish time in the barrier step increases the number of hotspots to 13945. (c) The optimal process setting is to change the pressure in the touchdown step, reducing the number of hotspots to 156.

Design Optimization With Metal Fill

Another way to improve thickness uniformity and reduce the number of hotspots is through design optimization. Previous approaches have used either density based [8-9], rule based [10], or physics model based [11] criteria to predict the CMP topography variation for metal fill. These techniques may not provide the accuracy or speed that is required for applications at advanced process nodes. We propose a novel CMP metal fill tool that is developed to incorporate the following items for metal fill optimization: 1. a hybrid model for fast topography and thickness simulation; 2. a field solver to estimate RC parasitic and timing impact for critical nets; 3. a mechanical model for ultra low k dielectric mechanical strength consideration.

Figure 5 describes our technique. Several parameters can be specified by the user. This includes process requirements for metal fill such as minimum and maximum fill size, minimum fill-to-signal space, and minimum fill-to-fill space. Additionally the user can specify criteria for electrical and mechanical requirements. Given these constraints, the algorithm can determine the best fill size and density for each small region or grid within the design such that it results in the most optimized solution.

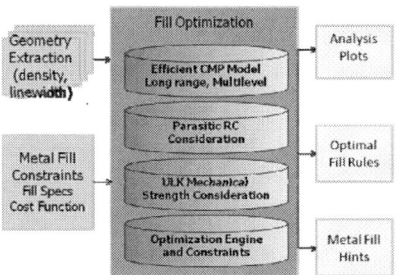

Figure 5. A novel metal fill optimization methodology. Process parameters, design information, and user defined constraints are all input to compute the optimal metal fill.

Figures 6(a)-(b) show the pattern density and distribution of a test chip without any fill and Figures 6(c)-(d) show the pattern density and distribution with the optimal metal fill. The results show that the standard deviation of density reduces from 29% to 11% with optimized metal fill, resulting in a much more uniform distribution. The corresponding copper thickness is shown in Figure 7 with and without metal fill. The standard deviation of copper thickness is reduced from 115 Angstroms to 50 Angstroms, as predicted by the model.

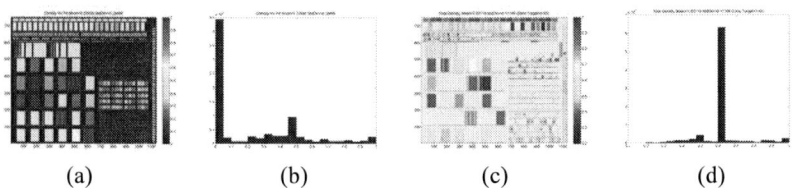

| (a) | (b) | (c) | (d) |

Figure 6. The pattern density of the test chip without (a),(b) and with (c),(d) optimized metal fill. The standard deviation of density is reduced from 29% to 11% with optimal fill.

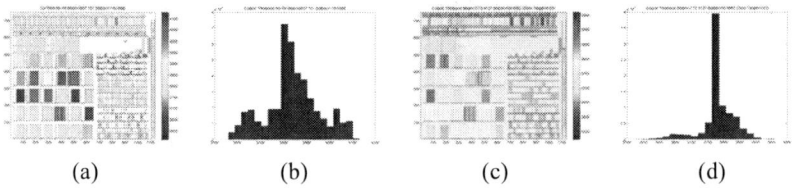

| (a) | (b) | (c) | (d) |

Figure 7. The corresponding copper thickness as predicted by the model both without (a),(b) and with (c),(d) optimized metal fill. The standard deviation of copper thickness is reduced from 115 Angstroms to 50 Angstroms with optimal fill.

Additional parameters may also be specified, such as the maximum acceptable change in capacitance to preserve the timing. A field solver is used to compute the capacitance to adjacent signal wires without any metal fill. The net segment capacitance is first simulated based on the geometry (nominal metal thickness, average linewidth inside the grid) and film stack parameters (dielectric thickness, dielectric constant). Based on the user specified tolerance for capacitance increase, the acceptable buffer distance is determined for each grid. The use of the hybrid model enables us to quickly simulate the CMP thickness. Each iteration computes the capacitance and resistance with the predicted copper thickness. If the percent RC change due to metal fill is greater than the user specified value, the buffer distance is increased until the optimal value is found.

Another constraint that can be specified by the user is minimal via density to preserve the mechanical strength, which may be an issue for low k dielectrics. One technique to improve the via density is to insert dummy via between two adjacent metal layers where there is already existing metal fill to ensure that there is no shorting between signal wires. Our metal fill methodology can estimate the percent increase in via density by considering the probability of metal fill overlap between two adjacent metal layers and include this additional constraint (the minimum percentage increase in via density) as part of the optimization. Figure 8 shows an example of the multivariate optimization algorithm that includes thickness uniformity, electrical impact, and mechanical strength consideration.

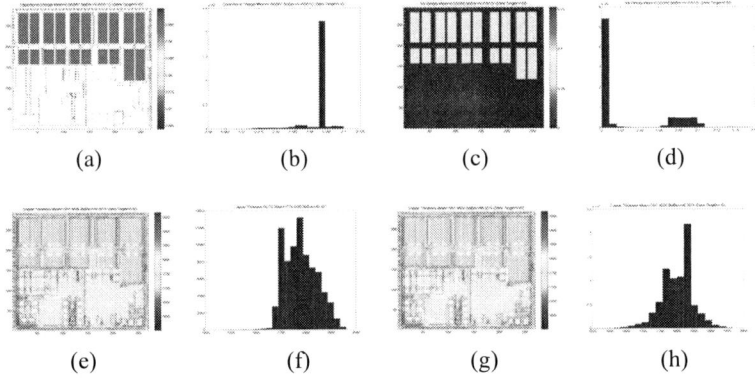

(a)   (b)   (c)   (d)

(e)   (f)   (g)   (h)

Figure 8. (a), (b) The capacitance change for a design with logic and memory, with a mean increase of 9% (target is a maximum increase of 10%). (c),(d) The via density with a mean of 2.5% (target is a minimum of 2%). (e), (f) The copper thickness without metal fill, with 62 Angstroms standard deviation. (g), (h) The copper thickness with metal fill, with 46 Angstroms standard deviation.

In addition to improving thickness uniformity, the model based approach can help reduce the number of other hotspots such as density, thickness and surface height gradients, topography variation, etc. While typical minimum density rules for CMP are generally around 15% or 20%, the optimal target density for thickness uniformity may be much higher. For a logic chip of size 6mm by 6mm, our results indicate that the optimal target density is 68%. Figure 9(a) compares the number of copper thickness hotspots with the density design rule vs. our optimization algorithm. The number of hotspots is reduced from over 7000 to just 100.

When provided, the topography of the incoming surface may also be input (e.g. the M3 surface when running the M4 metal fill algorithm). In Figure 9(b), the number of surface height hotspots is compared with the density design rule vs. our optimization algorithm. In this case, choosing the minimum density design rule makes a big difference in the number of surface height hotspots (over 2500 with 15% vs. 500 for 20% rule). However, the number of rule based hotspots is still much higher than with the automated optimization algorithm (150).

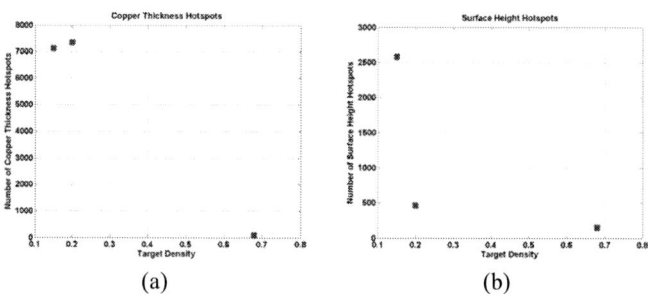

(a)   (b)

Figure 9. The number of hotspots is compared for copper thickness (a) and surface height (b) using 15% and 20% minimum density design rules vs. the metal fill optimization algorithm. Copper thickness hotspots are reduced to 100, surface height hotspots to 150.

Having general design rules for CMP does not guarantee that we will have the best CMP thickness uniformity or will eliminate all hotspots. We have seen that the optimal target metal fill density depends on several factors, including the process, design, metal level, and performance targets. Figure 10 shows the target metal fill density for different types of designs such as a test chip, logic chip, and logic/memory combination. These results indicate that a single target design rule for metal density is not adequate for high performance design since the optimal target density varies.

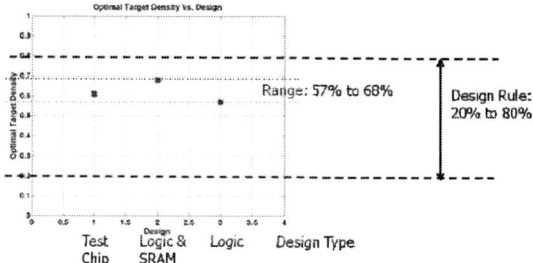

Figure 10. The optimal target density for different types of designs varies based on design type, process, metal level, and user defined constraints.

Summary

A hybrid CMP modeling methodology has been presented. We have shown that our model accuracy is comparable to that of a physics based model with significantly shorter simulation times of minutes compared with hours. We have shown that this methodology can be used for both process and design optimization with the implementation of the hybrid model in an automated software tool. The use of wafer level data can allow us to generate a full wafer level simulation at high resolution within minutes. With intermediate prediction results for thickness uniformity and hotspot simulation, our technique can be used to determine optimal process polish time and pressure. Our results indicate that a process change in an intermediate step may be required to determine the optimal process settings. Finally, we have developed a design optimization methodology through the use of metal fill insertion. The multivariate optimization technique ensures that all the aspects of IC design related to CMP are covered when developing fill rules. Simulation results show chip design using this metal fill technique yields much fewer hotspots while preserving the timing and mechanical strength.

References

1. L. Song , K. H. Chen , Taber Smith, "The Role of CMP Model in DFM", *ISTC* (2008).
2. X. Zhang, L. He, V. Gerousis, L. Song and C. Teng, "Case Study and Efficient Modeling for Variational CMP", *IET Circuit, Device and Systems* (2007).
3. N. Rodriguez, L. Song, S. Shroff, K. H. Chen, T. Smith, W. Luo, "Hotspot Prevention Using CMP Model in Design Implementation Flow", *ISQED* (2008).
4. H. Liao, L. Song, N. Jakartdar, R. Radojcic, "Integration of CMP Modeling in RC Extraction and timing flow", *CICC* (2007).
5. C. Hui, X.B. Wang, H. Huang, U. Katakamsetty, L. Economikos, M. Fayaz, S. Greco, X. Hua, S. Jayathi, C.-M. Yuan, L. Song, V. Mehrotra, K. H. Chen, T. Gbondo-Tugbawa, and T. Smith, "Hotspot Detection and Design Recommendation Using Silicon Calibrated CMP Model", *Proc. of SPIE* (2009).
6. L. Economikos, T. Gbondo-Tugbawa, A. Gower-Hall, K. H. Chen, T. Smith, L. Song, F. Raquel, H. Xiang, R. Puri, S. Greco, C. Truong, X. Chen, W. Tseng, P. O'Neil, "Framework for Using CMP Model in Evaluating Processes and Designs", *ICPT* (2008).
7. B. Lee, E. Drege, W. Luo, R. Pyke, and L. Song, "More than Modeling: The Use of CMP Modeling For Design Optimization," *24th International VLSI/ULSI Multilevel Interconnect Conference* (2007).
8. A. Kahng, G. Robins, A. Singh, A. Zelikovsky, "Filling Algorithms and Analyses for Layout Density Control," *IEEE Trans. Computer-Aided Design* (1999).
9. Y. Chen, A. Kahng, G. Robins, A. Zelikovsky, "Practical Iterated Fill Synthesis for CMP Uniformity," *ACM/IEEE Design Automation Conference*, (2000).
10. B. Stine, D. Boning, J. Chung, L. Camilletti, F. Krupa, E. Equi, W. Loh, S. Prasad, M. Muthukrishnan, D. Towery, M. Berman, A. Kapoor, "The Physical and Electrical Effects of Metal-Fill Patterning Practices for Oxide Chemical-Mechanical Polishing Processes,", *IEEE Trans. on Electron Devices* (1998).
11. S. Sinha, J. Luo, C. Chiang, "Model Based Layout Pattern Dependent Metal Filling Algorithm for Improved Chip Surface Uniformity in the Copper Process," *Asia and South Pacific Design Automation Conference* (2007).

Ge- and III/V-CMP for Integration of High Mobility Channel Materials

P. Ong, L. Witters, N. Waldron, L.H.A. Leunissen

Imec, Kapeldreef 75, 3001 Leuven, Belgium

For devices beyond the 15 nm generation the introduction of high mobility channel materials is foreseen. Ge has a very high hole mobility and is thus an interesting candidate for pMOS transistors. III/V-materials have very high electron mobility and thus provide interesting candidates for nMOS transistors. In this work we present a process-flow for the integration of Ge and III/V-materials as channel materials on bulk silicon wafers. The Ge- and III/V-CMP process is a critical step for this integration approach, as excessive material needs to be removed and the surface needs to be smoothened. Ideally, the use of high mobility channel materials, in combination with a high-k material as gate oxide, will result in a significantly improved transistor performance compared to devices with Si channels. In this paper we show results of the development for the Ge- and III/V-CMP.

Introduction

It is expected that the implementation of high mobility channel materials into CMOS devices will be needed to enable further scaling of devices beyond the 15 nm generation. Ge is a promising candidate as high mobility channel material for the pMOS transistors. And for the nMOS transistors III/V-materials offer several choices. The mobility for electrons and holes are 1600 cm^2/Vs and 430 cm^2/Vs for Si (1) while for Ge the hole mobility is much higher with 3900 cm^2/Vs and for e.g. GaAs the electron mobility is much higher with 9200 cm^2/Vs. The excellent channel performance for a Ge-based pMOS has been demonstrated recently: an intrinsic hole mobility that is a factor of 2.4 above the universal Si curve and a clear increase in device current were achieved (2).

In this paper we focus on a particular integration scheme for the introduction of the Ge and the III/V-materials as high mobility channel materials on regular Si wafers and on how CMP is a critical element for this approach. Regular Si wafers are used and the Ge and III/V-materials are deposited locally, which is a very cost-effective approach compared to starting with bulk Ge wafers. For III/V-materials there are no bulk wafers with diameters of 200 or 300 mm. A sketch of the involved process steps for this integration scheme is shown in Figure 1. After completion of the standard Shallow Trench Isolation (STI) integration on these Si wafers, the Si in the active area is recessed selectively to a depth of ~300 nm. Then, with a selective epitaxial growth process, the Ge or III/V material is grown in the trenches between the STI oxide.

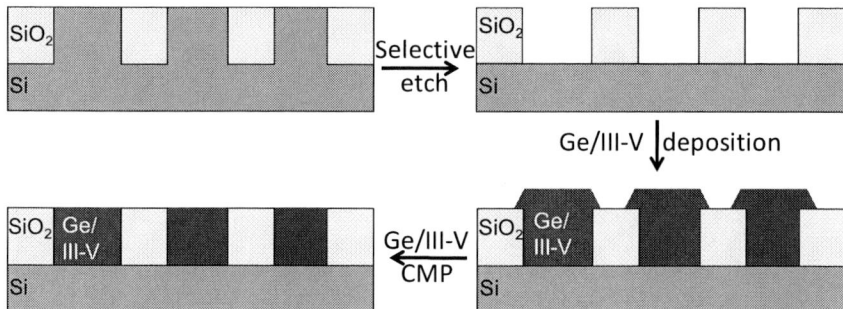

Figure 1. Integration scheme for implementing Ge and III/V-materials as high mobility channel materials on bulk silicon wafers

Due to the lattice mismatch between Ge or the III/V-material and Si there will be threading dislocations (TD) in the deposited film. To enable better devices the threading defect density (TDD) in the film needs to be as low as possible. For the integration of III/V-materials an InP buffer layer is grown on a Ge seed layer (3). One of the advantages of the local deposition is that for narrow trenches the confinement of the epitaxial material within the STI oxide results in TD trapping to the sidewalls (4) and thus the Ge or InP closer to the surface is to a certain extent defect free. To ensure proper filling and good coverage at the interface to the STI oxide a certain overgrowth of the epitaxial material is applied. A thicker overgrowth is also beneficial for reducing the TDD in wider features, where the TD trapping is not effective (5, 6). The overburden is removed with a CMP step and at the same time the surface, which is relatively rough after deposition, is smoothened to meet the requirements for device integration with ultra-thin high-k gate dielectrics. The difficulty is to obtain a defect and contamination free Ge surface with good uniformity across structures of different sizes.

This paper illustrates the development of a 2-step Ge CMP process with good uniformity for structures of different size for 300mm Ge-STI wafers. For III/V-CMP we present first results for patterned 200 mm wafers and discuss some Environment, Health and Safety (EHS) related issues.

Experimental Results

To polish the Ge overgrowth and to stop on the STI oxide we are using commercially available slurry, which was originally developed for W-CMP. This slurry was selected because it has a low oxide removal rate and it gave us good results after the initial screening of different process options. The 300 mm STI wafers have HARP oxide as STI oxide. Initial process development was already previously reported (7) and we chose to add a H_2O_2 concentration of 0.17% to the W-slurry to boost the Ge removal rate and to avoid pitting (8). The selectivity between Ge and HARP oxide of the selected mix is relatively low with 3.5:1 and therefore we need to monitor closely the oxide loss on the Ge-STI wafers. The post-CMP (PCMP) clean is done with diluted ammonia, which in comparison with DI-water helps to significantly improve the defectivity on blanket Ge wafers and to reduce metal contamination from the Ge surface (7).

By implementing a second CMP step with oxide slurry the roughness of blanket Ge wafers can be reduced to values as low as 0.11 nm (Figure 2), which is comparable to blanket Si wafers. This second CMP step also helps to further reduce the amount of defects by another order of magnitude compared to the situation after the first CMP step with diluted ammonia for the PCMP clean. To assist fast process development regarding the roughness of the Ge wafers, we have established a correlation between haze measurements on the defect metrology tool (SP2) and time consuming roughness measurements from atomic force microscopy (AFM). This correlation can be seen in Figure 3.

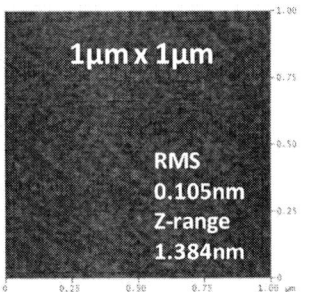 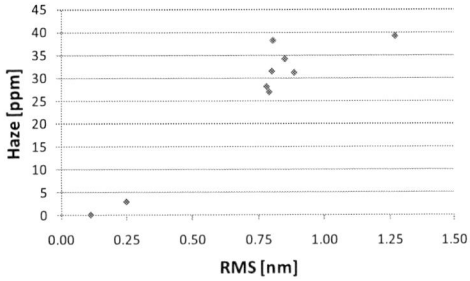

Figure 2. AFM of blanket Ge wafer polished with 2-step CMP process

Figure 3. Correlation between haze measured by SP2 and roughness as measured by AFM for blanket Ge wafers

For the patterned Ge-STI wafers, where the Ge is grown selectively in the trenches between the STI oxide, it is important that the trenches are completely filled. The deposition of the Ge needs to be optimized so that the overfill for structures of different sizes is comparable or even matches the CMP behavior. Further details about the Ge growth can be found in (5). For the results presented herein the Ge deposition was optimized to a quite uniform Ge overgrowth for structures of different sizes and the overgrowth is roughly 1000 nm thick. This high value for the Ge film thickness allows for the best filling performance. Thinner Ge layers result in a deficient filling in corner areas of Ge structures. As a consequence, slurry particles and other polishing debris remain after CMP in these recessed areas causing issues in later process steps.

In Figure 4 the results of step height measurements on three different structures for different polishing times are shown. Measurements are done across large Ge areas separated by a 30 µm, across a 10 µm wide Ge capacitor and across a feature with 3 µm and 1 µm wide Ge areas. In addition to the step height results also the measurements of the oxide loss in a large oxide pad (60 µm by 60 µm) are included in Figure 4. These structures are spread over the die and the overall effects of local pattern density variations on the topography still need to be investigated further.

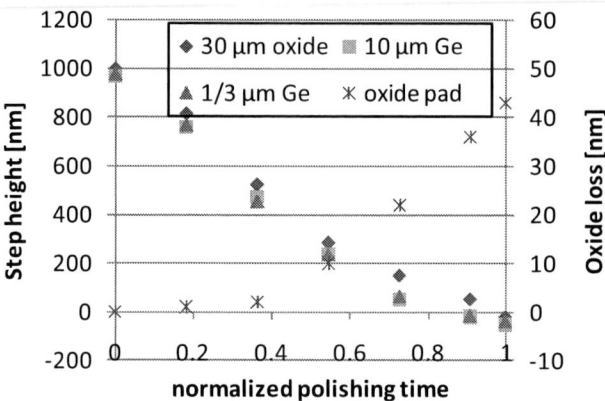

Figure 4. Step height measurements on different structures and oxide loss for different polishing times

The initial overburden is quite uniform. It can be seen that the Ge overgrowth in the 30 μm structure clears later than the Ge on the smaller structures. The step height reduction for these larger areas is slower than for the smaller features. This effect disappears when the wafers are polished long enough in which case the Ge is recessed uniformly in the different structures. With the selectivity of ~3.5:1 on blanket wafers there is an oxide loss of ~45 nm when the Ge overburden is completely removed in all structures. Even with this significant oxide loss, the remaining height of the STI oxide is sufficient to work as isolation between the different devices. The roughness of the Ge after the polishing step with the W-CMP slurry is still quite high with 0.75 nm RMS. To reduce the roughness and to have an option to tune the topography a second CMP step with oxide slurry is applied. While we consume 20-40 nm of oxide with this second CMP step it results in an improved roughness of 0.26 nm RMS and indeed provides the option to further improve the uniformity between the different features and to tune the topography. And based on the results from the blanket Ge wafers the developed 2-step approach is also beneficial regarding defectivity and metal contamination. Based on initial electrical data it seems to be desired that the oxide region is slightly recessed, which can be achieved easily with the 2-step CMP approach. Figure 5 shows a typical X-SEM picture with the final result.

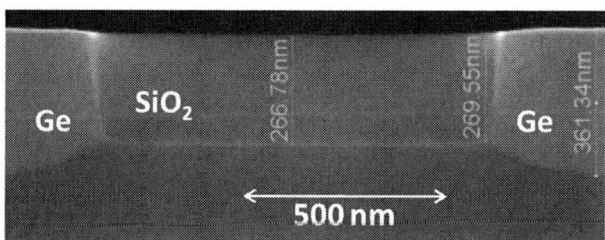

Figure 5. XSEM of 300 mm Ge-STI wafer after polishing with 2-step CMP approach

For the integration of the III/V-material a CMP process of the InP buffer layer is required. As this is a similar approach as for the Ge-STI CMP we also use the same W-slurry. We have achieved good results with this slurry (Figure 6). But due to the acidic pH (2-3) we anticipated EHS related issues. Due to the presence of phosphor we monitored for gas creation during the CMP process setup as we considered the possibility of phosphine creation. During polishing of InP wafers with the selected slurry, a high concentration above 800 ppm of phosphine is detected during processing inside the polishing tool by Fourier Transform InfraRed spectroscopy. The threshold limit value of phosphine is 300 ppm. As the CMP tool is equipped with an exhaust the phosphine is extracted and no phosphine is detected outside the tool. However, the generation of phosphine is a very serious issue and needs to be avoided. During tests with different basic slurries we have never seen indications for phosphine generation, but the process results are not meeting the requirements yet. Further studies and evaluations are ongoing. Another EHS aspect of the III/V-CMP process that is under investigation is the Indium contamination inside the tool, which is toxic in solid form and seems to accumulate in the tool. These kind of ESH aspects need to be taken into account for future developments, especially also when As containing III/V-materials (e.g. InGaAs) will need to be polished.

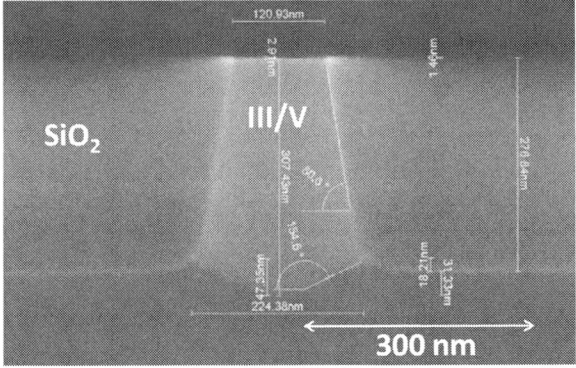

Figure 6. XSEM of 200 mm InP-STI wafer after CMP

Summary

The motivation for the implementation of high mobility channel materials is presented and a specific integration approach for the integration of Ge and/or III/V-materials on bulk Si wafers is proposed. Results from our process setup for Ge and InP CMP were presented. The resulting new CMP processes were successfully implemented and enabled the proposed integration scheme for the introduction of Ge and III/V-materials as high mobility channel materials. EHS aspects still need to be addressed further for the processing of the III/V-materials. Future work can be done with optimized slurries and also for the co-integration of the Ge and III/V-materials.

Acknowledgments

Parts of this work have been done within imec's industrial affiliation program on (sub-) 22 nm CMOS Technologies. Imec wants to thank their partners for the support of these activities.

References

1. Takagi et al., *presentation at INC4* (2008)
2. J. Mitard et al., *Symposium on VLSI Technology*, Digest of Technical Papers, p. 82 (2009)
3. G. Wang et al., *ECS Transactions,* **27**, (1), p. 959-964 (2010)
4. J.S. Park et al., *Appl. Phys. Lett.* **90**, 052113 (2007)
5. R. Loo et al., *Journal of The Electrochemical Society*, **157** (1), H13-H21 (2010)
6. G. Wang et al., *Thin Solid Films* **518**, p. 2538–2541 (2010)
7. P. Ong et al., *2010 International Conference on Planarization/CMP Technology* (2010)
8. J. Hydrick et al., *ECS Transactions*, **16** (10), p. 237-248 (2008)

Advanced Direct-polish Process on Organic Non-porous Ultra Low-*k* Fluorocarbon Dielectric on Cu Interconnects

Xun Gu [1], Takenao Nemoto [2], Yugo Tomita [2], Ricardo Duyos Mateo [2], Akinobu Teramoto [2], Shin-Ichiro Kuroki [1], Shigetoshi Sugawa [1] and Tadahiro Ohmi [2, 3]

[1] Graduate School of Engineering, Tohoku University
Aza-Aoba 6-6-10, Aramaki, Aoba-Ku, Sendai, 980- 8579, Japan
[2] New Industry Creation Hatchery Center, Tohoku University, Japan
[3] World Premier International Research Center, Tohoku University, Japan

A direct-polish process has applied to the ultra low-k dielectric without the etch stop layer to reduce effective dielectric constant in damascene interconnects. In this study, the direct-polish process on organic non-porous dielectric, fluorocarbon (*k*=2.2), was demonstrated and an optimum direct-polish process condition was investigated. Mechanical effect, not chemical effect, on fluorocarbon degraded electrical properties by changing of chemical structure of fluorocarbon. A surface plasma treatment of fluorocarbon before polishing was applied to avoid the degradation of electrical characteristics during direct-polish and this result revealed that the surface plasma treatment of fluorocarbon is a practical technique in advanced Cu interconnects.

Introduction

In order to reduce signal propagation delay and power consumption, an Cu interconnect with ultra low-k (ULK) dielectric is crucial. An organic non-porous ULK material, fluorocarbon (*k*=2.2), has been integrated into Cu damascene interconnects as an indispensable dielectric to avoid integration process induced damage in LSI fabrication [1-3], while the widely discussed ULK material, porous carbon doped silicon oxide (p-SiCO:H), was investigated for such challenges [4, 5]. A direct-polish process has applied to the ULK without etch stop layer to reduce an effective dielectric constant in damascene interconnects [6]. However, a porous low-k SiCO:H film often suffers from chemical or mechanical damage during chemical mechanical polishing (CMP) because slurries and cleaning chemicals penetrates into its pores [7, 6]. In this study, the direct-polish process on organic non-porous fluorocarbon was demonstrated and an optimum direct-polish process condition by nitrogen plasma treatment (NPT) on its surface was investigated. The mechanism of the damage induced by the direct-polish on the non-porous fluorocarbon film was also discussed.

Experimental

The organic ULK fluorocarbon film was deposited by using a new concept of microwave, 2.45 GHz, exited dual-shower -plate structure plasma reactor, which shows much higher frequency than that by the conventional radio frequency (~100 MHz) plasma. The microwave plasma produced by the radial line slot antenna indicates lower electron temperature (around 1-2 eV) and it prevents ion bombardment damage to the substrate sufficiently [2, 3]. The microwave plasma enhanced chemical vapor deposition (MWPE-CVD) method consequently enables to produce a high quality film.

Cu / fluorocarbon damascene lines were prepared for electrical evaluation. A damascene process flow with cross-sectional images in this experiment is illustrated in Fig. 1 (a). A 500 nm-thick fluorocarbon film was deposited by the MWPE-CVD on a carbon doped silicon-nitride (SiCN) film, followed by SiCO:H film formation as a hardmask layer. A photo lithography process was run by KrF stepper. After formation of 0.28 / 0.20 μm width / space trenches in the fluorocarbon dielectric stack, Cu was formed by electroplate deposition method on physical vapor deposition Cu / TiN. An overburden Cu and TiN layers were polished with the Araca APD-800 polisher. Over-polishing time of TiN, which was detected by monitoring shear force variation [8, 9], was varied from 0 to 40 sec. A Citric acid with additives, which has been reported as the suitable cleaning solution without damage generation for the fluorocarbon film, was used as a post-CMP cleaning solution [10, 11]. Optimum brush scrubbing process with high brush rotation rate in low down pressure was applied as a post-CMP cleaning process [12, 13]. Finally a passivation film was deposited, then an annealing treatment was applied at 150 °C. The same structure of TiN films was also prepared on the blanket fluorocarbon film, which is used as metal on insulator structure after TiN polishing as shown in Fig 1 (b).

Leakage current or dielectric constant was measured on line-line comb pattern of the damascene sample or by the mercury probe on the blanket sample. C1s and F1s photoelectron spectra with the high resolution X-ray photoelectron spectroscopy (XPS), which is a highly sensitive and high-resolution photoelectron spectrometer equipped with a monochromatic AlKα radiation source were measured to obtain chemical structures on the surface of blanket fluorocarbon film at the take-off angle (represents angle of an escape photoelectron) of 90° and a photon energy of 1486.6 eV by ESCA-300 [14, 15, 11].

Results and Discussion

Figure 2 shows the cross-sectional transmission electron microscopy (TEM) image of Cu / fluorocarbon damascene lines after 20 sec over-polish of TiN. Cu lines with fluorocarbon interlayer dielectric are successfully formed. Line-line leakage currents with a space of 0.20 μm for 20 and 40 sec over-polish of TiN in Cu-comb pattern are shown in Fig. 3. Almost one order of magnitude of leakage current increase is observed for 40 sec over-polish. It reveals that over-polish degrades the line-line leakage current during direct-polish on the fluorocarbon film in Cu-comb pattern. To clarify the effects of over-polish on degradation of leakage current, leakage current or dielectric constant was measured by the mercury probe on the blanket sample. Figure 4 (a) indicates leakage current of the blanket fluorocarbon film as a function of electric filed at various over-polishing time from 0 to 30 sec. A result of a blanket fluorocarbon sample dipped into slurry was also shown. Degradation of leakage current was observed as an over-polishing time increases and this is consistent to the result in damascene sample. Moreover, the result that no significant leakage current degradation in sample dipped into slurry revealed that degradation of leakage current is considered to be dominantly attributed by the mechanical effect. This result shows the different mechanism of degradation on electrical properties between porous low-k SiCO:H and non-porous fluorocarbon during direct-polish [7].

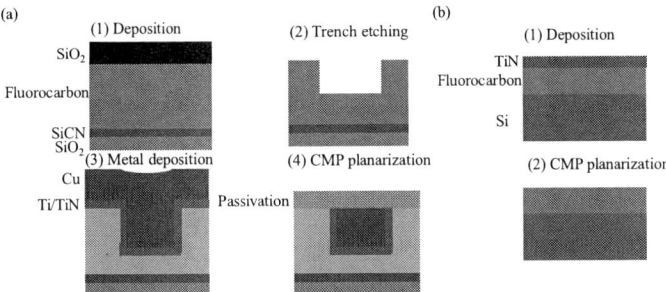

Figure 1 (a) Process flow of Cu / fluorocarbon damascene process and (b) Process flow of metal on insulator structure for blanket wafer.

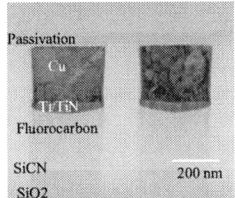

Figure 2 Cross-sectional TEM image of Cu / fluorocarbon damascene lines. Cu lines with fluorocarbon interlayer dielectric are successfully formed.

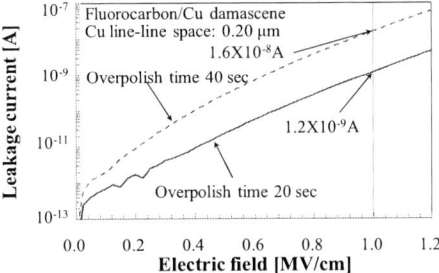

Figure 3 Cu line-line leakage current in Cu-comb pattern for 20 sec and 40 sec overpolishing time. Almost one order of magnitude of leakage current increase is found for 40 sec over-polish.

Figure 4 (b) summaries leakage currents at the electric field of 1MV/cm and dielectric constants of fluorocarbon as a function of over-polishing time. Both leakage currents and dielectric constants increase as over-polishing time increases. Almost 10% degradation of dielectric constant at 30 sec overpolishing time was found, which is much better than that for porous ULK SiCO:H material. It is considered that the fluorocarbon film is a robust material in terms of dielectric constant variation to CMP process.

To investigate causes of degradation of electrical characteristics, $C1s$ and $F1s$ photoelectron spectra of the fluorocarbon film were measured by XPS. Figure 5 (a) and (b) indicates $C1s$ and $F1s$ spectra of the fluorocarbon film at various over-polishing time from 0 to 30 sec. A blanket fluorocarbon sample dipped into slurry was also measured.

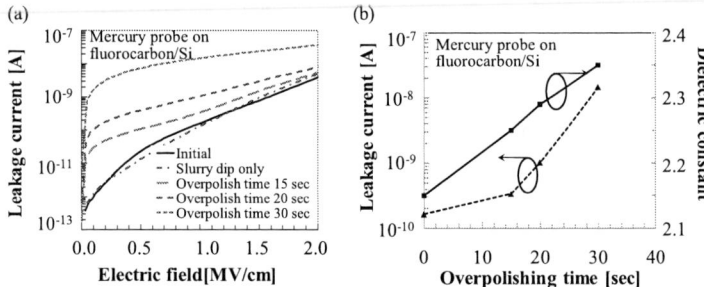

Figure 4 (a) Leakage currents of fluorocarbon for blanket fluorocarbon samples as a function of electric filed at various conditions of overpolishing time and slurry dip only treatment. (b) Leakage currents and dielectric constants of fluorocarbon as a function of overpolishing time.

Four separated C1s spectra are detected. These are considered as $-C-F_3-$, $-C-F_2-$, $-C-F-$, $-C-CFx-$ on surface of the initial fluorocarbon film [10]. After over-polish of TiN, intensities of $-C-F_3-$, $-C-F_2-$, and $-C-F$-bonds were decreased significantly. Moreover, a $-C-C-$ bond is generated and it was increased with increasing in over-polishing time. Meanwhile, intensities of F1s spectra decreased with over-polish time, as shown in Fig. 5. (b). These results reveal that decomposition of the C-F bond by over-polish ends up decrease in F concentration and increase in $-C-C-$ bond. No significant difference of intensities of all bonds in fluorocarbon, however, was observed before and after slurry dip. This result was in a good agreement with that of leakage current, shown in Fig. 5 (a) that mechanical effect of over-polish is considered to be dominated to degrade the electrical properties of fluorocarbon during direct-polish.

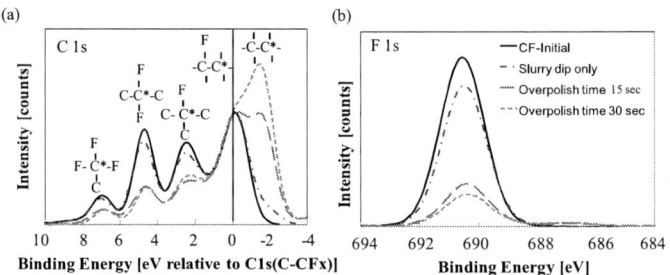

Figure 5 (a) C 1s and (b) F 1s spectra of the fluorocarbon film by XPS at take-off angles of 90° at various over-polishing time from 0 to 30 sec.

To avoid degradation of electrical properties during over-polish, an effective method by surface NPT technique was applied to the fluorocarbon film. NPT method was applied on the surface at the film deposition process between fluorocarbon film and TiN layer. Figure 6 shows (a) leakage current of the blanket fluorocarbon film and (b) C1s photoelectron spectra at various over-polishing times with and without the surface NPT method. Degradation of leakage current was avoided distinctly by the surface NPT method, and no clear change of leakage current was found even at the long over-polish time of 30 sec. This is due to the surface NPT method, which can improve the stability of chemical structure of fluorocarbon to prevent from damage of over-polish for direct-polish on fluorocarbon as shown in Fig. 6 (b). Figure 7 summaries leakage currents at the electric field of 1MV/cm and dielectric constants of fluorocarbon as a function of over-polishing time with and without NPT. Almost two orders of magnitude of leakage current decrease were found to be improved in the overpolishing time of 30 sec by NPT method. Meanwhile, degradation of dielectric constant was also improved. A model of NPT is illustrated in Fig. 8. After direct-polish process without NPT method, damages on the surface were introduced. These damage sites may result in degradation of electrical properties, as shown in Fig. 4. A nitride fluorocarbon layer by NPT on the surface of fluorocarbon is considered to be a protective layer with stable chemical structure. This layer enables to avoid damage induced by direct-polish process.

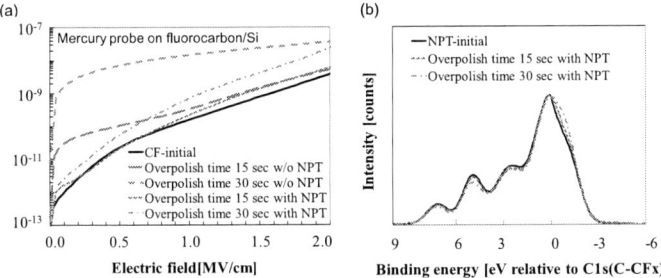

Figure 6 (a) Leakage currents of fluorocarbon and (b) C 1s spectra by XPS at take-off angle of 90° at various conditions of overpolishing time with and without the surface plasma treatment.

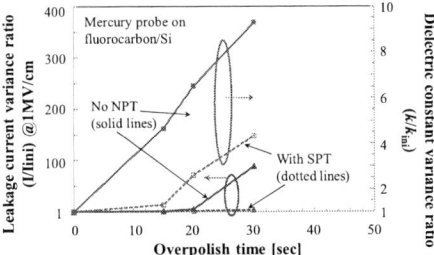

Figure 7 leakage currents at the electric field of 1MV/cm and dielectric constants of fluorocarbon as a function of over-polishing time with and without NPT.

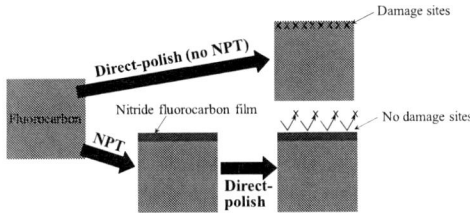

Figure 8 Schematics model with and without NPT method. A nitride fluorocarbon layer by NPT is considered to be a protective layer on the surface of fluorocarbon film.

Conclusions

A direct-polishing process was developed with organic non-porous ULK dielectric, fluorocarbon, to reduce the effective dielectric constant in Cu damascene interconnects. The mechanism of the degradation of electrical properties induced by direct-polish on the fluorocarbon film was investigated and over-polish of TiN is found to dominantly attribute to degrade electrical properties by changing of chemical structure of the dielectric fluorocarbon film. A mechanical damage, not a chemical damage, is considered as the dominant causes of degradation of electrical characteristics. The surface nitrogen plasma treatment of fluorocarbon is a practical technique to avoid the degradation of electrical characteristics during direct-polish in advanced non-porous ULK dielectric fluorocarbon / Cu interconnects.

References

1. X. Gu, T. Nemoto, Y. Tomita, K. Miyatani , A. Saito, Y. Kobayashi, A. Teramoto, S. Kuroki, T. Nozawa, T. Matsuoka, S. Sugawa, and T. Ohmi, *Proc. of Advanced Metallization Conference 2010: 20th Asian Session (ADMETA)*, p 54 (2010).
2. T. Ohmi, M. Hirayama, and A. Teramoto, *J.Phys. D: Appl. Phys.*, 39,1 (2006).
3. A. Itoh, A. Inokuchi, S. Yasuda, A. Teramoto, T. Goto, M. Hirayama, and T. Ohmi, *Jpn. J. Appl. Phys.*, 47, 2515 (2008).
4. M. Ueki, M. Tagami, F. Ito, T. Onodera, I. Kume, N. Furutake, H. Yamamoto, J. Kawahara, N. Inoue, K. Hijioka, T. Takeuchi, S. Saito, N. Okada, and Y. Hayashi, *IEDM Tech. Dig.*, 26.7 (2008).
5. A. Ishikawa, Y. Shishida, T. Yamanishi, N. Hata, T. Nakayama, N. Fujii, H. Tanaka, H. Matsuo, K. Kinoshita, and T. Kikkawa, *J.Electrochem Soc.*, 153(7),692 (2006).
6. S. Kondo, K. Fukaya, N. Ohashi, T. Miyazaki, H. nagano, Y. Wada, T. Ishibashi, M. Kato, K. Yoneda, E. Soda, S. Nakao, K. Ishigami, and N. Kobayashi, *Proc. of IITC*, p 164 (2006).
7. S. Kondo, B.U. Yoon, S.G. Lee, S. Tokitoh, K. Misawa, T. Yoshie, N. Ohashi, and N. Kobayashi, in *VLSI Technical Digest*, p. 68 (2004).
8. X. Gu, T. Nemoto, Y. Sampurno, J. Cheng, S. Theng, A. Philipossian, Y. Zhuang, A. Teramoto, T. Ito, S. Sugawa, and T. Ohmi, *Mater.Res.Soc.Symp.Proc.*, Vol.1157, p 157 (2009).
9. Y. Sampurno, X. Gu, T. Nemoto, Y. Zhuang, A. Teramoto, A. Philipossian, and T. Ohmi, *Jpn. J. Appl. Phys.*, 49, 05FC01 (2010).,
10. X. Gu, T. Nemoto, A. Teramoto, T. Ito, and T. Ohmi, *J.Electrochem. Soc.*, 156 (6) 409 (2009).
11. X. Gu, T. Nemoto, A. Teramoto, R. Hasebe, T. Ito, and T. Ohmi, *Solid State Phenomena Vols.*, 145-146, pp 381 (2009).
12. X. Gu, T. Nemoto, A. Teramoto, T. Ito and T. Ohmi, *ECS Trans.*, **19**(7), 103 (2009).
13. X. Gu, T. Nemoto, Y. Tomita, A. Teramoto, S. Sugawa and T. Ohmi, *Proc. of ADMETA*, p 160 (2010).
14. U. Gelius, B. Wannberg, P. Baltzer, H. Fellner-Feldegg, G. Carlsson, C. G.Johansson, J. Larsson, P. Munger, and G. Vergerfos, *J. Electron Spectrosc. Relat. Phenom*, 52(1) 747 (1990).
15. X. Gu, T. Nemoto, Y. Tomita, R. Duyos Mateo, S. Sugawa and T. Ohmi, *Proc. of International Conference on Planarization/CMP Technology (ICPT)*, p 51 (2010).

Effect of Slurry Application/Injection Methods and Polishing Conditions on Bow Wave Characteristics

X. Liao[a], Y. Sampurno[a,b], Y. Zhuang[a,b], F. Sudargho[b], A. Rice[a], and A. Philipossian[a,b],

[a] Department of Chemical and Environmental Engineering, University of Arizona, Tucson, Arizona 85721, U.S.A.
[b] Araca, Inc., Tucson, Arizona 85718, U.S.A.

In this study, the effect of slurry application/injection methods and polishing conditions on slurry flow dynamics at the retaining ring bow wave was investigated. Two slurry application/injection methods (standard slurry application method and novel slurry injection method) were used. For each method, an ultraviolet enhanced fluorescence system was implemented to measure the slurry film thickness at bow wave for a polyetheretherketone (PEEK) retaining ring at different sliding velocities, slurry flow rates and ring pressures. Results indicated that the novel slurry injection method generated significantly thicker slurry film at the bow wave than that of the standard slurry application method. For both methods, slurry film thickness at the bow wave increased with increasing flow rate and ring pressure while it decreased with increasing sliding velocity.

Introduction

Chemical mechanical planarization (CMP) is widely used in integrated circuit (IC) manufacturing industry to achieve both local and global surface planarity through combined chemical and mechanical forces. For most current commercial CMP polishers, slurry is applied onto the pad center area. As the pad rotates during polishing, however, a large amount of fresh slurry flows directly off the pad surface without entering the pad-wafer interface due to the centrifugal force, resulting in very low slurry utilization (1). In this paper, a novel slurry injection system is designed to efficiently introduce fresh slurry into the pad-wafer interface. Slurry film thickness at the bow wave for a PEEK retaining ring is measured at different sliding velocities, slurry flow rates and ring pressures using an ultraviolet enhanced fluorescence system. Results are compared between the standard slurry application method and novel slurry injection method.

Experimental

All experiments were performed on an Araca APD-800 polisher and tribometer. A detailed description of the apparatus can be found elsewhere (2). Two slurry application/injection methods used were as follows: (a) slurry was applied onto the pad center area as is the case in most state-of-the-art CMP processes and referred to here as the standard slurry application method; (b) slurry was injected onto the pad surface through a novel slurry injector referred as novel slurry injection method. In this case, fresh slurry was introduced through one slurry inlet from the slurry tank which then traveled inside an internal channel and flowed out through the multiple holes in the trailing edge of the inject bottom slit. The injector was placed adjacent to the wafer on the surface of the pad as shown in Fig. 1. For each method, the slurry film thickness at bow wave was measured under two ring pressures (1.4 and 2.8 psi), two sliding velocities (0.6 and 1.2 m/s), and two slurry flow rates (150 and 300 ml/min). An embossed 31-inch Politex pad was used with in-situ pad conditioning (with a 3M PB32A brush) at a conditioning force of 3 lb_f during the tests.

For each slurry application/injection method, an ultraviolet enhanced fluorescence system (UVIZ-100™ manufactured by Araca, Inc.) was implemented to measure the slurry film thickness at the retaining ring bow wave for a PEEK retaining ring designed for 300-mm wafer CMP process. Coumarin dye was added to Fujimi PL-7103 slurry at the concentration of 0.5 g/L. During the tests, ultraviolet (UV) light was projected on the retaining ring bow wave area to excite the dyed slurry and the emitted fluorescence was captured by a high resolution camera. Fifty bow wave images were taken during a 10-second period at each polishing condition. Each raw image was converted to a gray-scale image and the average brightness of the gray-scale image was obtained. The average brightness of the gray-scale image was directly correlated with slurry film thickness: the higher the gray-scale brightness value, the thicker the slurry film. Prior to the tests, a calibration curve was established to correlate the slurry film thickness with the average brightness of the grey-scale images.

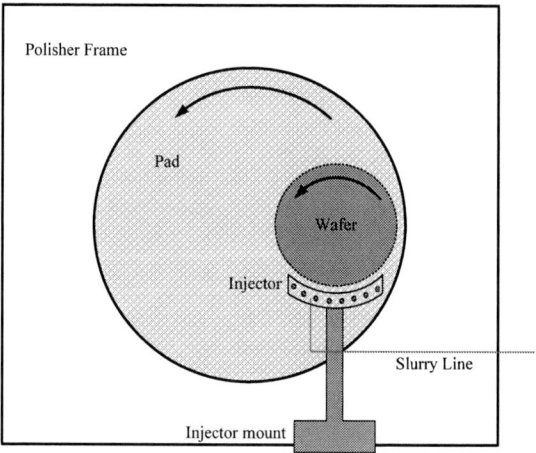

Figure 1. Schematic diagram of novel slurry injector device setup on a CMP polisher.

Results and Discussion

Figure 2 compares the average slurry film thickness at the retaining ring bow wave of the two slurry application/injection methods at different polishing conditions. Error bars represent the standard deviation of the fifty bow wave images taken at each polishing condition. At the same polishing condition, the slurry film thickness at the retaining ring bow wave of the novel slurry injection method is on average 45% higher than that of the standard slurry application method. This is due to the fact that as the pad rotates during polishing, a large amount of fresh slurry flows directly off the pad surface leading to a thinner bow wave around the retaining ring. In comparison, the novel slurry injector is placed close to the retaining ring, allowing most of the fresh slurry to be delivered to the retaining ring thus causing a thicker bow wave.

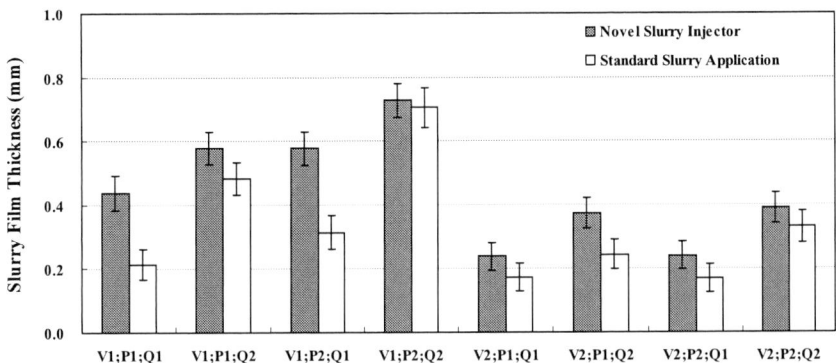

Figure 2. Effect of slurry application/injection methods on bow wave slurry film thickness.
(P1 = 1.4 psi, P2 = 2.8 psi, V1 = 0.6 m/s, V2 = 1.2 m/s,
Q1 = 150 ml/min, Q2 = 300 ml/min)

Figure 3 shows the effect of sliding velocity on the bow wave slurry film thickness for the two slurry application/injection methods. For both methods, slurry film thickness at the retaining ring bow wave decreases significantly with increasing pad/retaining ring sliding velocity. This is due to higher centrifugal force at the higher sliding velocity whereby the slurry is pushed towards the edge of the pad resulting in thinner bow waves.

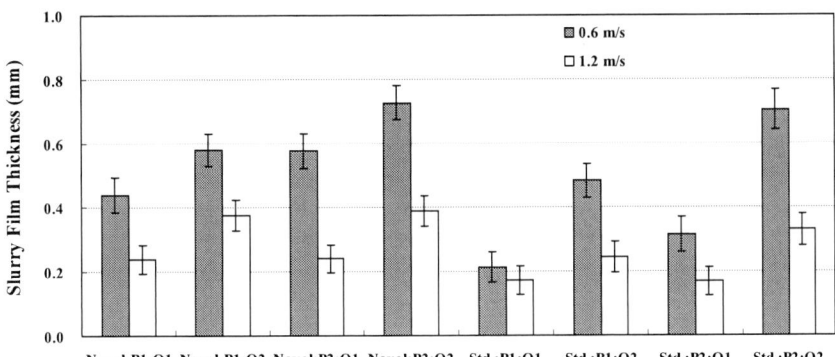

Figure 3. Effect of pad/retaining ring sliding velocity on bow wave slurry film thickness.
(P1 = 1.4 psi, P2 = 2.8 psi, V1 = 0.6 m/s, V2 = 1.2 m/s,
Q1 = 150 ml/min, Q2 = 300 ml/min)

Figure 4 shows the effect of retaining ring pressure on the bow wave slurry film thickness for the two slurry application/injection methods. For both methods, when pressure is increased from 1.4 to 2.8 psi, the bow wave slurry film thickness increases significantly except at the sliding velocity of 1.2 m/s and slurry flow rate of 150 ml/min. This is due to the fact that as retaining ring pressure increases, it is more difficult for the slurry to flow through the retaining ring-pad interface rendering thicker bow waves.

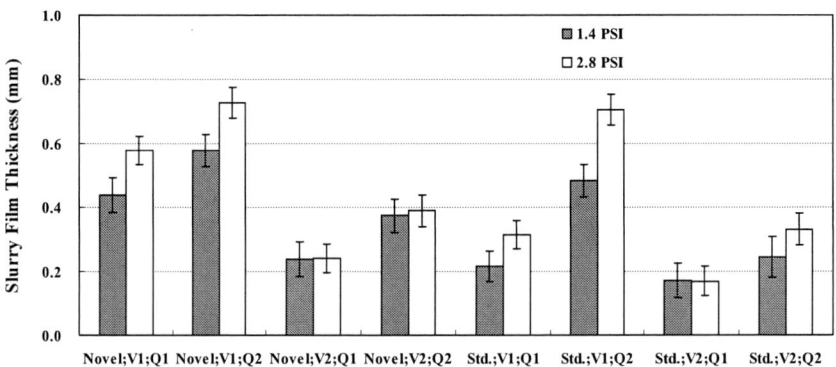

Figure 4. Effect of retaining ring pressure on bow wave slurry film thickness.
(P1 = 1.4 psi, P2 = 2.8 psi, V1 = 0.6 m/s, V2 = 1.2 m/s,
Q1 = 150 ml/min, Q2 = 300 ml/min)

Figure 5 shows the effect of slurry flow rate on the bow wave slurry film thickness for the two slurry application/injection methods. For both methods, when slurry flow rate is increased from 150 to 300 ml/min, the bow wave slurry film thickness increases significantly at different sliding velocities and retaining ring pressures.

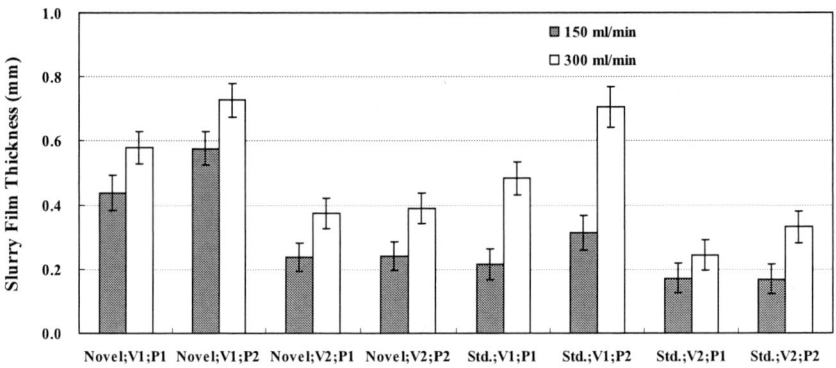

Figure 5. Effect of slurry flow rate on bow wave slurry film thickness.
(P1 = 1.4 psi, P2 = 2.8 psi, V1 = 0.6 m/s, V2 = 1.2 m/s,
Q1 = 150 ml/min, Q2 = 300 ml/min)

Conclusions

In this study, the effect of slurry application/injection methods and polishing conditions on slurry flow dynamics at the retaining ring bow wave was investigated. At the same pad/retaining ring sliding velocity, slurry flow rate, and retaining ring pressure, the slurry film thickness at the retaining ring bow wave of the novel slurry injection method was significantly larger than that of the standard slurry application method. For both slurry application/injection methods, slurry film thickness at the retaining ring bow wave increased with increasing flow rate and ring pressure while it decreased with increasing sliding velocity.

References

1. A. Philipossian and E. Mitchell, Jpn. J. Appl. Phys., **42**, 7259 (2003).
2. http://www.aracainc.com/products/apd-800.

664

Evolution of Post CMP Cleaning Technology

Gautam Banerjee

Air Products and Chemicals, Inc.

7201 Hamilton Blvd., Allentown, PA 18195, USA.

In the sub 40 nm technology nodes, new generation of materials are being incorporated into the complex film stack prior to planarization using CMP. The challenge of cleaning wafers after CMP step has evolved over the years from cleaning simply silicon dioxide to now cleaning porous ultra low-k film as well as various metallic films such as Cu, Ta, Al, Ru, Co, Mn, and sometimes alloy films such as GeSbTe(commonly called GST). It is no longer possible to have one common clean formulation working for all technologies. In this paper, evolution of post CMP cleaning technology will be discussed, primarily focused on Cu process but will address the post CMP cleaning challenges with some of the emerging materials applications.

Introduction

One important side effect of Moore's Law is that technical questions in the semiconductor industry are never definitively settled. Materials and processes that are perfect fit for current manufacturing nodes may be inadequate in a few years. Ideas that are very complex or expensive now may be essential in the future [1]. The problem stems from the industry's steady reduction in feature sizes. Behavior in narrow lines can be substantially different from the bulk properties of the material. Reducing the cross section of bulk wires increases the resistance (R) according to

$$R=(l*\rho)/A$$

where l = wire length, ρ=resistivity=constant and A = cross-sectional area. As line-widths drop below 100nm, however, scattering at interfaces and grain boundaries becomes more important than bulk behavior and the resistivity increases exponentially [2]. Any reduction in the copper cross-section thus has a double impact, increasing both bulk resistance and the resistivity multiplier. The resistance is more important for long lines, while capacitance has a greater impact on the performance of shorter lines [3]. Though resistance remains cross section-dependent, larger wires are less susceptible to the double impact of size effects. Other circuit constraints make it impractical to simply increase the wire size. Instead, manufacturers would like to use thinner barrier layers, allowing more copper to fit within a given lithographic feature.

New barrier layer – New Integration challenges

In order to reduce barrier layer thickness, two important considerations are; a) preventing copper diffusion into the dielectric and b) ensuring strong adhesion of the copper layer. There are competing process priorities in the above two requirements. In order to prevent copper diffusion into the dielectric, barrier layer must be thick enough, conformal and without any hole although for many applications at 32nm and below nodes, the use of porous low-k dielectric layer is now becoming common feature. The thick barrier layer adds to the increase of net resistance.

To improve adhesion at the interface, a barrier layer must form strong interfacial bond with Copper on one hand and with the low-k dielectric layer on the other hand. If a single barrier layer cannot achieve the goal, then the current trend is to use a bi-layer such as a combination of Ru or

Co or Mn below Cu (called glue layer) and then use another layer of TaN beneath the glue layer. To some extent, the use of bi-layer allows for reduction of thickness of the barrier layer compared to the use of a single barrier layer.

New post CMP Cleans for Cu Processes

To develop post CMP clean formulation for Cu process, therefore, the complexity arises from the presence of multiple dissimilar interfaces stacked together in a pattern wafer which get exposed to the post CMP clean solution when a wafer is cleaned in a scrubber after CMP process. The different galvanic couples present corrosion problems in the solution that may otherwise be difficult to predict using blanket wafer studies. Wang et al have described the challenges of using Co and Ru in the Cu interconnects integration for both CMP and post CMP cleaning formulation development. For post CMP cleaning of Co seed enhancement layer(SEL) and Ru barrier layer, an alkaline pH post CMP clean solution was found to be the ideal choice as shown in figure 1 below [4].

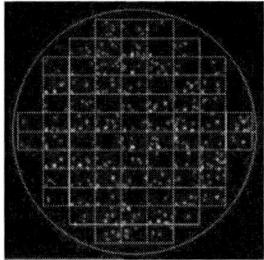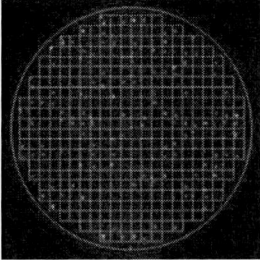

Fig. 1: Defect maps for test pattern wafers using new cleaning method, (a) Ru barrier; (b) Co SEL

Among the various post CMP cleans, there could be an issue of dendrite formation after the cleaning in high density Cu line areas of the pattern wafers. With an appropriate formulation, the dendrite formation could be avoided as shown in figure 2 below (for wafers with 32nm feature size – images courtesy a major OEM customer of Air Products and Chemicals, Inc.).

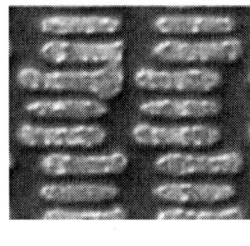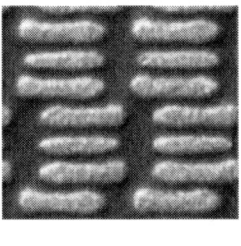

(a)   (b)

Fig. 2: Dendrite formation seen after post CMP clean with a non-optimized formulation (a) and no-dendrite even after 7 days with CoppeReady® CP98 (b).

In addition to different metals being introduced as barrier or seed enhancement layer, the porous ultra low-k film used as dielectric layer may be susceptible to shift in k value when exposed in the post CMP clean formulation. If this shift is due to the change in chemical nature of the dielectric film, then the process is deemed unstable and the formulation needs to be modified.

CoppeReady®CP98 has demonstrated to be suitable for use with porous low-k film as shown in the table 1.

Table 1: Effect of post CMP clean on K value of PDEMS 2.2

PDEMS 2.2	K	RI
Before treating with CP98	2.32	1.35
After optimized treatment w/CP98	2.32	1.35

One other issue with hydrophobic ultra low-k dielectric film integration in the Cu interconnect process is the appearance of water mark after post CMP clean step. In order to eliminate water marks from the wafer surface after cleaning and drying, an optimized formulation definitely is a big step towards that goal. However, it also requires an optimized process development [5]. The optimized process includes use of IPA vapor dryer based on Marangoni principle.

Since almost all the barrier CMP slurries contain some type of film forming Cu corrosion inhibitor, it is necessary to formulate a clean solution that would remove all organic films from the Cu surface after scrubber cleaning step. One of the most common Cu corrosion inhibitors is benzotriazole which forms a monolayer of Cu(I)-BTA on Cu. With an appropriate clean formulation, that monolayer should be removed as can be seen in the TOF-SIMS result of CoppeReady®CP98 clean performance given in figure 3 below.

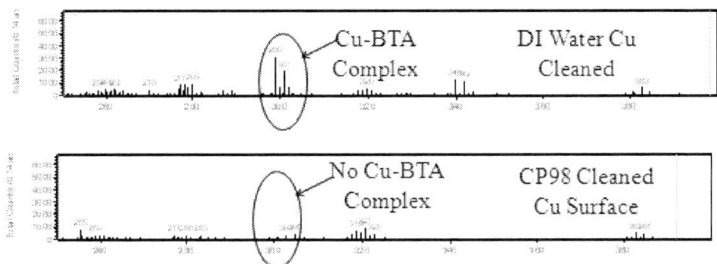

Fig. 3: TOF-SIMS results showing comparison of Cu surface cleaning efficiency of CP98.

New post CMP Cleans for Al Processes

Beside the Cu CMP process, specially formulated post CMP clean is also being used in Al-CMP process as well as for Silicon Nitride/Poly silicon CMP processes. For High K Metal Gate application (HKMG), Al CMP step has now become standard since Intel implemented the same at 45nm node [6].

The metal gate height uniformity and defectivity controlled by Al CMP across whole wafer are crucial to influence the device and yield performance of HKMG with replacement metal gate structure products [7]. The dimensional tolerance of the Al CMP is more challenging (10 times tighter) than conventional CMP process because the metal gate height is only several hundred Angstroms [8]. The product yield of the HKMG structures is particularly sensitive to the defectivity, including fall on particle, micro-scratch and corrosion defect types. All of these defect types could be influenced by post CMP cleaning and therefore need an optimized cleaning process and chemistry. For Al post CMP cleaning, beside particle removal, preventing corrosion is a big challenge. Depending on the integration schemes, either pure Al or Al-Cu alloy film may be used. If an alloy film is used, an inherently mismatched galvanic couple is present in the microstructure that acts as a source of pitting corrosion (figure 4a). For pure Al film, while classical pitting is not a possibility, galvanic corrosion at the interface of Al and other underlying film is a distinct possibility as shown in the figure 4b below.

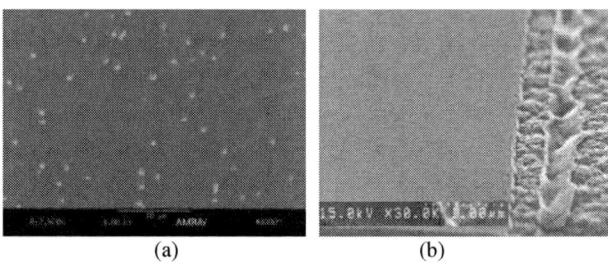

(a)   (b)

Fig. 4: Pitting on Al-0.5Cu film after post CMP clean (a) and Interfacial galvanic corrosion of Al at the TiN barrier layer junction (b)

As it appears above, whether true interfacial galvanic corrosion at the interface of Al and TiN or the pitting corrosion within the microstructure of Al-0.5Cu matrix by dissolution of Al around Cu (which is a representation of galvanic couple of Al and Cu), it is necessary to prevent the galvanic corrosion when using formulated post CMP clean solution. Figure 5 shows the Al lines without any galvanic corrosion at the interface with TiN and figure 6 shows pit free surface after post CMP clean step.

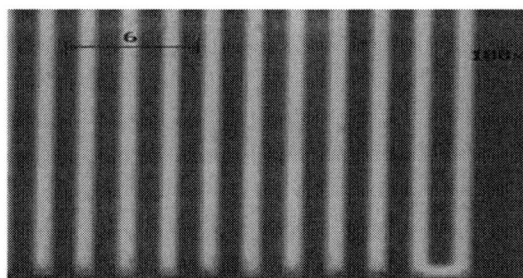

Fig. 5: Galvanic corrosion free interface of Al/TiN lines using CoppeReady®CP72B

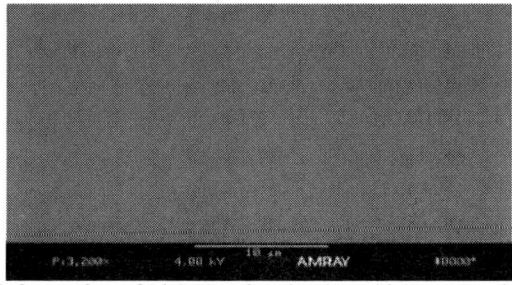

Fig. 6: Pit free surface of Al-0.5Cu after cleaning with CoppeReady®CP72B

Figure 7 shows the SP2 defect map of Al blanket wafers with higher defect counts using a competitor's clean formulation (left) and lower defect count with CoppeReady[R]CP72B (images courtesy Applied Materials Inc., Santa Clara, CA, USA).

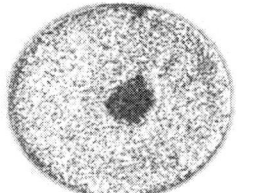

Fig. 7: SP2 defect maps of Al blanket wafers (left: competitor clean and right: CP72B)

New post CMP Cleans for GST Processes

Beside the above types of post CMP cleaning needs, another emerging area is the cleaning of Germanium Antimony Telluride film (Ge2Sb2Te5 or GST) after CMP. GST is a promising material for Phase Change Memory application (PCM) due to its multiple bit operation, scalability and very fast switching speed [9-11]. The first challenge for a GST CMP process is defectivity control. As shown in Table 2, the GST alloy is considerably softer and also more fragile than Cu metal [12].

Table 2: The physical properties of the GST alloy and Cu film

		GST	Cu
Young's Modulus,	GPa	120	130
Sheer Modulus,	GPa	46	48
Hardness,	GPa (10nm in thickness)	2.7	6.5

As a result, it is more difficult to planarize GST without scratching the surface [13] or causing film de-lamination in localized areas. The GST alloy is a ternary IV-V-VI compound with electro-negativity values on the Pauling scale for Ge, Sb and Te of 2.01, 2.05 and 2.1, respectively [14]. They will therefore have dissimilar chemical reactions in an electrochemically active solution such as a post CMP cleaning solution and potentially show corrosion/leaching out of one alloying element compared to other, if the formulation is not optimized for alloy stability. As shown by Liu et al [15], with the optimization of the CMP slurry formulation for GST CMP process, an optimized post CMP clean formulation is also required in order to prevent de-alloying of the GST film and thus the material property of GST film necessary for PCM application. Figure 8 shows the typical defects found after post CMP cleaning of GST wafers.

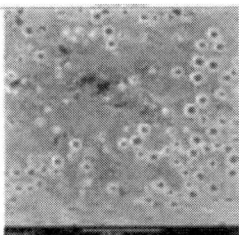

Fig. 8: Various defects found on GST wafers after post CMP cleaning

Figure 9 shows the defect counts using various post CMP clean formulations from Air Products and from another company. CoppeReady® CP88G has provided the best result.

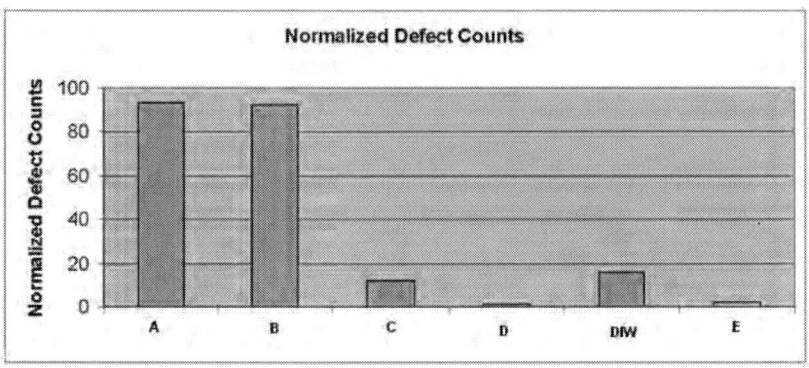

Fig. 9: Comparison of various post CMP cleaning formulations for GST film (D = CoppeReady® CP88G from Air Products)

Summary

CMP has become an established enabling technology for planarizing wafers and its scope has been broadening steadily to include varieties of new films. It has thus become necessary to formulate specific post CMP clean solutions to achieve lowest defects on different films without damaging the other exposed films in the stack. Significant advances are being made by the researchers in academia and industry to solve various technical challenges. Some of the examples given in this paper show the advances made by Air Products to address some of the technical challenges facing the post CMP clean application with new interconnect integration.

Acknowledgments

This paper is made possible through the contributions of several colleagues and collaborators. The author specifically acknowledges various discussions with Dr. D.C. Tamboli and Dr. M. B. Rao of Air Products and Chemicals Inc, Dr. Y. Chen of Applied Materials Inc. and Dr. R. Rhoades of Entrepix Inc.

References

1. K. Derbyshire, *Solid State Tech*, Oct. 2007.
2. G. Lopez et al, *IITC* 2007, pp. 40-42.
3. M. Aimadeddine et al. *IITC* 2007, pp. 175-177.
4. Y. Wang et al, ECS Trans., 2010, *33* (12), pp. 147-155.
5. Y. Chen et al, 215th ECS Symposia, May 2009, E2, Paper#742.
6. K. Mistry et al, *IEEE* 2007, p. 247.
7. J. M. Steigerwald, IEDM Tech Dig., 2008, p.247.
8. P. Packan et al, IEEE 2009, p. 28.4.1.
9. S. Lai, IEEE, 2008, 1-4244-2377-4/08, 01-02.
10. D.H. Im, IEEE, 2008, 1-4244-2377-4/08, 09-02.
11. B. J. Choi, *J. Electrochem. Soc.*, **156** 1, H59-H63, (2009).
12. J. Zhang, Proc. MRS, 2003, Vol. 734, B9.51.1 (2003).
13. M. Zhong, *J. Electrochem.Soc.*, **155**, 11, H929-H931, (2008).
14. A.G. Sharpe, Inorganic Chemistry, 1981, p287, Longman Group limited, New York.
15 F.Q. Liu et al, ECS Transactions, 2009, 19 (7) pp.73-79.

Cleaning aspects of novel materials after CMP

R. Vos[a], M. Wada[b], S. Arnauts[a], H. Takahashi[a,b], D. Cuypers[a],
H. Struyf[a] and P.W. Mertens[a]

[a]Imec, Kapeldreef 75, 3001 Leuven, Belgium
[b]Dainippon Screen Mfg. Co., Ltd., 480-1, Takamiya, Hikone, Shiga 522-0292, Japan

High mobility channel materials such as Ge and III-V compound semiconductors, are explored for the sub-22nm technology node. In order to allow process integration and full scale manufacturing of these materials, they need to be introduced by epitaxial growth in narrow trenches on silicon carrier wafers. Consequently, CMP is needed to flatten the surface before constructing the gate stack. Similarly to silicon, trace contamination is expected to have a detrimental effect on the devices and therefore, an appropriate cleaning of the wafer consisting of areas with mixed substrates is an absolute requisite. This involves the removal of particulate, metallic and organic contamination while controlling surface etching of these novel materials integrated on the silicon wafer. This will be addressed in this paper and allows the development of efficient post-CMP cleaning strategies.

Introduction

The successful downscaling of integrated circuits according to Moore's law and the scaling guidelines summarized in the ITRS roadmap is one of the biggest technological achievements of the last 40 years. This feat is mainly attributed to the good properties of the bulk Si and its native oxide. The continued reduction of the physical oxide thickness demanded by the scaling requirements has necessitated the replacement of the SiO_2 dielectric and poly-Si electrode in the heart of the transistor by a gate dielectric with higher dielectric constant and a metal gate, as has been realized in 32 nm technology node. However, the performance enhancement for scaled devices beyond the 22nm technology node cannot be guaranteed anymore due to intrinsic mobility problems. New channel materials with higher carrier mobility (μ) might be needed to meet the performance specifications. Ge is a promising candidate for the pMOS transistors since it has a very high hole mobility. For nMOS transistors, different III-V compound semiconductor materials such as InAs, InGaAs, InSb, ... having a higher electron mobility compared to silicon are being explored.

In addition, new device structures are being considered as well. Implant free Quantum Well - High Electron Mobility Transistors (QW-HEMT) have emerged to replace classical MOSFET devices [1]. Such QW-HEMT can be realized using Ge as PMOS channel material and a III-V compound semiconductor (e.g. InGaAs) as channel for nMOS in combination with a suitable heterojunction buffer material [2].

These new high-mobility channel materials are integrated on silicon carrier wafers inside narrow trenches using selective area epitaxial growth (SAG) [3] as schematically depicted in Figure 1.

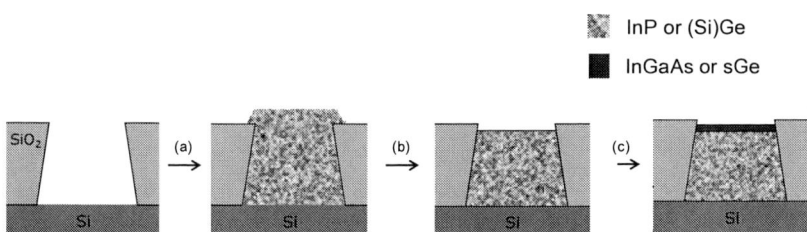

Figure 1. Schematic representation of process flow for introduction of high-mobility channel materials on Si carrier wafers.

Narrow trenches of widths ranging from 0.1 to 100 μm in SiO$_2$ on silicon wafers are made with fairly standard Shallow Trench Isolation patterning technologies. In these trenches, high quality crystalline epitaxial layers are selectively grown *(a)*. After chemical mechanical polishing and controlled etch back of the buffer material *(b)*, this substrate serves as the starting template for fabrication of high mobility devices by epitaxial growth of the channel material *(c)*. After deposition of the channel, the gate stack with the gate dielectric and electrode can be constructed and the raised source/drain areas are fabricated.

The development of a CMP process with good uniformity for structures of different size has been presented by Ong *et al.* [4]. Prior to channel deposition, the presence of a clean and well-ordered surface on an atomic level is necessary. Besides the chemically inhomogeneous native oxides, also unwanted contamination from the CMP step such as particulates and/or metallic contamination should be removed. Therefore a wet cleaning step with well-controlled and limited substrate loss is indispensable to remove all these contaminants. In this paper, the material loss of (Si)Ge and InP in various cleaning solutions is summarized and critical aspects for removal of particulate and metallic contamination are highlighted. This allows the selection of appropriate post-CMP cleaning sequences.

Experimental results and discussion

The etch rates of SiGe and Ge in different aqueous cleaning mixtures are summarized in Table I. No significant substrate etching is observed in non-oxidizing chemistries such as DIW and HF/HCl mixtures. In other acids, such as HNO$_3$ and H$_2$SO$_4$, a low but measurable etching is observed. In aqueous oxidizing environments such as ozonated DIW (O$_3$-DIW, see Table I) and NH$_4$OH/H$_2$O$_2$/H$_2$O (APM) mixtures (see Figure 2), significant etching of Ge occurs due to high solubility of the formed Ge-oxide in aqueous solutions. In the most concentrated APM, Ge is dissolved very rapidly and upon dilution of the APM mixtures, a decrease in the etch rate can be observed. Figure 2 also confirms that for SiGe, the etching is decreasing if the germanium content is lowered. In addition, it has been reported that a Si-rich oxide is growing for SiGe in APM mixtures while the GeO_2 and suboxides are dissolving [5]. Table I also shows that in diluted NH_4OH, Ge is being slowly etched with an etch rate lower than the one for Si in diluted NH_4OH (~0.4nm/min in 1:10000 diluted NH_4OH at RT). However, opposite to Ge, the etching of Si in NH_4OH is inhibited when the surface is covered with a native oxide. When SiGe is treated with diluted NH_4OH, a decrease in SiO_2 related bonds can be observed (see Figure 3) while no change is observed in the Ge-O bonding region. Detailed identification of the bondings in these areas is difficult due to overlap with the C-H bending and other peaks related to organic contamination which is found to be fairly constant for all samples as can be concluded from the C-H stretch intensities around 2900 cm^{-1}. Contrary to the diluted NH_4OH treated conditions, the samples were hydrophobic after diluted HF and the ATR-FTIR spectra show that the Si-O bondings are further reduced while also a new peak arises at 2010 cm^{-1} which can be attributed to Ge-H stretching. This indicates also further removal of Ge-oxides. Furthermore, the samples after short/long diluted NH_4OH followed by dHF result in comparable spectra, indicating similar surface conditions.

Table I. Etch rates of SiGe and Ge (nm/min) in different mixtures at RT.

	SiGe			
	$Si_{0.55}Ge_{0.45}$	$Si_{0.45}Ge_{0.55}$	Ge	Reference
$HCl:H_2O$ (1:8)	<0.017	<0.05	<0.04	[5]
$HF:HCl:H_2O$ (1:10:70)	<0.086	<0.16	<0.08	[5]
2 w-% HF	<0.090	<0.099	<0.11	[5]
1 M HNO_3	-	-	0.14±0.013	[6]
96 w-% H_2SO_4	-	-	0.16±0.006	[6]
H_2O	<0.022	<0.049	<0.036	[5]
H_2O (O_2-bubbling)	-	-	0.005	[6]
$DIW-O_3$	0.15±0.23	0.24±0.28	1.1±0.1	[5]
pH 11 NH_4OH (1:100)	-	-	0.29±0.007	[6]

'-' : not measured

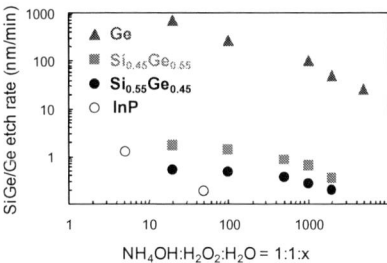

Figure 2. Ge, SiGe and InP etch rate as a function of APM dilution at RT.

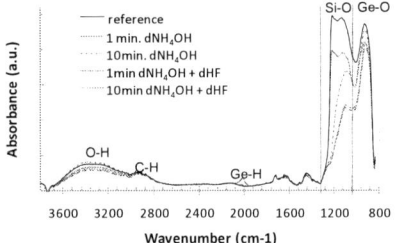

Figure 3. ATR-FTIR spectrum of SiGe55% after UV/O_3 based pre-clean followed by 1:100 diluted NH_4OH and 2% HF.

In Table II, an overview is given of the etch rates of InP in typical cleaning mixtures used in semiconductor manufacturing. It is shown that InP is etched in acidified solutions. Especially in concentrated HCl, a high etch rate is observed. However there is a risk for production of the toxic gas PH_3 in HCl containing solutions as has been reported previously [7,8]. This was also confirmed in this work. Table II also shows that, at low pH, the dissolution rate of InP is further increased if an oxidizer such as O_3 is present. At neutral pH with an oxidizer, such as DIW-O_3 mixtures, the etching is limited. Also in alkaline solutions, such as diluted NH_4OH, the InP is not etched unless an oxidizer is added. Figure 2 includes the etch rates of InP in APM mixtures at different dilutions and as shown in this graph, only for the most concentrated 1:1:5 mixture, a significant etching is observed. Similarly to SiGe, there is a risk for InP to alter the surface composition by using oxidizing chemistries due to the differences in solubility of the formed oxides.

Table II. Etch rates of InP (nm/min) in different cleaning chemistries.

	InP
37 w-% HCl	10721±349
3.7 w-% HCl (1:10)	2.5±0.2
0.037w-% HCl (1:100)	<0.05
2 w-% HF	<0.2
69 w-% HNO_3	0.68±0.09
96 w-% H_2SO_4	5.56±0.31
DIW-O_3	<0.4
0.037 w-% HCl + O_3	0.75±0.41
1M NH_4OH	<0.21

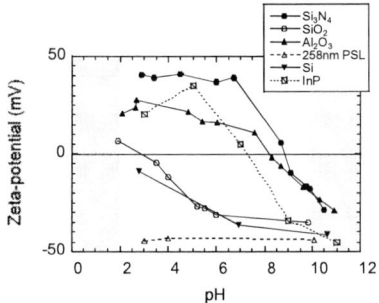

Figure 4. Zeta-potential measurements of different materials afo pH. (Values for InP are taken from reference [9].)

In order to select cleaning solutions with optimal particle removal efficiency, the chemistry needs to be tuned in order to eliminate electrostatic attraction forces between the wafer substrate and any particulate contamination. For wafers covered with areas consisting of different materials, this can be quite challenging since the diverse surfaces bear different charges depending on the pH. Figure 4 summarizes the surface charge as measured using zeta-potential measurements of typical materials that might be encountered in a manufacturing environment. Polystyrene latex particles (PSL) are included as model system of typical organic particulates. Germanium has an iso-electric point around 2 similar to Si [10] and is negatively charged over a broad pH range. The pH of the cleaning solution has to be choosen in a region where all particles/substrates have similar surface charges resulting in electrostatic repulsion, *i.e.* at pH < 2 or pH > 10. However at high ionic strength, such as I > 10^{-2} M, the electrostatic repulsion forces are shielded and van der Waals attraction forces become dominant. Therefore, the pH range 10 – 12 is favorable for particle removal.

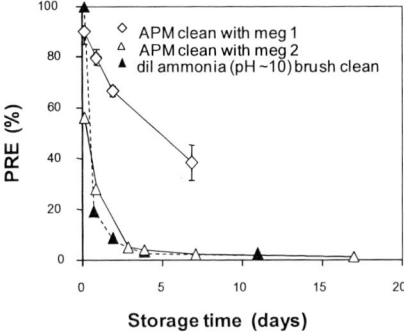

Figure 5. Particle removal efficiency of 30 nm SiO_2 slurry particles using megasonic or brush scrubbing clean from resp. Si and Si_3N_4 substrates as function of storage time (data taken from reference [11] and [12]).

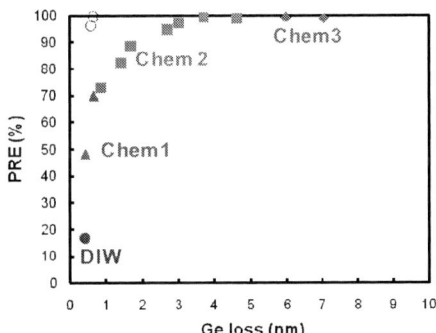

Figure 6. Removal of 30 nm SiO_2 slurry particles from Ge after storage as function of substrate loss using various chemistries with or without physical force.

It is well known that the removal of particulate contamination from flat substrates decreases if the cleaning is delayed. This is due to the increased adhesion forces attributed to particle deformation and/or formation of additional bonds as function of aging time. Figure 5 shows that this effect is very pronounced for the removal of small SiO_2 slurry particles even when high pH solutions such as APM or diluted NH_4OH are used in combination with physically assisted cleanings. This illustrates again the need for in-situ particle cleaning steps. As the physical forces cannot overcome the increased adhesion forces due to aging of the particles on the wafer substrate, the aged particles can only be removed from the wafer using substrate etching. Figure 6 shows that minimal ~3 nm substrate loss is needed for complete removal of 30nm slurry particles from Ge substrates. These results are in agreement with previously published results on Si substrates [13] and on Ge with larger sized particles [10].

In addition to particles, metallic contamination can be present after CMP and needs to be removed from the wafer substrate. For silicon, the best cleaning efficiency is obtained with HF based cleans [14]. This can be explained by etching and removal of the silicon hydroxyl groups that act as binding sites for metal cations [15]. However, metals as Fe, Ca and Zn are also removed to some extend using diluted HCl [16]. For Ge, it has also been reported that HF-based cleans perform better compared to diluted HCl [17].

Summary

(Si)Ge and III-V materials when used in future device processing result in new cleaning challenges. For removal of particulate contamination, either low pH (pH <2) or high pH (pH > 10) solutions such as diluted NH_4OH are recommended. High pH solutions cause no significant etching of the substrate at least when no oxidizer is present. Low pH solutions with HCl may impose some serious ES&H issues when used for cleaning of InP due to hydride evolution. However, diluted acids and especially HF-based cleanings are very efficient to remove metallic contamination from the substrate.

Acknowledgments

This work has partly been done within imec's industrial affiliation program on (sub-) 22 nm CMOS Technologies. Imec wants to thank their partners for the support of these activities.

References

1. G. Dewey, R. Kotlyar, R. Pillarisetty, T. Rakshit, H. Then, R. Chau, *IEDM 2009*.
2. Shinohara K., Yamashita Y., Endoh A., Hikosaka K., Matsui T., Mimura T. and S. Hiyamizu, *Jpn. J. Appl. Phys.*, **41**, L437 (2002); G. Hellings, G. Eneman, B. De Jaeger, J. Mitard, K. De Meyer, M. Meuris, M. M. Heyns, *2009 Silicon Nanoelectronic Workshop Proc.*, 33 (June 2009).
3. R. Loo, G. Wang, L. Souriau, J.C. Lin, S. Takeuchi, G. Brammertz and M. Caymax, *J. Electrochem. Soc.*, **157**, H13-H21 (2010); G. Wang, M. Leys, D. Nguyen, R. Loo, G. Brammertz, O. Richard, H. Bender, J. Dekoster, M. Meuris, M. Heyns and M. Caymax, *J. Electrochem. Soc.,* **157**, H1023 (2010).
4. P. Ong and P. Leunissen, to be presented at CSTIC (2011).
5. M. Wada, H. Takahashi, J. Snow, R. Vos, T. Conard, P.W. Mertens and H. Shirakawato, *Solid State Phenomena*, to be published (2011).
6. D. Brunco, B. De Jaeger, G. Eneman, J. Mitard, G. Hellings, A. Satta, V. Terzieva, L. Souriau, F.E. Leys, G. Pourtois, M. Houssa, G. Winderix, E. Vrancken, S. Sioncke, K. Opsomer, G. Nicholas, M. Caymax, A. Stesmans, J. Van Steenbergen, P.W. Mertens, M. Meuris and M. M. Heyns, *J. Electrochem. Soc.*, **155**, H552 (2008).
7. P.H.L. Notten, *J. Electrochem. Soc.*, **131**, 2641 (1984).
8. H.F. Hsieh and H.C. Shih, *J. Electrochem. Soc.*, **137**, 1348-1353 (1990).
9. A. Hachigo and T. Nishiura, US7569493.
10. S. Sioncke, M. Lux, W. Fyen, M. Meuris, P. Mertens and A. Theuwis, *Solid State Phenomena*, **134**, 173 (2008).
11. G. Vereecke, J. Veltens, K. Xu, A. Eitoku, K. Sano, S. Arnauts, K. Kenis, J. Snow, C. Vinckier and P. Mertens, *Solid State Penomena*, **134**, 155 (2008).
12. X. Kaidong, PhD dissertation, KULeuven (2008).
13. T. Hattori, Ed., in *Ultraclean Surface Processing of Silicon Wafers*, Springer (1998).
14. R. Vos, E. Kesters, S. Garaud, D. De Waele, K. Kenis, M. Lux, H. Kraus, J. Snow, D. Shamiryan, G. Catana, W. Deweerd, T. Schram, S. Degendt and P. Mertens, *Solid State Phenomena*, **103-104**, 241 (2005).
15. L.M. Loewenstein and P.W. Mertens, *J. Electrochem. Soc.*, **145**, 2841 (1998).
16. T. Q. Hurd, H.F. Schmidt, A.L.P. Rotondaro, P.W. Mertens, L. H. Hall and M.M. Heyns, in *Cleaning Technology in Semiconductor Device Manufacturing IV*, R.E. Novak and J. Ruzyllo, Editors, **PV 95-20**, p. 277, The Electrochemical Society Proceedings Series, Pennington, NJ (1995).
17. S. Sioncke, B. Onsia, K. Stuys, J. Rip, R. Vos, M. Meuris, P. Mertens and A. Theuwis, *ECS Transactions* **1(3)**, 220 (2005).

Study on the Ring Type Crater Defect Reduction in Cu CMP Process

Jin-Hai Xu[a], Paul-Chang Lin[a,b], Charles Xing[a,b], Pei Li[a], Zhi-Yong Ma[a]

a. SMIC, 18,Zhangjiang Rd, Pudong, Shanghai, 201203, PRC.
b. Department of Microelectronics and Solid-State Electronics, Fudan University.

Abstract:

Copper has been used as inter-wiring in deep-submicron integrated circuit (IC) for its lower resistivity and potentially higher resistance to electro migration. Copper Chemical–mechanical polishing/planarization (Cu CMP) is used to remove the redundant Cu and Ta/TaN/oxide to achieve dual damascene interconnect structure. Crater defect is one of the most dangerous killer defects in Cu CMP process. In this research, we will focus on study the forming mechanism of one typical ring type crater defect. Also the solution is studied and carried out to eliminate this ring type crater defect.

Introduction

Chemical mechanical planarization (CMP) has become the most effective method of achieving local and global planarity across a wafer surface for high-volume semiconductor manufacturing. CMP uses both mechanical and chemical means to selectively remove material from the surface of a wafer to result in a flat surface, which is required for subsequent processing steps [1, 2]. Copper is widely used in deep-submicron integrated circuit(IC) because of its valuable physical and mechanical properties. Because of the different performance of copper and aluminum, the Damascus process have been introduce to semiconductor manufacturing for copper as interconnect line. The Cu CMP is very important for Damascus process.

Crater is one of the most dangerous yield killer defects post Cu CMP process, an example of crater defect is shown in Figure1.This defect may cause the IC open circuit or impact device performance reliability. There are various of factors inducing crater defect. It is reported that process environment contamination and the concentration of some specific chemical will contribute to crater defect in both Cu ECP (electro chemical plating) and Cu CMP [3-6,8]. For example, if VOC (volatile organic compound) concentration in the air is higher, the crater defect is worse [8]. In this paper we are dedicated on studying one typical ring type crater defect forming mechanism, in final the solution is studied and carried out to eliminate this ring type crater defect.

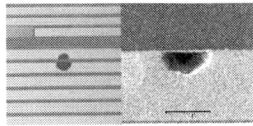

Figure 1.Crater image in Cu CMP

Experiment

It is known that the crater which is random or clusters on the wafer in semiconductor manufacturing is caused by the Cu seed surface contamination in BS/ECP process or the process environment during CMP [8].

One typical crater defect with ring pattern is often found after Cu CMP process, as Figure 2 shown. In order to find the root cause, we designed experiment to verify.

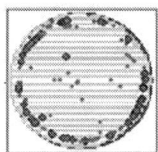

Figure2. Ring type crater defect

Design of experiment:

3M blocky disk with 4.25-inch diameter, polyurethane IC1010 polishing pad, CMP Titan head, colloidal silica based slurry (PH ~5, peroxide as oxidizer) and post CMP clean citric acid are used in this experiment. During this experiment, ~10s high pressure DI water rinse is performed after main polish step and no other inhibitor is added in process. We named the waiting or delay time after process finish on platen1/2/3 as P1/P2/P3 WT respectively. We use 200mm pattern wafers with ECP electrochemical planting about 16kÅ Cu film. Split table is shown as Table1.

Condition	Platen	Waiting time(second)	Platen	Waiting time(second)	Platen	Waiting time(second)
A	P1	<10	P2	<10	P3	<10
B_1	P1	150	P2	<10	P3	<10
B_2	P1	200	P2	<10	P3	<10
C_1	P1	<10	P2	150	P3	<10
C_2	P1	<10	P2	200	P3	<10
D_1	P1	<10	P2	<10	P3	150
D_2	P1	<10	P2	<10	P3	200

Table 1 DOE table for different waiting time and platen

Equipment applied in the experiment
1. Novellus ECP SABRE NExT system in standard process condition is used for Cu film deposition.
2. The polish is performed by AMAT 200mm CMP Mirra Mesa, Titan head.
3. The defect is inspected by KLA-Tencor inspection tool 2360.
4. The LEICA-INS 3000R and SEM (HITACHI S-5500) are used to review the defects.

Result and Discussion

Total 6 pieces pattern wafers are used in the experiment and wafers No. are named A-D. The wafers were processed in Mirra-polisher with standard process condition. The results are shown in Figure 3. Based on the results, we can find that condition A, D_1 and D_2 defects are good, but other conditions all suffered crater defect. And we can find that if the waiting time is longer, the defect is worse. This phenomenon also applies to Condition B_1 and B_2. And for the same waiting time, C_1 and C_2 crater defect performance is worse than B_1 and B_2's. It implies the waiting time on different platens is relative to the ring type crater defect. The results indicate P3 WT won't induce crater defect, but P2 WT and P1 WT will induce ring type crater defect, and P2 WT impact ratio is higher than P1 WT. As the WT increase, the ring type crater becomes worse. As we known, the chemical reaction is the major factor for metal erodes [3, 7]. The process on P1/P2 is to remove Cu metal and P3 is to mainly remove oxide film, the metal is easily eroded on P1/P2. Based on the trend shown in Figure 3, we can know that wafer has risk to suffer crater defect if the waiting time is more than 100s either on P1 or P2, and waiting time on P2 is the most critical.

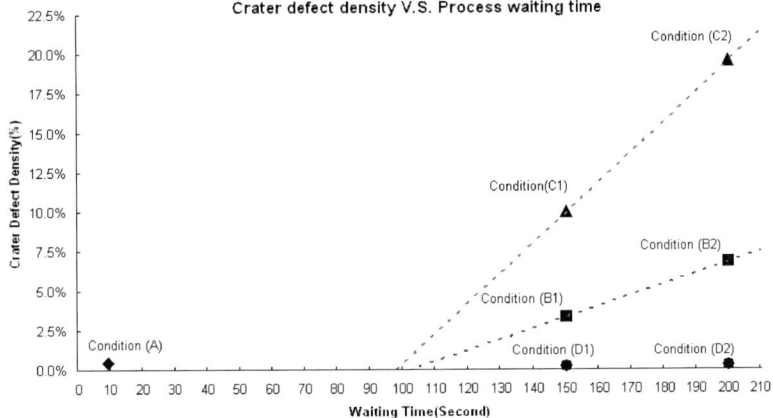

Figure3.Crater defect density of samples

Also after observing defect map, we find the craters are exactly formed in one circle if we composite all crater defect maps as Figure 4 shown. After inspection across the wafer diameter, the distance d between the circle and wafer edge is 17.5mm. As we known, the wafer carried head is Titan head which has three chambers as Figure 5 shown. The distance L of inner tube to the inboard edge of retaining ring is 17.5mm. It exactly matches the ring type position on wafer. This indicates the inner tube pressure may cause the ring type crater defect. After looking into the detail process step, we can find when Cu CMP process finished, the head was raised up by vacuum, but the inner tube presses the wafer to finish de-chuck action. This may induce the wafer locally tiny bend, the position of the wafer pressed by inner tube is easy to contact the chemical on pad, when the contact time is long enough and the chemical reaction will erode the Cu to form corrosion/crater or metal damage defect

[4,7, 8]. Further more, the chemical on wafer surface will flow to the bend position of wafer and it accelerates the local chemical reaction.

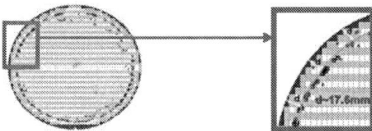

Figure4. Ring type defect map

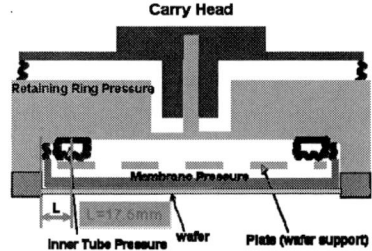

Figure5. Carrier head structure

In order to verify the correlation between inner tube pressure and ring type crater defect, we design another experiment as table 2 shown. We have known that waiting time on platen 2 is the most critical, so firstly we ensure all conditions listed in Table 2 waiting for 200s after process finish on P2, then 4 split conditions on inner tube pressure were done to check corresponding ring type crater defect density, experiment results are shown in Figure 6. Based on this diagram, we can find reducing inner tube pressure is an effective way to reduce ring pattern crater defect. But we can't reduce inner tube pressure to 0 since wafer de-chuck action can't be completed without inner tube pressure, we can reduce the inner tube pressure value in equipment constant to a minimum value by which wafer de-chuck can be performed without tool alarm.

Condition	P2 Inner tube pressure(psi)	P2 waiting time(s)
E	3.6	200
F	3	200
G	2.4	200
H	1.8	200

Table2. DOE table for different inner tube pressure (data is normalized)

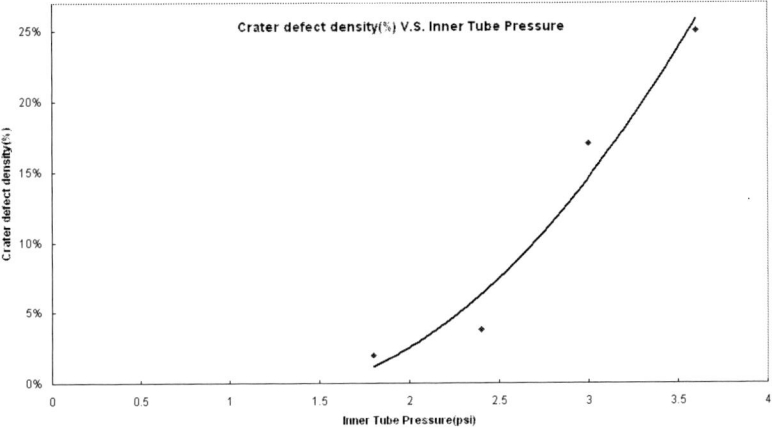

Figure6.Crater defect density V.S. inner tube pressure(data is normalized)

Conclusion

Based on above analysis, we can come to some conclusions. Wafer will have ring type crater defect if wafer waiting time on P1 or P2 is over 100 seconds, and waiting time on platen 2 is more critical. It is attributed to long time chemical reaction on the wafer local area which matches with inner tube position in Titan head. By optimizing inner tube pressure in equipment constant, we can effectively reduce ring type crater defect. Also P1 and P2 polish time balance well is a good way to prevent from waiting time on P1 or P2 longer. Especially for some products with extremely thick ECP plating (more than 30K), it is necessary to introduce dummy wafer between pattern wafers to reduce waiting time on P1 and P2.

References

1. P. Singer, Semicond. Int., 6, 91 (1998);
2. J. Xina, Wear 268 (2010) 837–844;
3. Taek-Soo Kim, Tomohisa Konno,ect, Acta Materialia 57 (2009) 4687–4696;
4. Jeng-Yu Lin, Yung-Yun Wang,etc, Electrochemical and Solid-State Letters, 10 (1) H23-H26 (2007);
5. Lucia D'Urzo,Benedetto Bozzini, Mater Electron (2009) 20:666–670;
6. J. Hayon, C. Yarnitzky,ect, Journal of The Electrochemical Society, 149 (7) B314-B320 (2002);
7. Jorn Plagmann,Martin Dicks,CMP-MIC conference,79,(2007);
8. Liang,Chen,Kan Wu,ect, ECS Trans. / Volume 27 / Issue 1 / Thin Film, Etch and Plating, CSTIC 2010;

New application of optical endpoint system: In situ Cu residue detection

Weifeng Zhang[a], Xucheng Wang[a], Changxing Tan[a], Shan Wang[a], Walters Shen[b], George Ge[c]

[a] Applied Materials China
[b] Applied Materials USA
[c] Semiconductor Manufacturing International Corp

Based on the detection method of optical inspection tool, an in-situ Cu residue detection method was developed to check all the wafers through the optical endpoint system. This method will help to block the wafer which suffered Cu residue defect flowing to the next process step and monitor the hardware variation, consumable variation and process variation. It will also help to improve manufacture efficiency.

Introduction

Cu CMP has been widely adopted in BEOL process as a key process of Cu Damascene technology, and made a rapid progress in recent years. For Cu CMP process, Cu residue is a major yield killer defect which is caused by hardware variation, consumable variation or process variation[1, 2]. It can be eliminated through reworking if it can be detected before the wafer processes the following process step.

The Cu residue can be detected by optical inspection tool, but not all the wafers will be scanned after Cu CMP in the production line due to low throughput of inspection tool, delayed feedback. Therefore some of the wafers which suffered Cu residue will flow to the next process step without reworking, and the yield will be impacted[2]. A more efficient Cu residue detection method is needed to check all the wafers.

Based on the principle of the detection method of optical inspection tool and the optical endpoint system of CMP tool, an in-situ Cu residue detection method is developed to check all the wafers during Cu CMP process, which can detect the Cu residue without throughput impact and hardware retrofit.

This method can also be used to monitor the CMP tool hardware variation, consumable variation and process variation, and it is helpful for Cu CMP process control and yield improvement.

Mechanism of in-situ Cu residue detection method

1. Principle of the detection method of optical inspection tool
In the visible range of radiation, due to the inherent properties of conductors, metal films exhibit high reflectivity to incident light. The amount of the reflective intensity is quantified by reflectance, R, which is given as

$$\mathbf{R} = I_r / I_i \qquad [1]$$

Where I_r is the intensity of the reflected beam and I_i is the intensity of the incident beam.

For a monochromatic illumination and a fixed geometry, the magnitude of the reflected intensity is directly dependent on the layer's reflectance ($I_r = R*I_i$). During defect inspection, optical inspection tool will scan the wafer surface to collect the reflected intensity. If the reflected intensity of some area is higher than that of other area, the inspection tool will define that there is Cu residue defect, because the reflectance of the Cu residue area is higher than that of normal area, see Figure 1.

Figure 1. Optical image of Cu residue defect

2. Introduction of optical endpoint system on CMP tool

A monochromatic laser is used as incident beam in optical endpoint system, its assembly is shown in Figure 2.

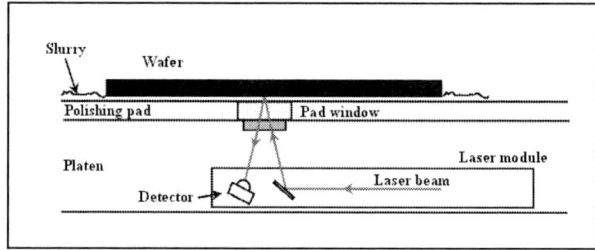

Figure 2. Platen assembly of optical endpoint system

Optical endpoint system has high sensitivity to the reflected intensity change, it can detect signal variations resulting from the exposure of different layers, even for metal layers with very similar reflectivity. For a stack of metal layers deposited on a silicon substrate wafer (Figure 3-a), the signal variation during wafer polishing is shown in Figure 3-b.

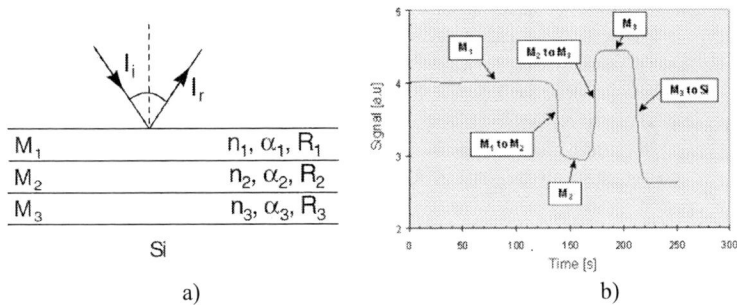

a)   b)

Figure 3. Signal variation during a stack of metal layers polishing ($R_3>R_1>R_2$)

Optical endpoint system has high spatial resolution across wafer's surface. It is able to collect the raw reflected intensity data across the wafer, and the measured signals can be easily correlated to the radius on the wafer. Normally the average intensity of one scan on wafer is plotted on an endpoint trace, characteristic changes on the trace will be used to determine endpoint time. At the same time, the reflected intensity at the special radius is able to be collected through a special algorithm.

3. Segmented Algorithm setting of in-situ Cu residue detection method

Normally Cu CMP is divided into the following 3 step polishing (3 platens polishing), see Figure 4. Based on optical endpoint system of CMP tool, the in-situ Cu residue detection method is set to detect the Cu residue at the end of step 2 (Cu film clearing). During this period, most of the wafer's surface should be barrier layer (Ta), if there is Cu residue at some area, the reflected intensity of this area should be higher than other area, because the reflectance of Cu layer is higher than Ta layer ($I_r = R*I_i$, $R_{Cu} = 0.947$, $R_{Ta} = 0.56$).

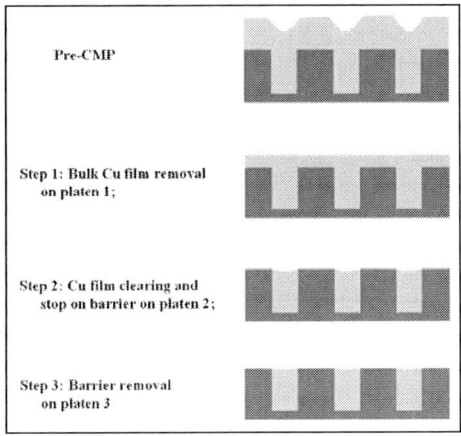

Figure 4. Cu CMP Process sequence

In this method, a segmented algorithm is used to calculate the average reflected intensity across the wafer (I_{avg}) and the reflective intensity at the selected radius (I_{sel}) from the raw data of optical endpoint system, then I_{sel} and I_{avg} will be compared.

If $I_{sel} \approx I_{avg}$, the algorithm will define that there is no Cu residue at the selected radius. And the wafer will continue the following steps normally.

If $I_{sel} \gg I_{avg}$, the algorithm will define that there is Cu residue at the selected radius, then a warning message will be sent to CMP tool from optical endpoint system, which includes where the residue is detected (Depend on your setting), and then the correction actions will be taken accordingly. The chart of this method is shown in Figure 5.

The sensitivity of this method can be adjusted through the setting of the algorithm to detect the tiny Cu residue (i.e. adjust the threshold of the difference of I_{sel} and I_{avg}).

The endpoint time of step 2 (Cu film clearing) is determined by the signal change of I_{avg}, the Cu residue will be detected at the step of over polishing, so this method does not impact throughput. And this method no needs any hardware retrofit.

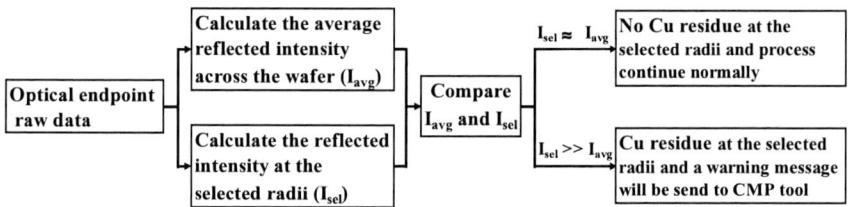

Figure 5. The algorithm of the in-situ Cu residue detection method

Examples of application

In this section, the raw data of two wafers (one suffered wafer edge Cu residue, another without Cu residue) are reprocessed with the in-situ Cu residue detection method to verify the function of this method. These two wafers are polished on Reflexion LK 300mm CMP tool.

Example 1: Wafer edge suffer Cu residue due to zone1 pressure shift of polishing head
The optical endpoint data is shown in Figure 6-a, it is obvious that wafer edge has Cu residue (residue defect size is 20um – 200um). The raw data is reprocessed with the in-situ Cu residue detection method. In this method, the reflected intensity of wafer edge (I_{edge}) and the average reflected intensity across the wafer (I_{avg}) are plotted on the endpoint trace (Figure 6-b), and the magnitude of them are compared to determine wafer edge has Cu residue or not. For this wafer, the I_{edge} is much higher than I_{avg} (Figure 6-b), which indicates that there is Cu residue at wafer edge, and then a warning message is sent to CMP tool from the optical endpoint system (Figure 6-c).

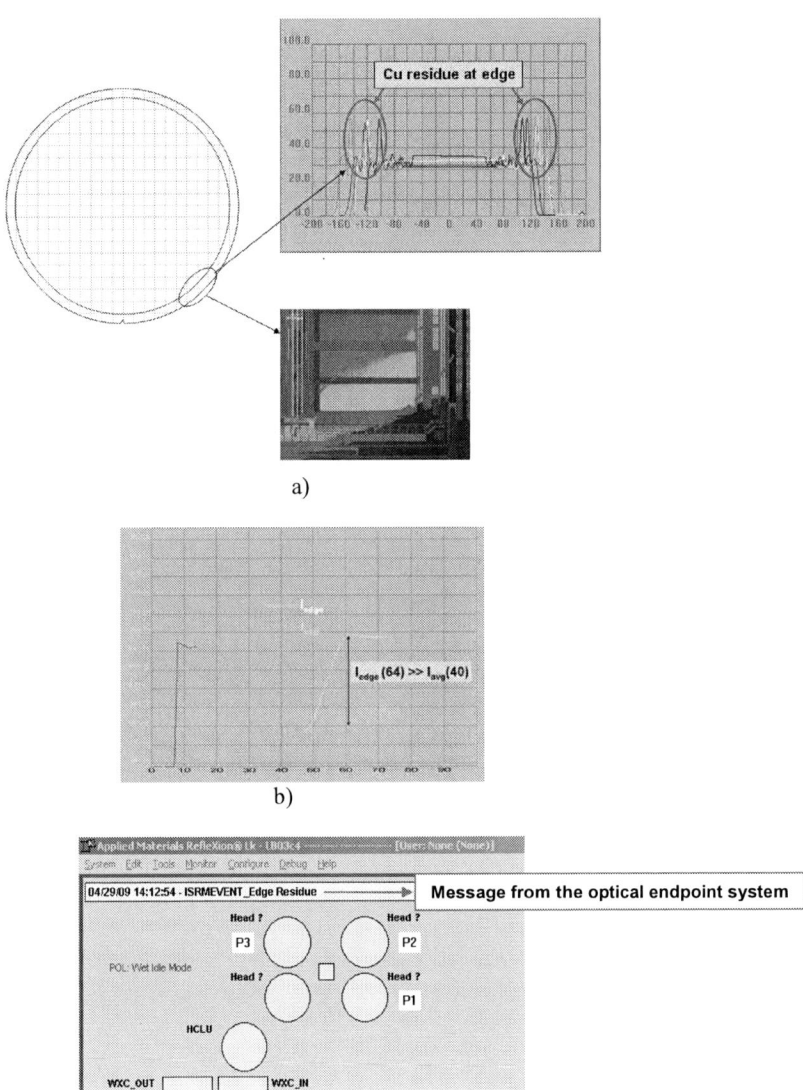

a)

b)

c)

Figure 6. Reprocess of the optical endpoint data of the wafer suffered edge Cu residue

Example 2 : Wafer without Cu residue at wafer edge
The optical endpoint data of this wafer shows that there is no Cu residue at wafer edge (Figure 7-a). The raw data is also reprocessed with the in-situ Cu residue detection method, and the endpoint trace shows that I_{edge} is similar to I_{avg}, which indicates that this wafer has no residue at wafer edge (Figure 7-b).

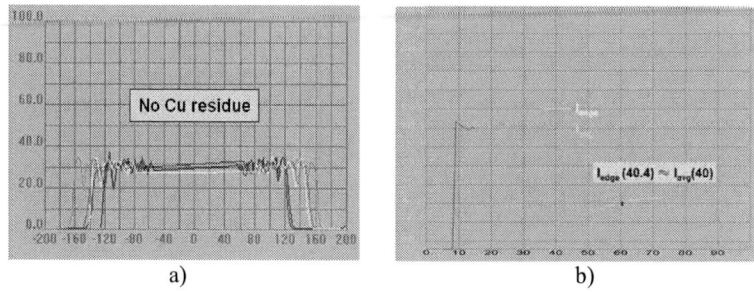

a)   b)

Figure 7. Reprocess of the endpoint raw data of the wafer without Cu residue

Work flow chart comparison

After compare the work flow chart which is used in production line (Figure 8-a) and the work flow chart with the in-situ Cu residue detection method (Figure 8-b), it is obvious that in-situ Cu residue detection method has the following advantage:
1) All the wafers will be checked, while only few of wafers are checked currently;
2) Wafer which suffered Cu residue can be rework as soon as possible;
3) Be able to feedback hardware / consumable / process variation on time;
4) Most of work is finished with in CMP tool.

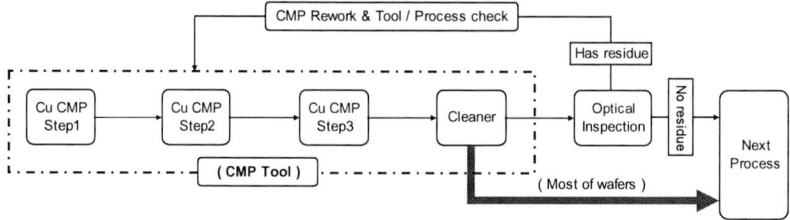

a) The work flow chart which is used in production line currently

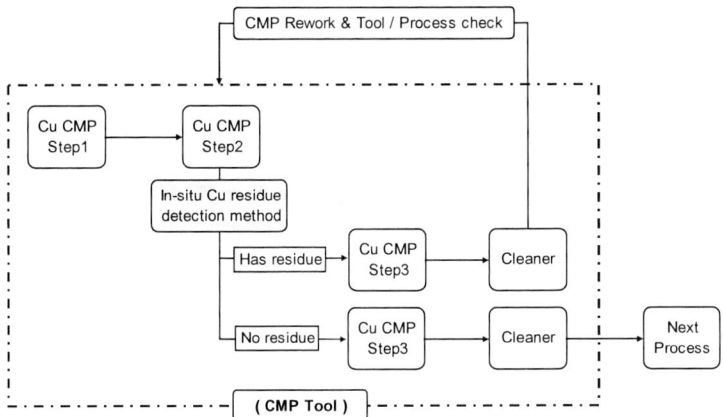

b) The work flow chart with in-situ Cu residue detection method

Conclusion

Based on the principle of the detection method of optical inspection tool and the optical endpoint system of CMP tool, an in-situ Cu residue detection method is developed. It is able to check all the wafers during Cu CMP process without throughput impact and any hardware retrofit. It can also be used to monitor the CMP tool hardware variation, consumable variation and process variation, which is helpful for Cu CMP process control and yield improvement. It is also helpful to improve the manufacture operation efficiency.

Acknowledgements

The authors would like to thank all the members of Applied materials China CMP technology group and Applied materials USA technology center for their support during the method development.

References
1. H. Chen, T. Tsai, Y. Huang, C. Huang, C. Chen, et al., *Proceedings of IEEE Int. Interconnect Technology Conf.*, 6, 21(2001).
2. J. Tony Pan, et al., *Proceedings of CMP-MIC Conference*, 2, 11(1999).

ECS Transactions, 34 (1) 691-697 (2011)
10.1149/1.3567659 ©The Electrochemical Society

The Mechanism of Organic Base and Surfactant in Silicon Wafer CMP Process

Li Weiwei

School of Information Engineering, Hebei University of Technology, Tianjin, 300401, China

In silicon substrate polishing, slurry has important influence on polishing rate and surface quality. Organic base and surfactant are the key aspects which decide slurry quality. This paper researches action mechanism of different organic bases in polishing and the influence of different surfactants on polishing rate. The results show that, polishing rate of slurry with polyamine is higher than that of slurry with monoamine, polyhydroxyl non-ionic surfactant can improve the stability of slurry system, and speed up the quality transmission, which is benefit to improve polishing rate. Slurry with polyamine and polyhydroxyl non-ionic surfactant has higher polishing rate, good pH stability, while no surface defects, such as scratches and corrosion pits.

Introduction

With shrinking feature size and growing substrate area of integrated circuits continuously, billions of devices can be integrated on the chip surface. High quality silicon substrate material is required in order to ensure the yield and electrical characteristics of integrated circuits[1]. Chemical mechanical polishing (CMP) is widely used to achieve a high level planeness on the substrate surface, while the polishing slurry plays a crucial role to determine the polishing quality directly. As we all know, CMP is an interworking of chemical action and mechanical action. Silicon CMP slurry mainly includes abrasive particles playing mechanical role; pH regulator, surfactants, buffers playing chemical role. In order to achieve high-level, low damage, low roughness, chemistry-oriented is proposed to approach to minimize the mechanical action. But how to enhance the chemical action further and match the chemical and mechanical action organically is the focus of current study[2,3].

In CMP process, pH regulators and surfactants play a very important role. The former adjusting the pH of polishing system, is the key of chemical role; the latter reducing the surface tension of CMP slurry, improve the consistency of mass transfer and the surface quality after CMP. Song Xiaolan et al. used non-ionic surfactant Triton X-100 and anionic surfactant SDBS compound to improve the stability of CMP slurry significantly[4]. W.R. Morrison used tetramethylammonium hydroxide as a pH regulator to improve the selectivity and achieved better processing results[5]. Ttriethanolamine was also suggested to be used as a pH regulator improving the selectivity and CMP quality[6]. In other studies, adding polyamines organic base and FA/O surfactant could reduce CMP scratches and improve the polishing rate[7~10]. PARK Jin-Hyung investigated the effect of an alkaline agent, tetramethyl ammonium hydroxide (TMAH) in a slurry with colloidal silica abrasives, on the polishing rate of nitrogen-doped $Ge_2Sb_2Te_5$ (NGST) film[11]. However, the impact of different pH regulators and surfactants on the removal rate, slurry stability, circulation ability and surface quality is lack of a systematic research.

691

In this paper, the effect of different organic bases and surfactants in silicon substrate polishing process were studied, as a result, the best choice and proportion were determined to achieve increased processing efficiency.

Experimental

Adding different organic bases to the colloidal silica respectively with 1.5wt% concentration and mixing uniformity to compare experiments effect. Basic components in slurry: organic base 1.5wt%, colloidal silica 35wt%, surfactants 0.05wt%; when polishing, diluting with water according to the ratio of 1:15(mass ratio),and then mixing well. Slurry parameters: pH = 10.5-12, the specific gravity = 1.278.Each sample was polished in the uniform parameters: pressure is 0.4Mpa; polishing speed is 100rpm; slurry flow is about 50ml/min; polishing time is 40min at a time, one sample polishing twice. pH value was measured every 20 minutes. Considering the low initial temperature of polishing lap and slurry, the slurry was heated to 25-26 ℃ before polishing, in order to make the temperature of slurry as uniform as possible before and after polishing. After polishing surface roughness was detected by atomic force microscopy after polishing.

Results and discussion

The effect of different organic bases in silicon substrate CMP process

The choice of alkali is very important to alkaline slurry. In this process, some factors must be taken into account, such as the pH stability, toxicity and volatility. Potassium hydroxide, sodium hydroxide and ammonia are commonly used as pH regulators. Sodium hydroxide or potassium hydroxide can adjust slurry pH efficiently, but the alkali metal ions is very easy to enter the substrate, causing the local penetration through device, leading to leakage current increasing and other effects in the polishing process. The ammonia, with the disadvantages of unstable pH and volatile, also causes environment pollution on the clean room. Therefore, the organic base is now widely used as pH regulator[12].

In this paper, several aliphatic amines with the similar molecular structure were selected as slurry pH regulators, and the polishing speed was compared at the condition. The results were shown in Fig 1.

The result of the polishing experiments shows that the polishing rate of polyamines is much higher than that of monoamine. This is because the polyamine in the polishing process has played a buffer role. When a local change of the slurry pH value occurs, lone pair electrons on the polyamine nitrogen atom seize H^+, and release OH^-, at last, to maintain a stable slurry pH. The chemical mechanism of silicon wafer polishing process is as follow:

First, Si reacts with OH^-, namely:

$$Si+4OH^- \rightarrow SiO_4^{4-}+2H_2\uparrow \text{ (redox reaction)} \qquad (1)$$

SiO_4^{4-} generated rapidly reacts with water further:

$$SiO_4^{4-}+4H_2O \rightarrow Si(OH)_4+4OH^- \qquad (2)$$

$$2Si(OH)_4+4OH^- = 2SiO_3^{2-}+6H_2O \qquad (3)$$

Thus, OH⁻ and water were consumed in the reaction, multi-amine organic base promotes the ionization of water:

$$R-NH_2 + H_2O \rightarrow R-NH_3^+ + OH^- \tag{4}$$

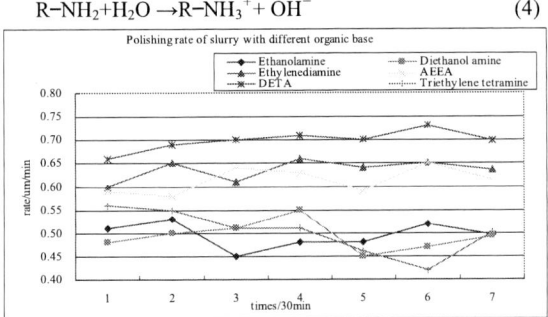

Fig.1 Polishing rate curve of slurry with different organic bases

With CMP chemical reaction, OH⁻ reduces and reaction (4) moves to the right, continuously providing OH⁻, which ensures a high polishing rate cycle.The capacity of providing OH⁻ is proportional to the basicity of organic bases. Organic base, different with inorganic base, is consistent with Arrhenius acid-base theory. The basicity is the ability to give electronic, a manifestation of this capacity is the charge density, more denser more stronger basicity[13]. As we known, −CH₃ and−H are electron donating group, while −OH is electron withdrawing group. Because of electron-donating capability: ethanolamine > diethanol amine, alkalinity is also the case. AEEA can be perceived as H replaced by −CH₂CH₂NH₂, so its basicity is obviously increased. Because of low molecular weight volatile and volatility, ethylenediamine is not preferred.

Because of containing multi-lone pair electrons of nitrogen atoms, the alkali and buffer capacity of polyamine is optimal. However, when polishing, as triethylenetetramine has a high molecular weight, reaction products can not be clear away in time, which covering on the wafer surface, and it impedes further reactions, resulting in polishing rate decreasing. As strong alkaline, diethylenetriamine has good buffering capacity, proper molecular weight, and non-volatile, and its reaction products can be taken away in time in polishing process, so polishing rate is the best.Therefore, we selected diethylenetriamine as the slurry pH regulator.

The effect of surfactants in silicon substrate CMP process

In CMP process, a large number of nano-particles, which produced in slurry and wafer surface under the action of chemical and mechanical roles together, on one hand, lead to agglomeration of abrasives, and on the other hand, left a lot of dangling bonds on the wafer surface. In order to reduce the surface energy, the wafer surface is very easily to absorb these nano-particles. These nano-particles diffuse on the wafer surface and crosslink corresponding atomics, then forming a stable adsorption state desorbed difficultly. These adsorbates are very difficult to clean, thus this will affect the subsequent steps, leading devices failure. This is a very critical problem in wafer CMP.

Anionic surfactant a, non-ionic surfactant b and cationic surfactants c, respectively were added to the slurry A (silica sol +2wt% diethylenetriamine) according to the proportion of 0.05wt%, then mixed well. Polishing with different slurries at the same parameters and comparing the polishing rate, the result was shown in Fig 2.

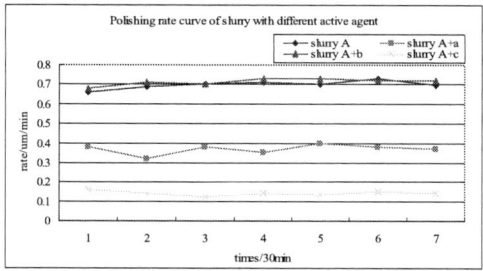

Fig. 2 Polishing rate curve of slurry with different active agent

According to the mechanism of surfactant, it needs to add one or more surfactant with special nature in the polishing slurry, and makes it absorb on the particle surface, which changes the surface properties of particles and greatly enhances the exclusion of particles, so that system dispersion performance significantly improves[14]. This will avoid scratches caused by particle agglomeration in polishing process.

Under alkaline condition, the surface of SiO_2 colloidal particles takes negative electricity, dispersed by electrostatic repulsion. When surfactant with charges was added into the system, the cationic surfactant produces gravity, which causes particles agglomeration together. Anionic surfactant has negative charge, due to Brownian motion, Zeta potential of the system changes rapidly, which makes the system in a more unstable condition. Non-ionic surfactant with high stability is not in the ionic state and it doesn't affect the colloid surface charge. The dispersion state of Non-ionic surfactant on the particle surface is shown in Fig.3. As steric role of the macromolecules enhances the stability of the system and speed up the quality transmission of process, to some extent, it improves the polishing rate.

Hydroxyl and amide group in non-ionic surfactant molecules are hydrophilic and good water-solubility, so only the Zeta potential changes of slurry added surfactants different with hydroxyl number was analyzed.

Fig.3 Dispersion state of Non-ionic surfactant on particle surface

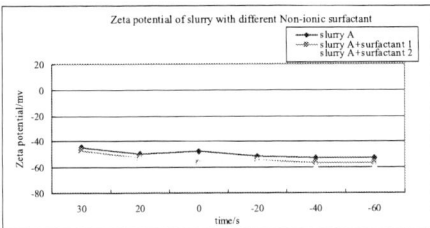

Fig.4 Zeta potential of slurry with different Non-ionic surfactant

Surfactant 1 is monohydroxyl surfactant, Surfactant 2 is multi-hydroxyl surfactant.

Fig.4 shows that zeta potential of the slurry with polyhydroxy surfactant was significantly higher than that without surfactant, while little higher than that with monohydroxy surfactant. This is because when there is a large number of hydroxyl, one of the hydroxyl forms hydrogen bond and adsorbs with the surface of colloidal particles, a number of hydrogen bonds left, which makes the number of hydroxyl groups increased on particle surface. This lead charge density and quantity and Zeta potential increased too, slurry system is more stable.

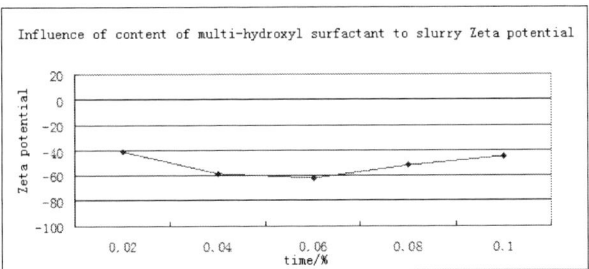

Fig.5 Influence of content of multi-hydroxyl surfactant to slurry Zeta potential

Fig. 5 shows that the absolute value of Zeta potential increases firstly, and then decreases with the polyol surfactant concentration increases. This is because when the concentration is low, the particle surface is not covered effectively, which makes low Zeta potential. The particles without adsorbing dispersant aggregate because of Brownian motion, this makes poor dispersion. When the concentration gradually increases, the particle surface coverage is improved, Zeta potential increased, thus dispersion stability increases. However, when the concentration is over high, the particle surface adsorption has reached saturation, the excess free dispersant molecules connects with each other leading to flocculation, so dispersion stability is deteriorated.

In conclusion, polyol nonionic surfactant self-synthesized was selected in this article, the addition is 0.05wt%.

Silicon (polished) surface quality inspection

As seen from Fig.6, scratch and corrosion pits on silicon surface disappeared after polish.

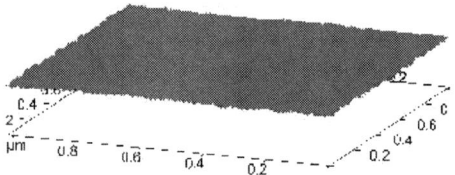

Fig.6 Atomic force microscope image of silicon surface

Conclusion

(1) The slurry added diethylenetriamine shows the highest polishing rate compared with other organic bases and can be cleaned easily. This indicates that polyamine playing good buffer role, whose polishing rate is higher than that of monoamine.

(2) Adding polyhydroxy non-ionic surfactant into slurry can increase Zeta potential, and keep slurry system stable, so it can speed up the mass transmission in polishing process and is beneficial to improve the removal rate.

(3) Slurry, added polyamine and polyol non-ionic surfactant, shows high polishing rate and stable pH. The silicon surface after polished has no scratches, corrosion pits and other defects.

References

1. H. Fusstter. Impact of chemomechanical polishing on the chemical composition and morphology of the silicon surface. Mat. Res. Soc. Symp. Proc, 1995, 386(97).
2. Kevin Cooper, z Jennifer Cooper, b Johannes Groschopf, etc. Effects of Particle Concentration on Chemical Mechanical Planarization. Electrochemical and Solid-State Letters, 2002, 5 (12): 109-112.
3. Mahadevaiyer Krishnan, Jakub W. Nalaskowski, Lee M. Cook. Chemical Mechanical Planarization: Slurry Chemistry, Materials, and Mechanisms. Chem. Rev., 2010, 110 (1): 178–204.
4. Song Xiaolan, Research of Chemical mechanical polishing and application in nano-silica slurry. Dissertation, Hunan: Central South University, 2008: 61~64.
5. W.R. Morrison, K.P. Hunt. European Patent 0853335, 1998.
6. R. Manivannan, S. Ramanathan. Role of abrasives in high selectivity STI CMP slurries. Microelectronic Engineering, 2008, 85:1748~1753.
7. Zhang Kailiang, Liu Yuling, Wang Fang, Research of nano abrasive used for chemical mechanical polishing ULSI silicon substrate. Semiconductor journal, 2004, 25(1):115~119.
8. Liu Yuling, Tan Baimei, Zhang Kailiang, Substrate material properties and processing test technology project of VLSI, Metallurgical industry press, 2002.
9. Yuling Liu, Kailiang Zhang, Fang Wang, Weiguo Di. Investigation on the final polishing slurry and technique of silicon substrate in ULSI. Microelectronic

Engineering, 2003, 66: 438~444.

10. Takeo Katoh, Hyun-Goo Kang, Ungyu Paik, etc. Effects of Abrasive Morphology and Surfactant Concentration on Polishing Rate of Ceria Slurry. Jpn. J. Appl. Phys. 2003, 42: 1150-1153.

11. PARK Jin-Hyung , CHO Jong-Young, HWANG Hee-Sub, PAIK Ungyu,etc. Effect of Alkaline Agent on Polishing Rate of Nitrogen-Doped $Ge_2Sb_2Te_5$ Film in Chemical Mechanical Polishing. Electrochemical and solid-state letters, 2008, 11: 288-291.

12. Liu Ruihong, CMP slurry development and performance study of silica dielectric layer. Dissertation, Dalian: Dalian University of Technology, 2009 : 49~53.

13. E. Matijevi´c , S.V. Babu, Colloid aspects of chemical–mechanical planarization. Journal of Colloid and Interface Science, 2008, 320: 219–237.

14. Kurtis D. Hartlen, Aristidis P. T. Athanasopoulos, Vladimir Kitaev.Facile Preparation of Highly Monodisperse Small Silica Spheres(15 to >200 nm) Suitable for Colloidal Templating and Formation of Ordered Arrays. Langmuir, 2008, 24: 1714-1720.

ECS Transactions, 34 (1) 699-704 (2011)
10.1149/1.3567660 ©The Electrochemical Society

Modeling Copper Chemical Mechanical Polishing Processes Using Linear System Method

Lixiao Wu, and Changfeng Yan

School of Mechanical & Electronical Engineering, Lanzhou University of Technology, Lanzhou, Gansu 730050, China

Abstract: In this paper, a solid-solid contact model is established for Cu CMP. The effects of different pattern structure on dishing and erosion during Cu CMP are investigated by the linear system method. The simulation results of the evolution of different pattern structure are given and are compared with experimental data. The effects of down force on the evolution of different pattern structure during overpolishing stage are also discussed. The simulation results show low down force will reduce dishing and erosion during Cu CMP.

Introduction

Copper (Cu) CMP is a necessary processing step in the fabrication of copper interconnects. Copper CMP clear the overburden copper and remove the barrier on top of the dielectric spaces separating the copper interconnect lines. Copper CMP is known to suffer from pattern dependent problems such as dishing and erosion, which cause increased line resistance (1). Dishing and erosion depend on layout patterns and polishing process parameters. Therefore, predictive pattern dependent models of copper CMP are highly desirable. Such predictive model will provide help on layout pattern design and specifying CMP process settings.

Efforts have been made on modeling Cu CMP. Cu CMP involves the simultaneous polishing of multiple materials. In this sense, it is similar to STI CMP processes. Tamba E.G. Tugbawa (1) developed a semi-physical chip-scale pattern dependent model for copper CMP processes. The model accounts for the temporal evolution of the bulk copper thickness and the temporal evolution of dishing and erosion. Zhan Chen (2) presented an empirical relation of dielectric erosion in metal CMP as a function of pattern density and metal line width. Ed Paul (3) gave a model of Cu CMP based on methods of chemical kinetics which includes both chemical and mechanical processes. Lai (4) developed contact mechanics models to explain the role of pattern geometry on the variation of material rate during Cu CMP.

In this paper, the linear system method is used to model the effect of pattern geometry on evolution of dishing and erosion during Cu CMP. The influence of different polishing parameters such as down force and polishing time on the evolution of dishing and erosion are also investigated.

Model Construction

Model established in this paper is same as the model developed for STI CMP (5-8). In the model, the effects of slurry flow and pad/wafer relative speed on the CMP processes are neglected. The contact between the wafer and the pad in the CMP is considered as solid-solid contact. The wafer is considered as a rigid punch and the pad as an elastic half space in the model (5-8). In the pattern structure studied in this paper, the length of the line is much larger than its lateral dimensions. Because of this, the contact between the wafer and the pad is considered as 2-D rigid punch contact problem.

As shown in Fig. 1, in the frame of reference in this paper, the boundary surface of the wafer and the pad is the x-y plane and the z-axis is directed into the solid. The stresses and deformations in the pad (elastic half-space) loaded one- dimensionally over a narrow strip. The loaded strip in the pad lies parallel to the y-axis and has a width in the x-direction; it carries normal tractions which are a function of x only.

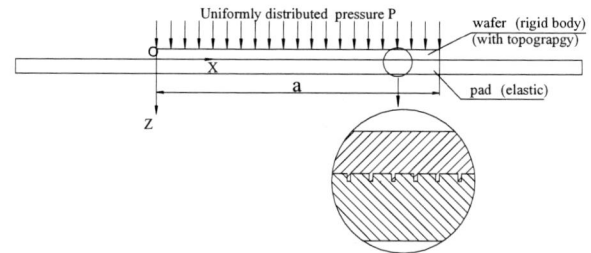

Figure1 Schematic of the rigid-elastic contact model of CMP (7)

If the wafer surface is in continuous contact with the pad surface, the contact pressure distribution between the wafer and the pad can be expressed as Eq. 1 (7).

$$p(x) = p_1(x) + p_2(x)$$

$$= (\frac{E}{2(1-v^2)})\{(\frac{2}{\pi})^{1/2}\int_0^\infty[(\frac{2}{\pi})^{1/2}\int_0^\infty \omega f_1(x)\cos(\omega x)dx]\cos(\omega x)d\omega + a_0\frac{a}{2}[\frac{a^2}{4}-(x-\frac{a}{2})^2]^{-1/2}\}$$

where $p_1(x) = (\frac{E}{2(1-v^2)})(\frac{2}{\pi})^{1/2}\int_0^\infty[(\frac{2}{\pi})^{1/2}\int_0^\infty \omega f_1(x)\cos(\omega x)dx]\cos(\omega x)d\omega$

$$p_2(x) = (\frac{E}{2(1-v^2)})a_0\frac{a}{2}[\frac{a^2}{4}-(x-\frac{a}{2})^2]^{-1/2}$$

$$\int_0^a p(x)dx = P \qquad\qquad [1]$$

where P is the applied force and a is width of the loaded strip on the pad; $f_1(x)$ is the expression of the wafer surface profile and a_0 is a constant.

Since the size of pattern feature is much smaller than the wafer diameter, $p_2(x)$ can be approximately taken as a constant p_2 for simplification. Eq. 2 can be obtained.

$$p(x) = (\frac{E}{2(1-v^2)})(\frac{2}{\pi})^{1/2}\int_0^\infty[(\frac{2}{\pi})^{1/2}\int_0^\infty \omega f_1(x)\cos(\omega x)dx]\cos(\omega x)d\omega + p_2 \qquad\qquad [2]$$

We can also have Eq. 3.

$$P_1(\omega) = \frac{E}{2(1-v^2)}(\frac{2}{\pi})^{1/2}\int_0^\infty \omega f_1(x)\cos(\omega x)dx = \frac{E}{2(1-v^2)}\omega F_1(\omega) \qquad\qquad [3]$$

Where $P_1(\omega)$ is the cosine Fourier transform of $p_1(x)$; $F_1(\omega)$ is the cosine Fourier transform of $f_1(x)$. According to Eq. 3, the CMP system is considered as a linear system. The pressure $p_1(x)$ can be calculated by the inverse cosine Fourier transform of $P_1(\omega)$. Therefore, the contact pressure between the wafer and the pad $p(x)$ is obtained.

According to Preston's law, the removal rate is calculated by Eq. 4.

$$MRR(x) = kp(x)V \qquad [4]$$

where $MRR(x)$ is the material removal rate, $p(x)$ is the contact pressure, V is the relative velocity over the wafer-pad interface and k is a constant representing the effect of other remaining parameters. Since the difference of the relative velocity is small, the relationship between the pressure and the removal rate is approximately linear as shown in Eq. 5.

$$MRR(x) = Kp(x) \qquad [5]$$

where K is a constant. The contact pressure $p(x)$ is obtained by Eq. 2 and Eq. 3.

The evolution of the wafer topography during Cu CMP is simulated by an iterative method. Polish time is discretized and the time interval is taken as ΔT. The wafer topography $f_1(x)$ is discretized and is denoted as $f_m(n\Delta T)$ in the calculation. Here $f_m(n\Delta T)$ is the discrete wafer topography after polishing time $n\Delta T$. The denotation m is the number of the samples in x direction and $n\Delta T$ is the polishing time.

$$f_m(n\Delta T) = f_m((n-1)\Delta T) - MRR_m(n\Delta T)\Delta T$$
$$n = 1,2,3... \qquad [6]$$

where $MRR_m(n\Delta T)$ is the material removal rate between polishing time $(n-1)\Delta T$ and $n\Delta T$. It is calculated by Eqs. 2, 3 and 5. The evolution of wafer topography $f_1(x)$ during Cu CMP can be simulated by Eq. 6.

Effects of Wafer Pattern Structure on Cu CMP Processes

In this section the evolutions of different pattern structures are given. One pattern studied in the paper is an array with a fixed line width of $20\mu m$ and line spaces varying from $1\mu m$ to $100\mu m$. This pattern is denoted as pattern I. Another pattern structure is periodic structure with line width of $9\mu m$ and line space of $1\mu m$ and is denoted as pattern II. The copper and oxide removal rate used in the simulation are not based on the experimental results from blanked wafer polishing. The reason is the blanket wafer surface is not a completely flat surface. The copper and oxide removal rate obtained by blanket wafer polishing is not the removal rate needed in the model. Appropriate copper and oxide removal rates and other parameters in the model are taken based on the simulation results of the CMP system and the experimental data of patterned wafer.

In this paper, only overpolishing stage during Cu CMP is simulated. It is assumed that at the beginning of the overpolishing stage bulk copper has been cleared and the topography $f_1(x)$ on the wafer surface is zero. When the overpolishing begins, the initial patterns on the wafer surface are pattern I and pattern II with small relative heights due to different removal rate for different material during the Cu CMP. The initial wafer pattern structures are shown Fig. 2 (a) and Fig. 3(a). The evolutions of the pattern I and pattern II during the overpolishing stage are shown in Figs. 2(c) and Fig. 3(c). The topography after polishing time 9s for pattern I is shown in Fig. 2(b).

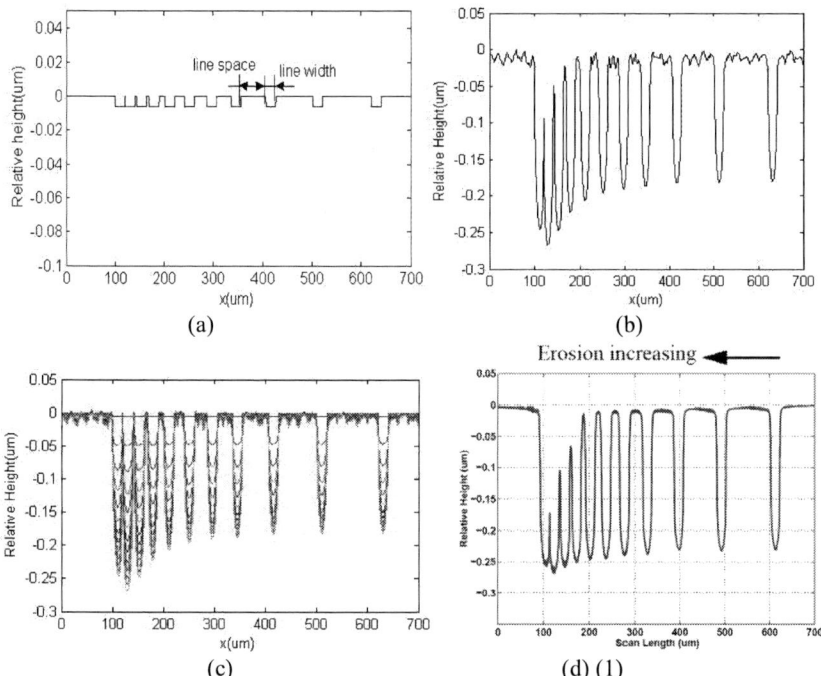

Figure 2 Simulation parameters are Cu removal rate of 636Å/s and oxide removal rate of 12Å/s and down force of 4 psi and time interval of 0.015s. (a) Wafer pattern at beginning of the simulation, (b) Simulated wafer pattern after polishing time 9s, (c) Simulated evolution of wafer pattern during polishing process, (d) Experimental data of array (1)

In Fig. 2 for pattern I , it is clear that the smaller the line space, the higher the erosion. The topography after polishing time 7.5s for pattern II is shown in Fig. 3(b). In order to verify the model, experimental data are used to compare with the simulated results. Fig. 2(d) and Fig. 3(d) show the profilometer scans of pattern I and II during overpolishing stage during Cu CMP. The similarity of the simulation results and the experimental data verify the model developed in this paper. However, the values of dishing and erosion of simulation in Fig. 2(b) and Fig. 3(b) are not exactly same as the values in Fig. 2(d) and Fig. 3(d). The reason is the simulation in this paper is only for overpolishing stage. At the beginning of the overpolishing stage the topography on the wafer is assumed as a flat surface. The experimeatal data shown in the paper is obtained through bulk copper clearing stage, barrier clearing stage and overpolishing stage. The topography at the beginning of the overpolishing stage may be not a flat surface during experiment.

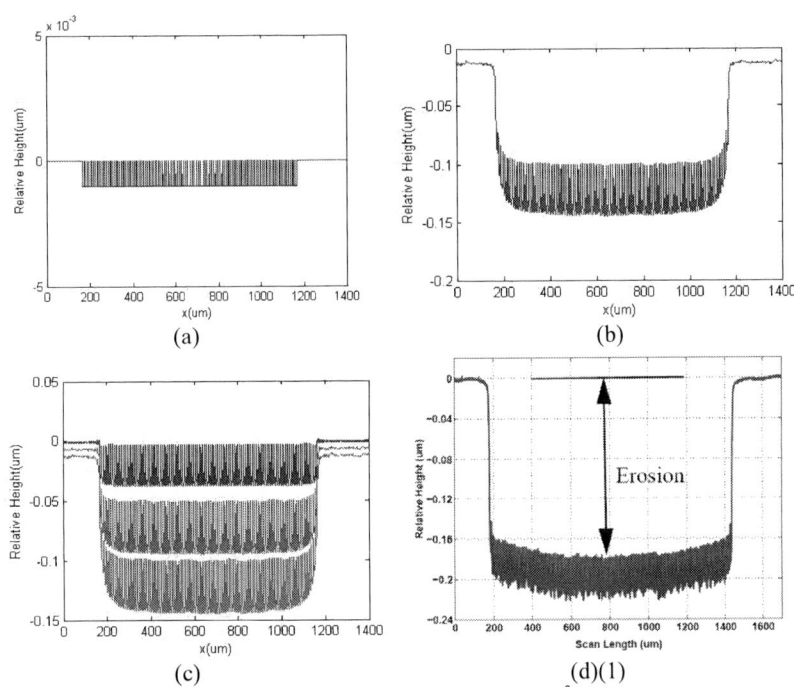

(a)   (b)

(c)   (d)(1)

Figure 3 Simulation parameters are Cu removal rate of 1272Å/s and oxide removal rate of 16Å/s and down force of 8 psi and time interval of 0.005s. (a) Wafer pattern at beginning of simulation, (b) Simulated wafer pattern after polishing time 7.5s, (c) Simulated evolution of wafer pattern during polishing process, (d) Experimental data of array (1)

Effects of Down Force on Cu CMP Processes

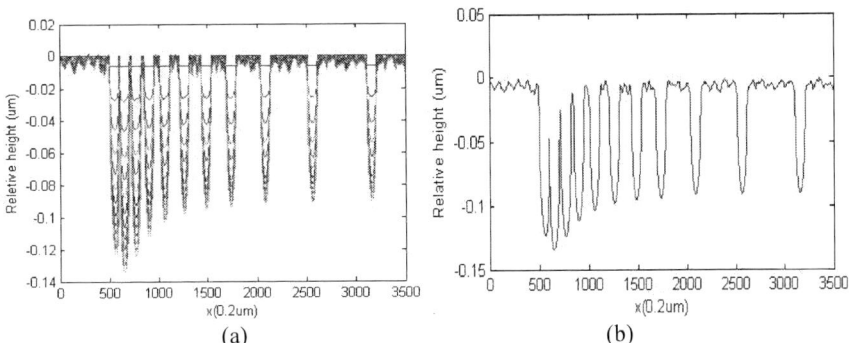

(a)   (b)

Figure 4 Simulation parameters are Cu removal rate of 318Å/s and oxide removal rate of 6Å/s and down force of 2psi and time interval of 0.015s. (a) Simulated evolution of wafer pattern during polishing process, (b) Simulated wafer pattern after polishing time 9s

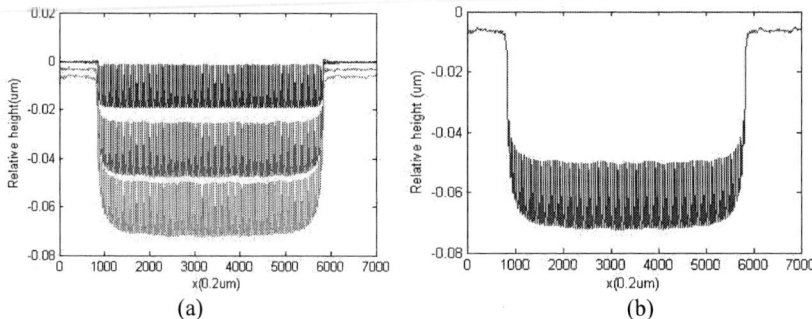

(a)   (b)

Figure 5 Simulation parameters are Cu removal rate of 636Å/s and oxide removal rate of 8Å/s and down force of 4 psi and time interval of 0.005s. (a) Simulated evolution of pattern during polishing process, (b) Simulated wafer pattern after polishing time 7.5s

In this section, the effects of down force on the evolution of dishing and erosion during Cu CMP processes are investigated. When small down force is applied, the evolutions of the pattern I and pattern II during the overpolishing stage are simulated and are shown in Figs. 4 and 5. It is clear that the smaller the down force, the lower the dishing and erosion. From the results it is seen that during the overpolishing stage the small down force should be applied to get small dishing and erosion.

Conclusion

A linear system model is established for Cu CMP in this paper. The evolution of different pattern structure is given and is compared with experimental data. The effects of wafer topography on Cu CMP are shown. The evolution of dishing and erosion for different pattern structure during the Cu CMP is simulation. The effects of down force on the evolution of dishing and erosion during Cu CMP processes are investigated.

Acknowledgments

This work is supported by the National Natural Science Foundation of China under Grant No.60806049

References

1. T. Tugbawa, Chip-Scale Modeling of Pattern Dependencies in Copper Chemical Mechanical Polishing Processes, PhD thesis, MIT, (2002)
2. Z. Chen, F. Sun, and R. Vacassy, *J. Electrochem. Soc.,* 153 (6) G582 (2006).
3. E. Paul, F. Kaufman,V. Brusic,, *J. Electrochem. Soc.,* 152 (4) G322-328 (2005).
4. J. lai, N. Saka, and J. Chun, *J. Electrochem. Soc.,* 149 (1) G31-G40 (2002).
5. L. Wu and C. Yan, *J. Electrochem. Soc.,* 154, H596 (2007).
6. L. Wu, *IEEE Transactions on Semiconductor Manufacturing,* 20, 439 (2007).
7. L. Wu, "An analytical Model of Contact Pressure caused by Wafer Topography in Chemical-Mechanical Polishing Process", *IEEE Transactions on Semiconductor Manufacturing* (submitted)
8. L. Wu and C. Yan, "An analytical model for contact height and contact pressure in chemical mechanical polishing (CMP) for different pattern structure," *IEEE Transactions on Semiconductor Manufacturing* (submitted)

Effect of pH on CMP of VOx Thin Films for RRAM

Yin Liguo, Zhang Kailiang*, Wang Fang, Wei Xiaoying, and Zhang Taofeng
School of Electronics Information Engineering,
Tianjin Key Laboratory of Film Electronic & Communication Device,
Extension of South Hongqi Road, Naikai District,
Tianjin University of Technology, Tianjin, China, 300384
*corresponding author, kailiang_zhang@163.com

Resistive random access memory has attracted enormous attention
as next generation high density nonvolatile memory for flash
memory, due to its low voltage operation, high programming speed,
and simple fabrication. And Chemical mechanical planarization
technology of the novel memory materials is also the necessary
work for its application in RRAM devices, especially for TiO_2,
NiO and VOx films which are the focused RRAM materials
recently. In this paper, CMP of VOx films was investigated firstly,
and the effect of slurry pH on MRR and surface roughness was
discussed. MRR results show that high polishing rates can be
observed with low pH (acidic slurry) and high pH (alkaline slurry),
which are more than 250 nm/min. However, lower surface
roughness can be achieved only when alkaline slurry was used
(0.12 nm vs 2.6 nm). In conclusion, the alkaline slurry with silica
abrasives is the optimum candidate for CMP of VOx films.

I. INTRODUCTION

Resistive random access memory (RRAM) is a kind of quickly developing non-volatile memory devices (1). The key materials for RRAM are transition metal oxides (TMO) (2, 3). Vanadium oxide is a traditional TMO material which is compatible with standard silicon processing. As you known, the ability of VOx to switch its conductance between two distinct states (insulator and metal, respectively) provides the basis for many electronic devices such as "smart windows" for energy saving and comfort, Mott field-effect transistors, electrical-optical switching devices (4, 5). Therefore, RRAM based on VOx is being studied as the promising candidate for next generation non-volatile memory.

In order to reduce the reset current, increase write/erase cycles and density, the cell size has to be shrinked to nanometer scale. Meanwhile, high lithography resolution process requires high degree of planarization for wafer surface to meet the shallow depth of focus. Chemical mechanical polishing (CMP) is the optimal and necessary planarization process for the IC manufacture with the feature size that is less than 0.25 microns. In addition, the gap fill structure of RRAM needs to be fabricated by means of damascene technology, in which CMP is an inevitable process for VOx. Last but not least, planarization solutions for new materials utilized in non-volatile memory are required according to ITRS 2007. Thus it is important to do some research on the CMP of VOx for the fabrication of RRAM devices. However, few attempts have been done on this work.

It is generally known that several process parameters including equipment and consumables (pad, backing film and slurry) can optimize and improve the CMP performance (6). Among the consumables for CMP process, especially, the slurry and its properties play a very important role in the removal rates and planarity for the global planarization ability of the CMP process. There are several slurry properties that affect the material removal procedure such as slurry chemicals, size and hardness of abrasive particles, slurry viscosity, and stability of the abrasive suspension in the slurry, etc. Potential of hydrogen (pH) is one of the most important factors that affect the material removal process.

In this paper, CMP of VOx films was investigated firstly, and the effect of slurry pH on MRR and surface roughness was studied.

II. EXPERIMENTAL

A 1000-nm film of VOx was deposited directly on 4-inch silicon substrates by using RF magnetron sputtering. The original surface observed by the atomic force microscope (AFM) is shown in **Fig. 1** and the RMS-roughness is 2.6 nm.

In the polishing experiments, a Strasbaugh n-Spire 6EC polisher with IC1000 polishing pad (Rohm and Hass) was operated at platen/carrier rotating speeds of 50 rpm/50 rpm and downward pressure of 3 psi. During CMP, the slurry was delivered to the gap between VOx and the pad with a flow rate of 150 ml/min. CMP processing time was 2 min. Commercial grade colloidal silica abrasives were used to prepare the slurries. The solids concentration in the slurry was 5% by weight and the silica particle's diameter was about 120 nm. The pH values of the solutions were adjusted by adding variable amounts of HNO_3 or KOH. The slurries were mixed with a magnetic mixer as the abrasives and chemical additives were added to the slurry supply tank. To maintain good dispersion, the slurry in the supply tank was continuously stirred with a magnetic mixer, as it was fed to the polish tool. After polishing, samples were ultrasonically cleaned in DI water for 3 min to remove SiO_2 particles from the surface. All experiments were conducted at room temperature.

The polishing removal rate was determined from the thickness difference before and after polishing. The thicknesses of the films were measured using a profilometer (Dektak 150) with a deviation of 0.1 nm. And the reported data were obtained by averaging over three experiments in each case. Before and after CMP, the surface morphology of VOx was characterized by AFM (Agilent 5600LS) in tipping mode, and sample areas of 2 × 2 μm were used to measure RMS-roughness of the sample surfaces.

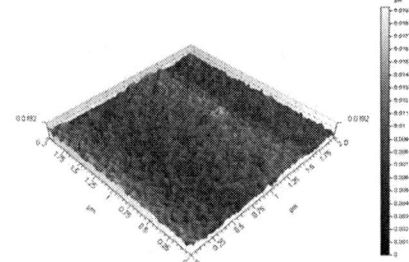

Figure 1. Original surface morphology of VOx observed by AFM.

III. RESULTS AND DISCUSSION
1. VOx CMP removal rate

Fig. 2 shows the removal rate of VOx film polished by silica slurry in 2 min with the different pH value. pH value was split from 3 to 11 at interval of 2. The acidic slurry with pH 3 was found to have a removal rate of 320 nm/min. The neutral pH slurry had a non-removal characteristic with very low removal rate. However, the alkaline slurry with pH 11 revealed an excellent removal rate of 278 nm/min. This implies that both acidic and alkaline slurry are suitable for VOx-CMP application in terms of removal rate.

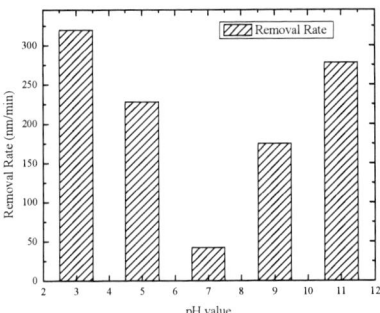

Figure 2. VOx polishing material removal rates in various slurry pHs.

During the CMP, the chemical reaction on VOx surface is believed to be as follows:

$$V_2O_5 + 2HNO_3 \rightarrow 2VO_2NO_3 + H_2O$$
$$V_2O_5 + 6KOH \rightarrow 2K_3VO_4 + 3H_2O$$

In highly alkaline solutions with pH > 10, vanadium oxide dissolves as VO_4^{3-}. On the other hand, in acidic solutions with pH < 4, there is the possibility of forming VO_2^{+}. Lower pH means that the hydrogen ion molar concentration increased in the acidic slurry. Meanwhile, higher pH means that the hydroxyl ion molar concentration increased in the alkaline slurry. From the above chemical equations, it can be seen that the reaction rates of VOx thin films increased due to more diffusion of hydrogen (H^+)/hydroxyl (OH^-) ions into the surface by the increase of hydrogen (H^+)/hydroxyl (OH^-) groups. Consequently, the dissolution rate increased. Therefore, the removal rate increased.

A passivation layer forms on the VOx surface. The brittle layer is removed by the abrasive particle in the slurry and is dissolved in the slurry by the chemical etching. During the CMP process, the formation and removal of the passivation layer is repeated. The differences in removal rates at all slurry pH values indicate that the formation of a protective layer inhibits material removal to various degrees. The magnitude of the removal rate at pH 3 was higher than at other pH values, indicating that the surface layer formed at pH 3 has less passivation characteristics.

For rapid CMP removal, the slurry chemicals must react with the VOx surface to form a weakened layer, which can be removed by the abrasive and shear forces within the CMP slurry. No such reaction occurs at the VOx surface at neutral pH.

2. VOx surface quality following CMP

Following removal rate studies, the VOx films were measured using AFM to compare the surface quality in various pH value slurries. After deposition, VOx had high surface roughness due to hills of deposited material, as shown in **Fig. 1**. Following 2 min of CMP, the surface hills were eliminated, and a smooth surface was obtained.

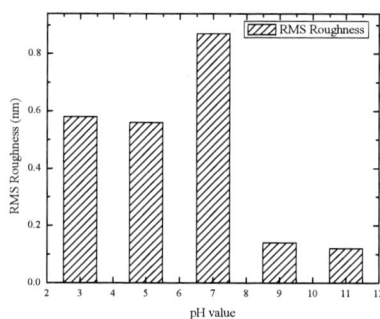

Figure 3. RMS roughness as a function of different slurry pHs.

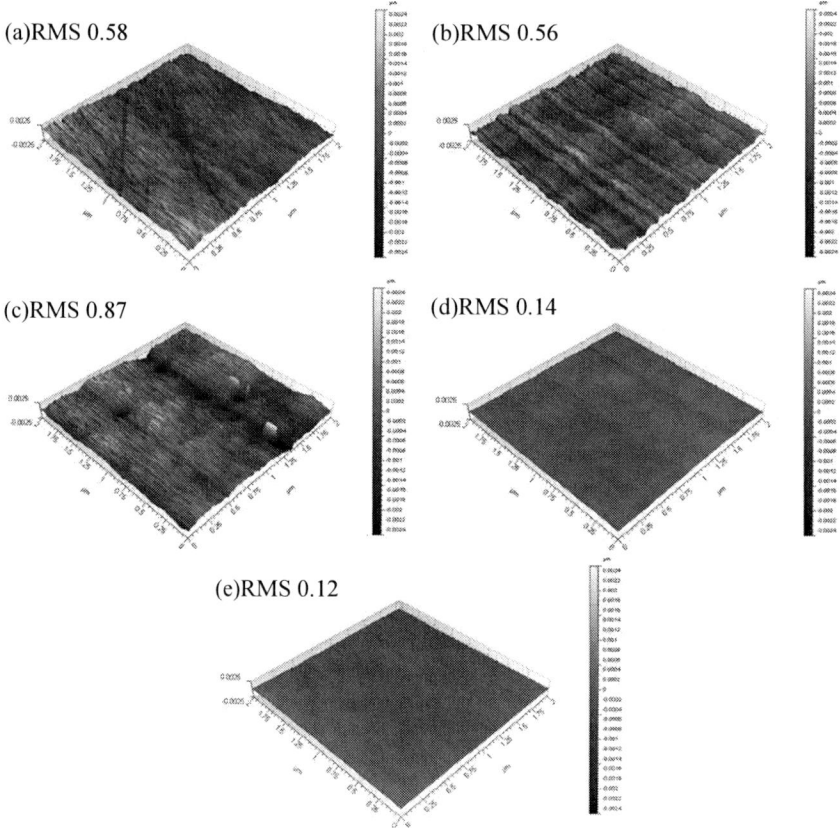

Figure 4. AFM images of the polished surfaces of
(a) pH 3, (b) pH 5, (c) pH 7, (d) pH 9, and (e) pH 11.

Fig. 3 shows the RMS roughness of the polished VOx thin films as a function of different pHs. And **Fig. 4** presents the corresponding AFM surface morphology. The excellent surface roughness was achieved at the alkaline pH.

The reason may be that at the alkaline pH, the silica particles have large potential with the same sign so they repel each other without agglomeration. Hence, when the alkaline solution is used rather than acid one, the VOx surface will be fine since the abrasion occurs with small abrasive particles during CMP.

Following CMP with neutral slurry (pH 7), the RMS roughness of VOx measured 0.87 nm, and deep pits were observed in the material surface. We attribute the less desirable surface to the instability of neutral slurry at a pH of 7. At this pH, the colloidal suspension of the silica particles is degraded compared to acidic and alkaline slurries, resulting in more agglomeration and settling of slurry particles, which causes damage to the VOx thin film.

Overall, the polishing solution with pH 11 gives the best result in terms of both VOx removal rate and surface roughness.

IV. CONCLUSION

Chemical mechanical polishing of VOx films with silica slurry has been examined as a function of slurry pH. VOx is removed more rapidly in acidic and alkaline slurries than in neutral slurry due to the slurry reactivity. The removal rate increases with the concentration of H^+ ions in acidic slurry or OH^- ions in alkaline slurry. Surface quality was characterized with atomic force microscopy. CMP results in smooth VOx surfaces over a range of alkaline pH, but at neutral pH the decline in the abrasive particle suspension results in higher roughness following CMP. Slurry instability also results in a number of post-CMP surface defects at acidic pH. Controlling pH values of the slurry will influence the formation and dissolution of the passivation layer significantly, thus resulting in different CMP performance.

In conclusion, based on the present study, alkaline slurry with silica abrasives at high pH value could contribute to CMP polish rate and surface quality and is recommended for the VOx damascene structure to achieve the best VOx CMP efficiency. The fabrication of damascene device structure will be further investigated so that the CMP process can become an effective patterning process for RRAM materials.

ACKNOWLEDGMENTS

We acknowledge the financial support provided by the National Natural Science Foundation of China under Grant No 60806030, and Tianjin Natural Science Foundation under Grant No 08JCYBJC14600, No 10SYSYJC27700 and Tianjin Science and Technology Developmental Funds of Universities and Colleges under Grant No ZD200709.

REFERENCES

1. C. Y. Lin et al., *J. Electroceram.*, **21**(1-4), 61 (2008).
2. R. Waser and M. Aono, *Nat. Mater.*, **6**, 833 (2007).
3. H. Akinaga and H. Shima, *Proc. IEEE.*, **98**(12), 2237 (2010).
4. E. M. Heckman et al., *Thin Solid Films.*, **518**(1), 265 (2009).
5. F. B. Dejene and R. O. Ocaya, *Curr.Appl.Phys.*, **10**(2), 508 (2010).
6. A. Vijayakumar et al., *IEEE Potentials.*, **27**(1), 28 (2008).

710

Study of Inhibition Effects on Copper CMP Slurry Performance

Jianfen Jing[a], Zhiyong Ma[b], Pei Li[b], Chen Lu[b], Paulchang Lin[b],
Jian Zhang[a], Xinyuan Cai[a]

[a]Anji Microelectronics (Shanghai) Co., Ltd., Shanghai, 201203, China
[b]Semiconductor Manufacturing International Corporation, Shanghai, 201203, China

Abstract

With the development of the products with more advanced technology nodes, new challenges continue to emerge in CMP processes. For example, there is a strong need to use lower down force in polishing process while maintaining a high removal rate to avoid low-k/ultra-low k dielectric film damage. The requirements for dishing, overpolish window, Cu residue clearance capability and surface defectivity, especially corrosion related defects become more stringent as copper lines become narrower and narrower. In this paper, we share our research efforts and results in developing a novel copper CMP slurry to overcome these performance challenges in Cu CMP process. In short, a well designed inhibitor system was applied in this slurry. The impacts of these inhibitors on the slurry performance including blanket removal rate/profile, static etch rate, dishing, Cu residue clearance capability and corrosion were evaluated through polishing experiments, electrochemistry evaluations, and laboratory tests simulating CMP aggressive conditions. With a synergetic effect of these inhibition systems, the slurry shows the excellent CMP performance

Introduction

Chemical mechanical planarization (CMP) of copper (Cu) is a critical step in Cu dual damascene process. It is well known that in Cu CMP, slurry chemistry and polishing process significantly affect CMP performance. Cu CMP slurry typically includes complexing agent, corrosion inhibitor, oxidizer, abrasive, and other desired components. Chemical additives in the slurry formulations are utilized to enhance the removal rate of Cu being polished and/or to passivate the low lying regions of the wafer. To effectively planarize Cu surface, a balance between dissolution rates of the surface using an oxidizing agent, formation of thin passivation layer with the help of an inhibitor, and removal of excess Cu with complexing agents is required. Many recent studies on Cu CMP have been mainly focused on chemical nature of Cu CMP process [1~3].

As technology node advances, the industry moves to lower k dielectric materials that promise much faster devices with lower power consumption [4], resulting in more stringent requirements in Cu CMP. The challenges include more demanding requirements for Cu dishing and erosion performance while maintaining high throughput and low defectivity like corrosion, scratches and surface contamination. Cu CMP process is becoming more chemical than mechanical in order to perform CMP at lower down force to avoid low-k material delaminating and thin Cu line deforming. However, increased chemical component strength normally results in higher dishing, erosion and corrosion. Hence, the selection of inhibitors in slurry formulation is becoming more important in advanced Cu CMP. The inhibitors should not only have the good inhibition capability to minimize dishing, erosion and corrosion, but also give high Cu removal rate at bulk removal stage. In this paper, we will investigate the effects of inhibitors in Cu slurry formulation on Cu CMP performance.

Experimental

Polishing experiments
The polishing experiments were carried out using an 8" Mirra Mesa polisher, with IC1010 pads (Rohm & Haas) and 3M A165 diamond disk. Cu blanket wafers and Semitech 854 patterned test wafers were used for all the polishing experiments. The down force used ranged 1 psi to 3 psi for polishing head and 6 lbs for in-situ conditioning disk. The platen rotational speed was set at 93 rpm. The slurry flow rate was 100 ml/min. Automatic four-point probe was used to measure Cu thickness.

Copper static etch rate experiments
Cu static etch rate was evaluated on a coupon Cu blanket wafer. The Cu blanket wafer was immersed in CMP slurry in a beaker for 30 minutes, and the Cu wafer was subsequently rinsed in DI water and air dried. The static etch rate was calculated based on pre and post soaking Cu thickness.

Electrochemical studies
Potentiodynamic polarization tests were carried out by using Electrochemical Instrument CHI660b. The Cu electrode (working electrode) used in these experiments was attached to a Teflon rod. Saturated calomel electrode and a platinum electrode were used as the reference and counter electrodes, respectively. The copper electrode was first allowed about 900 seconds to attain a stable open-circuit potential. Tafel plot was then obtained by scanning the potential at a rate of 10 mV/s from -1.0 V to 1.0 V.

Results and Discussions

Cu removal rate and static etch rate
The Cu CMP slurries were formulated with an abrasive, a complexing agent, inhibitors, and an oxidizer. Inhibitors A and/or B were used in the slurries. As shown in Fig. 1, both Cu removal rate (RR) and static etch rate (SER) decreased with the addition of inhibitor A and/or B. Inhibitor B itself has very strong inhibition capability which suppresses RR and SER much more than that of inhibitor A. As commonly accepted, low SER is useful for dishing improvement, but the passivation film formed with inhibitor B is too hard to be removed. When inhibitor A is used, the RR did not decrease as much compared with that of inhibitor B, but the SER is also higher. With the slurry formulation containing both inhibitors A and B, the slurry can provide high RR and low SER at the same time.

A typical Cu CMP process involves three steps: bulk Cu CMP (P1), remaining Cu CMP (P2, a step to reach Cu and barrier layer interface), and barrier layer CMP (P3) steps. The effects of inhibitor concentration [A] and [B] on CMP performance were further studied. As shown in Fig. 2, with the exception of combination of [A] and 2[B], the rest of the slurries with different A and B concentrations and ratios have similar RR at higher down force (> 2.0 psi). However, it shows much different behavior at lower down force. When inhibitor A concentration is low, Cu RR at low down force decreases significantly with the increase of inhibitor B, and it shows obvious non-Prestonian behavior, which always means a wider over-polishing window. However, Cu RR at low down force decreases little with the increase of inhibitor B when inhibitor A concentration is high.

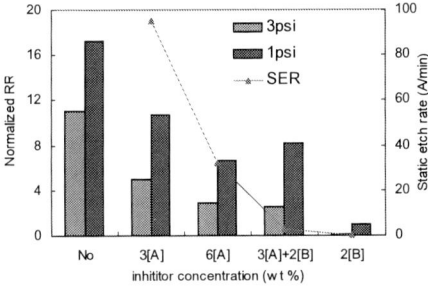

Fig. 1 Inhibitors effects on Cu removal rate and SER

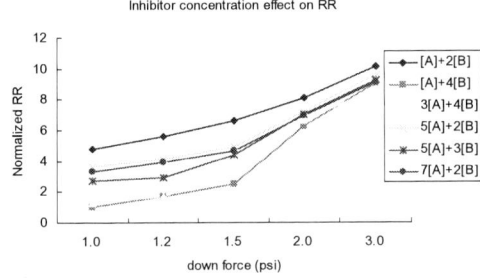

Fig. 2 Removal rate as a function of down force for Cu slurries with different combinations of inhibitors A and B

Performance on patterned wafers

The patterned wafers were next polished with these slurries. The post P2 polishing performances of patterned wafers were shown in Table I and Fig. 3. As shown in Table I and Fig. 3, if the inhibitor A concentration was too low, the passivation film was not so efficient to prevent corrosion no matter how much inhibitor B was used (Fig. 3(a)). A sufficient protective film was formed when a higher level of inhibitor A was used (Fig. 3(b)). Increasing inhibitor A and/or B concentration is good for dishing and overpolishing window. The slurries have very good dishing performance when inhibitor B concentration is higher than 2[B], but there is Cu residue even after over- polishing by 50%. However, the Cu residue decreases with the increase of inhibitor A.

TABLE I. Inhibitor concentration effect on patterned wafer performance.

Normalized inhibitor concentration	Dishing @ 120 um pad at end point (A)	Dishing rate (A/s)	Cu residue after over polishing 50%	Corrosion
[A]+2[B]	800@ 1 psi	15	None	Yes
[A]+4[B]	136 @ 1.5 psi	3.5	Extensive	Yes
3[A]+4[B]	95 @ 1.5 psi	2.9	A Few	No
5[A]+2[B]	435 @ 1 psi	7.8	None	No
5[A]+3[B]	20 @ 1psi	2.4	Some	No
7[A]+2[B]	364 @ 1 psi	6.1	None	No

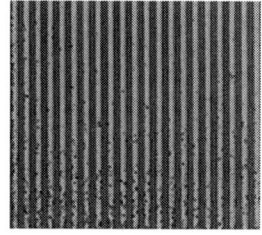 

a) with inhibitor [A]+4[B]   (b) with inhibitor 7[A]+2[B]

Fig. 3 post P2 polishing defects on patterned wafers

Electrochemical behavior

The electrochemistry methodology was used to evaluate the inhibition behavior of inhibitors A and B. As shown in Fig. 4, with the increase of [A] and [B], the corrosion potential (Ecorr) shifts to the higher potential region, this indicates a better protection on Cu surface. The corrosion potential of slurries with low inhibitor A concentration ([A]+2[B] & [A+4[B]) is lower than that of others, which results in corrosion on patterned wafer. There is no obvious correlation between corrosion potential and inhibitor B concentration, indicating that low dishing and wide over-polishing window of slurries with higher [B] was caused by the unique Cu surface protection mechanism of inhibitor B.

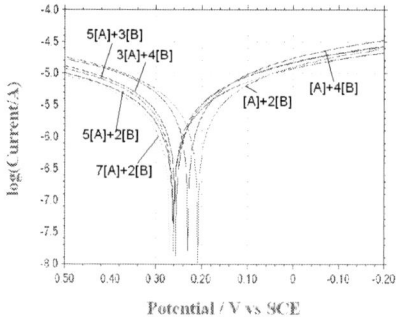

Fig. 4 Tafel plot of Cu slurries with different additive combinations

Patterned wafer soaking test

In order to investigate the corrosion control capability of slurry formulation in this work, patterned wafer polished using Cu slurry with the inhibitors combination of 5[A]+2[B] were soaked for 30 minutes in the slurry. As shown in Fig. 5, good Cu surface is obtained even after 30 minutes of soaking in the slurry. This result illustrates that the slurry has a stronger corrosion control capability.

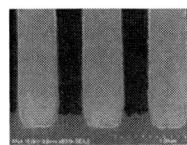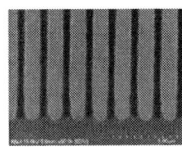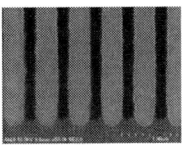

Fig. 5 Patterned wafer SEM micrographs after polishing and soaking in Cu slurry with inhibitor combination of 5[A]+2[B]

Conclusions

In this work, inhibitor A and B were investigated in the Cu CMP slurry. Inhibitor B has a strong inhibition capability, which suppresses Cu RR and SER much more than inhibitor A. By tuning the combination of inhibitor A and B, the slurry can not only provide high Cu RR, low SER, low dishing and dishing rate, but also has robust Cu residue clearance capability and strong corrosion control capability. The electrochemical analysis results indicate that the inhibitor B can form a unique passivation film on Cu surface which helps minimizing dishing.

References

1. S. Seal, S.C. Kuiry, B. Heinmen, Effect of Glycine and Hydrogen Peroxide on Chemical-Mechanical Planarization of Copper. Thin Solid Films 423(2003)243-251
2. Y. Luo, Slurry Chemistry Effects on Cu Chemical Mechanical Planarization. (2004)
3. M. Surya Sekhar, S. Ramanathan, Characterization of Copper Chemical Mechanical Polishing (CMP) in Nitric Acid-Hydrazine Based Slurry for Microelectronic Fabrication. Thin solid Films 504 (2006) 227-230
4. M.R. Oliver, Chemical Mechanical Planarization of Semiconductor Materials, Springer, New York, 2004, p. 1.

CHAPTER 6

MATERIALS AND PROCESS INTEGRATION
FOR DEVICE AND INTERCONNECTION

SiON Gate Dielectric Optimization for NBTI Improvement

Yong Chen, Yonggen He, Wenbo Wang, Rui Guo, Zhaoyun Tang, Jialei Liu, Jingang Wu, Jianhua Ju

Semiconductor Manufacturing International Corp., No.18 Wenchang Avenue, Economic-Technological Development Area, Beijing, P.R.C

> Plasma nitride SiON films have been widely used as gate dielectric in advanced CMOS device fabrication since 90nm technology node. As a replacement material for conventional silicon dioxide, it can provide increased dielectric constant and serve as an effective boron barrier. Post nitridation anneal (PNA), a critical step for plasma nitride SiON formation, is used to stabilize plasma incorporated nitrogen radicals and ions through form thermal stable Si-N banding. Furthermore, PNA can repair the damaged Si/SiO2 interface by interface substrate re-oxidation under oxygen contained ambient. In this work, several PNA conditions were studied. Their effect on transistor performance and PMOS NBTI (Negative Bias Temperature Instability) lifetime were evaluated. It demonstrates that NBTI performance can be improved 20% by PNA process optimizing, while keeping the same Tinv thickness and Ioff/Ion performance.

1. Introduction

In the past decade, the semiconductor industry gradually moved from using pure "conventional" SiO2 to nitrided SiON(oxynitride) film as the gate dielectric. The main advantage of SiON is its better ability to block boron diffusion and to reduce gate leakage. Unfortunately, the incorporation of nitrogen also results in considerably higher threshold voltage shift of pFETs when it biased in inversion voltage at elevated temperature [1]. The Negative Bias Temperature Instability (NBTI), first observed almost 40 years ago [2], has thus become the number one reliability concern as CMOS technology downscaling. During the NBTI stress, interface states and charged defects in the gate dielectric are created. These defects decrease the mobility, reduce the drain current, and induce a threshold voltage shift.

The correlation between NBTI performance and the nitrogen incorporated in the oxide is widely reported in literature [3-6], results shown both nitrogen concentration and its distribution in dielectric film will impact device performance and reliability. So in one hand we need increase dielectric consistent and reduce gate leakage by much more nitrogen incorporation, and in other hand we must keep nitrogen in oxide surface as shallow as possible. Sometimes, it will be a trade-off to balance device Ion/Ioff performance and reliability behavior because more nitrogen incorporation result in more nitrogen accumulate at Si/SiON interface. Decoupled plasma nitridation (DPN) provide a alternative method to form SiON gate dielectric, which consisting of three steps. Firstly, a thin SiO2 film is grown using in-situ-stream generation oxidation or rapid thermal oxidation. The next step consists of exposing the oxide to a high density, plused-RF N2 plasma, during which nitrogen is incorporated into the dielectric film. A final stabilization step, the so-called post-nitridation anneal (PNA), consists in a high temperature anneal, in an O2 atmosphere. All of the three process steps need to be optimized to target the specifications of each transistor node, while PNA acts as a key factor on NBTI, for the reason that it can repair the plasma damaged Si/SiO2 interface by interface substrate re-oxidation under oxygen contain anneal ambient. In this paper, we report the results on NBTI improvement by PNA optimization. A baseline PNA (PNA-1) and 2 weaker PNA (PNA-2, PNA-3) were studied, to evaluate its effect on device NBTI.

2. Device Fabrication

300mm bare wafers and device wafers were fabricated for unit process optimization and device characterization respectively. Unit process is optimized through pure SiO2 growth, N2 plasma nitridation control and post anneal. Structure wafers were processed through a standard CMOS flow as depicted in the following sections. Following shallow trench isolation (STI) formation, different Silicon oxynitride(SiON) gate dielectric processes were formed. And then, poly deposition, gate etch, source/drain extension implantation, spacer formation, sources/drain ion implantation and spike activation were performed. Finally, nickel-based silicide films were formed on gate, followed by metallization process.

3. Experimental Results

3. 1 PNA process and Gate dielectric process

Baseline condition (PNA-1) is using a higher temperature, higher pressure, and longer anneal time process. In order to avoid driving N into SiO2 and close to Si/SiO2 interface, weaker PNA was selected to modulate N profile. Table-1 listed 3 PNA processes settings. Weaker PNAs (PNA-2, PNA-3) using a lower temperature, lower pressure and shorter anneal time correspondingly. Comparing to PNA-1 baseline process, PNA-2 and PNA-3 are much weaker.

Table-1. PNA process settings

PNA Process	Temp (C)	Ambient	Pressure (Torr)	Anneal Time (sec)
PNA-1 (BL)	T	N2/O2	P	t
PNA-2	T-50	O2	P/100	1/4 t
PNA-3 *	T-50	N2/O2	P/10	1/2 t + 1/4 t

* PNA-3 using a 2-step process with different N2/O2 flow ratio.

Different PNA lead to different N concentration and profile in SiON film. In order to match baseline device Tinv and keep same device performance, base oxide and DPN process were tuned correspondingly, as Fig. 1 showed. For weaker PNA, to match baseline final Tinv, thicker base oxide and shorter DPN are necessary. It also indicated that PNA has re-oxidation behavior based on process condition. Higher temperature, higher O2 partial pressure, longer process time may result in more re-oxidation. Meanwhile, it may also drive N into the silicon oxide more and close to silicon interface more.

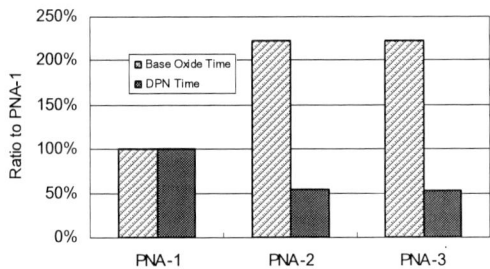

Fig 1. Base oxide and DPN process time to gain the comparable Toxi

3. 2 Unit film properties

Film thickness and N Dose have been measured by XPS and summaried in Fig. 2. It shows, for same device Tinv, comparing with baseline process, weaker PNA process (PNA-2, PNA-3) has thinner thickness and higher N dose. It indicated weaker PNA process owns a little bit higher N concentration in film.

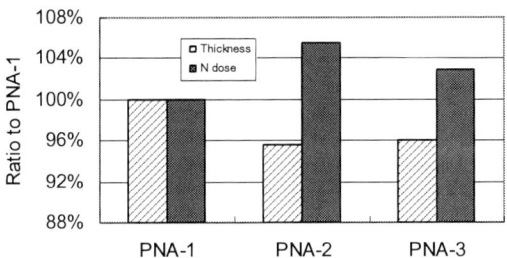

Fig 2. Film properties for different PNA process

Film stability was checked as well. Fig-3 shows the film thickness and nitrogen dose change after queue 5 days. Comparing to baseline process, weaker PNA process has a much big N dose lose (2.5E14 at/cm2, or 5%, 4X N dose lose higher than baseline). It indicates that more stable nitrogen can be got by strong PNA (higher temperature and high O2 pressure).

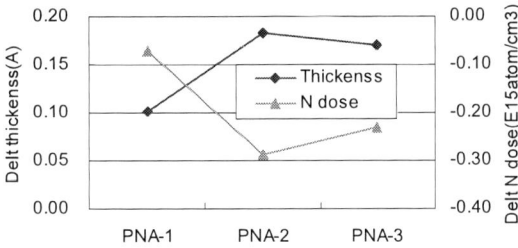

Fig 3.Thickness and N dose change after 5 days queue time.

3. 3 Device characteristics and NBTI performance

Among two PNA split condition, PNA-2 was chosen for device check and NBTI performance evaluation with a 65nm node CMOS test vehicle. Fig4 shows the Ion/Ioff performance of devices fabricate with baseline PNA condition and PNA-2, it indicates weaker PNA (PNA-2) match well with baseline (PNA-1). Other electrical parameters, such as Rs, Cg0, Vt···, also shows good matching between these two PNA conditions (result not shown here).

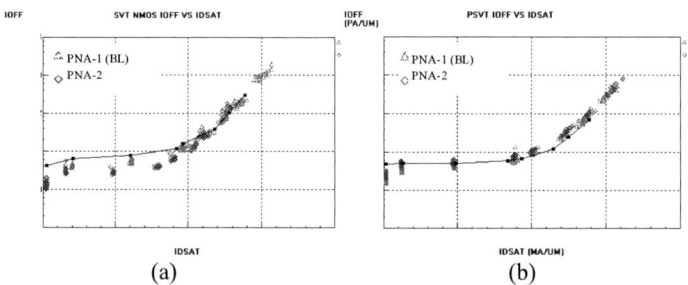

(a)   (b)

Fig.4. Ion/Ioff performance with different PNA condition: a) NMOS, b) PMOS

Fig 5 is the NBTI result for PNA-1 and PNA-2. The devices were stressed 6000sec under Vg=-2V and 125C ambient. It is clear that the PNA-2 gives a smaller Idsat shift than the PNA-1, with 20% Idsat shift reduction. Extrapolate NBTI lifetime was improved by 2 yrs based on baseline lifetime.

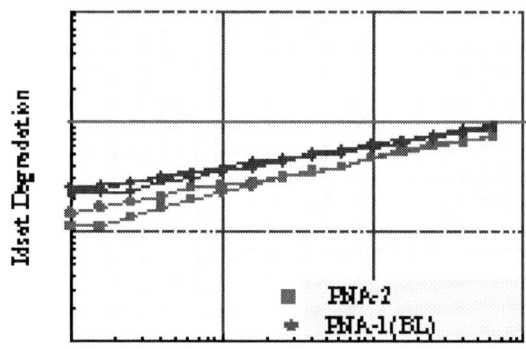

Fig-5. The Idsat shift comparison after 6k sec NBTI stressing at Vg=-2.0V and 125C

4. Discussion

NBTI is influenced by several factors. One key factor is N peak concentration and N profile in Gate SiON [7]. Both concentration and profile are decided not only by DPN, but also by PNA. PNA affects N peak concentration and profile through thermal process drive-in and re-oxidation in the O2 ambient.

A stronger (higher temperature, longer process time) PNA normally produce more re-oxidation at Si/SiON interface, which will benefit on channel mobility and NBTI. But, on the other hand, a stronger PNA will result in more N drive-in during the thermal process, lead to N close to Si/SiON interface, accordingly, degrade NBTI. Therefore, there is a trade-off for re-oxidation and N drive in to obtain better NBTI performance.

In this study, in order to get same device Tinv, for different PNA, DPN process time is determined by final N dose. A weaker PNA needs a thicker base oxide and a shorter DPN, as showed in Fig 1. SIMS analysis was carried out on SiON films processed by PNA-1 and PNA-2 respectively. As shown in Fig.7,. Compare to baseline gate dielectric film, the SiON film formed by weaker PNA (PNA-2) shows three features: 1) the peak N concentration is 1.14x higher than basline, 2) N concentration at Si/SiON interface is 36% lower, 3) N peak is much closer to top surface. All these features favor NBTI performance improvement.

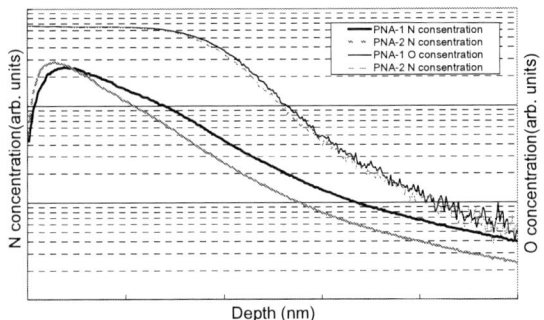

Fig 6. SIMS profile

Table. 3. Profile comparison data (Normalized).

PNA Process	SiON Thickness (A)	N concentration at Si/SiON interface (atom/cm3)	Rp (A)	Rc (atom/cm3)
PNA-1	1.00	1.00	1.00	1.00
PNA-2	0.97	0.64	0.72	1.14

5. Summary

A weaker PNA process is developed, which need combine thicker base-ox and less DPN to get same inversion thickness as baseline. Though the SiON film formed by a weak PNA is less stable as verified by 5 days queue time test, it indicates comparable device performance and superior NBTI lifetime as compare with baseline. The NBTI performance gain is contributes to weak PNA less thermal dirve-in of plasma doped nitrogen, as it confirmed by SIMS profile.

Acknowledgments

The authors would like to thank our colleagues who supported this study.

References

1. G. Chen et al., "Dynamic NBTI of PMOS transistors and its impact on device lifetime," in proc., *Int. Reliability Physics Symp.*, p.196(2003).
2. E. N. Kumar, et. al., "Material dependence of NBTI physical mechanism in silicon oxynitride (SiON) p-MOSFETs: A comprehensive study by ultra-fast On the-fly (UF-OTF) IDLIN technique," in proc., *Int. Electron Device Meet.*, pp. 809-812(2007).
3. T. Gjani, et, al, "Scaling challenges and device design requirements for high performance sub-50nm gate length planar CMOS transistors", *Symposium on VLSI technology*, p.174-175(2000).
4. M. Yamamura, et al., "Improvement in NBTI by catalytic-CVD silicon nitride for hp-65nm technology", *Symposium on VLSI technology*, p.88-89(2005)
5. F. Arnaud, et al., "Low cost 65nm CMOS platform for low power& general purpose applications", *Symposium on VLSI technology*, p.10-11(2004)
6. M. Tanaka, et al., "NBTI immune first plasma nitridation SiON with multiple single-wafer tools for 45nm node gate dielectrics", *14th International conference on advanced thermal processing of semiconductors-RTP 2006*, p. 127-130
7. C.C. Liao, et al, "Factors for Negative Bias Temperature Instability Improvement in Deep Sub-Micron CMOS Technology" ©2008 IEEE

Analysis of the Temperature Dependence of Trap-Assisted-Tunneling in Ge pFET Junctions

M.B. Gonzalez[1,2], G. Eneman[1,2,3], G. Wang[1,2], B. De Jaeger[1],
E. Simoen[1] and C. Claeys[1,2]

[1]Imec, Kapeldreef 75, B-3001 Leuven, Belgium
[2]EE Dept., KU Leuven, B-3001 Leuven, Belgium
[3]Postdoctoral fellow of the Research Foundation-Flanders

In this work, the temperature behaviour of trap-assisted tunneling in Ge pFET junctions selectively grown in STI substrates is evaluated. The experimental results are compared with the TAT model for Si proposed by Hurkx *et al.* (1), where several parameters (such as the effective mass and trap level values) are adapted in order to obtain the best agreement between the model and the experimental results.

Introduction

In the frame of channel engineering, Ge pMOS devices are currently being investigated as a potential candidate to extend the technology roadmap below the 22 nm node. For the most recent technology, the implementation of high-k gate dielectrics, which reduces the low-field mobility due to phonon scattering, leads to the necessity of searching for new channel materials to achieve a significant carrier mobility enhancement and higher drive current. This offers an excellent opportunity for other materials than Si. For example, Ge for pFETs and Ge or III-V materials for nFETs. Germanium has a 4.4 times higher hole mobility than the conventional Si channel pMOSFET. This can enhance the circuit speed, through an increase of the transistor drive current or I_{ON} (2).However, in order to integrate germanium in the silicon process tools, Ge transistors need to be fabricated in Ge thin layers processed on a Si handle or carrier wafer, which raises new concerns. One such issue is the presence of extended defects at the Ge/Si hetero-interface, (owing to the 4% lattice mismatch between Si and Ge), which can significantly degrade the mobility enhancement, reducing the on-state current. Furthermore, the presence of defects can increase the leakage current in the sub-threshold region or I_{OFF} state leakage, through the drain-to-substrate junction leakage (1). Moreover, one of the key concerns for Ge is the narrow bandgap (0.66 eV) compared to unstrained Si (1.1 eV). For advanced CMOS nodes, where higher halo implantations are used and high electric fields are present, the smaller bandgap can considerably increase the drain-substrate junction leakage for the shorter channel MOSFETs, through field assisted mechanisms like Trap-Assisted-Tunneling (TAT) and Band-To-Band Tunneling (BTBT). The high drain-to-substrate leakage will increase the off state current and the power consumption. A good defect control with an acceptable leakage current level, requires a better understanding of the field mechanisms occurring in the Ge drain-substrate junction.

Experimental Conditions

The Ge pFET junctions were fabricated on 200 mm diameter (100) n-type Czochralski silicon wafers. Active diode regions were defined by shallow trench isolation (STI) followed by P implantations and an n-well anneal. After a standard wet clean and HF dip, an in-situ 850°C H_2 bake was done in order to remove the native oxide. Subsequently, the Si substrate was etched out with an etch depth around 330 nm by in-situ HCl vapor phase etching. Selective epitaxial growth of 330 nm undoped Ge was performed using an ASM Epsilon® 2000 reactor. The layers were grown with GeH_4 as precursor in two steps (at 450°C with H_2 as carrier gas, and secondly at 350°C with N_2 carrier gas). A post growth anneal at 850°C for 3 min in a N_2 ambient was done in some of the splits. After the Ge growth and post growth anneal, a phosphorus n-well implantation and anneal were done, resulting in an active donor doping density of $N_D \sim 2 \times 10^{17}$ cm^{-3}. Subsequently, halo, extensions, pre-amorphization and HDD regions were implanted followed by a 550°C anneal to activate the dopants. The resulting active acceptor concentration was $\sim 2 \times 10^{20}$ cm^{-3} (3) resulting in a junction depth around ~ 100 nm. The process was continued with a nickel germanidation of the contacts and TiN/Al backend. More detailed information about the processing conditions can be found in reference (4). The threading dislocation density (TDD) in the Ge layer was evaluated by High Resolution Transmission Electron Microscopy (HR-TEM), where a TDD $\sim 1.6 \times 10^{10}$ cm^{-2} was observed for the non-annealed split, while a reduction of the TDD down to 4.2×10^8 cm^{-2} was obtained for the Ge layers processed with the 850°C post growth anneal.

Simulated Electric Field Enhancement Factor (Γ_{Dirac})

For the Trap-Assisted-Tunneling leakage mechanism Hurkx *et al.* (1) defined an adjusted lifetime for Si pn junctions, which could be described as $\tau^* = \tau_g / 1 + \Gamma_{Dirac}$, with τ_g the Shockley-Read-Hall (SRH) generation lifetime and Γ_{Dirac} given by (1, 5-6) :

$$\Gamma_{Dirac} = \frac{\Delta E}{kT} \int_0^1 \exp\left[\frac{\Delta E}{kT} u - \frac{4}{3}\frac{\sqrt{2m^*}(\Delta E)^{3/2}}{q\hbar|E|}u^{3/2}\right] du \qquad [1]$$

where k is the Boltzmann constant, $\hbar$ is the reduced Planck constant, E is the electric field and ΔE is the position of the trap level which is equal to half of the bandgap ($E_g/2$) for midgap traps. The variable u is associated to the energy of the carriers after tunneling, and m* is the tunneling mass. The m* values for silicon and germanium are typically taken as 0.36 m_0 and 0.19 m_0, respectively (7), where m_0 defines the electron rest mass. It is important to note that the electric field enhancement factor Γ_{Dirac} decreases exponentially with increasing temperature (5) and is strongly dependent on the parameter values implemented in the model such as the effective mass m* and trap level position. Figure 1 shows the simulated sensitivity of Γ_{Dirac} (implemented in MATLAB) on the effective mass (a,b) and trap level (c,d) variations at -40°C and 145°C. For electric fields lower than $1\cdot10^5$ V/cm, Γ_{Dirac} is considerably smaller than 1. At this low field regime the SRH generation mechanism is dominant, while at higher electric field $\sim 10^6$ V/cm, BTBT is expected to be dominant and different models need to be considered.

About three decades of leakage current variations were observed when changing the effective mass values at fixed temperature. The value of the electric field, where the trap level variations start to be significant for Γ_{Dirac}, is reduced by considering smaller effective mass and temperature values.

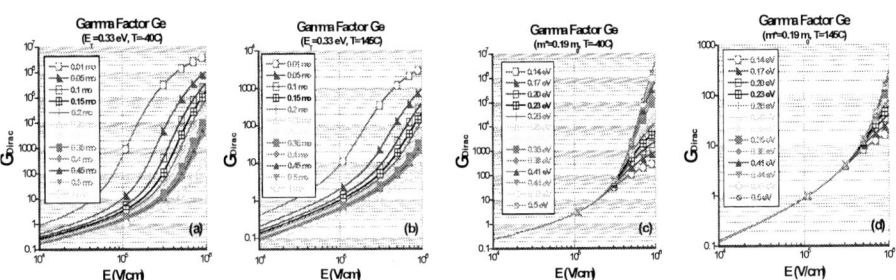

Figure 1. Simulated electric field enhancement factor (Γ_{Dirac}) for Ge based on Hurkx model (1). Effective mass m^* variations are considered between $0.01m_0$ and m_0 at -40°C (a) and 145°C (b), while trap levels variations are assumed in the range 0.15 eV < ΔE <0.5 eV at -40°C (c) and 145°C (d).

(Γ_{Dirac}+1)R_{SRH} Curves Association Model and Experiment: m^* Extraction

In order to compare the experimental results with the simulated Γ_{Dirac}, the experimental area current density (J_A) derivative vs. depletion depth (W) has been calculated combining current voltage (I-V) and capacitance voltage (C-V) characteristics by the method described in Ref. (8), for temperatures ranging from -40°C to 145°C and a reverse voltage up to 1 V. Taking into account the diffusion (J_{diff}), the SRH and the trap-assisted-tunneling contributions, is defined as:

$$\frac{dJ_A}{dx}\bigg|_{x=0}^{x=W} = \frac{dJ_{diff}}{dW} + (1+\Gamma(x))R_{SRH}(x) \qquad [2]$$

where the first term $dJ_{diff}/dW \approx 0$ and $R_{SRH} = qn_i/\tau_g$, with n_i as the intrinsic carrier concentration and q the electron charge. The temperature behaviour of the experimental results of Eq. [2] is studied for a non-annealed and 850°C post growth annealed case. A voltage range $V_R < 1$ V is chosen in order to avoid the BTBT contribution. The maximum electric field has been calculated using $E = qN_DW/\varepsilon$, where ε is the permittivity of the semiconductor. The depletion width is assumed constant with temperature. The considered donor concentrations for the non-annealed and 850°C cases are ~ 1.3×10^{17} cm$^-$^{3} and 1.8×10^{17} cm^{-3}, respectively (evaluated by C-V characteristics).

The experimental curves are compared with a simulated (Γ_{Dirac}+1)R_{SRH} curve with varying R_{SRH}=[1,1x10^7] A/cm^3 in 70 steps, m*=[0.01,0.5] m_0 in 50 steps and E_T=[0.15,0.45] eV in 4 steps, where a strong dependence of the best fitting parameters for R_{SRH} and E_T on the effective mass value is obtained, as is seen in Fig. 2, where the error is defined as the sum of the relative distances between the experimental curve and the simulated one. Therefore, in order to evaluate the R_{SRH} variations with temperature, the effective mass should be fixed for all temperatures, as only minor thermal variations of m* are expected (9). For all the studied temperatures of both splits, effective mass values lower than 0.1 m_0 have been generally obtained, while a full range m*=[0.01,0.5] m_0 using an 0.01 m_0 as step has been tested. An average value for the best fit of <m*>=0.05 m_0 was obtained, which is significantly lower than the physical values of 0.19 m_0 reported in literature (7). However, this result is closer to the ~0.025 m_0 value, which was recently obtained combining TAT experimental junction leakage in Ge p$^+$n junctions with MEDICI simulations (6).

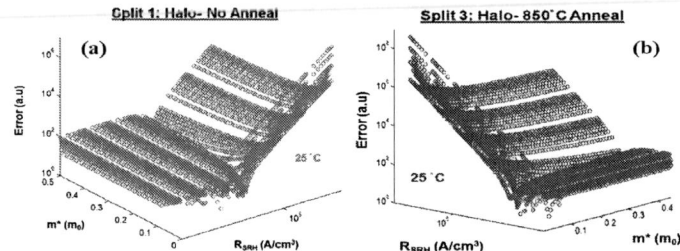

Figure 2. Error between the experimental curve and the simulated one as a function of R_{SRH} and effective mass variations for a non-annealed (a) and 850°C post growth annealed (b) case at 25°C.

$(\Gamma_{Dirac}+1)R_{SRH}$ Curves Association Model and Experiment: $R_{SRH}(T)$, τ_g and $n_i(T)$ Extraction

In order to evaluate the R_{SRH} dependence on temperature, the simulated $(\Gamma_{Dirac}+1)R_{SRH}$ curve at fixed $<m^*>=0.05\ m_0$ is compared with the experimental data with varying $\Delta R_{SRH}=1$ and $E_T=[0.15, 0.5]$ eV in 36 steps. The best fitting conditions are shown in Table I, and the resulting $(\Gamma_{Dirac}+1)R_{SRH}$ curve comparison with the experimental data are shown in Fig. 3, where a good agreement between the experimental data and the model is observed. While, no clear correlation of E_T with temperature is obtained for the 850°C post growth annealed case, the non-annealed split, which has a higher density of G-R centers in the depletion region, shows an 0.07 eV energy decrease with increasing temperature from -40°C till 145°C. This decrease could be affected by the bandgap dependence on temperature, where a ~0.072 eV wider Ge bandgap is predicted between -40°C and 145°C (10-12).

TABLE I. Comparison of experimental and best fitting $(\Gamma_{Dirac}+1)R_{SRH}$ curves obtained by varying $\Delta R_{SRH}=1$ and $E_T=[0.15, 0.5]$ eV in 36 steps. An effective mass of $m^*=0.05\ m_0$ has been used.

	Split 1: Halo- No Anneal				Split 3: Halo- 850 °C			
Temp (°C)	R_{SRH} (A/cm³)	m^* (m_0)	E_T (eV)	Error	R_{SRH} (A/cm³)	m^* (m_0)	E_T (eV)	Error
-40	382	0.05	0.22	4.36	26	0.05	0.26	4.41
-25	837	0.05	0.22	4.55	59	0.05	0.26	2.18
-5	2087	0.05	0.22	4.30	156	0.05	0.26	1.16
5	3141	0.05	0.22	4.40	243	0.05	0.26	2.04
25	6889	0.05	0.22	4.08	494	0.05	0.27	3.30
45	14039	0.05	0.21	4.55	963	0.05	0.31	2.23
65	26645	0.05	0.20	4.08	1813	0.05	0.28	3.43
85	49058	0.05	0.18	3.91	3148	0.05	0.50	4.90
105	86463	0.05	0.16	3.63	5274	0.05	0.50	2.44
125	144893	0.05	0.15	4.60	8745	0.05	0.27	4.29
145	223676	0.05	0.15	5.12	12990	0.05	0.50	3.71

The best agreement between the experiment and the model leads to a SRH generation rate R_{SRH} dependence on T shown in Fig. 4(a). From the extracted R_{SRH} (T) it is possible to obtain the intrinsic carrier concentration dependence on temperature n_i (T) and the generation lifetime assuming $R_{SRH}(T) = q n_i / \tau_g$. The experimental n_i variations with temperature are represented in Fig. 4(b), and compared with two experimental trends reported for n_i in Ge (13-15):

$$n_i = 1.76x10^{16} \ T(K)^{3/2} \exp(-0.3925(eV)/kT) \qquad [3]$$

$$n_i = 7.3x10^{20} \ \exp(-0.44(eV)/kT) \qquad [4]$$

τ_g values of $4.89x10^{-10}$ s for the non-annealed and $6.82x10^{-9}$ s for the 850°C post growth annealed splits have been employed.

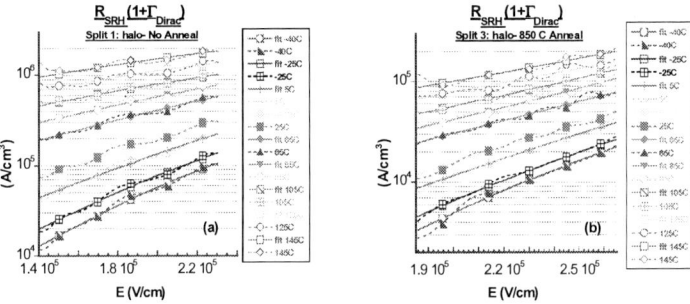

Figure 3. Comparison of experimental and best fitting $(\Gamma_{Dirac}+1)R_{SRH}$ curves vs. electric field by varying $\Delta R_{SRH}=1$ and $E_T=[0.15, 0.5]$ eV in 36 steps for a non-annealed (a) and 850°C post growth annealed (b) cases. An effective mass of m*=0.05 m$_0$ has been used for the simulated curve.

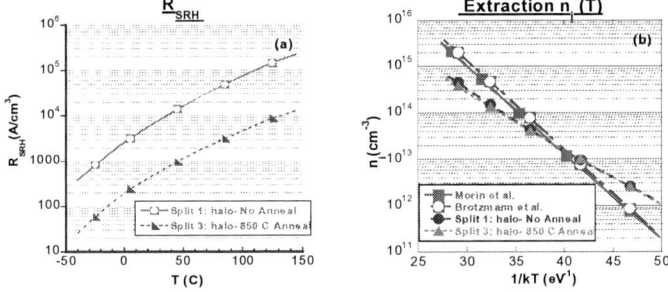

Figure 4. Temperature dependence of the best fitting condition of the SRH generation rate for the studied Ge pFET junctions (a) and the extracted intrinsic carrier concentration compared to values of $n_i(T)$ reported in the literature for Ge (14, 15) (b).

The same 1/kT dependence has been observed for both splits (Fig. 4b), however possible discrepancies between the literature values (14-15) and this work may be affected by the lifetime dependences on temperature (16) and depletion depth. The extracted generation lifetimes show a factor ~14 reduction for the non-annealed case compared to the annealed one. The origin of this reduction is due to the higher trap density (N_T), which can be attributed to the presence of non-annealed point-defect clusters and/or the higher TDD in the junction depletion region. It is interesting to mention that a factor of 38 higher TDD is found in the non-annealed samples while only a factor of ~14 in lifetime reduction is observed. This is fully in agreement with previous work (1, 17), reporting a factor of 2 in leakage reduction for a factor of 5 in TDD reduction.

Conclusions

The impact of the electric field and the threading dislocation density (TDD) on the TAT temperature behaviour of Ge pFET junctions selectively grown in STI substrates has been studied for temperatures ranging from 233 K till 418 K and a reverse voltage up to 1 V. The results are compared with the TAT model for Si proposed by Hurkx *et al.*, and a good agreement between the model and experimental data has been observed. However, an effective mass value of m*=0.05 m_0, which is 3.8 times lower than the value reported in the literature has been employed in order to obtain the best agreement between the experimental results and simulations. Moreover, a factor of 38 higher TDD is observed in the non-annealed samples compared to the splits which received a post epi anneal at 850°C for 3 min in N_2 ambient. This TDD increase has a marked impact on the lifetime, where a reduction by a factor of 14 was observed.

References

1. G.A.M. Hurkx, D. B. M. Klaassen and M. P. G. Knuvers, *IEEE Trans. on Electron Devices*, **39**, 331 (1992).
2. G. Eneman, R. Yang, G. Wang, B. De Jaeger, R. Loo, C. Claeys, M. Caymax, M. Meuris, M. M. Heyns and E. Simoen, *Thin Solid Films*, **518**, 2489 (2010).
3. D.P.Brunco, B. De Jaeger, G. Eneman, J. Mitard, G. Hellings, A. Satta, V. Terzieva, L. Souriau, F. E. Leys, G. Pourtois, M. Houssa, G. Winderickx, E. Vrancken, S. Sioncke, K. Opsomer, G. Nicholas, M. Caymax, A. Stesmans, J. Van Steenbergen, P. W. Mertens, M. Meuris and M. M. Heyns, *J. Electrochem. Soc.*, **155**, H552 (2008).
4. G. Wang, F. E. Leys, L. Souriau, R. Loo, M. Caymax, D. P. Brunco, J. Geypen, H. Bender, M. Meuris, W. Vandervorst and M. M. Heyns, *ECS Trans.*, **16**(10), 829 (2008).
5. E. Simoen, F. De Stefano, G. Eneman, B. De Jaeger, C. Claeys and F. Crupi, *IEEE Electron Device Lett.*, **30**, 562 (2009).
6. G. Eneman, M. B. Gonzalez, G. Hellings, B. De Jaeger, G. Wang, J. Mitard, K. De Meyer, C. Claeys, M. Meuris, M. Heyns, T. Hoffman and E. Simoen, *ECS Trans.*, **28**(5), 143 (2010).
7. A. G. Chynoweth, W. L. Feldmann, C. A. Lee, R. A. Logan, G. L. Pearson and P. Aigrain, *Phys. Rev.*, **118**, 425 (1960).
8. E. Simoen, M.B. Gonzalez, G. Eneman, P. Verheyen, A. Benedetti, H. Bender, R. Loo and C. Claeys, *Mater Sci: Mater Electron*, **18**, 787 (2007).
9. S. Richard, N. Cavassilas, F. Aniel and G. Fishman, *J. Appl. Phys.*, **94**, 5088 (2003).
10. J. Vanhellemont and E. Simoen, *J. Electrochem. Soc.*, **154**, H572 (2007).
11. C. D. Thurmond, *J. Electrochem. Soc.*, **122**, 1133 (1975).
12. S. Sze, *Physics of Semiconductor Devices*: John Wiley & Sons, New York, 1981.
13. E. Simoen and J. Vanhellemont, *J. Appl. Phys.*, **106**, 103516 (2009).
14. F. J. Morin and J. P. Maita, *Phys. Rev.*, **96**, 28 (1954).
15. S. Brotzmann and H. Bracht, *J. Appl. Phys.*, **103**, 033508 (2008).
16. A. Poyai, E. Simoen, C. Claeys, A. Czerwinski and E. Gaubas, *Appl. Phys. Lett.*, 78, 1997 (2001).
17. G. Eneman, E. Simoen, R. Yang, B. De Jaeger, G. Wang, J. Mitard, G. Hellings, D. P. Brunco, R. Loo, K. De Meyer, M. Caymax, C. Claeys, M. Meuris and M. M. Heyns, *ECS Trans.*, **19**(1), 195 (2009).

eSiGe global and micro loading effect study in high performance 45nm CMOS technology

Yonggen He[a], Huojin Tu, Jing Lin, Hualong Song, Jun Wang, Guiyin Ma, Weizhong Xu, Bin Ye, Tzuchiang Yu, Jingang Wu

Technology R&D Center, Semiconductor Manufacturing International Corporation
No.18, Zhangjiang Road, Shanghai, P.R.China
[a]Allan_He@Smics.com

The embedded silicon germanium (eSiGe) is widely applied in advanced CMOS device fabrication to boost PMOS channel mobility. Beside selectivity, defect control and thermal compatibility, one big challenge of epitaxy growth SiGe process is loading effect between different product and different features. In this work, two precursors of SiH_4 and SiH_2Cl_2 (DCS) were applied for SiGe epitaxy growth respectively. Their impact on embedded silicon germanium global and micro loading was investigated also. The results reveal some solutions to minimize the loading effect such as proper precursor selection, partial pressure optimization and silicon open space constraint.

Introduction

There has been an explosive increasing of interest in embedded SiGe epitaxial applications on the recessed source/drain (S/D) areas beyond 90nm CMOS process during the past years (1-4). Because eSiGe help us to create a compressive strained channel to enhance carrier mobility, and makes it possible to fabricate a faster transistor without shrinkage of channel length. In the embedded S/D application, PMOS source/drain area is recessively etched and filled with SiGe, which gives two important benefits in terms of transistor performance improvement. The first benefit is hole mobility improvement due to SiGe induced compressive strain in the device channel since Ge has a larger lattice constant than that of Si. The second benefit of eSiGe is it can reduce source/drain electrical resistance and favor an abrupt junction formation. SiGe can incorporate more dopant than Si, furthermore boron diffusion in SiGe is retarded and comparatively lower than that of Si (5). However, with device size further scaling down to 45nm and beyond, process integration challenges have imposed more stringent requirements on SiGe epi process. Among these challenges, the SiGe global and micro loading effect is a big barrier for volume production. Furthermore the spice model accuracy in development phase can not really reflect the volume production as traditional technology without considering the pattern layout dependence effects (LDE) (6).

In this work, the study of the global and micro loading effect of embedded silicon germanium for advanced CMOS process is presented. The global loading effect compares SiGe thickness and Ge concentration between blanket wafer and pattern wafer, also the different pattern wafers, while the micro loading effect is defined as SiGe thickness and Ge concentration difference in same test structure on different pattern wafer or in different test structure on same wafer. The process factors, such as precursor type, reaction gas partial pressure and silicon open space were investigated to find their actual impact on loading effect.

Experimental

Fig.1 lists key process steps used in the 45nm technology fabrication sequence. After active area definition by shallow trench isolation, channel implants were used to determine device threshold voltage (Vth). Then, a thin plasma nitrided SiON gate dielectric were formed followed by poly-Si deposition and patterned. PMOS S/D area is opened by an anisotropic etch, while poly top and NMOS area is protected by hard mask or photo resist. After a RCA and HF clean (7), SiGe film is selectively growth in the recessed PMOS S/D area. After LDD engineering, main spacer is deposited, followed with S/D junction engineering, and nickel silicidation, which are formed subsequently. To further enhance the carrier mobility, dual stressed contact etch stop layers (DSL) were deposited. The metal process is followed for electrical characterization.

Bare wafer with SiGe film was also fabricated for process optimization and film characterization respectively, which use the same epi process condition as that of the pattern wafer except the higher temperature (1000~1100C) hydrogen pre-baking is used for bare wafer.

The SiGe thickness and germanium concentration were measured by SE (Spectra Ellipsometer) on blanket wafer and pattern wafer with 55*50um2 monitoring pad. The optical measurement was calibrated by SIMS (Secondary Ion Mass Spectroscopy) and XRD (X-ray Diffraction). The TEM (Transmission Electron Microscope) was performed for micro-loading comparison between different features and different products.

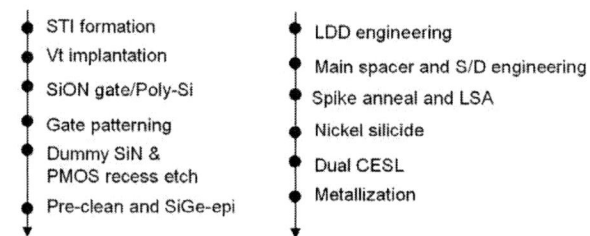

Fig.1 45nm process flow illustration used for pattern wafer preparation

Results &Discussion

Global Loading

The SE measured SiGe thickness and Ge concentration is used to be the index to study the global loading effect. As shown in Table1, though the same SiGe growth condition was used both for bare wafer and pattern wafer, the higher growth rate and Ge concentration was achieved in pattern wafer both for SiH_4 and DCS precursor. The reason is that bare wafer gets full exposure of Si surface, while for pattern wafer, only PMOS recess S/D area was exposed which result in higher growth rate and higher Ge concentration. Same mechanism can explain the SiGe thickness and Ge concentration difference between different product (product1 vs. product2, abbreviate as P1 &P2 hereunder) and different reticles (reticle1 vs. reticle 2, abbreviate as R1 &R2 under this).

Beside silicon transmission ratio's impact on SiGe growth rate and Ge concentration, different Si precursors also exhibit different loading effect as shown in Table 1. SiH_4 based SiGe demonstrate less global loading effect than that of DCS. One proposed model is that the DCS has much more chlorine by-products which will impact the epitaxial growth process.

TABLE 1. SiGe global loading effect of different Si precursor

SiGe process	Product2/ Bare wafer		Product2/ Product1		Reticle2/Reticle1*	
	Thickness	Ge	Thickness	Ge	Thickness	Ge
SiH4 based	1.56	1.30	0.90	0.97	1.05	1.05
DCS based	2.43	1.54	0.87	0.97	1.14	1.05

*Different reticles on same product1.

Micro Loading

Cross-section TEM is used to quantify SiGe overfill or underfill micro loading on different testkeys with varied silicon open area, pattern density and different products or reticles(same testkey). The Ge concentration difference is not compared in this part because it's difficult to get the accurate value from TEM or other analytical method within nanometer scaled devices.

P1 and P2 wafers were processed with same SiGe condition using SiH_4 and DCS respectively. On each of product, there are two sets of testkey (testkey 1 and testkey2, abbreviate as T1 &T2 hereunder) with different silicon open area. For same testkey, the Δ T(thickness difference) of SiGe layer between P1 and P2 is used as an index to characterize loading effect. As shown in Fig.2 (a), T2 gets much worse micro loading compare with T1 as it has higher growth rate due to less silicon open area. It is also found that SiH_4 based SiGe exhibits much better micro loading as compared to that of DCS. Same phenomena were observed when comparing the micro loading of same product (P1) with different AA (active area) masks (fig.2 (b)). It is interesting that different micro loading trends for T1 &T2 were demonstrated when we compare Fig 2 (b) to Fig. 2 (a). The loading effect seems get improved based on T1, whereas for T2, the micro-loading even get worse. As showing in Fig. 3, the DCS base SiGe even change from under-fill to over fill. One possible model is that the Si transmission ratio reduce from reticle 1 to recticle 2 induce overall growth rate increase as observed in table1, which favor to reduce the loading for PMOS dense area such as SRAM (T1 like), while it will cause a drawback for PMOS iso area such as scribe lane (T2 like).

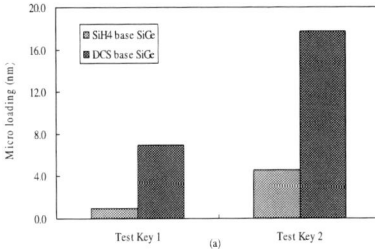 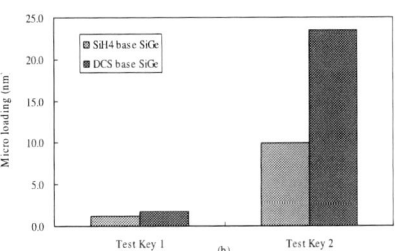

Fig.2 Micro loading of different products (a) (P2 vs. P1); and reticles (b) (R2 vs. R1)

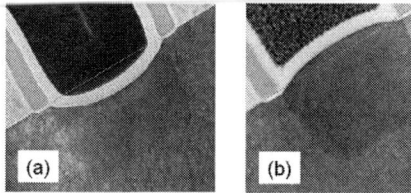

Fig.3. AA mask change induces DCS-SiGe profile change for same test key T2: (a) under-fill for reticle 1; (b) over-fill for reticle 2

The micro loading effect between neighbored features is also checked by TEM. In this case, the poly CD is keep same, while increase spacing from 0.7X 45nm rule to 7X 45nm rules. Fig.4 lists the cross-section TEM of on rule pitch and maximum pitch embedded SiGe profile. It shows that the SiH4 base SiGe get very flat profile and thickness increasing with pitch size up, while DCS base SiGe profile change from round to shoulder with slight thickness variation. The shoulder comes from eSiGe thickness

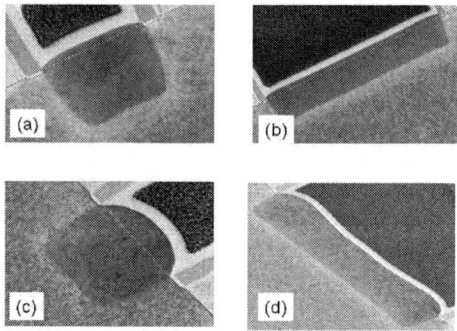

Fig.4. e-SiGe profile and thickness with 45nm on rule (a & c) and 7X (b & d) rule poly spacing: (a), (b) SiH₄ base; (c), (d) DCS base

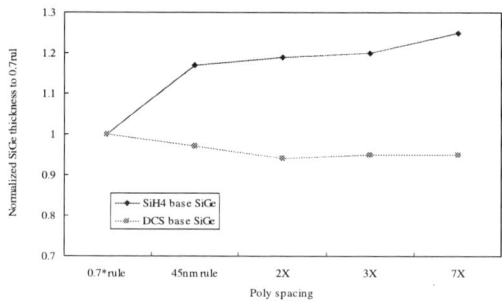

Fig.5. Through-pitch behavior of SiH4 and DCS based SiGe

difference within one larger pitch AA area which was called "shoulder" effects due to DCS-SiGe growth more depends on crystal plane and can grow from sidewall (8),while SiH$_4$-SiGe is more a bottom_up process compare to that of DCS. Other pitch size eSiGe are also checked and plot in Fig 5. Different from global loading and micro loading effect discussed above, DCS based SiGe demonstrate much better through pitch performance than SiH$_4$ based SiGe if it doesn't take into account of the profile change. The over-all thickness variation for DCS base SiGe is within 10%, while more than 25% fluctuation observed for SiH4 base SiGe.

Minimize global and micro loading effect

    eSiGe is a very complex process module, with many factors such as recess profile, Ge concentration, doping, channel proximity which will impact final PMOS performance (2-4, 9). In our work, we found SiGe thickness is a very sensitive parameter which will affect PMOS device performance. As shown in Fig.6, about 200A thickness difference will result in over 30% Idsat difference. So minimize SiGe global and micro loading is one of big challenges for eSiGe process. As discussed in above sections, proper precursor selection, silicon open space constrain include dummy pattern optimization are effective methods to reduce SiGe global and micro loading. Beside these knobs, we also found reduce HCl gas partial pressure or ratio is another effective way to reduce global and micro loading. Fig.7 shows two different products and testkeys' SiGe thickness and Ge

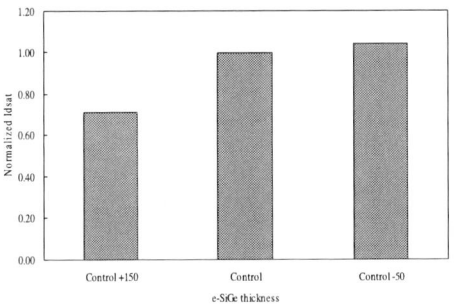

Fig.6. e-SiGe thickness impact on device performance

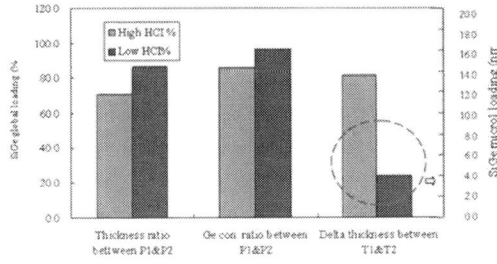

Fig.7. HCl partial pressure impact on global and micro-loading performance

concentration difference with varied HCl gas ratio. Reduce HCl flow, can minimize the thickness and Ge% difference between two products, it also can reduce the micro loading of different test structures within same wafer. It needs pay attention to optimize HCl flow since too low HCl will induce side-effect of non-selective growth.

Summary

Embedded SiGe's global and micro loading is very critical with device shrinkage which requires complex process engineering and very tightening process control. This work indicates SiH_4 is much better than DCS as eSiGe process precursor to reduce SiGe global and micro loading, while DCS demonstrate better through pitch performance than SiH_4. Silicon transmission ratio both in wafer level and die level is another important factor to affect eSiGe loading effect. Beside these methods, decrease HCl partial pressure is favor to reduce loading effect, while it need optimized to balance between SiGe loading effect and selectivity.

Acknowledgments

The authors would like to thank SMIC FA lab for their support of TEM data collection, also acknowledge Dr. Jiongping Lu & Dr. Jianping Wang for their useful discussion.

References

1. T. Ghani, et al, "A 90nm High Volume Manufacturing Logic Technology Featuring Novel 45nm Gate Length Strained Silicon CMOS Transistors", *IEDM,* p.978-980 (2003).
2. Ohta, H. et al, "High performance 30 nm gate bulk CMOS for 45 nm node with /spl Sigma/-shaped SiGe-SD", *IEDM*, p.240 (2005).
3. Yasutake, N. et al, "Record-high performance 32 nm node pMOSFET with advanced Two-step recessed SiGe-S/D and stress liner technology", *VLSIT*, p.48-49 (2008).
4. Tamura, N. et al, "Embedded silicon germanium (eSiGe) technologies for 45nm nodes and beyond", *IWJT*, p.73-77 (2008).
5. H. Okamoto, et al. "A Study on Aggressive Proximity of Embedded SiGe with Comprehensive SDE Engineering for 32 nm-node high-performance pMOSFET Technology", *IEDM* ,p.326-329 (2008).
6. Cliff Yung-Chin Hou, "Design Challenges and Enablement for 28nm and 20nm Technology Nodes", *VLSI*, p.225-226 (2010).
7. Chin-I Liao, et al, "Effective Surface Treatments for Selective Epitaxial SiGe Growth in Locally Strained pMOSFETs", *IDEM* (2006).
8. J. R. Holt, et al, "Selective Epitaxy: Morphology and thickness control for high performance CMOS technology", *ECS Transaction,* p.475-483 (2008).
9. J. P. Liu etc. "Loading effect of selective epitaxial growth of silicon germanium in sub micrometer-scale silicon (001) windows", *Electrochemical and solid state letters,* p.58-59 (2009).

Investigation of laser spike anneal dwell time and it's compatibility with embedded-SiGe

Yonggen He[a*1], Yong Chen[a], Jiongping Lu[a], Jingang Wu[a], Chuanjin Xu[a], Tzuchiang Yu[a], David M. Owen[b], Yun Zhang[b], Shrinivas Shetty[b]

[a]Semiconductor Manufacturing International Corp., No.18, Zhangjiang Road, ShangHai, P.R.C
[b]Ultratech, 3050 Zanker Road, San Jose, CA, U.S.A
[1]Allan_He@Smics.com

Laser spike anneal (LSA) is applied in advanced CMOS device fabrication to achieve efficient dopant activation without excess dopant diffusion. Dwell time, defined as the ratio of full width at half maximum of laser beam and beam scan speed, is found to be a critical and effective knob in LSA processes. In this work, different dwell time of LSA and its impact on dope activation, junction profile, and compatibility with embedded SiGe (e-SiGe) was studied. Dwell time in the range of 800us to 275us has been investigated. LSA processes with different dwell time result in similar dopant activation and almost identical doping profile. On the other hand, reduced dwell time resulted in significant improvement in wafer warpage and overlay performance, when LSA is applied on CMOS devices with e-SiGe S/D pMOSFET. With electrical results consistent with offline characterization, this work indicated that lower dwell time LSA is more flexible and compatible with e-SiGe process for advanced CMOS devices.

Introduction

Application of millisecond annealing (MSA) to CMOS fabrication has recently been reported (1-4). Compared to conventional spike annealing, it offers higher annealing temperatures at much shorter durations to increase dopant activation without additional diffusion, which will degradate short channel effects obviously. Furthermore, the MSA techniques can be easily integrated into conventional CMOS flow as a directly replacement of RTA or combination with spike-RTA. Laser spike annealing (LSA), as one of MSA techniques, has been widely used for its unique features. It adopts a long and single-wavelength laser (10.6um) incident?? at the Brewster angle, which allows for negligible local temperature variation on patterned wafers (5). With device continue scaling down, besides implantation engineering, we need increase LSA temperature to achieve much more dopant activation and form an ultra shallow, abrupt, and low resistivity junctions. However, the introduction of new materials in advanced CMOS device, puts more stringent requirements on thermal budget management. For example, e-SiGe is typically used to introduce channel strain and boost the performance of PMOS. As Ge concentration increases, wafer deformation becomes a serious issue which limits the maximum annealing temperature.

Dwell time, defined as the ratio of full width at half maximum of laser beam and beam scan speed, is found to be a critical and effective knob in LSA processes beside temperature and scan overlap percentage. In this work, different dwell time of LSA and their impact on dopant activation, junction profile was studied. It was found that the dwell time in the range of 800us to 275us result in similar dopant activation and almost identical doping profile. When applying the low dwell time LSA on CMOS devices with e-SiGe S/D pMOSFET, we found a significant improvement of wafer warpage and overlay performance, while the Ion-Ioff performance is comparable between low dwell time and control. This work indicated that lower dwell time LSA is more flexible and compatible with e-SiGe for advanced CMOS devices.

Experimental

Two sets of blanket wafers were prepared with simulated NSD and PSD implantation respectively. An identical spike annealing was performed on blanket wafer just before LSA to achieve same thermal budget as pattern wafer, while different temperature and dwell time LSA split matrix were done on these bare wafer sequential. Four point probe (4PP) and Second Ion Mass Spectrum (SIMS) were used to characterize dopant activation and diffusion respectively. Another set blanket wafer with epitaxial growth SiGe film was prepared also, the wafer warpage and stress were measured by CGS-300 system (5) before and post LSA process.

To investigate the real impact of LSA dwell time on transistor performance, a state of art 45nm CMOS flow with e-SiGe were used to fabricate devices with varied LSA condition. Fig.1 lists key process steps used in the 45nm technology fabrication sequence. After active area definition by shallow trench isolation, channel implants were used to determine device threshold voltage (Vth). Then, a thin plasma nitrided SiON gate dielectric were formed followed by poly-Si deposition and patterned. PMOS S/D area is opened by an anisotropic etch, while poly top and NMOS area is protected by hard mask or photo resist. After a RCA and HF clean, SiGe film is selectively growth in the recessed PMOS S/D area. After LDD engineering, main spacer is deposited, followed with S/D junction engineering, and nickel silicidation, which are formed subsequently. To further enhance the carrier mobility, dual stressed contact etch stop layers (DSL) were deposited. The metal process is followed for electrical characterization. Beside electrical test, the pattern wafer warpage and stress were characterized by CGS and optical ellipsometer also.

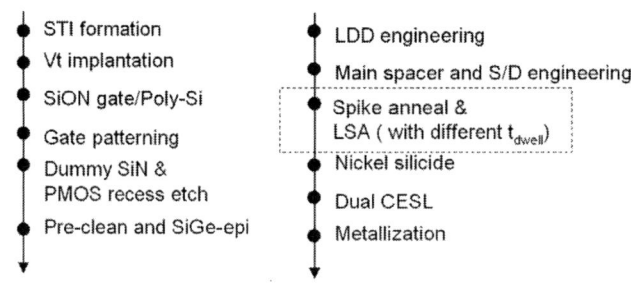

Fig.1 45nm process flow illustration used for pattern wafer preparation

Results &Discussion

Activation and diffusion

One of the main considerations in LSA processing is the activation of the dopants. In this study, the level of activation is measured directly by the sheet resistance. Figure 2 shows the measured sheet resistance using different dwell times versus temperature. Over a temperature range of ~175°C it is clear from Fig.2 that sheet resistance has different dependence on dwell time in different temperature scope. Below 1200°C, both N and P-type dopant demonstrate very weak dependence on temperature and dwell time. When temperature is above 1200°C, it become a very sensitive factor to sheet resistance both for long dwell time (800us) and short dwell time (275us). The slope of sheet resistance versus temperature almost identical for dwell time ranged from 275us to 800us, the Rs gap between 800us and 275us dwell time is slightly, around 15°C temperature difference. It is interest that 2X 275us LSA process even get much higher activation compared to that of 800us LSA process under same peak temperature.

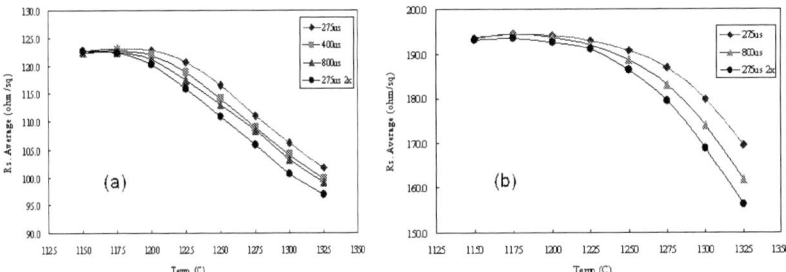

Fig. 2 Rs change with temperature under different dwell time with: (a) NSD IMP; (b) PSD IMP

Beside activation, the diffusion of dopant with varied dwell time also checked by SIMS. Different with the activation, the dwell time seems to has negligible impact on dopant vertical diffusion. As shown in Fig. 3, both phosphorus and arsenic profile seems identical when we varied the dwell time from 800us to 400us and 275us while keep LSA peak temperature as the same. The same case for boron SIMS profile change versus LSA dwell time, as indicated in Fig. 4.

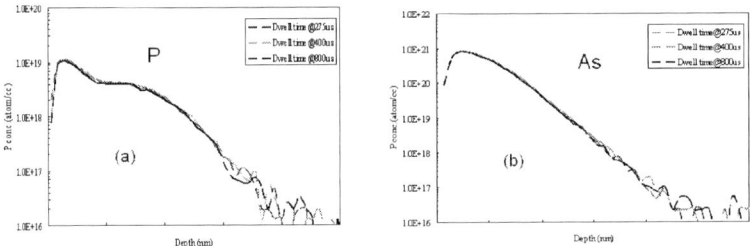

Fig. 3 NSD junction profile with different dwell time LSA (a) P; (b) As

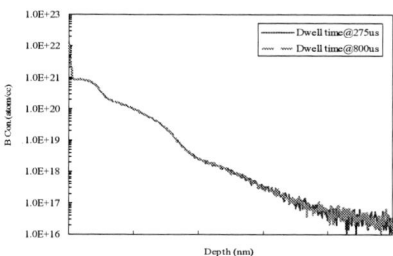

Fig. 4 Boron profile with different dwell time LSA

Warpage and overlay errors

e-SiGe is widely used to introduce compressive strain in PMOS channel and boost device performance. However, interfacial dislocations and slip usually occur during annealing, and this problem becomes exacerbated with the Ge concentration increasing (5, 6). In this work, the blanket wafers with epitaxial growth SiGe film were processed with LSA with different dwell time and measured for warpage and deformation. Though the actual warpage value on blanket wafer is less than 10um, the deformation difference between varied dwell-time LSA is obviously as shown in Fig.5. The PV (peak-to-valley, the index of wafer topography) value decreased from 3.92um to 1.02um when LSA dwell time changed from 400us to 275us. Furthermore, the within wafer topography uniformity also get improved.

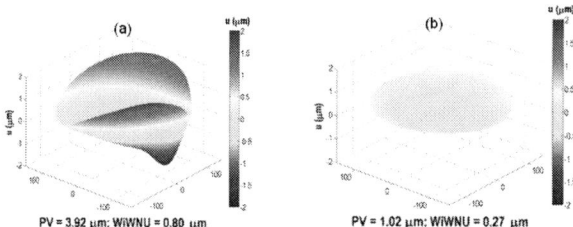

Fig. 5 CGS measured SiGe wafer deformation induced by LSA process: (a) dwell time 400us; (b) dwell time 275us

The dwell time impact on e-SiGe wafer warpage is also checked on real pattern wafer. The wafer bow was measured just before and post LSA process, the delta value of bow is used to characterize wafer deformation. As shown in Fig.6 (a), the wafer warpage is dramatically increased when dwell time increase from 400us to 700us and 800us. The bow change with 800us LSA is almost 2times to that of 400us LSA. It is found that low dwell time also has wide process window. For 400us LSA process, even increase peak temperature 20°C, the warpage still much lower than that of higher dwell time (700us or 800us). The warpage induced by LSA process will impact sequential process, also on litho overlay performance. Fig.6 (b) exhibit CT overlay errors for different LSA conditions. Consistent with warpage behavior, lower dwell time LSA results in much less CT overlay errors and has much wide peak temperature range.

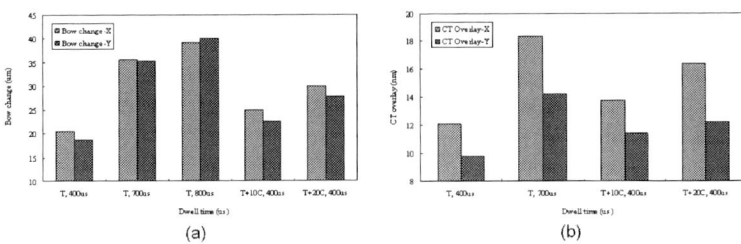

(a)   (b)

Fig. 6 Different LSA dwell time and peak temperature impact on: (a) wafer warpage; (b) CT overlay errors

Device performance

Pattern wafers were processed with the state of art 45nm flow described in Fig.1. At LSA step, they are split into two conditions of 400us and 700us dwell time respectively. As shown in Fig. 7, the device performance of different dwell time LSA is comparable both for PMOS and NMOS. For NMOS, the 400us LSA demonstrate much better within wafer uniformity both for Ion and Ioff. One suspect model is low dwell time LSA result in more uniform stress cross wafer during SMT. Sheet resistance of different dwell time LSA are also checked for AA (active area) and poly respectively. Consistent with what we have observed on blanket wafer, short dwell time LSA gets little higher sheet resistance, it can easily be compensated by temperature adjusting.

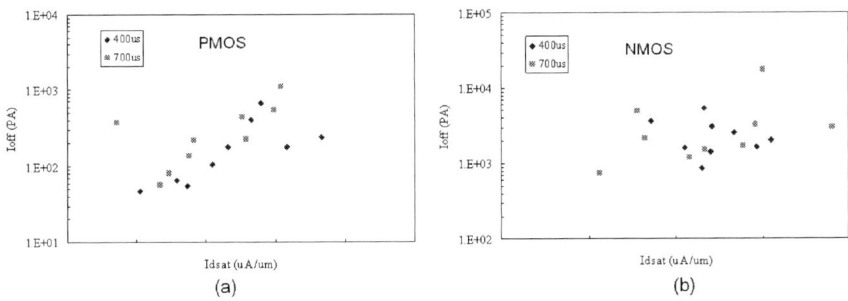

(a)   (b)

Fig. 7 Ion/Ioff performance of different dwell time LSA: (a) PMOS; (b) NMOS

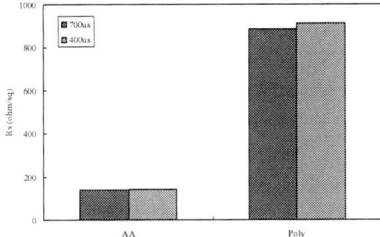

Fig. 8 Sheet resistance comparison between different dwell time LSA

Summary

Laser spike annealing (LSA), as one of the major MSA techniques, is very critical to achieve ultra shallow junction with high activation and less diffusion. Dwell time, as one of key tuning knobs of LSA process, was investigated in this work. Based on the offline and inline characterization of activation, diffusion, warpage, CT overlay errors and device performance, it indicated that lower dwell time LSA is more flexible and compatible with e-SiGe for advanced CMOS devices process.

Acknowledgments

The authors would like to thank Dr. Jianping Wang for his useful discussion.

References

1. Yamamoto, T. et al, "Advantages of a New Scheme of Junction Profile Engineering with Laser Spike Annealing and Its Integration into a 45-nm Node High Performance CMOS Technology", *VLSI*, p.122-123 (2007).
2. Shima, A. et al, "Laser annealing technology and device integration challenges", *ICSICT*, p.454-457 (2006).
3. Hoffmann, T. et al, "Laser Annealed Junctions: Process Integration Sequence Optimization for Advanced CMOS Technologies", *IWJT*, p.137-140 (2007).
4. Yun Wang, et al, "Laser spike annealing for advanced CMOS devices", *IWJT*, p.126-130 (2008)
5. Shetty, S. et al, "Impact of laser spike annealing dwell time on wafer stress and photolithography overlay errors", *IWJT*, p.119-122, (2009).
6. Fujii, O. et al, "Sophisticated methodology of dummy pattern generation for suppressing dislocation induced contact misalignment on flash lamp annealed eSiGe wafer",*VLSI*, p. 156-157, (2009).

A Robust Shallow Trench Isolation High Density Plasma Chemical Vapor Deposition Void Free Process for 0.13μm CMOS Technology

Grace Ning[1], Paul-Chang Lin[1,2], Charles Xing[1,2], Allen Bian[1,2]
, Hong-Bo Zhao[1], Ya-Lu Cao[1]

1. SMIC, 18,Zhangjiang Rd, Pudong, Shanghai, PRC, 201203
2. Department of Microelectronics and Solid-State Electronics, Fudan University, Shanghai

ABSTRACT

Shallow Trench Isolation(STI) is widely used in advanced CMOS technologies. This paper describes a shallow trench isolation for 0.13μm CMOS technologies development which utilizes AMAT Ultima Plus High Density Plasma (HDP) CVD oxide process to fill 0.18μm wide and 0.5μm deep trenches with void free. Through optimizing source/bias RF power, process gas flow and cross section verification, as a result, we got a robust gap-fill recipe with void free.

INTRODUCTION

Shallow Trench Isolation(STI) is a device isolation technique for integrated circuits. As the semiconductor industry moved to sub 0.25um CMOS technology there is a need for creating very small void free gaps on the wafer sub-layer[1,2,3,4]. STI is a mainstream isolation method for advanced logic, DRAM, SRAM and flash memory. Shrinking features of STI request stringent process control. The challenge is providing void-free, seamless gap fill, especially for high-aspect-ratio trenches. STI gap fill will directly impact device performance, induce transistor to transistor leakage or transistor to substrate leakage[5,6]. Now the circuit densities further increase and hence the width of these gaps decreases causing the gap aspect ratios (AR) to increase, filling these narrower gaps becomes more difficult. and normal defect inspection procedure can be used to monitor STI gap fill performance, which is also known as STI void(Figure1).

In this paper, The poor STI gap fill mechanism is studied, and we developed a robust process recipe with optimizing source/bias RF and gas flow, finally we got STI with void free and excellent device isolation.

EXPERIMENT

Design of Experiment

In 0.13um CMOS technology, The main STI process machine type is AMAT Ultima-HDP. and STI void defect usually distributes at wafer edge(Figure1).Based on the Ultima-HDP chamber hardware configuration, wafer edge plasma density is much lower than wafer center(Figure2) , which makes wafer edge gap fill performance much worse. To improve wafer edge gap fill capacity, We focus on recipe source/bias

RF and gas flow setting to develop a new STI process recipe to improve gap fill performance.

Experimental Method

    1) AMAT Ultima-HDP Centura 5200 in standard HDP process condition.

    2) KLA-TENCOR KLA-2351 Bright filed

    3).KLA-TENCOR F5X film thickness measurement

    3) FEI TF08 in TEM test.

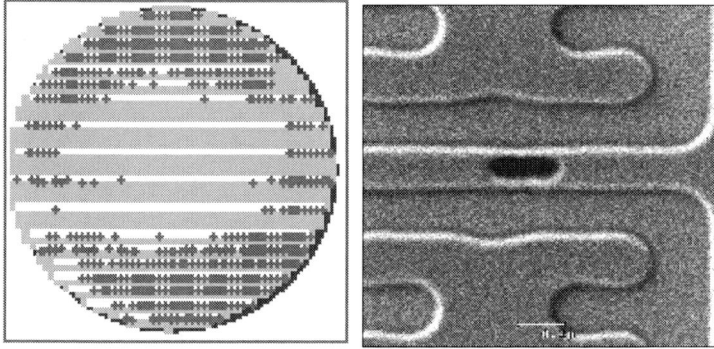

Fig.1 Wafer sort map showing wafer edge STI void map and defect SEM image

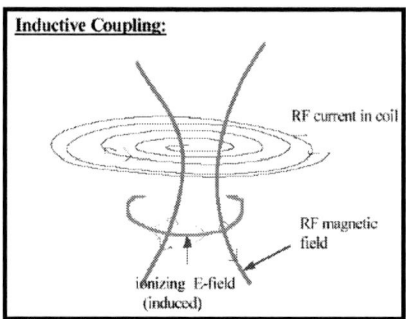

Fig.2 The illustration for Ultima-HDP plasma distribution

Result and Discussion

    In STI HDP process, The gap-fill capability for a given aspect ratio depends on the ratio between the two processes: deposition and sputtering. Too much deposition and too little sputtering will generate cuspidal profile of deposited oxide, resulting in poor gapfilling and formation of void (Figure 3).Too much sputtering relative to deposition can result in corner cutting (Figure 4).Thus, an equilibrium between deposition and sputtering has to be established. The D/S ratio is a key process factor, which can to be adjusted to improve the gap-fill demands.

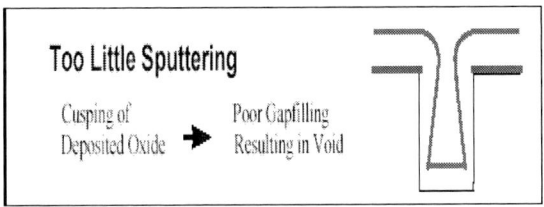

Fig.3 The illustration for tool much deposition and too little Sputtering

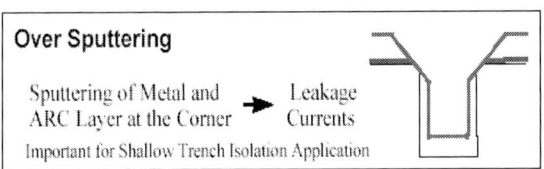

Fig.4 The illustration for tool much Sputtering and too little deposition

High deposition plasma process is used to fill the high aspect ratio gaps with Silicon dioxide as the deposition layer. A typical STI HDP oxide deposition has a gas mixture containing Oxygen, Silane and inert gases, like Argon or Helium, to achieve simultaneous deposition and etching and hence the unwanted bonding of Silane with hydrogen and with hydroxyl is also minimized by choosing the right gas configuration. A RF bias is applied to the wafer substrate in the process chamber. The ions of some molecules, especially Argon gas, are formed by ionization in the plasma and accelerate toward the wafer surface as RF bias is applied to the substrate. When these heavy ions hit the surface, the material present on the wafer gets sputtered (etched or removed) due to this striking. Hence, dielectric material deposited on the wafer is sputter etched at the same time to keep the gaps open during the deposition process.

And source RF including top RF and side RF which are used to tune chamber plasma density.

Condition No.		1	2	3	4	5	6
Top RF	Setpoint	1300	1300	1300	1300	1300	1300
	Actual	1340	1270	1200	1280	1290	1240
	Delta	40	-30	-100	-20	-10	-60
Side RF	Setpoint	3100	3100	3100	3100	3100	3100
	Actual	3150	3120	3010	3060	3030	3110
	Delta	50	20	-90	-40	-70	10
Bias RF	Setpoint	3500	3500	3500	3500	3500	3500
	Actual	3510	3480	3420	3480	3600	3420
	Delta	10	-20	-80	-20	100	-80
Total RF Delta		100	-30	-270	-80	20	-130

Table 1 Process source and bias RF power splits for HDP STI

In the 0.13um STI HDP recipe standard setting, The power setting is 1.6:1.2:1 (Top:Side:Bias).and gas setting is 14:8:1(O2:SiH4-Side:SiH4-Top) Here we choose 6 conditions for different source and bias RF,(Table 1). Based on CP data, We found serious CP loss when source RF and bias RF total decay were higher than 100W. The CP loss map matched STI void defect map.(Figure 5)

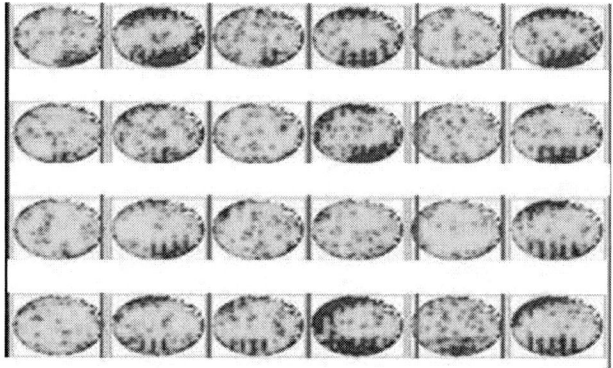

Fig.5 Condition 3 CP map data

From Table 1 and Figure 5. It shows worse condition had more RF decay which caused decreased ionization, especially at wafer edge, it caused wafer edge gap-fill performance worse.

To study source RF power & wafer edge thickness correlation, we found wafer edge plasma density is very sensitive with RF decay at low power setting (Figure 6). So we choose the best source power setting to avoid RF decay impact. And we do not recommend higher bias power to improve film gap fill capacity for the already high power input to the chamber. Low bias powers are also not recommended as lower the sputter rate. As a result, lower deposition rate recipe for a fixed D/S ratio. And new recipe power setting is 1.8:1.3:1(Top:Side:Bias) and gas setting is 12:6.8:1(O2:SiH4-Side:SiH4-Top).

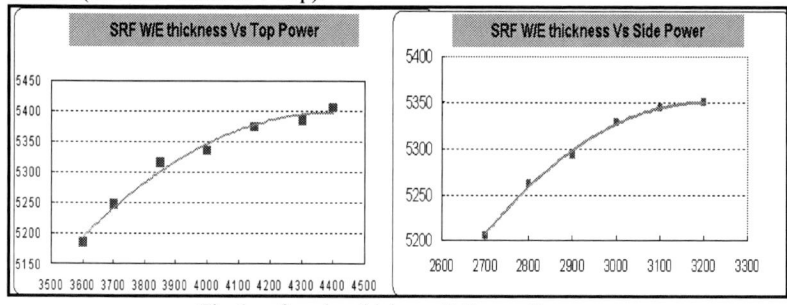

Fig.6 wafer edge thickness& Source Power

Then we did FEM to check new recipe capacity for different AR by increasing . AA-CD(Figure7).From the X-SEM image, standard recipe can get void free in W/E with AR:2.6(Figure10), but new recipe can get void free in W/E with AR:2.9 and the profile become flat(Figure8).

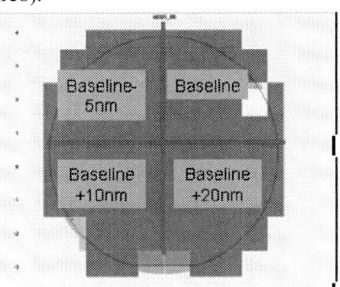

Fig.7 FEM shows AA Etch CD Matrix

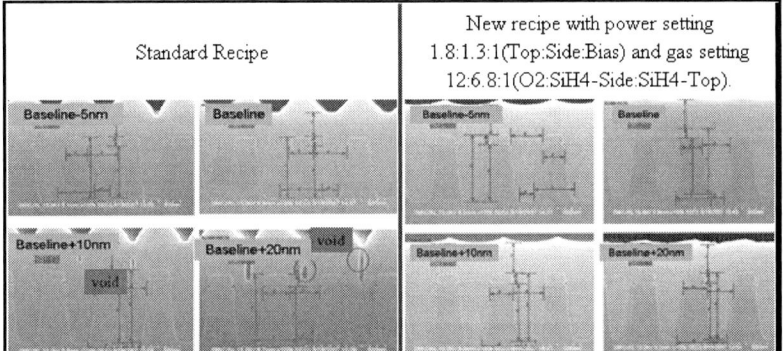

Fig.8. XTEM analysis pictures showing new recipe has good STI gap-fill and profile

CONCLUSION

Shallow trench isolation (STI) standard gap-fill capacity is marginal and usually suffer void on wafer edge. In this experiment, The correlation between STI void and source/bias RF was studied. CP and FEM data indicated that new recipe with power setting 1.8:1.3:1(Top:Side:Bias) and gas setting 12:6.8:1(O2:SiH4-Side:SiH4-Top) could greatly enlarge STI gap-fill capacity.

References
1. M. Nandakumar, A. Chatterjee, S. Sridhar, K.Joyner, M. Rodder and I. 4. Chen, "Shallow trench isolation for advanced ULSI CMOS technologies," in IEDMTech. Dig., p. 133, 1998.

2. P. VanDerVoorn, D. Gan, and J. P. Krusius,"CMOS shallow-trench-isolation to 50 nm channel width," IEEE Trans. Electron Devices,
3. Amerasekera, I.-C.Chen, "A Shallow Trench Isolation for Sub-0.13um CMOS Technologies," IEDM Tech.Dig., 1997. P.657
4. C. H. Li, K.C. Tu, H. C. Chu, I. H. Chang..etc." A Robust Shallow Trench Isolation (STI) with SiN Pull-Back Process for Advanced DRAM Technology, 2002 IEEElSEMl Advanced Semiconductor Manufacturing Conference,P.21~26
5. Jony Indahwan*, Z.G. Song...etc. Failure Analysis on Wafer Edge Issue in 0.13,um Technology.ICSE2004 Proc. 2004, Kuala Lumpur, Malaysia. P. 268~270
6. APPLIED MATERIALS.Ultima HDP-CVD Centura Process:Optimization & Troubleshooting,Version D,10/13/2000

ECS Transactions, 34 (1) 749-754 (2011)
10.1149/1.3567668 ©The Electrochemical Society

CMP-less Planarization Technology with SOG/LTO Etchback for Low Cost 70nm Gate-Last Process

Huaxiang Yin, Lingkuan Men, Tao Yang, Gaobo Xu, Qiuxia Xu, Chao Zhao and Dapeng Chen

Integrated Circuit Advanced Process Center, Institute of Microelectronics, Chinese Academy of Sciences, Beijing 10029, China
yinhuaxiang@ime.ac.cn

A novel 70nm gate-last process featuring with CMP-less spin-on glass/low-temperature oxide (SOG/LTO) etchback method for developing high performance low cost logic platform is demonstrated. The special 3-step sacrificial etchback process all in one RIE chamber is developed to realize the pseudo global planarization on the composite SOG/LTO 2-layer structure. The finally fabricated PMOSFETs demonstrate nice electrical characteristics as well as similar uniformity distribution to that of planarization process.

Introduction

Recently, the novel gate stack structure with high-k/metal gate (HK/MG) materials is implemented into Metal Oxide Semiconductor Field Effect Transistors (MOSFET) to promise conventional scaling of the high-performance Complementary MOS (CMOS) process down to 45/32nm node (1). HK/MG structure enables thinner equivalent oxide thickness (EOT) than previous oxide/poly-silicon (poly-Si) structure while at a much lower gate leakage current level, since the former architecture introduces a thicker physical dielectric-film thickness as well as the suppression of poly-gate-depletion effect. However, two completely different integration schemes, namely gate-first and gate-last, for HK/MG structure integration are proposed (2). Gate-first is compatible with conventional planar process flow but suffers from thermal instabilities and etching issues of HK/MG structure. Gate-last demonstrates minimum thermal effect on HK/MG integration and subsequently brings the best film quality and the control ability to the threshold voltage (V_{th}) of transistors. Besides the limited design-rule, the main disadvantages of gate-last are the process complexity and cost due to a special integration flow and device structure involving more process steps and more challenging process techniques.

For the gate last integration scheme of Intel (3), the normal transistor structure with poly-Si dummy gate-electrode is initially formed, and then an inter-layer-dielectric zero (ILD0) layer is deposited followed by a chemical mechanical planarization (CMP) step to planarize ILD0 layer and expose dummy gates. After that, the poly-Si dummy gate is removed by a special etch-method with high selectivity to ILD0. Then, multi-layer MG materials for V_{th} control and gate fill are deposited in the gate followed by another CMP process for removal of extra MG materials distributed on ILD0 layer. Since the gate stack is the heart of the transistor at nanometer scale, extreme control is necessary for all these gate processing steps to ensure proper device function and less parameters variation. Specially, CMP planarization ILD0 layer is critical for the final formation of HK/MG structure with good process uniformity. However, controlling CMP at the atom level with

749

various materials and pattern sizes on the same wafer is relatively challenging (4). To save process development cost, the sacrificial etch-back is another proper choice due to the excellent processing ability of dry-etching approach to control dimensions at the nanometer level.

This paper proposes one special CMP-less planarization technology with Spin-On-Glass (SOG)/Low Temperature Oxide (LTO) etchback for gate-last MOSFET integration.

SOG is an excellent planarization dielectric material in multi-level interconnection (5). With matured dry-etch approach, sacrificial etchback method can realize complete ILD planarization on metal lines with different pattern density (6). In contrast to CMP, the conventional SOG etchback process demonstrates a lower cost, lower complexity and better compatibility with FEOL process, while maintaining similar planarization result. This makes SOG etchback especially meaningful for the ILD0 planarization in gate-last process for sub-45nm CMOS integration.

Experiment Method

To evaluate our creative proposal, the experiment is performed on the 4-inch test fab-line in Integrated Circuit Advanced Process Center, Institute of Microelectronics of Chinese Academy of Sciences (IMECAS). The starting materials are normal p-type 4-inch Si wafers with resistivity equal to $15\Omega\cdot cm$. First, the device isolation and normal gate stack with different gate length size and pattern density are formed on the wafer. The test wafers are classified into two groups: one is the engineering group only with gate stack structure for etchback planarization technology development; another is the device group with full flow of source-drain doping and activation process for device characteristics evaluation. The dummy gate material is poly-Si film with thickness of 180nm deposited by LPCVD. The initial gate dielectric film is oxide film obtained by dry oxidation method with thickness of 2.5nm. The offset-spacer/spacer structure is applied on the gate lines after fine etch on gate stack. In following steps, a thick layer of 800nm CVD LTO plus a layer of 300nm SOG with low viscosity for better step coverage and robust gap filling are deposited on the gate stack.

Different etchback approaches as well as various reactive-ion-etching (RIE) parameters are investigated to realize the complete planarization on ILD0 layer and expose the dummy gate rightly. The main RIE recipe is based on CF_4/CHF_3 mixture and carried out on a LAM Rainbow 4520 RIE tool. All ILD RIE processes are performed in one chamber without any chamber transfer step. The planarization extent is evaluated by thickness distribution of post-etching ILD film across whole wafer and local gate-line structure. The film-thickness measurement tools are thickness meter of KLA-Tencor NANOSPEC/AFT and cross-sectional SEM view of JEOL 4500F.

Results and Discussion

Figure 1 describes different etchback approaches for composite SOG/LTO 2-layer structure on normal gate stack structures. This first approach is the reference method, with a so called name of "Conventional 3-Step Etchback", provided by LAM corporation. The second one is the special one developed for gate-last process with a so called name of "Recessed 3-Step Etchback". In this approach, the 1st-step etch is designed for etching SOG layer to form a recess profile across whole wafer. The 2nd-step etch is for adjusting the recess extent in previous step as well as trimming interface-profile between SOG and LTO layers. The 3rd-step etch is for the LTO etching and reducing the LTO thickness to the designed value, which should be smaller than that of dummy gate height so that the poly-Si materials can be successfully removed in the following step. "Recessed 3-Step Etchback" is expected to show a better ILD thickness uniformity within wafer and within die than the first method. The process results of ILD profile distribution by different etching approach are summarized in Figure 2. The figure shows a convex profile for ILD etching by "Conventional 3-Step Etchback", where the residual ILD thickness at wafer center is much larger than that around wafer edge. Meanwhile, the thickness variation from center to edge is approximately linear and the slope is larger than 1e-6. To compare etch quality of different approaches in quantity, we defines the effective wafer diameter (EWD) as the position where the ILD thickness is greater than 90% thickness at wafer center. "Conventional 3-Step Etchback" demonstrated 60mm EWD, corresponding to only 60% of original wafer size (100nm), which is a relatively poor result. It also easily leaves some residual oxide on top of gate stack due to a higher etch-rate at wafer center and then results in an inferior wet-etch result of poly-Si gate.

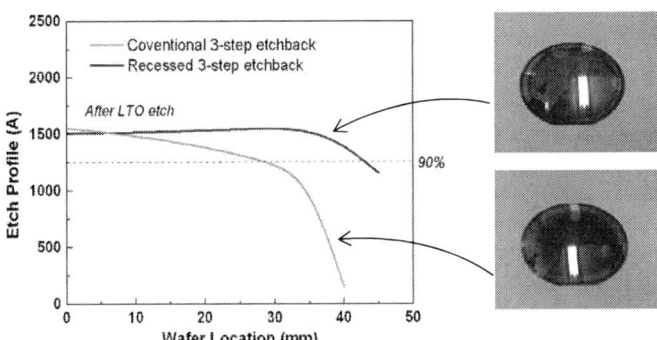

Figure 1. Process flows of "Conventional 3-Step Etchback" and "Recessed 3-Step Etchback" for composite SOG/LTO 2-layer planarization on gate stack structure.

Figure 2. Final profile distribution of ILD0 by different etchback approaches.

The convex ILD result by "Conventional 3-Step Etchback" method is due to a concave etch-rate profile of SOG RIE with the standard recipe. SOG is a kind of polymeric oxide obtained by solidifying sol-gel organic through curing process at $350^{\circ}C$ and demonstrates a smaller Si-O bind energy than normal thermal oxide. During RIE process, the etch by-products may be fragmented and re-deposited on the surface of the etched film. It results in a slower etch-rate than that for LTO in the same chamber. Moreover, the re-deposited polymer near wafer edge is more easily removed due to an enhanced electrical field and gas flow around circumjacent chamber sheath. This affects the remained amount of polymer around wafer edge and induces a faster SOG etch-rate at the edge of the wafer (so-called wafer edge micro-loading effect). As a result, the etch-rate of standard SOG RIE has serious dependency of wafer location.

Through modifying the etch parameters (reactive gas pressure, rf power, inserting additional step of Ar/O_2 polymer treatment during etching), a novel SOG RIE process with convex etch-rate profile distribution across the whole wafer is developed. Figure 3 presents the relationships between gas pressure and SOG etch-rate difference between center and edge. As gas pressure increases, the etch-rate difference between center and edge is changed from negative to positive. A stronger gas pressure is expected to increase the effect of chemical etching and decrease the effect from sheath electrical field. Furthermore, it may enforce a fully uniform coverage of reaction-polymer on the whole wafer. The following step of Ar/O_2 polymer treatment with normal gas pressure reduces more polymers at center than at edge. Therefore, the SOG at wafer center is etched faster than that around wafer edge and the concave profile of post-etching SOG is formed. The maximum thickness difference between wafer edge and center is as high as 50nm.

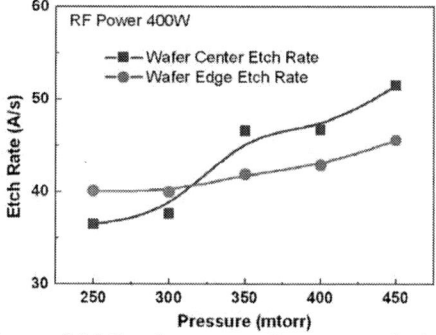

Figure 3. The dependency of SOG etch rate at wafer center and edge for gas pressure.

For "Recessed 3-Step Etchback" approach, the RIE parameters in the 2nd etch step are based on the standard SOG etch recipe and demonstrates a concave etch-rate profile, which is just opposite with the convex etch-rate profile of the 1st etch step. The 1st etch step together with the 2nd etch step is capable of smoothing the etch uniformity issue of "Conventional 3-Step Etchback". This is because of the compensation effect on the larger loss of film thickness near wafer edge for the serious micro-loading effect during SOG RIE and the thickness variations during SOG spin-coating. The 1st etch step produces a concave profile for etched SOG; the 2nd etch step trims the recessed depth of SOG in previous step and etches into beneath LTO layer to form a special structure; the 3rd etch step has no uniformity issue and is used for ILD thickness reduction. With this new "Recessed 3-Step Etchback" approach, the final ILD profile is like a flat-bottomed bowl across whole wafer. In contrast to "Conventional 3-Step Etchback" method, the EWD is increased by 30% and it is greater than 90% of original wafer size (100nm). The uniformity within the effective area is over 95% and slope less than 2.5e-7, which is much closer to the result of complete global planarization by CMP.

The SEM photos of engineering structures with gate lengths of 0.4µm and 70nm for "Recessed 3-Step Etchback" planarization on SOG/LTO composite layers are shown in Figure 4 and Figure 5, respectively. In Figure 4, both cross-sectional views of structure after etchback planarization and after dummy gate removal are presented. The figure indicates a relatively uniform degree of etchback planarization across isolated gate line and dense lines. The ILD thickness difference across a die is below 10%. The residual LTO thickness is 153nm and it is smaller than that of residual poly-Si gate (162nm). Moreover, a small recessed profile of ILD near each gate line is formed and the LTO thickness in the recessed region is 142nm and slightly smaller than general thickness of 153nm. This phenomenon, with a so called name of micro-concave effect, is due to the 2nd etch step trimming of the SOG/LTO interface. The standard SOG etch recipe has a higher etch-rate for LTO and the rate-ratio is about 2.1:1. During interface etch, the firstly exposed LTO on the top of gate lines is etched out instantly and subsequently forms a small recessed profile around dummy gate. This special gate structure with ILD micro-concave profile is of great benefit for poly-Si materials removal in TMAH wet-etch solution with a ultra-high etch selectivity to oxide, especially in nanometer level pitch etching.

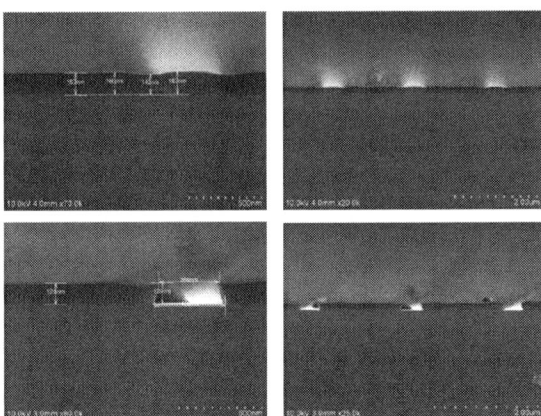

Figure 4. Cross-sectional view of ILD0 planarization on gate stack structure with feature length equal to 0.4µm by "Recessed 3-Step Etchback".

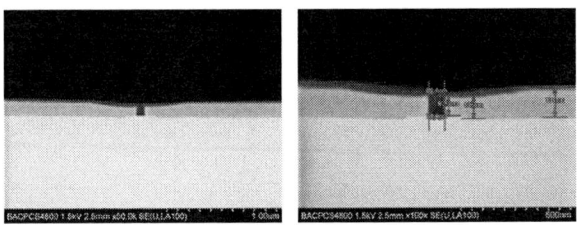

Figure 5. Cross-sectional view of ILD0 planarization on gate stack structure with feature length equal to 70nm by "Recessed 3-Step Etchback".

Figure 5 demonstrates the result of 70nm engineering structures for "Recessed 3-Step Etchback" planarization on SOG/LTO composite layers. In this figure, the dummy gate has been etched out and the uniform degree of planarization across die is similar to that with large gate lengths. The depth of micro-concave on dummy gate is increased to smoothly etch out poly-Si materials embedded in gate pitch by TMAH in short time.

The experimental gate-last HK/MG PMOSFET with "Recessed 3-Step Etchback" ILD planarization method is successfully fabricated. The gate length of transistor is 85nm and the integrated HK/MG structure is HfSiAlON/MoAlN. Figure 6 summarizes PMOSFET on-current (I_{on}) and V_{th} distribution across the wafer. The device demonstrates nice I-V characteristics at wafer center and the electrical parameters distributions of I_{on} and V_{th} across the wafer demonstrate a similar uniformity tendency to that of ILD0 profile. It indicates the final ILD profile has great effect on HK/MG integration quality in following process steps.

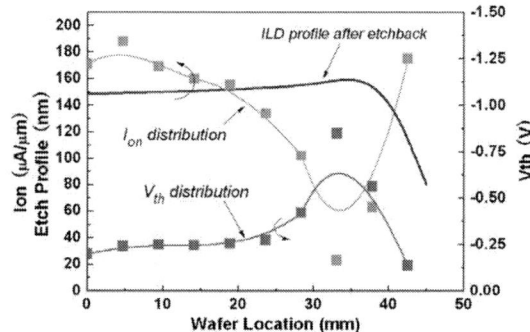

Figure 6. I_{on} and V_{th} distributions of Gate-last HK/MG PMOSFETs with "Recessed 3-Step Etchback" ILD planarization approach across whole wafer.

Conclusions

The CMP-less etchback technology for ILD0 planarization in gate-last MOSFET with nanometer-level gate length is successfully developed and it promised a low cost solution-platform for gate-last HK/MG CMOS process integration.

Acknowledgments

The authors would like to thank engineers in Integrated Circuit Advanced Process Center, IMECAS for their support in wafer processing.

References

1. P. Packan, S. Akbar, M. Armstrong, et al., IEDM Tech. Dig., p. 659, (2009).
2. Thomas Y. Hoffmann, *Solid-State Technology*, **53**(3), (2010).
3. Jie Diao, Garlen Leung, Jun Qian, et al., IEEE ASMC, p. 247, (2010)
4. Paul Feeney, *Solid-State Technology*, **53**(10), (2010).
5. Aric C. Madayag and Zhiping Zhou, Proceedings of the 14th Biennial University/Government/Industry Microelectronics Symposium, p. 136, (2001).
6. S. Wolf, *Silicon Processing for the VLSI Era: Volume 2-Process Integration*, p. 224, Lattice Press, California (1990).

Etch and Wet Clean Challenges and Joint Optimization

B. Yen, J. Lin, C. Lee, M. Hegarty, and P. Loewenhardt

Lam Research Corporation, Fremont, California 94538, USA

The emergence of new integration schemes and materials for advanced technology nodes has presented greater challenges to both etch and post etch wet clean. The mutual interactions between etch and clean processes must be considered, thus having both etch and clean process optimized together has become more essential and beneficial in order to provide the optimized process solution.

Introduction

For the 45 nm node and beyond in the Front-end-of-line (FEOL), novel high-k gate dielectrics (HK) are considered to allow further scaling of the gate dielectric. In order to prevent Fermi-level pinning, metal gates (MG) with the proper work function have to be used on the high-k dielectrics. The integration of metal gate with high k dielectric along with new FEOL applications like sigma shape recess introduce new challenges for FEOL etch and clean processes. For the back-end-of-line (BEOL), the porous low-κ materials used for the dual damascene process can be damaged by the etch chemistries. The dry etch and the subsequent cleaning processes and the adoption of metal hard mask (MHM) create new process and productivity challenges for both etch and clean as well. In this talk, we will discuss etch and wet clean challenges associated with new integration schemes and examples where etch and clean steps are optimized together.

Etch and Wet Clean Challenges

High K Metal Gate (HKMG) Formation

As logic technology transitions to high-k metal gate at the 32nm node, both etch and clean challenges became more stringent. For gate first approach, the usage of different high-k capping layers for both NMOS and CMOS presented challenges for etch to control footing, profile, and residue. Post etch residues usually contain metal and high-k elements, which can present queue-time challenges. This drives the need for an effective clean. A typical post HKMG etch surface is illustrated in Figure 1. It should be noted that post etch wet clean can also have profound impact to the final profile and CD. The trade-off between aggressive clean vs. profile erosion must be considered when developing etch and post etch clean unit processes. Figure 2 shows an example on how the wet clean can impact the final profile.

For gate last approach, one of the key steps that requires joint development between etch and wet clean/etch is the replacement gate dummy poly Si removal. Table I highlights the benefits of various approaches for this process step. Again, the dry and wet clean/etch steps are tightly correlated to each other. The optimized approach, however, are highly dependent on the integration details, and the capability of etch and clean tool. The best fabrication solution can be realized by integrating etch and clean together.

Figure 1. Post Etch Residue after HKMG Stack Etch

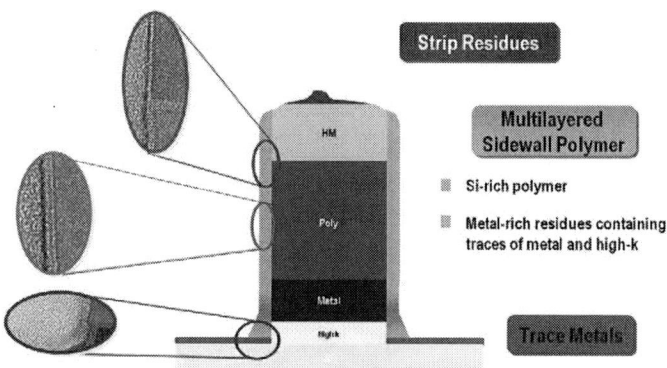

Figure 2. Control of HKMG CD Using Sequential Clean

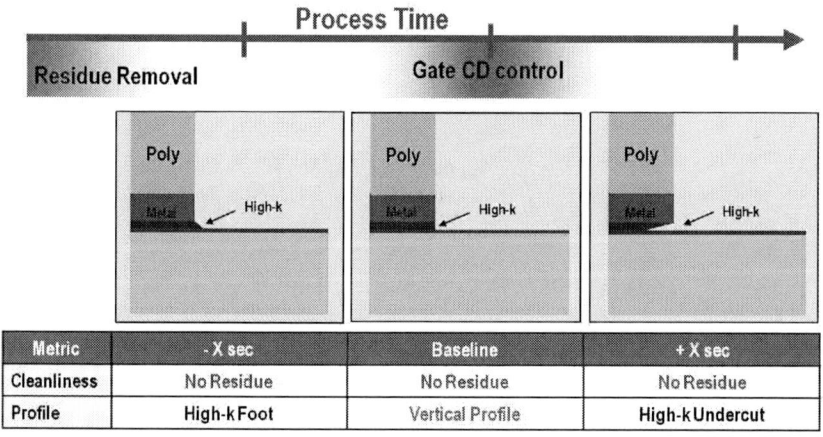

Metric	- X sec	Baseline	+ X sec
Cleanliness	No Residue	No Residue	No Residue
Profile	High-k Foot	Vertical Profile	High-k Undercut

TABLE I. Comparison of Various Approaches for Replacement Gate Poly Si Removal

Poly Si Removal	Benefits
Full Wet Etch	No plasma interaction with gate dielectric
Full Dry Etch + Wet Clean	Highest removal rate, independent of poly Si doping
Partial Dry Etch + Partial Wet Etch	Higher removal rate, no plasma interaction with gate

Sigma Shape Formation

Using embedded SiGe in the source/drain region of a transistor to increase channel stress has been incorporated since the 45nm technology node.[1] Traditional recess methods create either anisotropic or isotropic shape in the source/drain region; while this worked well, it does not stress the channel as effectively as a "sigma-shaped" recess. This shape is enabled by utilizing the unique nature of the Si crystal orientation. While dry etch behavior is relatively independent of crystal orientation, wet etch chemistries respond quite differently for <100> and <111> surfaces. Since both dry and wet etching will remove poly Si during this process, the final shape, including depth, top and bottom width, are greatly contributed by both steps. For example, a deeper and wider dry etch process will usually results in larger sigma shape even with a fixed wet etching process with good <100>/<111> selectivity. Therefore, while wet etch is the final step in defining the sigma shape, the interaction from the earlier dry step is very significant, and must be considered together to create the desired "sigma-shape" recess. In the Figure 3, the mutual dependent of dry and wet steps is shown. In addition, uniformity across the wafer for both steps (dry + wet) need to be tightly controlled and integrated to provide the overall uniformity.

Figure 3. Sigma Recess Shape Dependence on Dry and Wet Process

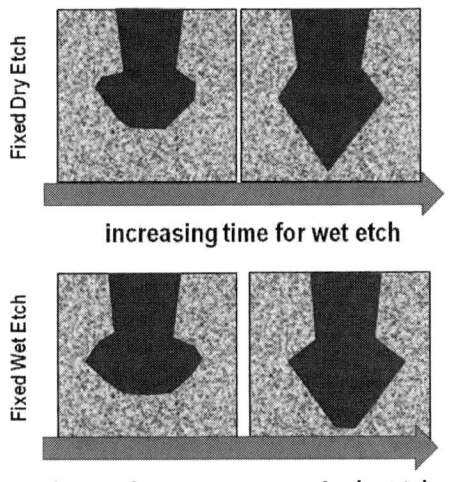

BEOL Interconnect Challenges

Cu line dimension shrink and the continuous reduction of dielectric constant (k) of the dielectric material between metal lines have brought major changes in the integration approach for BEOL interconnection. Traditional Via first photoresist scheme has been replaced with partial trench first – metal hard mask approach, which has become the dominate scheme for 28nm and below. In Via first scheme, via holes are opened first. After filling the holes with resist plugs, trench patterns are formed by plasma etching using a resist mask. The key challenges for the via-first process are to suppress resist poisoning, line-edge roughness, and ashing damage.[2] Resist poisoning is caused by a resolution failure due to amine penetration through the resist plugs.[3] The line edge roughness is caused by trench pattern distortion due to the poor plasma resistance of the resist material. The ashing damage is caused by the decomposition of methyl groups in low-k material due to resist ashing.[4] This becomes more serious as the feature size shrinks together with using lower-k value dielectric. To solve these problems, a trench-first scheme using a metal hard mask, typically TiN, has been proposed.

While the new scheme does provide several benefits to solving the problems, it has also brought new challenges to both etch and subsequent clean. In terms of challenges on wafer, using metal layer as mask during trench etch requires sufficient selectivity of dielectric layer over TiN in order to preserve enough mask layer with minimal mask corner erosion to maintain trench profile for high and reliable interconnect performance. The existence of metal mask layer during etch also implicates changes in plasma properties, thus the etch reactor and chemistry must be re-conditioned from the process parameters used for previous technology nodes to provide adequate ion density, ion energy, and desired species.

In terms of productivity, due to its non-volatile nature, TiF_x-containing polymer on wafer surfaces post etch needs be minimized to prevent defectivity and chamber matching issues. In addition, the low K damage is becoming a bigger concern as k values continues to decrease along with the shrinking critical dimension. Strip & etch plasmas can induce chemical changes that raise the k-value and consequently the capacitance of the low-k dielectrics. It has been studied that excess electron beam and UV cure increases k value of porous PECVD/SOD Low-k SiCOH[5]. Radicals inside etch reactor can also reduce k value of the dielectric by ion scattering and synergistic interaction with UV to deplete the carbon content of the low-k SiCOH and cause low-k damage during etch[6]. Furthermore, Cu loss/undercut also needs to be minimized during etch and wet clean. In Table I below, the etch challenges associated with each etch process step are listed. In Image I, on wafer and productivity challenges are illustrated.

TABLE II. Etch Challenges for BEOL DD Via and Trench with Metal Hardmask

Etch Step	Etch Challenges
Via Etch + PR Strip	Profile control, CD shrink, Low K damage
Trench Etch +Barrier Open	TiN selectivity, Profile control, Cu damage, low K damage
Post Etch Treatment	Cu surface control, Polymer removal, Residue control, Low K damage

Following the etch is the post etch wet clean process, which is designed to remove post etch polymer residue and preserve the profile. It is no surprise that the wet clean step will be highly dependent on how the etch process is finished, thus the etch and the clean process are often and best optimized together. In the scenario that the final etch step is polymerizing, the wet clean chemical should be tuned to effectively remove all polymers, which may contain metal to form non-volatile defects to impact device yield. If the etch step finishes with a relatively lean process, then the polymer removal becomes less of a critical factor during wet clean. In addition to polymer removal requirements, wet clean also has potential impact to Cu surface and can alter profile. For example, dHF will attack damaged low K and can cause an undercut trench profile which is not ideal for following metal fill process. An O_2 containing dHF will oxidize Cu and then CuO_x can be removed by dHF. The formation of Cu undercut and excess Cu loss due to this mechanism pose a significant challenge for Cu fill and impact reliability performance.

Figure 4. Etch Challenges for BEOL DD Via and Trench with Metal Hardmask

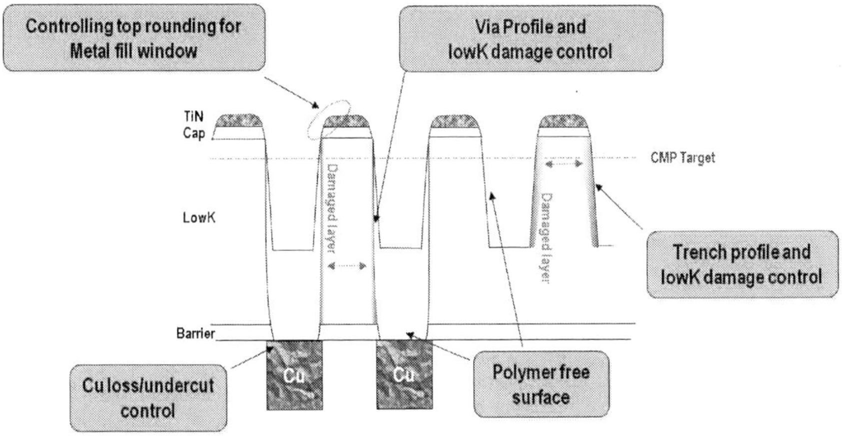

Conclusion

The successful implementation of the new high-k and metal gate materials in the different integration schemes, as well as the use of MHM for BEOL dual damascene, require a fundamental understanding and a joint optimization of the etch and clean integration. HKMG and Sigma shape are defined by both etch and clean process and must be tailored carefully to meet specific requirements. For BEOL dual damascene formation using Metal HM as mask, the post etch clean process must be tuned accordingly to the etch process and can vary significantly depending on the details of integration approach. In summary, etch and clean need to be optimized together, and etch and clean equipment providers must render integrated process solutions for both etch and clean for device makers to meet more stringent requirements of advanced technology nodes.

Acknowledgments

The authors like to thank Gowri Kamarthy, Tae Won Kim, and Anthony Ozzello for their joint work for etch and clean.

References

1. N. Tamura, Y. Shimamune *Applied Surface Science*, V254, issue 19, PP6067-6071
2. Takeshi Furusawa et. al. *Journal of The Electrochemical Society,* 153 _2_ G160-G163 _2006.

3. S. Lin, C. Jin, L. Lui, M. Tsai, M. Daniels, A. Gonzalez, J. T. Wetzel, K. A. Monnig, P. A. Winebarger, S. Jang, D. Yu, and M. S. Liang, in *Proceedings of the IEEE International Interconnect Technology Conference*, p. 146 _2001.
4. T. Furusawa, D. Ryuzaki, R. Yoneyama, Y. Homma, and K. Hinode, *Electrochem. Solid-State Lett.*, 4, G31 _2001.
5. JOURNAL OF APPLIED PHYSICS 101, 013305 2007 (M. A. Worsley et al.)
6. Japanese Journal of Applied Physics Vol. 47, No. 8, 2008, pp. 6923 (Yoshihisa Iba et. al.)

ECS Transactions, 34 (1) 761-768 (2011)
10.1149/1.3567670 ©The Electrochemical Society

Growth and Processing Defects in CMOS Homo- and Hetero-Epitaxy

E. Simoen[a], M. Bargallo Gonzalez[a,b], G. Eneman[a,b,c], E. Rosseel[a], A. Hikavyy[a],
D. Kobayashi[a,d,] R. Loo[a], M. Caymax[a], and C. Claeys[a,b]

[a] Imec, Kapeldreef 75, B-3001 Leuven, Belgium
[b] Department Electrical Engineering, K.U. Leuven, Kasteelpark Arenberg 10, B-3001
Leuven, Belgium
[c] also Post-doctoral Fellow of the Fund for Scientific Research-Flanders (FWO), 1000
Brussels, Belgium
[d] on leave from the Institute of Space and Astronautical Science, JAXA, Japan

The impact of different processing steps on the electrical properties
of homo- and hetero-epitaxial junctions deposited on silicon
substrates is described. In particular, the influence of the pre-epi *in
situ* cleaning, using a high temperature bake in H_2 is investigated.
It is shown that the removal of oxygen and carbon from the starting
surface is crucial in obtaining high-quality, low-leakage epitaxial
junctions. In addition, it is demonstrated that post-epi implantation
and anneal should be carefully optimized in order to maintain the
strain in SiGe layers and to control the defect formation.

Introduction

Epitaxial deposition has become an essential part of state-of-the-art CMOS processing.
For example, strained SiGe and Si:C epi layers are currently used as uniaxial stressors in
p- or n-channel transistors, respectively (1-3). Bi-axially strained Si or Ge films can be
deposited on strain-relaxed $Si_{1-x}Ge_x$ buffer layers (SRBs), yielding a higher electron or
hole mobility (1). Further down the ITRS roadmap, epitaxial Ge or III-V layers on a
silicon substrate are expected to yield so-called high-mobility devices, somewhere around
the 16-11 nm technology node. However, all these systems are characterized by a certain
lattice mismatch and a difference in thermal expansion coefficient with the underlying Si
substrate, which results in the build-up of elastic and thermal strain energy, increasing
with the thickness of the epitaxial layers. Beyond a critical thickness this strain will relax
by the formation of misfit dislocations, with threading arms reaching up to the surface. It
is well-documented in the literature that the presence of such extended defects in the
device active area leads to enhanced junction leakage current and, in some cases, to a
malfunctioning of the MOS transistor. Therefore, it is important to monitor and control
the formation of extended and, generally speaking, of electrically active defects during
hetero-epitaxy and subsequent CMOS processing.

It is the aim of this invited paper to overview the impact of different processing
steps on the formation of harmful defects. As will be shown, critical steps are the pre-epi
cleaning and pre-epi bake, the epitaxial deposition conditions, i.e., selective in Shallow
Trench Isolation (STI) regions and whether post-epi-growth thermal annealing is applied
or not. Results will be presented on the impact of different pre-epi bake temperatures on
the defectivity of Si and SiGe layers, as measured by junction leakage and recombination

lifetime. In addition, Deep Level Transient Spectroscopy (DLTS) has been utilized to assess the defects at the hetero-interface. In combination with Secondary Ion Mass Spectrometry (SIMS) one can demonstrate that the amount of remaining oxygen at the interface is directly related to the resulting lifetime of the wafers. However, for state-of-the-art Low Pressure Chemical Vapor Deposition (LPCVD), the pre-cleaning bake can be reduced to 800 °C without significantly compromising the electrical properties of the SiGe-Si hetero-junction.

In general, ion implantation followed by an activation and defect removal anneal is being performed on these epitaxial layers. As will be shown, this may introduce additional defects, especially when aggressive annealing schemes, like laser annealing are employed.

Impact of the pre-epi clean

$Si_{1-x}Ge_x$ or $Si_{1-y}C_y$ source/drain (S/D) stressors are selectively deposited by CVD in regions defined by STI boundaries, where the Si has been removed by dry etching. The pre-cleaning of the etched Si surface is of utmost importance for the subsequent epi layer quality and should, therefore, remove adequately the traces of C and O remaining on the starting Si surface (4,5). As a first step in the pre-epi cleaning procedure, the native oxide can be removed by an *ex situ* wet-chemical HF-dip (6). The duration should be minimized in order not to attack the oxide hard mask on the pMOS transistors (6,7). At the same time, the time interval between the HF-dip and the loading of the wafer inside the epi-reactor should be as short as possible in order to maintain a well-passivated Si surface. After the loading of the wafer into the reactor, an H_2 bake at high temperature is performed. The temperature should be sufficiently high to clean the surface by the volatilization reaction of silicon oxide (8):

$$SiO_{2(s)} + Si_{(s)} \leftrightarrow 2SiO_{(g)} \qquad [1]$$

However, in order to avoid dopant diffusion during the bake, its thermal budget should be kept under control. As shown by the SIMS results of Fig. 1, the oxygen concentration is at the detection limit for an H_2 bake at 800 °C or higher, while C is effectively removed in all cases at the hetero-interface (8).

The impact of the high-temperature pre-epi bake also translates in the electrical properties of the junctions, as shown in Fig. 2 for both Si and SiGe selective epitaxial layers on a Si substrate. It is clear that mainly the area leakage current density J_A is increased for lower T_{bake}, while the peripheral leakage current density J_P is hardly dependent on the bake temperature (8). An empirical relationship between J_A and T_{bake} has been derived (8), described by:

$$J_A = B \exp(0.011[850°C - T_{bake}]) \qquad [2]$$

for 750 °C < T_{bake} < 850 °C.

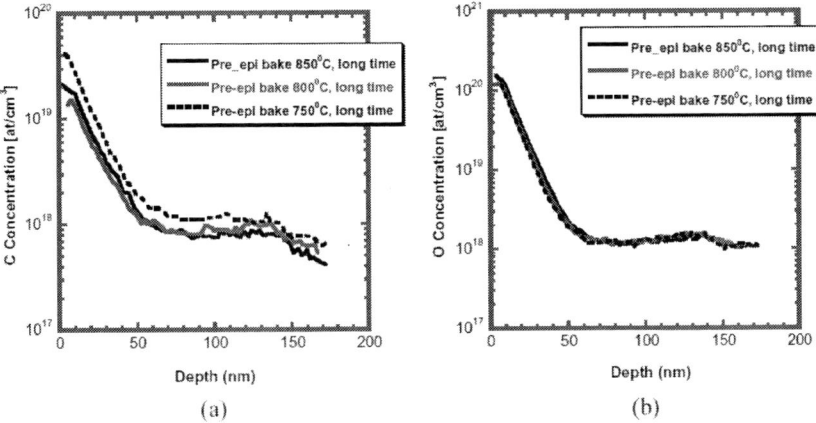

(a)   (b)

Figure 1. SIMS depth profile of C (a) and O (b) for junctions, with 140 nm thick Si epilayer for different pre-epi H_2 bake temperature, following a 90 min waiting time between the HF dip and the loading in the reactor (after Gonzalez *et al.* (8)).

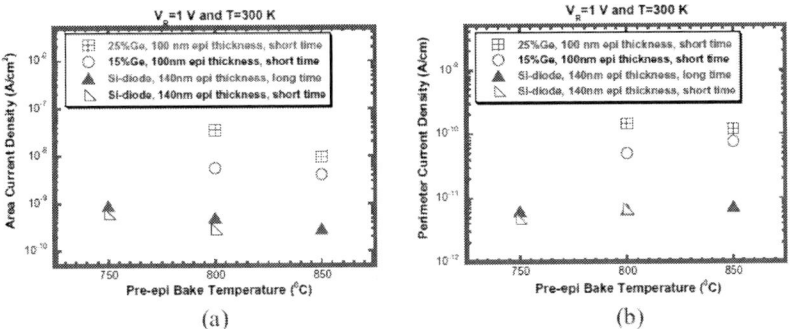

(a)   (b)

Figure 2. Area (a) and perimeter (b) leakage current density at -1 V for the junctions, consisting of an *in situ* highly B-doped Si or $Si_{0.85}Ge_{0.15}$ ($Si_{0.75}Ge_{0.25}$) epi layer on 300 mm n-type Cz Si wafers at 300 K (after Gonzalez *et al.* (8)).

The area leakage current density is inversely proportional with the Shockley-Read-Hall (SRH) generation lifetime τ_g, implying a reduction of τ_g for reduced bake temperature. This relationship has more recently been confirmed by Photoluminescence (PL) and microwave Photoconductance decay (μPCD) measurements on blanket 200 or 300 mm wafers with $Si_{0.85}Ge_{0.15}$/50 nm Si cap epi-layers (9). The minority carrier recombination lifetime data corresponding with 300 mm wafers measured by μPCD is represented in Fig. 3, showing a sigmodial behavior, with a threshold at 800 °C H_2 bake temperature. A similar optimal pre-bake condition has been derived by other groups as well (10,11).

The reduction of the lifetime for lower pre-epi bake temperatures suggests the presence of a higher density of electrically active defects, which mainly originate from the interfacial oxygen and carbon contamination (12). This can give rise to dislocation generation at the epi/Si interface and in the case of threading defects, deep levels are introduced in the depletion region of a junction. Interfacial defects in homo-epitaxial Si layers have been studied by Deep Level Transient Spectroscopy (DLTS), revealing trap states across the band-gap (13,14). It has been speculated that they were associated with oxygen and carbon at the interface (13). According to the data of Fig. 4, both the current-voltage (I-V) and DLT-spectra of 400 nm $1.2 \cdot 10^{17}$ cm^{-3} B-doped p-Si epi on 300 mm n-type Si substrates show a pronounced dependence on the pre-epi bake temperature. Overall, a lower leakage current and lower trap concentrations are observed for a higher T_{bake}. The main peak around 210 K in Fig. 4b is similar to literature data (14) and may be related to C- and O-induced deep levels at the epitaxial interface.

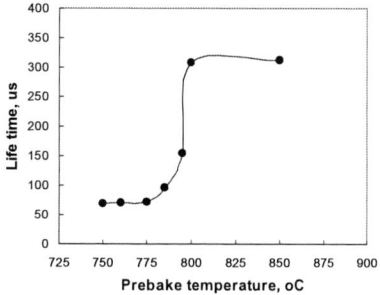

Figure 3. Correlation between carrier lifetime and pre-epi bake temperature as measured for 200 nm $Si_{0.85}Ge_{0.15}$/50 nm Si-cap epi-layers. The data shown has been obtained on 300 mm wafers. Similar results have been obtained on 200 mm Cz-Si wafers (after Loo *et al.* (9))

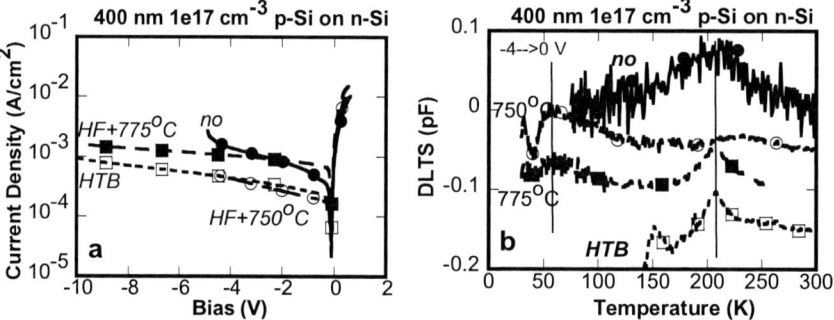

Figure 4. (a) Current density versus bias for a Si homoepitaxial p-n junction at room temperature, following an HF dip with different H$_2$ bake: no, High Temperature Bake (HTB=1000 °C), at 750 or 775 °C. (b) Corresponding DLT-spectra from -4-->0 V.

Impact of post-epi implantation and anneal

In situ doping of the epitaxial embedded SiGe source and drain regions should in principle lead to sharp and steep junctions, which is highly desirable from a scaling viewpoint. However, often a post-epitaxial ion implantation of the so-called Highly-Doped Drain (HDD) is performed for integration reasons (15). A high-dose ion implantation creates a significant amount of lattice damage and also strain relaxation in the SiGe layer (Fig. 5) (16), which need to be cured by a post-implantation anneal. The thermal budget of this anneal should be minimized in order to avoid excessive dopant diffusion, so that generally a high-temperature spike anneal is being used. The risk exists of course that not all damage is removed and that the strain is not completely recovered by incomplete recrystallization or by the interaction with the implantation-induced point defects. In fact, it has been shown that a so-called window-size effect exists in post-epi HDD implanted SiGe/Si junctions, whereby the perimeter leakage current density becomes smaller for smaller active areas (17-19). The origin of this effect is the interaction of the STI strain with the implantation-induced point defects (20-24), leading to a lower density of traps at the STI/Si interface. At the same time, it has been demonstrated by different stress-measurement techniques that elastic relaxation of the strain in the SiGe layers occurs for small active areas (25,26). However, the fact that both J_A and J_P remain constant for several window sizes in the case of non-HDD-implanted, *in situ* doped SiGe/Si heterojunctions (27) indicates that this has only a marginal impact on the leakage current. It emphasizes once more that the main factor in the leakage current are the residual, implantation-induced point defects and point-defect clusters and not so much the strain in the silicon substrate (28).

50 nm SiGe:B (25% Ge)

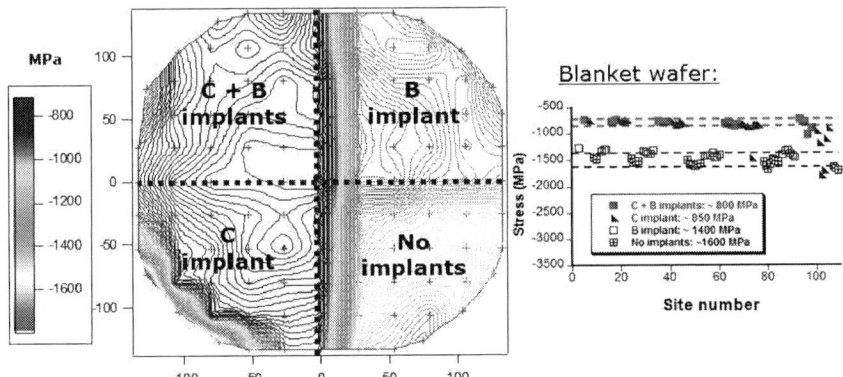

Figure 5. Stress profile and wafer map as measured by wafer curvature (bow) analysis, using two-dimensional laser beam scanning, for an *in situ* highly B-doped 50 nm $Si_{0.75}Ge_{0.25}$ epilayer on a 300 mm Si substrate after 12 keV C to a dose of $1 \cdot 10^{15}$ cm^{-2} (left side) and 3 keV B to a dose of $3 \cdot 10^{15}$ cm^{-2} (top side) implantations (after Gonzalez *et al.* (16)).

In order to reduce the thermal budget further, flash-lamp or laser annealing (so-called ms annealing - MSA) is becoming more and more popular (29,30). However, the high thermal stresses associated with the strong temperature gradients may result in strain relaxation of the SiGe layer when the laser annealing conditions (dwell time, maximum power) are not properly optimized (16,31-33). This yields an increase in the junction leakage current (32,33), as illustrated in Fig. 6, which is particularly pronounced when a C co-implantation is used. The latter is frequently employed to reduce the silicon interstitial-mediated transient enhanced diffusion of B, in order to better control the junction depth. A more in-depth study of the electrically active defects, combining high-frequency C-V and I-V junction characteristics leads to the conclusion that the laser annealing itself creates point defects in the junction depletion region (34). This is demonstrated by the result of Fig. 7, showing a MSA induced increase of J_A. At the same time, it has been demonstrated that by fine-tuning the C co-implantation parameters (energy and depth) one can perform defect engineering in such a way that the J_A for typical CMOS operation conditions (a reverse bias of e.g. -1 V) is reduced, compared with the case without C co-implantation.

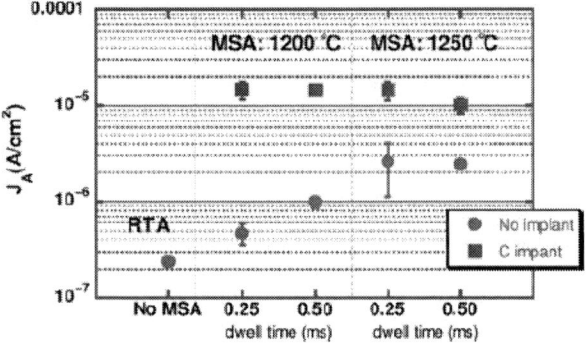

Figure 6. Area leakage current density at -0.1 V and 300 K after a 1035 °C spike anneal and additional MSA, for *in situ* highly B-doped $Si_{0.75}Ge_{0.25}/Si$ p^+-n junctions implanted with 3 keV B ($3 \cdot 10^{15}$ cm^{-2}) and with or without (no implant) 12 keV C ($1 \cdot 10^{15}$ cm^2) implant (after Bargallo *et al.* [33]).

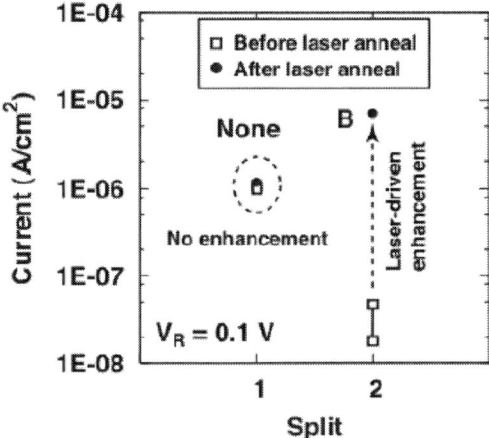

Figure 7. Comparison of the leakage currents before and after the MSA laser anneal at 1200 °C with a scanning speed of 150 mm/s. Split 1 did not receive and split 2 did receive a B implantation. The average values of 5 samples are plotted for the MSA case. For the case without laser anneal, two samples have been measured and their values are directly plotted on the graph (after Kobayashi *et al.* [34]).

Acknowledgments

The Authors would like to express their thanks to: P. Absil, T. Hoffmann, N. Thomas, P. Verheyen, F.E. Leys, M. Wada, N. Naka, M.K. Chowdry, B. De Vos, H. Dekkers, T. Fernandez-Lanas, A. Pacco, Y. Okuno, K. Hirose, B. Vissouvanadin, B. Van Daele, L. Geenen, L. Souriau, H. Bender, V. Machkaoutsan, P. Tomasini, S.G. Thomas, J.P. Lu, J.W. Weijtmans and R. Wise, for the use of co-authored results and for many encouraging discussions.

References

1. M.L. Lee, E.A. Fitzgerald, M.T. Bulsara, M.T. Currie and A. Lochtefeld, *J. Appl. Phys.*, **97**, 011101 (2005).
2. T. Ghani, M. Armstrong, C. Auth, M. Bost, P. Charvat, G. Glass, T. Hoffmann, K. Johnson, C. Kenyon, J. Klaus, B. McIntyre, K. Mistry, A. Murthy, J. Sandford, M. Silberstein, S. Sivakumar, P. Smith, K. Zawadzki, S. Thompson and M. Bohr, *IEDM Tech. Dig.*, 978 (2003).
3. P. Verheyen, G. Eneman, R. Rooyackers, R. Loo, L. Eeckhout, D. Rondas, F. Leys, J. Snow, D. Shamiryan, M. Demand, Th.Y. Hoffmann, M. Goodwin, H. Fujimoto, R. Cavit, B.-C. Lee, M. Caymax, K. De Meyer, P. Absil, M. Jurczak and S. Biesemans, *IEDM Tech. Dig.*, 907 (2005).
4. R. Loo, M. Caymax, I. Peytier, S. Decoutere, N. Collaert, P. Verheyen, W. Vandervorst and K. De Meyer, *J. Electrochem. Soc.*, **150**, 638 (2003).
5. M. Caymax and R. Loo, *Proc. Symposium on SiGe: Materials, Processing and Devices*, D. Harame *et al.* Editors, PV **2004-07**, The Electrochem. Soc. (Pennington, NJ), 815 (2004).
6. R. Loo, C. Walczyk, P. Verheyen, R. Rooyackers, F.E. Leys, G. Eneman, D. Shamiryan, P.P. Absil, T. Delande, A. Moussa, H. Bender, C. Drijbooms, L. Geenen, M. Caymax, J.W. Weijtmans, R. Wise, V. Machkaoutsan, P. Tomasini, C. Arena, J. McCormack, S. Passefort, H. Sorada, A. Inoue, B.C. Lee, S. Hyun, S. Jakschik and S. Godny, *ECS Trans.*, **3** (7), 453 (2006).
7. C.I. Liao, Y.C. Chen, P.L. Cheng, H.Y. Wang, C.C. Chien, C.L. Yang, K.T. Huang and S.F. Tzou, in *Conf. Dig. of the Third International Silicon Germanium Technology and Devices Meeting (ISTDM 2006)*, 166 (2006).
8. M. Bargallo Gonzalez, N. Thomas, E. Simoen, P. Verheyen, A. Hikavyy, F.E. Leys, Y. Okuno, B. Vissouvanadin, B. Van Daele, L. Geenen, R. Loo, C. Claeys, V. Machkaoutsan, P. Tomasini, S.G. Thomas, J.P. Lu, J.W. Weijtmans and R. Wise, *ECS Trans.*, **11** (3), 47 (2007).
9. R. Loo, A. Hikavyy, F. Leys, M. Wada, B. De Vos, A. Pacco, M. Bargallo Gonzalez, E. Simoen, P. Verheyen and M. Caymax, *Proc. of UCPSS, Diffusion and Defect Data Part B (Solid State Phenomena)*, **145-146**, 177 (2009).
10. M.S. Carroll, J.C. Sturm and M. Yang, *J. Electrochem. Soc.*, **147**, 4652 (2000).
11. A. Abbadie, J.M. Hartmann, P. Holliger, M.N. Séméria, P. Besson and P. Gentile, *Appl. Surf. Sci.*, **225**, 256 (2004).
12. M. Fukuda, Y. Shimamune, K. Tanahashi, K. Ikeda, M. Nishikawa, H. Maekawa, N. Tamura, T. Mori, A. Shimizu and M. Kase, *ECS Trans.*, **19** (1), 213 (2009).
13. D. Stievenard, X. Wallart and D. Mathiot, *J. Appl. Phys.*, **69**, 7640 (1991).

14. F. Lu, D. Gong, H. Sun and X. Wang, *J. Appl. Phys.*, **77**, 213 (1995).
15. H. Okamoto, A. Hokazono, K. Adachi, N. Yasutake, H. Itokawa, S. Okamoto, M. Kondo, H. Tsujii, T. Ishida, N. Aoki, M. Fujiwara, S. Kawanaka, A. Azuma and Y. Toyoshima, *Jpn. J. Appl. Phys.*, **47**, 2564 (2008).
16. M. Bargallo Gonzalez, T. Fernandez-Lanas, E. Rosseel, A. Hikavyy, H. Dekkers, G. Eneman, P. Verheyen, R. Loo, E. Simoen and C. Claeys, *ECS Trans.*, **25** (7), 217 (2009).
17. E. Simoen, M. Bargallo Gonzalez, B. Vissouvanadin, M.K. Chowdhury, P. Verheyen, A. Hikavyy, H. Bender, R. Loo, C. Claeys, V. Machkaoutsan, P. Tomasini, S. Thomas, J.P. Lu, J.W. Weijtmans and R. Wise, *IEEE Trans. Electron Devices*, **55**, 925 (2008).
18. M. Bargallo Gonzalez, E. Simoen, B. Vissouvanadin, G. Eneman, P. Verheyen, R. Loo, C. Claeys, V. Machkaoutsan, P. Tomasini and S. Thomas, *Phys. Stat. Sol. C.* **6**, 1901 (2009).
19. E. Simoen, G. Eneman, M. Bargallo, D. Kobayashi, A. Luque, J.-A. Tejada and C. Claeys, *ECS Trans.*, **31** (1), 307 (2010).
20. R. Hull, J.C. Bean, J.M. Bonar, G.S. Higashi, K.T. Sort, H. Temkin and A.E. White, *Appl. Phys. Lett.*, **56**, 2445 (1990).
21. N.G. Rudawski, K.N. Siebein and K.S. Jones, *Appl. Phys. Lett.*, **89**, 082107 (2006).
22. K.L. Saenger, J.P. de Souza, K.E. Fogel, J.A. Ott, C.Y. Sung and D.K. Sadana, *J. Appl. Phys.*, **101**, 024908 (2007).
23. N. Burbure, N.G. Rudawski and K.S. Jones, *Electrochem. Solid-State Lett.*, **10**, H184 (2007).
24. J.P. Liu, J. Li, A. See, M.S. Zhou and L.C. Hsia, *Appl. Phys. Lett.*, **90**, 261915 (2007).
25. A. Hikavyy, N. Bhouri, R. Loo, P. Verheyen, F. Clemente, J. Hopkins, R. Trussell, *Proc. of the International Conf. on Silicon Epitaxy and Heterostructures (ICSI)*, 145 (2007).
26. M. Bargallo Gonzalez, E. Simoen, Y. Okuno, N. Naka, G. Eneman, A. Hikavyy, P. Verheyen, R. Loo, C. Claeys, V. Machkaoutsan, P. Tomasini, S.G. Thomas, J.P. Lu and R. Wise, *Mater. Sci. Semicond. Process.*, **11**, 285 (2008).
27. M. Bargallo Gonzalez, E. Simoen, B. Vissouvanadin, P. Verheyen, R. Loo and C. Claeys, *IEEE Trans. Electron Devices*, **56**, 1418 (2009).
28. A. Luque, M. Bargallo Gonzalez, E. Simoen, C. Claeys and J.A. Jiménez Tejada, *Proc. ESSDERC 2010*, 384 (2010).
29. M. Hane, *ECS Trans.*, **19** (1), 63 (2009).
30. S. Govindaraju, C.-L. Shih, P. Ramanarayanan, Y.-H. Lin and K. Knutson, *ECS Trans.*, **28** (1), 81 (2010).
31. D. Riley, H. Bu, A. Jain and R. Khamankar, *ECS Trans.*, **16** (10), 333 (2008).
32. M. Bargallo Gonzalez, E. Simoen, E. Rosseel, P. Verheyen, L. Souriau, J. Geypen, H. Bender, T. Hoffmann, R. Loo, P. Absil and C. Claeys, *ECS Trans.*, **13** (1), 23 (2008).
33. M. Bargallo Gonzalez, E. Rosseel, A. Hikavyy, T. Fernandez-Lanas, G. Eneman, P. Verheyen, R. Loo, E. Simoen and C. Claeys, *IEEE Trans. Semicond. Manufact.*, **23**, 538 (2010).
34. D. Kobayashi, M. Bargallo Gonzalez, E. Rosseel, A. Hikavyy, K. Hirose, E. Simoen and C. Claeys, *ECS Trans.*, **33** (11), 191 (2010).

ECS Transactions, 34 (1) 769-774 (2011)
10.1149/1.3567671 ©The Electrochemical Society

Precise Control of Spike Anneal Process for Advanced CMOS

Zhibiao ZHAO, Ji Yue TANG, Ganming ZHAO
Applied Materials China, Shanghai 201203, China

Aggressive scaling of planar bulk CMOS device has resulted in the need for ultra-shallow junctions (USJ) formation. This represents some special concern are not only integration solution, but also the individual process precise control to meet the evolutional requirements. In the USJ particular, as junction thickness (depth) decreases, the series resistance of the junction increases. Therefore, the dopants concentration must to be increased to improve the resistance. Dilemma, the diffusion is an important issue in shallow junction technology. The spike anneal process was widely adopted at advanced CMOS fabrication to meet the device requirements. In this paper, how to precisely control the processes of spike anneal is discussed. The two major interacted process factors, peak-temperature and residence-time must be separately tunable, not only to meet the requirements of device parameter, but also to achieve an operational precise controlling for process setup, optimization and maintenance.

Introduction

With CMOS downscaling, the device geometries become smaller and smaller. The Figure 1 shows some shrink factors of the device geometries. Correspondingly, the ultra-shallow

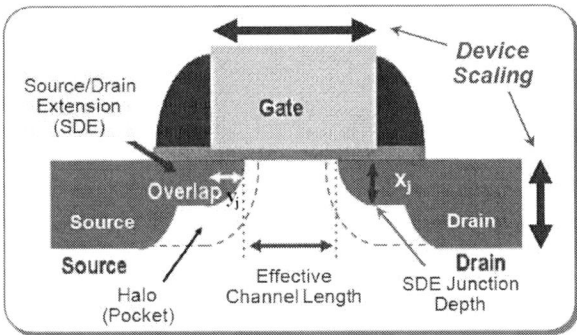

Fig. 1, Double arrows are shown the scaling factors of device geometries.

junction (USJ), shown as Source/Drain Extension(SDE) in Figure 1, was required to control both dopant activation and the diffusion depth (x_j and y_j, as shown in Figure 1). To lower the resistance, the higher dopant concentration of implantation had to be implemented for SDE, which results in the diffusion depth controlling become more challengeable. Thus, the necessary choice for dopant activation and diffusion limitation is to anneal at highest temperature with as possible as lower thermal budget, annealing with short time. Compared to conventional soak anneal, the spike anneal with increasing

769

temperature ramp-up and ramp-down rates can minimize the annealing time to improve the diffusion depth control [1]. In this paper, the precise control of Spike anneal process was discussed.

Spike Anneal Process Control

To form the USJ, the low energy implantation was used. The dopant implantation process inevitably results in implantation-induced damage, the transient enhanced diffusion (TED) must to be considered [2]. The extended defects are metastable, and subsequently dissolve. As long as extended defects exist, an interstitial supersaturation is maintained, resulting in a boron diffusion enhancement. The enhancement ends soon after the defects have dissolved; the diffusion enhancement is thus transient. A consideration of the interstitial supersaturation and of the time to defect dissolution allows the increase in junction depth, Δx_j, due to dissolution of interstitial-type defects, to be expressed as

$$\Delta x_j^2 \propto R_p \bullet \exp[-(-E_f + E_B - E_m)] \qquad [1]$$

where, R_p is the projected ion range, $E_f + E_m = 4.9 \pm 0.1 \text{eV}$, is the activation energy for Si self-diffusion, and E_B is the energy barrier for Boron diffusion [3]. Thus, for spike anneal process, the first factor must to be controlled precisely is the time of temperature ramp-up from lower to higher. This time will be limited by higher ramp-up-rate, therefore, the Δx_j will be minimized. Meanwhile, the R_p in equation-1 gave the reason why the lower energy implants result in reduced TED. On the other hand, for the low energy implantation, the projected range of dopants are very shallow, close to surface of substrate, the dopants out diffusion could be significant and could be affected by oxide thickness and non-uniformity [4]. As a result, the annealing ambient conditions must to be considered and controlled well.

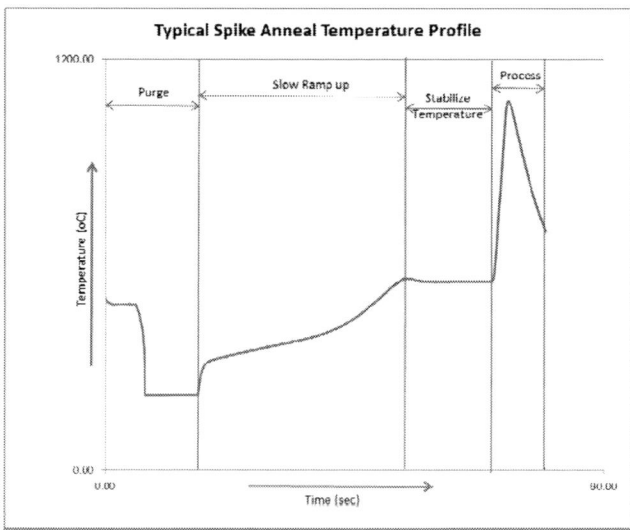

Fig. 2, Typical spike anneal temperature Profile

Figure 2 shows the temperature profile of Spike process, based on the *RadancePlus* Chamber of Applied Materials. At the process beginning, the purge steps were shown, to construct a repeatable with minimum O_2 environment. Then, as shown in Figure 2, there are "slow ramp-up" steps. It was reported that, the doping concentration has a strong effect on the heat up rate at low temperature [5]. Any dopant non-uniformity will effect on temperature non-uniformity due to free carriers, which generated by optical and thermal reason, has a strong effect on the absorption of radiation to heat the wafer. Therefore, the carrier recombination in bulk and surface can cause non-uniform absorption and result in non-uniform heating. So, followed by temperature stabilization step will help to minimum this effect. One key factor of spike anneal is the peak temperature variation of process. Some critical parameters of device, suck like, Cov, Vt roll-off, etc, cannot tolerate this variation too much. That's why the stabilization at lower temperature is a necessary step to minimum the variation of peak temperature. The major steps, process steps, will be focused to discuss at bellow.

Sharpness of Residence Time

The residence time of peak temperature is extremely critical for spike annealing. The process steps, which shown in Figure 2, were enlarged and redrawn in Figure 3. To characterize this factor, the Sharpness was defined as a time, the duration of temperature which is below 50°C of peak point. As shown in Figure 3, the Sharpness is the bottom width of the colored area which can be treated as the total thermal heated from peak temperature (T_{peak}) to T_{peak}-50°C.

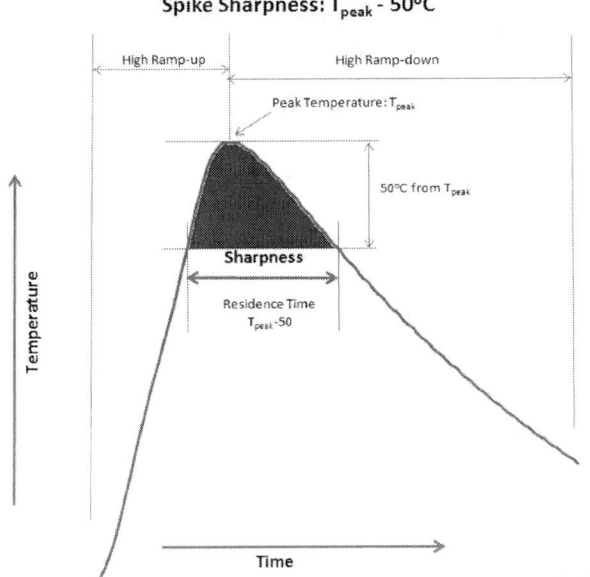

Fig. 3, Spike sharpness definition: the duration time of temperature below peak point 50°C.

Sharpness was considered as the key factor for both recipe setup and its maintenance, because it directly influences devices electronic characteristics which strong correlated with SDE performance. Figure 4 shows some experiments results of Sharpness splits on blank wafer with same low energy implantation processes. The recipes were set up with various residence times tuned by changing the cooling rate, since ramp-up rates above 200°C/sec do not change sharpness obviously. Different cooling rates were achieved by adjusting helium flow in the recipe. There were five residence times adopted to investigate the influence of junction depth and sheet resistance. From these experiments, about 15% of Sharpness variation will cause about 5% of resistance and 10% of junction depth shift, respectively. Therefore, to tune and control the Sharpness as well as the peak temperature, will be the necessary choices for recipe setup and maintenance.

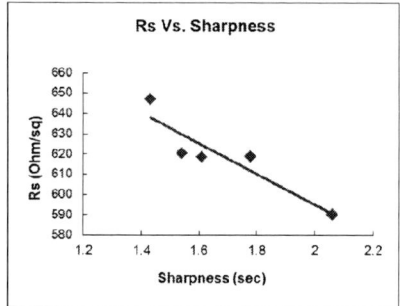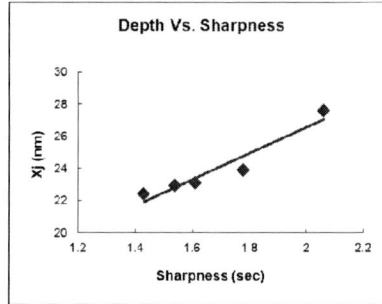

Fig. 4, Resistance and Junction Depth versus Sharpness with same processed substrate.

Precise Control of Residence Time

With CMOS continue scaling down to 65nm, even to 45nm, to precisely control the SDE lateral abruptness (defined as the distance over dopant concentration drops to 1/10, its unit is nm/decade) and SDE-to-gate overlap, shown as y_j in Figure 1, promote more challenge for annealing process. Figure 5 shows one Source/Drain resistance model, in which, four components of this series resistance included: the accumulation resistance, the spreading resistance (which affected by the abruptness of the SDE junction), the sheet resistance of the SDE region, and the contact resistance.

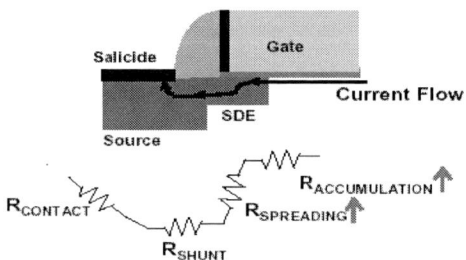

Fig. 5, Source/Drain resistance model: $R_{ext} = R_{co} + R_{sh} + R_{sp} + R_{acc}$

In order to meet the overall resistance requirements, the SDE sheet resistance value must be optimized together with the contact resistance and junction lateral abruptness.

For spike anneal process, to meet this aggressive requirements, not only the T_{peak} and Sharpness must be considered, but also the temperature shape below Sharpness need to be controlled, as color marked in Figure 6.

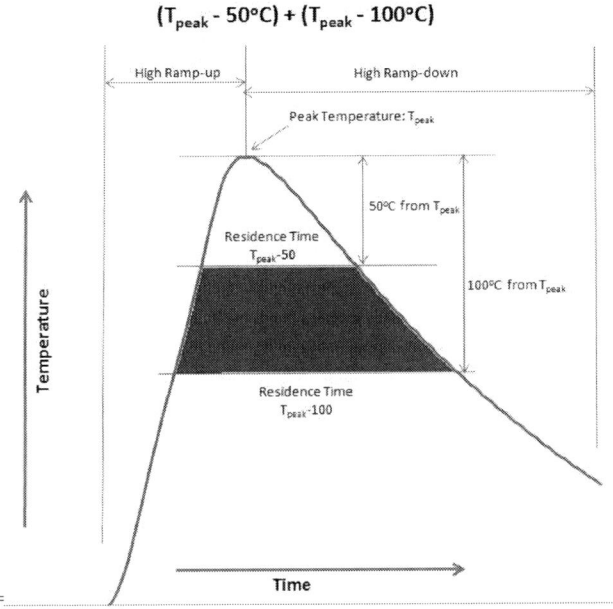

Fig. 6, Precise calculation: sum of two times, the duration of temperature below peak point 50°C and 100°C.

The sum of two values, time of T_{peak}-50°C (Sharpness) and time of T_{peak}-100°C, was treated as a controlled factor for spike process setup and maintenance. As shown in Figure 6, the trapezoidal area was defined by Sharpness (top-width) and time of T_{peak}-100°C (bottom width), therefore, the sum of these two time values, the colored area, indicate the total thermal heated between T_{peak}-50°C and T_{peak}-100°C. To precisely control this factor, the high ramp-up step, shown in Figure 3, has to be separated into two or more steps setting with different ramp-up-rates, for minor adjusted and optimization. Meanwhile, the cooling steps were fixed for easier maintenance.

With device further scaling down, the Laser-based anneal was introduced into industry.

Acknowledgments

The authors would like to thank the learning and knowledge sharing from Dr. Wang Chenyu, Dr. Zhou Qinggang and process expert Yang he.

References

1. D.F. Downey, et al., *7th Int. Con. on Advanced Thermal Processing of Semiconductor*, p. 229, 1999.
2. S.C. Jain, et al., *J. Appl. Phys*, p.8919-8941, Vol.91, No. 11, 1-June, 2002.
3. H.-J. Gossmann, *Semiconductor Silicon / 1998*, H.R.Huff, U.Goselle, and H.Tsuya, Editors, ESC Proc. Vol.98-1, 884, 1998.
4. H.-H. Vuong et al, *J.Vac.Sci.Technol*, p.428-434, B 18(1) Jan/Feb, 2000.
5. P.Timans, *4th Int.Con. on Advanced Thermal Processing of Semiconductors*, p. 145, 1996.

Improving Copper Interconnect Reliability via Ta/Ti Based Barrier

Xiao-Wen Hu[1], Paul-Chang Lin[1,2], Jenny Ma[1], Jian-Yong Jiang[1], Peng He[1]

1. SMIC, 18,Zhangjiang Rd, Pudong, Shanghai, PRC, 201203
2. Department of Microelectronics and Solid-State Electronics, Fudan University, Shanghai

As copper line width tightens to 100nm and below, except integration issues, reliability is another challenge. Barrier's performance is critical for the significant impact to interconnect speed and reliability. Ti based barrier is reported as an excellent barrier material from the standpoint of cost and performance, especially for the porous low-k ILD materials. While Ta based barrier is used mainly for its good adhesion, diffusion prevention. In this paper, we demonstrate a composite barrier with Ti adding to the Ta based bi-layer, which shown stable metal resistance while significantly improving reliability performance. Better interface is confirmed by TEM. Our study proved that such kind of Ti doped Ta barrier is better than the standard Ta N/Ta barrier.

Introduction

With copper metallization, the interconnect geometries are rapidly approaching nanoscale dimension, the barrier with comparatively higher resistance became one focus of research for its significant impact to interconnect speed and reliability. Numerous approaches have been taken to optimize the barrier. There are mainly two means: the engineering of the deposition process and the designing of the new type barrier. The former optimized the metal barrier's microstructure and morphology, and the later focused on developing new materials used as barrier.

To solve the integration problems of Cu interconnects and improve reliability performance, tantalum (Ta)-based barrier is typically used due to its good adhesion, Cu diffusion prevention, and Cu seed nucleation underlayer. While recently, titanium (Ti)-based barrier has regained attention for its compatible with porous ultra low-k in advanced logic products and significant lower cost in memory products applications compared to tantalum (Ta) barrier. And also several researches have focused on the integration issues associated with Ti (1,2,3). In this paper, we introduced a new type barrier which is a composite barrier with Ti adding to the Ta based barrier. Besides metal resistance, reliability is another important index for us to do comparison with the standard Ta barrier.

Experiment

Design of Experiment

Using dual-damascene copper metal lines in the 90nm technology node, the barrier process of both the Ti doped barrier and the standard TaN/Ta barrier were deposited by PVD. The different Ti doped percentages containing in the standard TaN/Ta barrier are listed in table 1 (split 1 is the standard TaN/Ta barrier, and split 2 to 4 are corresponding to the different Ti doped percentages in the standard barrier). The initial study was taken by four-point resistance measurements. The interfaces of the barrier with upper and lower layer were characterized by EM (Electro-migration) test. And also the samples were monitored with SEM (scanning electron microscopy) and EELS mapping.

TABLE 1. Process sequence splits for Ti/Ta barrier and Ta barrier

Split	Ti % *
1	0
2	0.11
3	1.15
4	2.05

*, weight %

Experimental Method

1) AMAT Endura5500 Bara&Seed system in standard PVD process condition.
2) KE DD-835V furnace system in standard alloy process.
3) FEI TF20 in TEM test.

Results and discussion

Interface of barrier and seed

As a metal barrier, ensuring a good interface with the copper seed is the first factor to concern. From TEM cross-section check (Figure 1), there is no significant difference among those splits which proved that such lower Ti doped percentage will not cause process problem

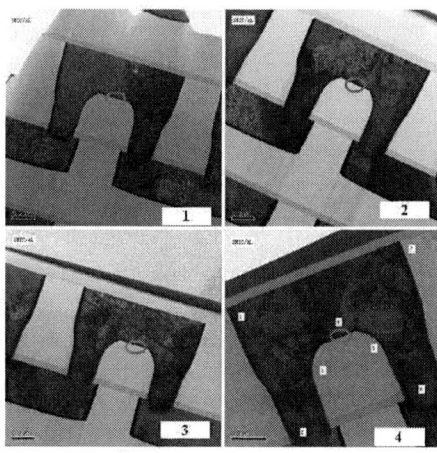

○ -- EDX analysis point

Figure 1 interface of Cu and barrier with different content of Ti

Resistance results

Considering Ti diffuses into Copper upon thermal treatment to cause excessive resistance increase, the resistance of via and metal were firstly chosen to compare. The metal sheet resistance of different splits is shown in Figure 2. Although the resistances value of all the splits are within spec, it trends up obviously along with the content of Ti increases, especially in split 4 when Ti content reaches to 2.05%, the resistance jumps significantly with larger standard deviation (listed in table 2).

Similarly, in the via resistance comparison (Figure 3), split 4 still exhibits a relatively higher resistance than other splits, while split 2 and split 3 are almost comparable with baseline split 1.

Based on those observations, we will dedicate on one of the split 2 or 3 to compare with baseline split 1 for the further study because split 4 may induce the R*C product with relatively larger resistance although within spec.

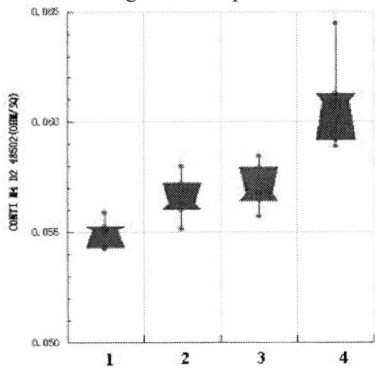

Figure 2 CONTI data of the splits with different content of Ti

TABLE 2. CONTI data in details corresponding to Figure 2

Split	Ti %	N	Avg	Std
1	0	5	0.0549	0.00069
2	0.11	5	0.0565	0.00110
3	1.15	5	0.0570	0.00111
4	2.05	5	0.0609	0.00222

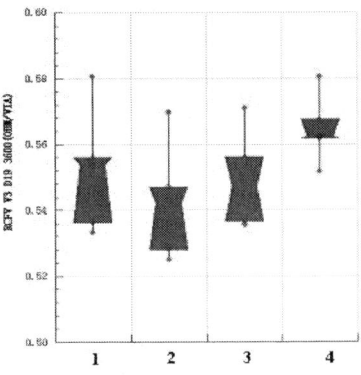

Figure 3 RCFV data of the splits with different content of Ti

TABLE 3. RCFV data in details corresponding to Figure 3

Split	Ti %	N	Avg	Std
1	0	5	0.5518	0.01896
2	0.11	5	0.5423	0.01787
3	1.15	5	0.5493	0.01470
4	2.05	5	0.5647	0.01053

Electro-migration Results

The package level EM test was performed with Qualitau system for via-terminated downstream structures (V2D). Based on the above resistance results of the splits, we chosen split 3 to do comparison with split 1. In figure 4, the MTTF of the sample with Ti doped barrier is more than two times larger than baseline, which has the same phenomenon with the study by W.Wu (1). Failure analysis has been done for the failed sample of split 3. FIB and TEM image show M2 void under cathodal via2 which is the normal EM failure location for Copper downstream structure (Figure 5). Moreover, EELS mapping analysis shown the obvious Ti segregation at the side wall, but not conspicuous on the bottom (Figure 6). We explain it for the re-sputter process to make the barrier very thinner at the bottom and the EQ limitation of the poor resolution for low concentration material during mapping. To the best of our knowledge, there two mechanisms are most likely to responsible to the improvement of split 3. The first is that Ti segregates at the Cu/SiC interface, gettering oxygen which can improve the adhesion of the interface (1). The second mechanism is as reported by Ueki et al (4), Ti/Cu adhesion is better than Ta/Cu adhesion through a better strengthened interface.

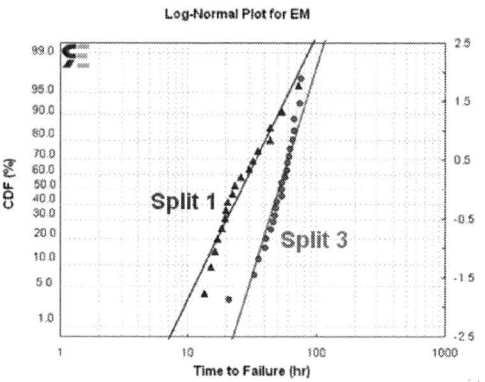

Figure 4 Lognormal distribution for EM data on V2D structure

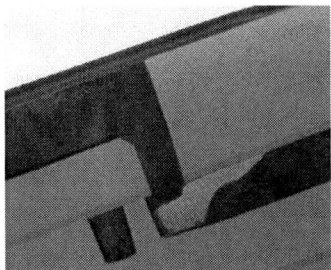

Figure 5 Cross-sectional TEM image show M2 void under cathodal via2 of split

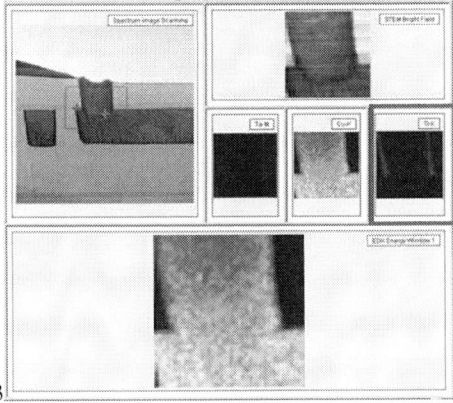

Figure 6 EELS Ti map on split 3. Ti segregation at the side wall is obvious.

Optical Microscope (OM) after Alloy

Once barrier layer is damaged, copper can diffuse to PEOX and diffuse along the interface of metal line. And then CuOx is visible from the top view by OM (Figure 7). This diffusion will be more serious especially after alloy stage because the high temperature (about 450°C) can exacerbate copper diffusion. In our experiment, the sample of split 3 was monitored by OM after once, twice and three times alloy process, respectively. Figure 8 shows that there is no copper diffusion after even three times alloy for 1.15% Ti concentration. This result provides another evidence for the good resistance of this new type barrier.

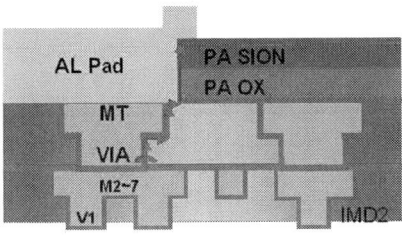

Figure 7 the flow of copper diffusion

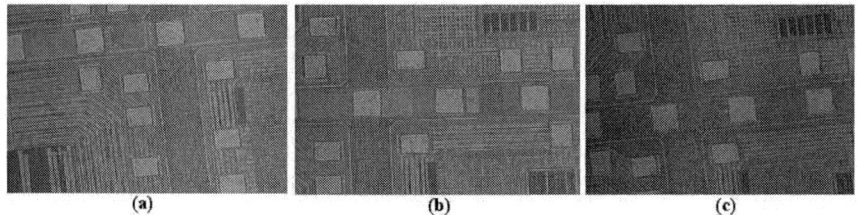

Figure 8 OM results of Split 3 after (a) one time alloy;
(b) two times alloy; and (c) three times alloy

Conclusion

In this experiment, the incorporation of Ti demonstrates a new type of composite barrier. Evidence indicates that this Ti containing barrier provides improvement in EM performance as compared with standard TaN/Ta barrier. And this barrier also showed good resistance even at high temperature.

References

1. W.Wu,et al, Ti-based Barrier for Cu Interconnect Applications, Novellus Systems Incorporation, IEEE Trans. Electron Devices, 2008, p202
2. Sridhar K. Kailasam, INOVA Copper Barrier/Seed for Memory Application, SEMICON Korea 2008, SEMI Technology Symposium (STS)
3. A.Sakata,et al, Reliability Improvement by Adopting Ti-barrier Metal for Porous Low-k ILD Structure, Toshiba Corporation, IEEE Trans. Electron Devices, 2006, p101
4. M. Ueki, et al, Suppression of strss-induced open failures between via and Cu wide line by inserting Ti layer under Ta/TaN barrier, IEDM 2002, pp. 749

Glue layer study of inter via between Cu and Al metal lines

John Chen ,Cheney Qiao, Linhong Yang, Kevin Chang

SMIC (Semiconductor Manufacturing International Corp) TJ
19,Xinghua Rd, Xiqing, Tianjin 300385,PRC

ABSTRACT

A 0.13um Cu-Al hybrid backend between a dual damascene process with Cu line and a conventional process with Al line is introduced in this paper. First 3 metal layers (M1~M3) interconnects with Cu line for process capability and circuit speed concern. Whereas following metal layers (M4~M7) interconnects with Al line for cost concern. The interconnection via between Cu and Al interface is the key process for Cu-Al hybrid process. This paper mainly focused on via process study between Cu and Al interconnection.

INTRODUCTION

Most 0.13um node and beyond technologies' interconnect process is Cu-based for lower RC delay concern, comparing with Al-based interconnect process. But some fabs are interest in Al based possess for cost concern; therefore Cu and Al hybrid interconnect process is worthy of developing.

Considering some IPs developed based on Cu-based interconnect process are used in 0.13um technologies and the first three metal layers(M1~M3) will be used for interconnect of IPs. We kept first three interconnect layers for Cu to ensure IPs' performances. Other metal layers will be used for global connect between IPs , we used Al-based interconnect process to form these metal layers in this process.

0.13um Cu-based process is matured and some 0.13um Al-based processes are also been published. So the key processes in 0.13um hybrid process is the interface between Cu and Al interconnect This paper focused on the interface process between Cu and Al interconnect process.

2. EXPERIMENTS

General 0.13um Cu-based interconnect process film structures of dual damascene process in Figure1 where Ta/TaN glue layer is used. General 0.13um Al based interconnect process film structures in Figure 2. where W plug with Ti/TiN glue layer is used.

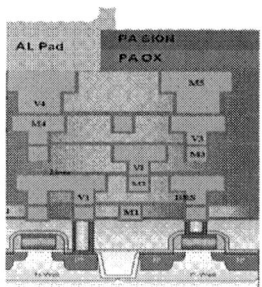

Figure1. Cu process film structure

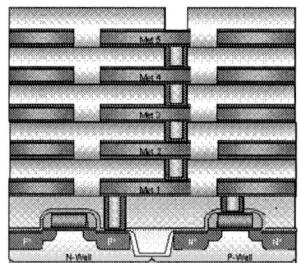

Figure2. Al process film structure

Designed Cu and Al hybrid interconnect process film structures as Figure 3. First there layers(M1 to M3) are standard Cu-based interconnect processes and other metal layers are standard Al-based interconnect processes. W with glue layers is used for interface between M3(Cu) and M4(Al).

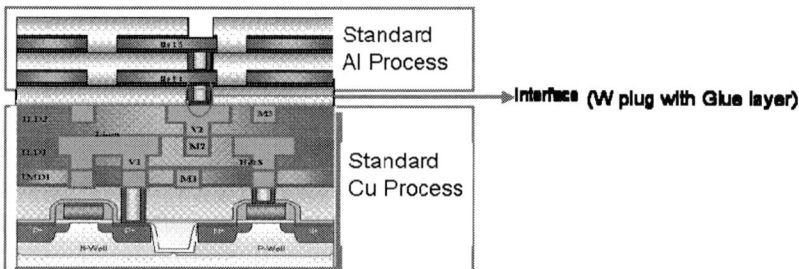

Figure 3. Hybrid interconnect process film structure

The most important film in this structure is the glue layer between W and Cu. Some experiments are designed to get a suitable glue layer.

Experiment design:

Ta/TaN is widely used as glue and barriar layer in Cu-based interconnect process. Its glue and barrier performance is widely reported [1] and verified by mass production. So Ta/TaN is selected for this experiment. TiN is a good barrier layer material for Al based interconnect process and has been widely used for this application, while it also performed well as a barrier layer for Cu. Its behavior as an adhesion-layer and seed-layer for Cu is not as good as Ta-based material. Hence, TiN deposited by either PVD or CVD hasn't been widely adopted for use in Cu interconnect metallurgy [2] .In this case, we just need TiN's barrier function to prevent Cu diffuing. So TiN is also selected as one optional glue layer material. Considering the gap filling performance, we selected CVD TiN.

In this test, we selected several combination as glue layers,

Choice 1: Ta/TaN, general Cu interconnect process glue layer

Choice 2: TaN/TiN, TiN is deposited with MOCVD method

Choice 3: TaN/Ta/TiN , TiN is deposited with MOCVD method

Choice 4: TaN/Ti/TiN, TiN is deposited with MOCVD method

Choice 5: Ti/TiN, general Al interconnect process glue laye. TiN is deposited with MOCVD method.

After the glue layer material is selected, the whole structure is designed as Figure4 to check the interconnect performance .

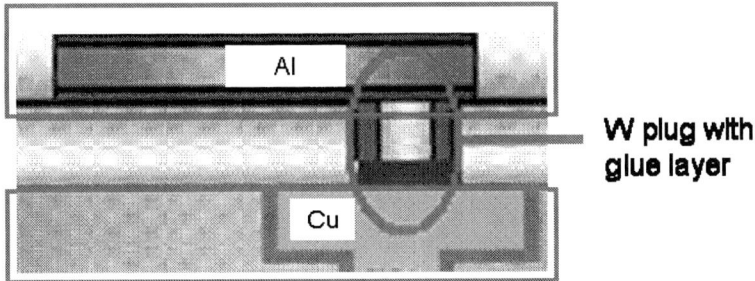

Figure 4. Test structure for hybrid interface layer

Experiment Result:

TEM cut result comparison between different glue layers was shown in Figure 5. Without TiN (Ta/TaN condition) show Cu void induced open while with TiN condition(TaN/TiN, TaN/Ta/TiN,TaN/Ti/TiN,Ti/TiN) TEM pictures show W and Cu contact well via glue layers. Without TaN condition (Ti/TiN) show smaller contact resistance while with TaN conditions show much bigger contact resistance.

Glue layer Condition	TEM		Contact Resistance
	TEM Pictures	Comment	
Ta/TaN	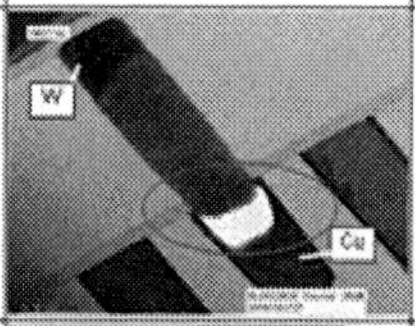	Cu Void	Open
TaN/TiN		OK	300 Ohm/sq
TaN/Ta/TiN	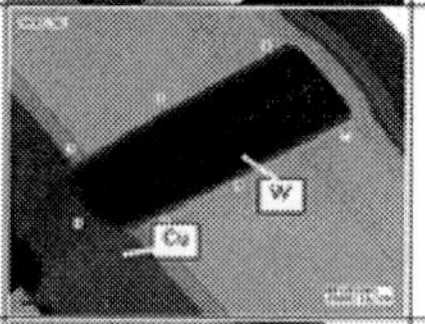	OK	40 Ohm/sq
TaN/Ti/TiN	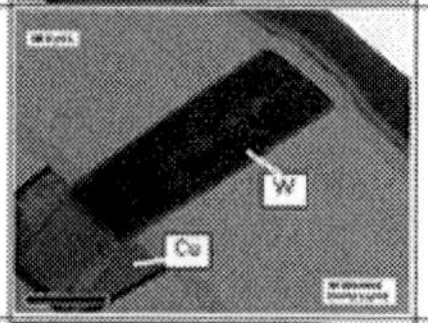	OK	45 Ohm/sq
Ti/TiN	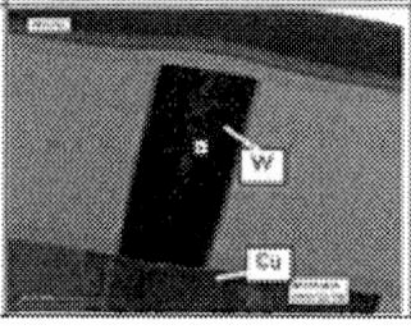	OK	4 Ohm/sq

Figure 5. Test structure contact resistance and TEM pictures

Ti/TiN performs well as a barrier layer to prevent Cu from diffusing out while it provided low contact resistance as glue layer.

CONCLUSION

To carry out Cu and Al hybrid process, the interface between Cu and Al process is the key factor. The paper focused on a structure with W plug and glue layer to connect Cu and Al. By TEM and electrical parameter test, Ti/TiN performs as a good barrier layer to prevent Cu from diffusing out and provides low contact resistance as glue layer. Thus provide a potential way to carry out Cu and Al hybrid process.

ACKNOWLEDGMENTS

Authors would like to thank our supervisors Tengkuet_Lo, Assistant Director of SMIC-TJ, for his encouragement and guidance. In additional, authors would give special thanks to SMIC-S1 ,for their technical support and Cu process wafers supply

REFERENCES

[1]   C.Ryu et al , Proc.Mtg. *Material Res. Soc*, Spring 1998. p.75
[2] C.Ryu et al *"Barriers for Copper interconnects"*, *Solid State Technology*, April.1999,p.53.

The Influence of The SIN Cap Process on The Voltage Breakdown and Electromigration Performance of Dual Damascene Cu Interconnects

Yalu Cao, Charles Xing, Neil Xu, Hua Zhou, Allen Bian, Paul-Chang Lin
SMIC, 18,Zhangjiang Rd, Pudong, Shanghai 201203,PRC

ABSTRACT

The influence of the SiN cap-layer deposition process including different pre-clean treatments on the voltage breakdown (VBD) and electromigration (EM) behavior of copper dual damascene metallization has been studied. A remarkable improvement for voltage breakdown and electromigration were revealed depending primarily on the pre-clean treatment before cap-layer deposition rather than the deposition process itself.

On one hand an "aggressive" pre-clean treatment yields improved Cu/SiN-interface properties with higher electromigration failure times and activation energies (1.23 ... 1.27eV). On the other hand these pre-clean treatments were found to induce voltage breakdown failures because of the redundant Cu/Cu^{2+} or Copper Silicide defects induced in the Dielectric (FSG)/SiN-interface during the plasma pre-clean treatment. The micro damage on surface and copper Silicide was found to increase with the pre-clean intensity. In contrast, No VBD risk is related to "less aggressive" pre-clean treatments.

The results indicate the need to adjust the SIN cap-layer process parameters with respect to both VBD&EM performance to meet the overall reliability requirements.

INTRODUCTION

The interface between the copper and the capping layer is to be the dominant diffusion pathway in copper dual damascene interconnects. Therefore, the properties of this particular interface play a key role for wear-out mechanisms such as electro-migration and voltage breakdown, limiting the life time of complex integrated copper interconnect systems.

SIN- and Sic-based films are widely used as capping materials for copper interconnects. It has been published in several studies that the EM life time depends on the adhesion behavior, i.e. the sticking coefficient between cap-layer and copper surface. Good adhesion and hence large EM life times are enabled by a tightly bonded interface which suppresses the migration along this pathway. In contrast, a poorly adhering interface would correspond with enhanced diffusion and lower EM life times [1,2].

One of the major tasks during process development is to find cap layer materials as well as pre-clean and deposition techniques that yield good interface adhesion to obtain adequate EM performance. It's found that P-SIC caps show improved EM life times compared to P-SiN[3]. In contrast, Martin et al. [4] observed a similar EM behavior for both SiN and SiCN caps, but obtained a significant improvement when applying a plasma pre-clean treatment before the SiCN cap-layer deposition. The importance of a pre-clean e.g. a NH, - or NH, M,-plasma treatment, has been highlighted also by Lloyd et al. [2,5,6]. The key point here is to remove a Cu-oxide film that has possibly formed on the copper surface before the cap-layer deposition [7].

Another way to improve EM resistivity is the integration of selectively-deposited cap-layers e.g. CoWP [2,8] or pure metals such as Pt, Pd, Ag, Au, Rh or Ir [9]. These materials promote the adhesion to the copper surface and allow a more effective suppression of Cu ion migration along this interface in comparison to conventional SIN- or Sic-based capping layers.

All mentioned publications including our earlier studies discuss the influence of the cap-layer deposition process only with respect to the EM behavior. In this paper comprehensive studies were performed to investigate both the EM and VBD performance dependent on the pre-clean treatment and the deposition technique of the SIN cap-layer, respectively. It will be shown that despite of excellent EM results some of the pre-clean treatments are not suitable as standard processes since they induce voltage breakdown failures because of the redundant Cu/Cu2+ or Copper Silicide defects in the Dielectric (FSG)/SiN-interface during the plasma pre-clean treatment .

EXPERIMENTAL

Copper dual damascene samples of a 0.13um technology with a TaN/Ta liner system were processed using different cap pre-clean treatment (PCT), gas flow and SIN cap deposition (DEP) conditions (Tab.1).This table is based on and refers to Taguchi-method. The pre-clean treatments PCT1 and PCT2, PCT3 are different mainly in terms of the plasma power. The power of the PCT1 is considerable larger compared to PCT2 and PCT2 is larger than PCT3. In addition, in PCT1 the SiH4 flow, leading gas for DEP, varied from "high" to "low". The processes DEP1, DEP2 and DEP3 represent different plasma techniques and SiH4 flow for the SiN cap-layer deposition. All other processes such as Cu plating, anneal and CMP are identical for the splits in this paper.

Experiment No.	Pre-Clean treatment power	Leading SiH4 in Pre-Clean treament	DEP treatment power	DEP Gas(SiH4)
A	PCT1	High	High	High
B	PCT1	Median	Median	Median
C	PCT1	Low	Low	Low
D	PCT2	High	Median	Low
E	PCT2	Median	Low	High
F	PCT2	Low	High	Median
G	PCT3	High	Low	Median
H	PCT3	Median	High	Low
I	PCT3	Low	Median	High

TAB.1 Pre-clean treatment and SIN-cap deposition process
for Electromigration and Voltage Break Down

Electromigration kinetic studies were carried out on package-level in a temperature range from 200 to 400°C applying current densities between 5 and 20mA/um^2. Test structures with a "fully landed" via (i.e. line width = 0.6um equals approx. 3x via diameter) were stressed with the electron flow direction from the uncritical metal line over the via into the critical metal line underneath (Fig.1). Here, the voiding occurs in the metal line under the via. This test structure is sensitive with respect to the cap-layer quality and therefore the test configuration is especially suitable to detect process-related changes of the Cu/SiN-interface properties [10].

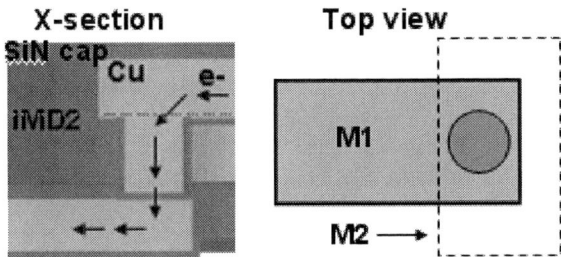

FIG.1 Test structure for EM studies(X-section and Top view)

In parallel to the EM tests, the voltage breakdown behavior was monitored. The structure consists of two groups of metal lines isolated by oxide and SiN cap-layer (Fig.2). A ramping up voltage is forced on the end of two groups (A and B). In case of detecting current between A and B, the voltage is regard as Voltage of Break Down.

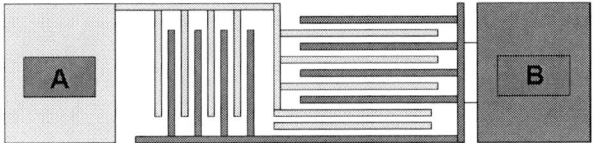

FIG. 2 Voltage Break down test structure. A ramping up
voltage is forced on A and B

Results and Discuss

The electromigration failure times turned out to be significantly higher on all those splits (G, H, I) which are subjected to pre-clean PC3 (Tab.2). In addition, the

Experiment No.	MTTF (mean time to failure)(hrs)	Ea(eV)	n
A	70	1.05	1.4
B	67	1.05	1.5
C	69	1.06	1.4
D	178	1.13	1.4
E	180	1.12	1.5
F	182	1.14	1.4
G	272	1.24	1.5
H	266	1.27	1.4
I	276	1.23	1.4

TAB.2 Eletro-Migration mean time to failure(MtTF), Activation Energies
result based on Pre-clean and SIN-cap deposition split table

activation energies (Ea=1.23 ... 1.27eV) are about 18% higher compared to splits (A, B, C) treated by PC1(1.05 ... 1.06eV). The current density exponents are almost independent of the pre-clean treatment. In contrast, no systematic correlation was found between EM performance and the SIN cap deposition process DEP1 vs. DEP2.

The VBD behavior was found to be influenced by the pre-clean treatment rather than the cap-layer deposition itself. VBD change (based on the average of all result) has no obvious correlation with deposition, including plasma power and gas flow (Fig.3). Whereas, pre-clean treatment change will obtain a tremendous improvement of VBD(Fig.4). It increase from -7.78% to 44.30% with the power of pre-clean treatment drop from median to low level. And the same trade-off was found on gas flow change. VBD trends to 61% from -1.8% when gas percentage reduces from median to low.

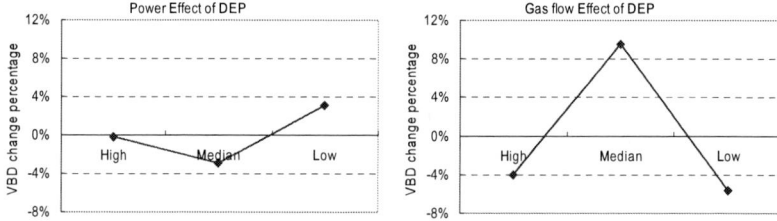

FIG.3 The effect of different deposition condition

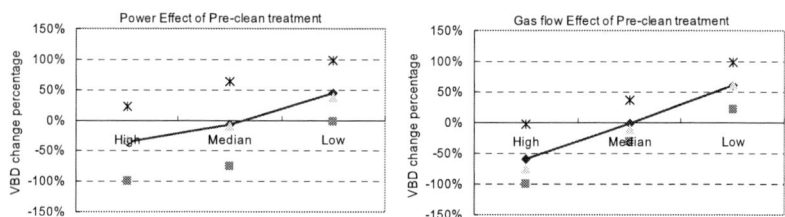

FIG.4 The effect of different deposition condition

Moreover, Based on this series of experiment, it was found that high SiH4 and plasma power on pre-clean treatment with very low VBD result often induced a new material copper silicide. This material, which formed by Cu/Cu2+ and SiH4, can be identified clearly SEM and TEM (Fig.5).

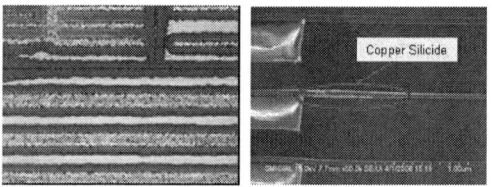

Fig.5 Copper Silicide top view SEM and X-section TEM

The formation of the copper Silicide is illustrated in Fig.6. At first, H or H-decomposed from NH3 under plasma assist, and reacted with CuO2, dioxided to Cu and H2O. Once CuO2 was eliminated drastically, Cu atoms in the surface will suffer ion bombard and get energy. If the energy is large enough, the Cu atom may sputter into plasma or onto dielectric neighboring Cu lines. And only small energy more will make Cu ion diffuse far from barrier interface into the interface between Cu line and SiN cap-layer. These dissociative Cu atoms will loss their electron and become to Cu ion while there are high electric negative element, such as fluorine in FSG. Furthermore, when SiH4 is enough and redundant in the reaction chamber, copper react with SiH4 under high plasma and copper Silicide come into being at last.

Before SiN cap-layer deposition copper is exposed on plasma, so higher plasma will sputter more Cu/Cu2+ to dielectric, which will add the probability of short between Cu lines. Lower plasma power sputter Cu less and copper Silicide less, so it was found that the variance of VBD is less under low plasma than under high plasma(Fig.4 left bottom chart). The rich SiH4 accelerate the formation of copper Silicide on dielectric and it will link the isolated Cu/Cu2+ into island, continuous film further. And the continuous copper Silicide induce VBD drop down significantly.

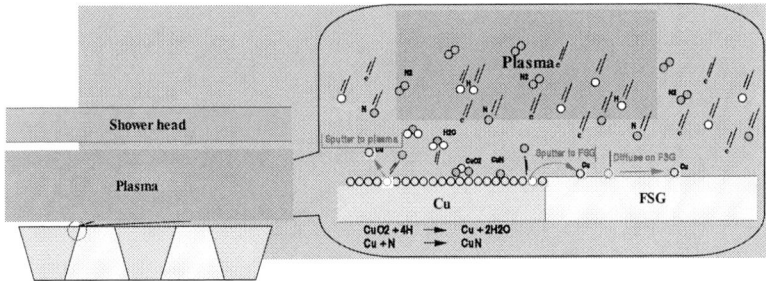

Fig.6 Copper Silicide formation mechanism

CONCLUSION

Comprehensive studies were performed to investigate both the EM and VBD performance dependent on the pre-clean treatment and deposition technique of the SiN cap-layer, respectively. The investigations revealed a trade-off between the EM and VBD performance primarily depending on the pre-clean treatment rather than the deposition process. On one hand an "aggressive" pre-clean treatment yields improved Cu SiN-interface properties with significantly higher electromigration life times and activation energies. On the other hand these pre-clean treatment were found to cause failures due to voltage break down because of Cu/Cu2+ and copper Silicide formation on dielectric. The studies show the importance to optimize the SiN cap-layer deposition process not only in terms of the electromigration performance but also with respect to the voltage break down behavior to meet the overall reliability target of the products.

ACKNOWLEDGMENT

The authors would like to express gratitude to all process engineers who do great work for this project in CVD, SMIC.

REFERENCES

1. N.E. Meier et al., "In situ studies of EM voiding in passivated Cu lines", Stress-Induced Phenomena 1999
2. J. Lloyd et al., "'Relationship between Interfacial Adhesion and Electromigration in Cu Metallization", proceedings IRW 2002
3. M. Hatano et al., "EM Lifetime improvement of Cu Damascene Interconnects by P-SIC Layer"; proceedings IITC2002, pp. 212-214
4. Martin et al., "Integration of SiCN as a low-k Etch Stop and Cu Passivation in a High Performance low-k Interconnect"; proceedings IITC2002, pp. 42-44
5. Parikh et al., "Defect and Electromigration Characterization of Two level Copper Interconnect", proceedings IITC 2001
6. R. Gonella, STM, "Cu quality and interconnect environment impact on Cu Dual Damascene electromigration performances", 2002 SEMI, Tutorial at the SEMICON Europe
7. Goldberg et al., "Interface Reliability of High Perfvrmance Interconnects", to be published in proceedings of the Advanced Metallization Conference 2002
8. P. Andricacos, "Copper On-Chip Interconnections", The Electrochemical Society Interface, Spring 1999
9. J.A. Cunningham, "Using electrochemistry to improve copper interconnects", Semiconductor International 2000
10. A.H. Fischer et al., "Electromigration Failure Mechanism Studies on Copper Interconnects", proceedings IITC 2002

Effect of RF Power on Carbon Nanotubes Synthesized at Low Temperature by RF PECVD

Xinyuan Lin, Kailiang Zhang*, Kai Hu, Xiaoyong Qiang, and Shiwei Wang

School of Electronics Information Engineering, Tianjin Key Laboratory of Film Electronic & Communication Devices, Tianjin University of Technology, Tianjin 300384, China, *corresponding author , kailiang_zhang@163.com

The growth of carbon nanotubes (CNTs) at low temperature is one of the key obstacles on currently application in electronic device interconnection. In this paper, CNTs were grown from acetylene and hydrogen gas mixture on Si substrate by RF-PECVD, in which Fe was selected as catalyst. The effect of RF power on CNTs at low temperature was mainly discussed. SEM and Raman spectra were used to characterize the morphology and defects of CNTs. Results showed that the optimum RF power for CNTs growth at low temperature (450°C) is 100W. When the RF power is too low, it can't provide enough energy for acetylene decomposition. When the RF power is too high, there are two aspects reasons that affect CNTs growth. One is that high density plasma can etch new-grown CNTs, and another reason is that lots of carbon atoms in the chamber are easy to make catalyst poisoned.

Introduction

With the development of integrated circuit (IC) process, the feature size of IC will become smaller and smaller, and current density of interconnects will surpass the maximum carrying capacity of Cu wire which is widely used. Electro migration and thermal effects caused by high current density will make the device invalid, so it is necessary for IC industry to explore new interconnect materials(1). Since carbon nanotubes (CNTs) were discovered in 1991(2), they have attracted worldwide attention and research because of their unique mechanical, electromagnetic and thermal properties. CNTs are widely used in electrical field due to their high electrical conductivity, high stability and other excellent characteristics(3-5). Some research has reported that multi-walled carbon nanotubes (MWCNTs) can continuously work more than 350 hours at $10^{10}A/cm^2$ and the current transport capacity is not degraded.

Though CNTs have been considered as a candidate for next generation interconnects application, there are many problems for the practical application of CNTs in IC interconnects. For example, CNTs with low bundle density or a number of defects are not suitable for IC interconnects, because they will lead to greater resistance than traditional interconnect materials. What is more, CNTs which are non-aligned cannot be applied to the integrated circuit interconnects, either. Therefore, preparation of aligned CNTs with few defects and high-density is the main problem that needs to be solved in IC interconnects(6-8).

The growth of CNTs is affected by many factors, such as substrate temperature, chamber pressure, RF power and so on. In this report, CNTs were synthesized by radio frequency plasma enhanced chemical vapor deposition (RF-PECVD), and effect of RF power on the growth of CNTs at low temperature was investigated.

Experimental

In this experiment, a catalyst was required in order to grow CNTs on Si surface (9). The catalyst (Fe, 5nm) film was deposited on polishing Si (100) substrates by ion beam sputtering system (FJL560, CAS Shenyang Scientific Instrument Center) at room temperature. Then, the samples were introduced into a RF-PECVD chamber (PECVD-II, CAS Shenyang Scientific Instrument Center). Firstly, the chamber vacuum was 1×10^{-3}Pa, and the substrate temperature was increased to the growth temperature with hydrogen (H_2) gas flow (40sccm), to prevent oxidation of the catalyst. Prior to the growth of CNTs, catalyst thin film was pretreatment under hydrogen plasma for 10 minutes (10). The RF power used was 300W and the pressure was 200Pa while the catalyst was pretreated. After the pretreatment H_2 flow increased to 60sccm, acetylene (C_2H_2) flow (20sccm) was admitted, chamber pressure increased to 900Pa and then the measurement of growth time started. The substrate temperature of the catalyst pretreatment and CNTs growth was both set 450°C, and was monitored by a pyrometer.

After the growth process was done, samples were taken out with the chamber temperature natural cooling to room temperature in the protection of H_2. The morphologies of CNTs were observed by scanning electron microscope (SEM, JEOL, JSM-6700F). A micro Raman spectroscopy (LRS-5) was used to analyze the graphitization degree of CNTs.

Results and discussion

Fig. 1 shows the SEM images of samples prepared under different RF power at the substrate temperature of 450°C. Fig. 1(a) shows the surface morphologies of the sample prepared at 50W. As seen, there is almost no CNTs existence on the sample surface, and catalyst particles can be clearly found. Morphology of the sample prepared at 100W is shown in Fig. 1(b). It can be seen that a large number of CNTs with small diameter disorderly distribute, and a small amount of impurities exist, as previously shown in Fig. 1(b). Fig. 1(c) shows the SEM image of the sample prepared at 200W. Under this RF power, CNTs of lager diameter appears and the density of CNTs is high. Amorphous carbon hardly exists on the sample surface. Fig. 1(d) shows the SEM image of surface morphologies of sample prepared at 300W. A large number of amorphous carbon are existent on the sample surface, while less CNTs exist.

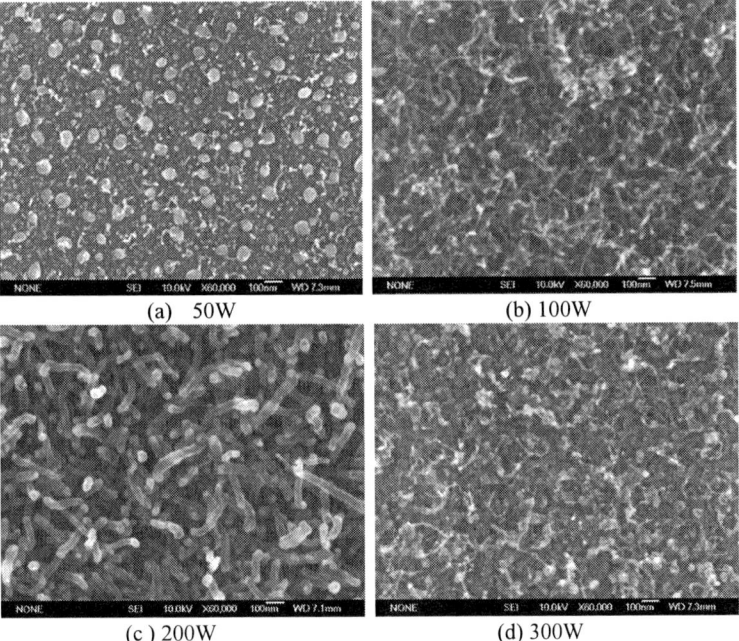

(a) 50W (b) 100W

(c) 200W (d) 300W

Figure 1.   SEM images of CNTs synthesized under different RF power

It can be seen that the morphology of samples is influenced by the RF power. At low temperature, C_2H_2 gas can be decomposed into C, CH, CH_2 and H in H_2 plasma, and the RF power dominates the number of C groups. The growth processes of CNTs can be equal to dissolution and separate out of carbon atoms in the catalyst. In the case of low RF power, such as Fig. 1(a), it can't provide enough energy for C_2H_2 gas decomposition and there is a handful of C groups existing in the chamber. The number of carbon atoms is not enough to separate out in the catalyst, CNTs can't grow. However, the RF power is not as larger as best, such as Fig. 1(c) and Fig. 1(d). When the RF power is too high, there are two aspects reasons that affect CNTs growth. One is that high density plasma can etch new-grown CNTs. CNTs with small diameter are easier to be etched in high density plasma, so there are only some large diameter CNTs existing in Fig. 1(c). And another reason is that lots of carbon atoms generated at high RF power in the chamber are easy to make catalyst poisoned. As seen in Fig. 1(d), a large number of carbon deposited on the surface of catalyst particles, CNTs can't grow. In this report, the optimum RF power for carbon nanotubes growth at low temperature is 100W. It can be seen that a large number of CNTs with small diameter existing in Fig. 1(b).

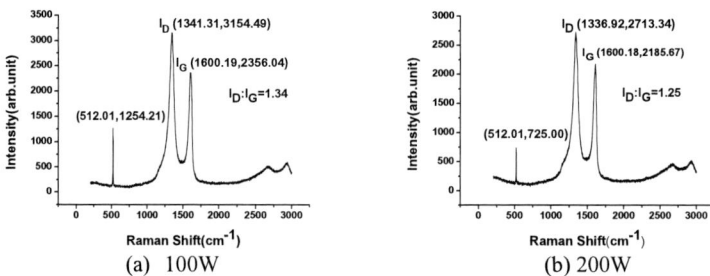

(a) 100W   (b) 200W

Figure 2.   Raman spectra of CNTs synthesized under different RF power

Fig. 2 shows the Raman spectra of CNTs prepared at different RF power. The peak near $500 cm^{-1}$ in Raman spectra indicates Si substrate. The G peak near $1580 cm^{-1}$ in Raman spectra which is generated from Raman active vibration modes E_{2g} indicates that the CNTs are multi-walled. The D peak near $1350 cm^{-1}$ in Raman spectra which is produced from the Raman non-active vibration mode A_{1g} represent the defect and impurity in the CNTs. The peak intensity ratio I_D: I_G represents the graphitization degree of CNTs. The ratio closer to zero indicates higher graphitization degree of CNTs. We can see that the ratio I_D: I_G of CNTs synthesized at 100W is higher than that at 200W from Fig. 2, which indicates that the graphitization degree of CNTs prepared at 200W is higher than that at 100W. That is because when the RF power is high, the amorphous carbon on the samples surface is easier to be etched in high density plasma. This is why the graphitization degree of CNTs synthesized at 200W is higher than that at 100W.

Conclusions

In summary, CNTs were prepared at low temperature (450°C) by RF-PECVD in this work. Results show that RF power has a large effect on the morphology of CNTs. When the RF power is too low (50W), it can't provide enough energy for acetylene decomposition and CNTs can't grow. When the RF power is too high, there are two aspects reasons that affect CNTs growth. One is that high density plasma can etch new-grown CNTs, and another reason is that lots of carbon atoms generated at high RF power in the chamber are easy to make catalyst particles poisoned. In conclusion, the optimum RF power for carbon nanotubes growth at low temperature is 100W.

Acknowledgments

We acknowledge the financial support provided by the National Natural Science Foundation of China under Grant No 60806030, and Tianjin Natural Science Foundation under Grant No 08JCYBJC14600, No 10SYSYJC27700 and Tianjin Science and Technology Developmental Funds of Universities and Colleges under Grant No ZD200709.

References

1. H. Li, C. Xu, N. Srivastava, K. Banerjee, *IEEE Sens. J.*, **56**(9), 1799(2009).
2. S. Ijima, *Nature*, **354**(7), 56 (1991).
3. Y.-Y. Zhang, G.-F. Zou, K. D. Stephen, *ACS NANO*, **3**(8), 2157(2009).
4. S. Hofmann, R. Blume, C. T. Wirth, *J. Phys. Chem. C*, **113**(5), 1648(2009).
5. D. H. Lee, W. J. Lee, S. O. Kim, *Chem. Mater.* **21**(7), 1368(2009).
6. Y.-S. Shi, Y.-C. Ding, H.-Z. Liu, *Appl. Surf. Sci.*, **255**, 7713(2009).
7. M. Meyyappan, L. Delzeit, A. Cassell, *Plasma Sources Sci. Technol.*, **12**, 205 (2003).
8. J.-M. Ting, K.-L. Liao, *Chem. Phys. Lett.*, **396**, 469(2004).
9. Y. M. Shin, S. Y. Jeong, H. J. Jeong, *J. Cryst. Growth*, **271**, 81(2004).
10. G. D. Nessim, A. J. Hart, J. S. Kim, *Nano Lett.*, **8**(11), 3587(2008).

Improving Yield with High-Performance Cables

Paul Warren

W. L. Gore & Associates, Landenberg, Pennsylvania 19350, USA

In the competitive solar and semiconductor industries, high manufacturing yields are very important. The type of cable system used with manufacturing equipment plays an important role in ensuring reliable performance, yet the impact of the cable system is often overlooked during the design process. Cable chain systems can affect yield in two ways: they have a direct impact on the equipment's precision, and they can increase particulation in cleanroom environments. Testing has indicated that eliminating the use of cable chains improved precision and reduced particulation.

Introduction

As the solar and semiconductor industries have become more competitive, higher manufacturing yields have become increasingly important. Yield is affected by both the rate of production and the amount of defective product from the manufacturing line. The production rate can be improved with higher machine speeds; however, precision of manufacturing equipment is often compromised as speed increases. In addition, particles generated by moving components increase the potential for defective product.

The type of cable system chosen plays an important role in ensuring reliable performance of manufacturing equipment, yet its impact is often overlooked during the design process. Selecting the right cable system for solar and semiconductor applications requires an understanding of the relationship between materials and the environment in which they will be used. To ensure that a cable delivers reliable electrical and mechanical integrity, three considerations should be evaluated: (1) the affect of cable systems on the equipment's precision; (2) the impact of continuous flexing on particulation; and (3) the importance of having the manufacturer test the cable system to ensure it will maintain electrical and mechanical integrity in the specific application.

Impact of Cable Systems on Positioning Accuracy

Higher through-put requires precise motion control, and a flex cable system has the most impact on motion control and precision. Increased positioning accuracy allows for higher through-put. A semiconductor manufacturer needed smooth linear motion of the stage but could not achieve the degree of precision required by the equipment. The controller experienced large swings from positive to negative force caused by vibration from a large cable chain with a large number of conductors for power, video, signal, and motion control. The manufacturer designed a test to compare the performance of two cable systems that carried all of the conductors:

- a cable chain system
- a GORE® Trackless High Flex Cable — a self-supported, four-layer planar cable that eliminates the need for a cable chain

<u>Test Setup</u>

The manufacturer designed a test protocol to record the amount of force required to move each system at both low and high speeds. An accelerometer attached to the cable stage measured the pull force required to move each cable system across the stage. Each cable system was tested at a speed of 200 millimeters/second (mm/s) with acceleration of 1 meter per second squared (m/s^2). The tests were then repeated at a speed of 400 mm/s with acceleration of 5 m/s^2.

<u>Test Results</u>

The tests of the cable chain showed significant variations in the force required to move the cable chain at both low speed (Figure 1) and at high speed (Figure 2). At low speed, 1.0 Newton (N) of force was needed to initiate movement. The force increased over the length of travel to nearly 10 N at 0.5 meters. At high speed, almost 3 N of force were required to initiate movement, and for the chain to travel 1.65 meters, the force varied between +6 N and -7 N. The swinging from positive to negative force was caused by the links in the cable chain.

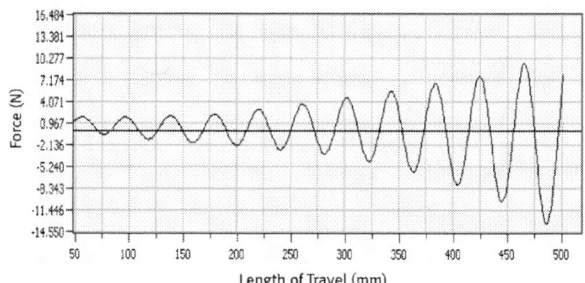

Figure 1. Force Required to Move Cable Chain at Low Speed

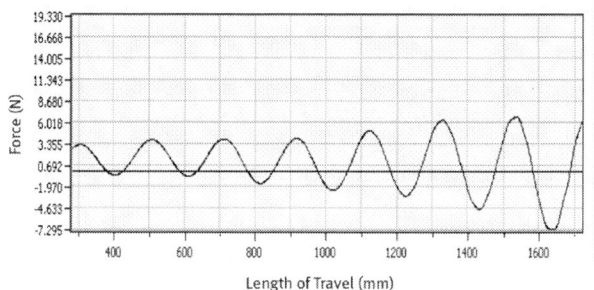

Figure 2. Force Required to Move Cable Chain at High Speed

The tests of the GORE® Trackless High Flex Cable indicated that the amount of force required to move the cable was predictable and linear at both low speed (Figure 3) and at high speed (Figure 4). The force required to continue moving this cable was consistent with no negative force, and the force remained linear as the distance traveled increased. At low speed, approximately 1 N of force was needed to initiate movement and approximately 8.5 N of force at 2 meters of travel. At high speeds, the force to move the cable remained consistent with even less spikes, with 2 N initiating movement and approximately 10 N for the cable to travel 1 meter.

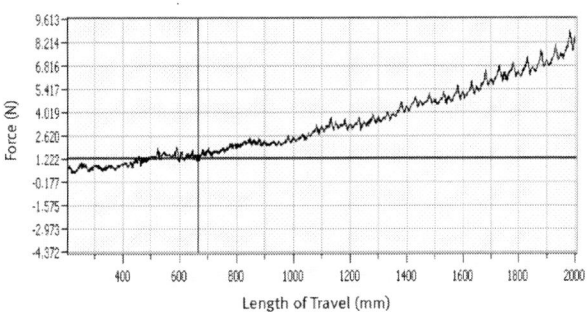

Figure 3. Force Required to Move GORE® Trackless High Flex Cable at Low Speed

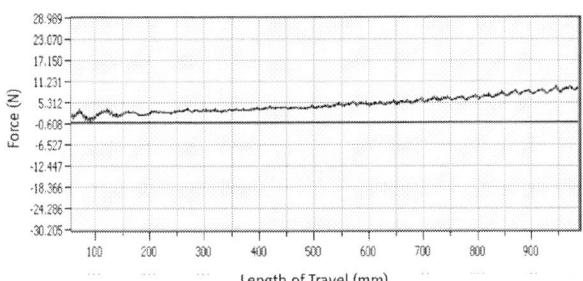

Figure 4. Force Required to Move GORE® Trackless High Flex Cable at High Speed

These tests showed that the pull force required to move the trackless cable increased almost linearly, but the force on the cable chain was very unstable. Consistent, linear increases in force resulted in easier control of the stage. Eliminating the use of the cable chain enabled the manufacturer to increase the speed of the stage without compromising positioning accuracy, and the result was an increase in through-put.

Particulation from Continuous Flexing

One of the major concerns with continuous flexing of cable systems is particle generation. This is particularly an issue for equipment used in cleanroom environments in semiconductor and solar manufacturing. In a recent study (1), W. L. Gore & Associates contracted with the Fraunhofer Institute for Manufacturing Engineering and Automation IPA in Germany to measure particulation of various cable systems.

Test Setup

To determine whether particulation was due to the cable or the cable chain, Gore tested identical flat cables in two different cable chains made by the same manufacturer. Gore selected two non-metal cable chains to use in the test — a low-vibration, quiet, and clean cable chain (chain A), and a conventional chain design with links and pins (chain B). Gore evaluated four cable systems:

- GORE® Trackless High Flex Cable
- Cable chain A containing two GORE® High Flex Flat Cables, positioned one on top of the other
- Cable chain B containing two GORE® High Flex Flat Cables, positioned one on top of the other
- Cable chain A containing two round cables with low particulation jackets, positioned beside one another without dividers. To achieve the lowest possible abrasion on these cables, Gore sized the chain with sufficient space so that the cables would not touch.

On each cable and cable chain combination, three optical particle counters were placed in the critical areas that most likely generate particles, and each counter recorded particles ranging in size from 0.1 to 5.0 micrometer. Each cable was tested for 100 minutes at velocities of 1.64 feet (0.5 meters), 3.28 feet (1.0 meter), and 6.56 feet (2.0 meters) per second, and airborne particles were recorded for 100 minutes. Using criteria set forth in Guideline VDI 2083 Part 9.1, Fraunhofer calculated the amount of particles at each measuring point and then determined the operating utility of each cable chain system. The operating utility was then used to determine the ISO classification.

Test Results

The test results indicated that the optical particle counters registered zero particulates at each measurement point for the GORE® Trackless High Flex Cable and for cable chain A with the flat cables. Cable chain B with the flat cables emitted particles at varying rates depending on the velocity — rates ranging from 0.1 to 1.7 particles per cubic foot. Cable chain A with round cables was also affected by velocity, emitting particles ranging from 0.0 to 2.5 particles per cubic foot.

Using calculations set forth in VDI Guideline 2083 and ISO 14644-1, Fraunhofer determined the probability of each cable and cable chain system emitting particulates (Table I).

TABLE I. Probability for Particulate Emission

Cable/Cable Chain	Probability
GORE® Trackless High Flex Cable	Less than 0.1%
Chain A with GORE® High Flex Flat Cables	Less than 0.1%
Chain B with GORE® High Flex Flat Cables	3%
Chain A with round cables	3%

Based on ISO guidelines, the Fraunhofer Institute determined the ISO 14644-1 cleanroom certifications based on the velocity that generated the most particulation (Table II).

TABLE II. ISO Certification Class

Cable/Cable Chain	ISO Certification Class
GORE® Trackless High Flex Cable	Class 1
Chain A with GORE® High Flex Flat Cables	Class 1
Chain B with GORE® High Flex Flat Cables	Class 5
Chain A with round cables	Class 4

These results showed that round cables with low particulation jackets in an ideal flex cable management condition can only meet ISO Class 4 requirements. Cable chains designed for low particulation applications with low vibration can meet ISO Class 1 with low friction, ePTFE jacketed planar cables. The study also demonstrated that GORE® High Flex Flat Cables and GORE® Trackless High Flex Cables generated less than 0.1 percent particulation, at all speeds, which exceeds the requirement for ISO Class 1 certification. Low-friction ePTFE flat cables provide numerous advantages, including

- reducing particulation with fewer points of friction in the cable chain and lower coefficient of friction in the jacket;
- eliminating the need for additional shelves and dividers in a cable chain;
- managing tubing, hoses, electrical cables, and optical fibers more easily;
- reducing the size and weight of cable chains, and possibly eliminating them completely with the lowest mass, self-supporting cables.

Verifying the Electrical & Mechanical Integrity of Cable Systems

As the results of these tests indicated, it is essential to evaluate the electrical and mechanical performance of a cable system under real-world conditions. Most manufacturers perform some level of electrical testing on every cable design before it is approved for delivery, so basic electrical requirements can be checked against the cable specifications. However, most cable systems used for semiconductor and solar manufacturing go beyond the basic requirements because signal integrity can be compromised by random, rolling, and torsion types of motion common in high-speed automation processes. Mechanical testing should be done to verify electrical performance while the cable is operating in conditions such as continuous and high-speed flexing, abrasion, and tight bending. The cable's electrical performance should also be measured while simulating the environmental operating conditions, such as variable temperatures, humidity, and pressure; vibration and acceleration; or exposure to liquids or gas. Particulation should also be monitored during high-speed flex testing to ensure that particulates are not released from the cable jacket or from the cable chain. Only with this type of testing can the cable system's integrity be validated.

Summary

Issues with positioning accuracy, particulation, and electrical performance of the cable system used with solar and semiconductor manufacturing equipment can have a significant impact on production rates. Testing done by a semiconductor manufacturer indicated that GORE® Trackless High Flex Cables significantly improved precise motion control and allowed for higher through-put. In addition, laboratory testing at the Fraunhofer Institute indicated that these same self-supporting cables and GORE® High Flex Flat Cables used in a non-metal, low-vibration chain were ISO-certified for Class 1 Cleanroom applications because they created sufficiently low quantities of particulates at both low and high speeds. The results of these tests support the importance of selecting cables specifically engineered for the challenging manufacturing environments of the solar and semiconductor industries.

References

1. W. L. Gore & Associates, Inc., *Cable Particulation Study in Cleanroom Environments* (2009).

Study on the Reliability of Fast Curing Isotropic Conductive Adhesive

Wenhui Du[1], Huiwang Cui[1,2], Si Chen[1,2], Zhichao Yuan[2], Lilei Ye[3] and Johan Liu[1,4]

[1]Key Laboratory of Advanced Display and System Applications, Ministry of Education and SMIT Center, School of Mechatronics Engineering and Automation, Shanghai University, Shanghai 200072, China
[2]Shangda Rui Hu Microsystem Integration Technology Co.Ltd, Room 101, Science &Technology Building, Shanghai University, No.149 Yangchang Road, Shanghai 200072, China
[3]SHT Smart High Tech AB, Fysikgränd 3, 41296 Gothenburg, SWEDEN
[4]SMIT Center, Department of Microtechnology and Nanoscience, Chalmers University of Technology SE-41296 Gothenburg, Sweden

With the development of semiconductor technology, electronic packaging technology with increasingly high integration density is needed. Solder is a traditional interconnect material widely used in electronic packaging industry, but now more attention is being paid to isotropic conductive adhesive (ICA) as an environmentally friendly interconnect material with advantages of low processing temperature, simple processing conditions and good manufacturability. However, compared with solder, reliability studies of ICA are still scarce. This paper aims to enrich the reliability study of ICA, by studying the reliability of a novel fast curing ICA which has a curing degree of 97.8% with 3 min of curing time at 150 ℃. It was found that after 85 ℃/85% RH humidity and heat test for 160 h, the bulk resistivity of ICA decreased about 82% and then remained stable. Besides, ICA embrittled as aging time extending.

Introduction

Sn/Pb based solder has been widely used in the electronic packaging industry as a reliable interconnection material for long[1-3]. But lead, one of its main components, can result in environmental pollution, endanger human health, and is diffcult to be recycled. It is said that at present only 25% of lead exploited can be recycled, so there is an urgent need to develop new material to substitute the traditional solder. Morever with the development of high density electronic packaging and enhancement of public environmental awareness, more studies have been carried out into isotropic conductive adhesive (ICA) as a green interconnection material with low processing temperature, simple processing conditions and good manufacturability. Now, ICA has become more and more widely used in liquid crystal display, smart card applications, flip-chip assembly, chip scale package, and ball grid array [4-6].

However the long curing time of ICA, one hour or longer, extends the process time in its application and increases the cost. Thus how to control its fomulation to achieve the aim of fast curing at lower temperature is on the agenda today. Morever, there are still some limitations and challenges to ICA, such as lower electrical and thermal conductivity and poor mechnical strength [7,8]. Because of the complex structure of ICA, many factors, such as heat stress and humidity may result in the failure of interconnection in application. Besides, compared with solder, the reliability studies of ICA are still scarce. Therefore, reliability test of ICA in severse environments has become most important for electronic packaging.

Based on previous work, we investigated material performance variation of ICA such as electrical conductivity and mechanical properties and the destruction of interfacing formed during service process. We gave a discussion of the analysis of failure mechanism by studing the performance change of a new fast curing ICA during accelerated aging test.

Experiment

A. Preparation of ICA

ICA consists of polymer matrix which would offer mechanical property and conductive fillers which would give electrical conductivity. The preparation procedure of ICA was divided into two parts, firstly preparing the matrix and secondly adding fillers. The preparation of matrix was a mixing process of expoy, curing agent, diluent and coupling agent. The mixing time was 20 min and the mixing speed was 3000 rpm.

B. Reliability Test

In the reliability test our study included two parts: one was to study the change of electrical conductivity of ICA itself and electrical connect during the humidity and heat test; and the other was to study changes in mechanical property after this aging process. In this paper the aging condition of 85 ℃/85 RH was selected. The definition of an actual 85-85 test is 85 ℃ and 85% RH for 1000 h. However as the aim of our experiment was to study the performance change of ICA during humidity and heat test and the schedule was intense, we changed the time of the test.

The electrical conductivity of ICA is characterized by bulk resistivity which is measured by four point method. The bulk resistivity testing samples were prepared by coating ICA with 76% silver flakes on a glass board. Adhesive tape is used to keep the shape of the sample. Finally all samples were firstly cured at 120 ℃ for 5 min and then kept for another 10 min at 150 ℃.

Electrical connect reliability is always studied by investigating the survival rate during humidity and heat test. In our study it was stipulated that if the resistance of one point is higher than 50 Ω, the connection is a failure. The sample of this test was given in Fig 1. Resistance of the two ends of the component was measured by an Ohmmeter. There are fifty points on every test board, designed by ourselves. The component matched with testboard was 1206-size capacitors. ICA with 76% silver flakes was used as binder.

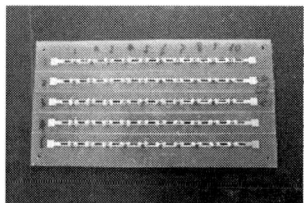

Fig 1. Sample for Electrical Connect Test

In our paper a kind of shear test for connection between substrates was studied. The shear test was conducted using an instron 5548 and the details of this shear test are shown in Fig 2. The dashed area in the picture is the bonding area, which was square in shape with a size of 8×8 (mm²). The test board was Copper FR4 substrate. The speed of test was 5 mm/min.

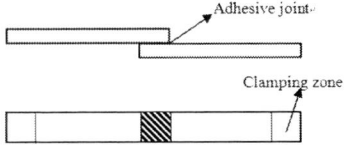

Fig 2. Shear test between substrates

In order to characterize the changes to ICA during the aging processs and the effect of humidity and heat on the material itself in more detail, we studied the change in impact strength of ICA after a humidity and heat test of 360 h. The samples for impact test were first cured in a copper mould at 120 ℃ for 5 min and then kept at 150 ℃ for another 10 min. At last the cured samples were polished into a cuboid of 80 mm long, 10 mm wide and 4 mm thick. To enhance the fillibility of ICA in the copper mould, the silver flake loading of ICA was reduced to 73%. The impact test was conducted in ZBC1400-2 Pendulum Impact Testing Machine.

Results And Discussion

A. Variation in Electrical Property

The variation in electrical property of ICA was divided into two parts to be studied in the paper, to investigate the failure mechanism of electrical connect after aging test. One is the variation in the material itself during the aging process and the other is the trend of electrical connect survival rate during the same aging process.

Fig 3 shows the variation trend of bulk resistivity of ICA with 76% silver flakes. From the diagram we can see that bulk resistivity decreased rapidly as aging time extending first and then remained stable. When the test continued for about 250 h, the bulk resistivity decreased by about 85.6% compared with the primary date. This indicates that the aging test had a good effect on the electrical conductivity of the material.

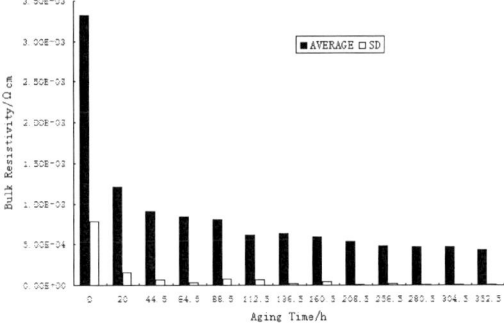

Fig 3. Variation of bulk resistivity during aging process

The change of electrical connect during the aging test is shown in Fig 4 and Fig 5. From Fig 4 we can see that as aging time extended the survival rate decreased in the mass but rose occasionally. Occasional increasing of survival rate may be due to accumulation of melioration of the electrical conductivity of some points. Thus we chose a number of points and respectively made a variation trend showing their resistance at the primary 168 h aging stage. The results are given in Fig 5. From the diagram we can see that all three randomly selected points showed inordinately the phenomenon of resistance occasional melioration.

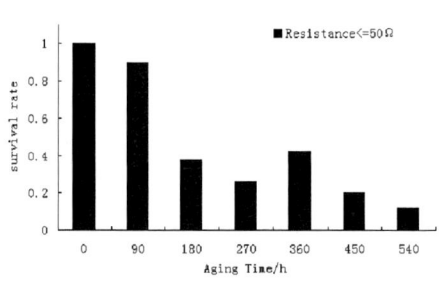

Fig 4. Variation of survival rate during aging process   Fig 5. Variation of points' resistance during aging

B. Variation of Mechanical Performance

Impact strength is usually used to characterize the ductility of material. In our paper it was used to characterize variation of ductility of ICA after aging. The impact strength of ICA with 73% silver flake is shown in Table 1.

TABLE 1. Impact Strength

Sample No.	Original (KJ/mm2)	Aging (KJ/mm2)
1	6.692	3.268
2	5.752	3.908
3	5.426	2.775
Average	5.957	3.317

Analyzing the original data and that gathered after aging, we can see that the impact strength of ICA dropped after aging and decreased about 44.3% after 360 h aging.

The shear strength gained from the shear test for connection between substrates is given in Table 2. From the table we can see that aging has a bad effect on the shear strength. After 650 h aging process, the shear strength reduced by 82.4%.

TABLE 2. Shear Strength of Test for Connection between Substrates

Sample No.	Original (MPa)	Aging (MPa)
1	5.71	0.988
2	6.22	0.831
3	6.68	1.53
4	5.26	0.828
5	5.38	0.957
Average	5.85	1.03

C.Micro Analysis of the Failure of Electrical and Mechanical Connect

From the results gained from electrical and mechanical performance, we concluded that conditions of humidity and heat had a good effect on the electrical conductivity of the material itself but weakened ICA's mechnical property. In order to explain the phenomenon more directly and vividly, changes in the micro structure of ICA during aging process were also studied. We selected some failpoints of electrical connection and fracture gained from impact test to make SEM observation. Fig 6 shows SEM pictures of the failpoints.

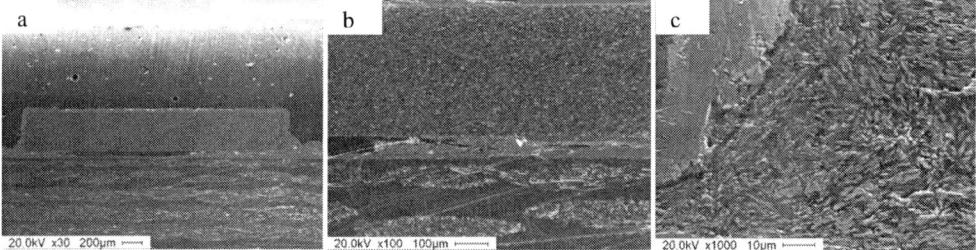

Fig 6. Micro Structure of the Interface; (a) is full view of connection between substrate and capacitor (b) are interfaces of ICA after 500 h aging (c) are these of ICA after168 h aging

Failure in adhesive bonded joints is generally divided into three types: cohesive failure, adhesive failure, and mixed failure [9,10]. From Fig 6, we can see that failure occurred in all pictures, but the one which existed in samples with 168 h aging seemed smaller than that appeared in samples after 500 h aging. The failure in samples with 168 h aging was cohesive failure and the one in samples with 500 h aging was adhesive failure. The three types of failure should all appear in the samples after aging, but in the random selection of samples, they were not all observed in one experiment.

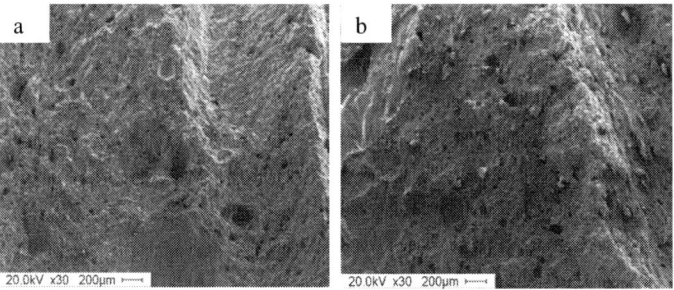

Fig 7. Fracture Apperance of Impact Test; (a) is the fracture apperance of ICA without aging, (b) is the fracture appearance of ICA with 360 h aging

From Fig 7 we can see that there were some voids in the cured adhesive. The existence of voids weakened its strength and resulted in lower impact and shear strength. The formation of voids happened in the process of adhesive preparation and services and was also related to the composition of adhesive. So the addition of foam breaker or vacuum solidification were considered to be effective methods to improve the strength of ICA. Comparing Fig. 7 (a) and (b), we see that some fine particles appeared on the fracture suface of (b) but didn't appear in (a). The fracture mode shown in (b) displayed a brittle rupture compared with (a). It indicated that the matrix of adhesive can be embrittled in humid and heat environment for some time.

CONCLUSION

The humidity and heat test improved the elelctrical conductivity of ICA itself, which can be explained by postcure theroy and shrinkage of ICA. In respect to reliability of the electrical connection, as the aging test proceeded, the reliabilty decreased and it was mainly expressed by the increase of resistance and greater instability of resistance value. For mechnical property, humidity and heat tests enbrittled material, possibly due to water absorptivity of the matrix. Water penetration inside the epoxy network would produce rupturing of interchain interactions, weaken the mechnical property and generate cracks.

Acknowledgments

This work was supported by the STC Torch program, contract no: 0903H195300, EU programs "Thema-CNT", "Mercure" and Nanopack. This work was also carried out within the Sustainable Production Initiative and the Production Area of Advance at Chalmers. This support is gratefully acknowledged. The authors are also grateful for the support of the Swedish National Science Foundation under the project "Nanointerconnect" (621-2007-4660) and the Vinnova Program on "Designade Material" through the contract No. 2009-03230.

References

1. Manko HH. Solders and soldering materials, design, production and analysis for reliable bonding, 2nd ed. McGraw-Hill: New York. 1979.
2. Wassink RJK. Electrochemical Publications. 1984.
3. Sinnadurai FN, Handbook of microelectronics packaging and interconnection technologies. UK: Electrochemical Publications.1985.
4. Li Y, Moon K, Wong CP. Science. 2005;308:1419.
5. Li Y, Wong CP. Proceedings of the IEEE polytronic fourth international conference on polymers and adhesives in microelectronics and photonics. Portland. 2004. p. 1–7.
6. Vona SA, Tong Jr QK, Kuder R, Shenfield D. Proceedings of the fourth international symposium and exhibition on advanced packaging materials, processes, properties and interfaces. 1998. p. 261.
7. Liu J, Lai Z. Proceedings of the third international conference on adhesive joining and coating technology in electronics manufacturing. 1998;184:1.
8. Kang SK, Buchwalter S, Tsang C. J Electron Mater. 2000;29(10):1278.
9. Davis M, Bond D. Int J Adhes Adhes. 1999;19(1):91.
10. Balkova R, Holcnerova S, Cech V. Int J Adhes Adhes. 2002;22(4):291.

The Effect of Functionalized Silver on Rheological and Electrical Properties of Conductive Adhesives

Qiong Fan[a], Huiwang Cui[a], Chune Fu[a], Dongsheng Li[a], Xin Tang[a], Zhichao Yuan[b], Lilei Ye[d] and Johan Liu[a,c]

[a]Key Laboratory of Advanced Display and System Applications, Ministry of Education and SMIT Center, School of Mechatronics Engineering and Automation, Shanghai University, Shanghai 200072, China
[b]Shang Da Rui Hu Microsystem Integration Technology Co. Ltd, Room 101, Science &Technology Building, ,Shanghai University, No.149 Yanchang Road, Shanghai 200072, China
[c]SMIT Center, Department of Microtechnology and Nanoscience, Chalmers University of Technology SE-412 96 Gothenburg, Sweden
[d]SHT Smart High Tech AB, Fysikgränd 3, 41296 Gothenburg, SWEDEN

This research used low molecular surface modifiers, and observed that chemisorptions took place through the formation of a bond between silver surface and an adsorbed molecule, which improved the dispersion of silver flakes in the organic resin. Several different functionalizers, such as thioglycolic acid, silane and di-acid, were used to functionalize the silver surface. Results of shear viscosity, bulk resistivity etc. showed that by using these low molecular organic functionalizers, isotropic conductive adhesives (ICAs) with lower shear viscosity and better electrical conductivity at high silver fillers content were obtained. The adipic acid had the greatest effect on the rheological and electrical property of ICAs, so its weight percentage in silver flakes was also optimized; ICAs displayed the maximum electrical conductivity when there was 0.5 wt% of silver flakes.

Introduction

Tin-lead solder alloy has been widely used in the electronics industry. However, there are several intrinsic problems associated with tin-lead solder that limit its future applications, such as environmental concerns and relatively high processing temperature [1-2]. Considerable advances have been made in the development of lead-free solders, and electrically conductive adhesives (ECAs) also provide a promising alternative. Due to a higher melting point than that of eutectic tin/lead solder, lead-free alloys cause some problems, such as higher stress generated in package, requirement of more expensive substrates, and more serious package pop-corning problem when the package is assembled on the printed wiring board (PWB). On the other hand, ECAs have the advantages of environmental friendliness, mild processing conditions, fewer processing steps, lower stress on the substrates, and finer pitch interconnect capability [3-4].

Isotropic conductive adhesives (ICAs) have been studied for many years as a lead-free alternative for the electronic industry, due to advantages of low processing temperature, simple processing conditions and good manufacturability [5]. However, the dispersion of silver particles in the organic resin greatly affect printing quality and production technology and also have a great influence on adaptability. In order to maximize the electrical conductivity, high conductive filler are needed in ICAs formulation. However, too high filler loadings usually deteriorate the mechanical property of ICAs.

Silver flakes coated with surfactant stearic acid (C-18 carboxylic acids), are widely used as conductive fillers in the formulations of most ICAs. The existence of such organic lubricants can affect the viscosity of conductive adhesive and prevent the agglomeration of silver flakes, but it also decreases the electrical conductivity of ICAs due to its insulation properties. This research used low molecular surface modifiers, and observed that chemisorptions took place through the formation of a bond between silver surface and an adsorbed molecule, which improved the dispersion of silver flakes in the organic resin [6]. In addition, small molecules on the silver surface had a better effect on the formation of conductive pathways than that with surfactant stearic acid, and improved the electrical conductivity of ICAs.

Experiment

Preparation of Materials

The silver flakes used in this work were supplied by Sino-Platinum Co. Ltd. Four types of functionalizers, including 3-Aminopropyltrimethoxysilane, adipic acid, (3-mercaptopropyl) trimethoxy silane, and thioglycolic acid, were used to functionalize silver flakes. The chemical structures of the functionalizers are shown in TABLE I. After the treatment of silver flakes, ICAs were prepared using epoxy, diluents, coupling agent, curing agent as matrix [7-8], treated and untreated silver flakes were respectively added in the matrix as conductive fillers. Samples filled with untreated silver flakes were marked as 0 for contrast.

TABLE I. Chemical Structures of the Functionalizers

Sample No.	Chemical name	Structure	Molecular weight
1	3-aminopropyl trimethoxysilane		179.29
2	adipic acid		146.14
3	3-mercaptopropyltrimethoxysilane		196.34
4	thioglycolic acid		76.12

Characterization

Shear viscosity. The viscosity of ICAs was measured by a viscometer (Brookfield DV II+ viscometer) with the spindle CPE-51. All viscosity was measured under different speeds from 0.5 rpm to 5 rpm at room temperature [9]. In this work, the viscosity at 5 rpm was chosen to characterize rheological properties of ICAs.

Fourier Transform Infrared Spectrometer (FT-IR). To characterize the treatment of functionalizers, FTIR of Silver flakes were studied before and after treatment using AVATAR 370 FT-IR from Nicolet, USA.

Electrical conductivity. Electrical conductivity of ICA is characterized by the bulk resistivity of cured ICA. Bulk resistivity was measured by the four-point method as shown in Figure 1. The bulk resistivity testing samples were prepared by coating ICA on a glass board [10]. Adhesive tape was used to control the shape of sample. The bulk resistivity can be calculated by:

$$\rho = \frac{VWt}{IL}$$ [1]

Where L=2.54 cm, W and t are the width and thickness of the sample, V and I are voltage, and constant current of being 0.2 A supplied by DC Power Supply (Velleman PS 613). Taking error of measurement into account, three samples were selected for every component.

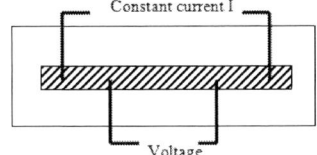

Figure 1. Four-point measurement method

Results and Discussion

The viscosity of ICAs filled with silver flakes treated with different functionalizers at the same level (1 wt% of Ag flakes) is shown in Figure 2. It can be seen that viscosity of ICAs filled with treated silver flakes was lower than that untreated except with thioglycolic acid. ICAs filled with silver flakes treated by adipic acid displayed the lowest viscosity. This phenomenon suggested that the functionalizers substituted the original surfactant stearic acid on silver flakes and a bond between silver surface and an adsorbed molecule was formed, this improved the dispersion of silver flakes in organic resin.

The effect of the functionalizers on electrical properties was characterized by the bulk resistivity of cured ICA. As shown in Figure 3, changes in bulk resistivity were similar to those in viscosity. Compared to others, the shear viscosity and bulk resistivity of ICAs with silver flakes treated by adipic acid ICAs showed the lowest viscosity and bulk resistivity, which indicated that the adipic acid had the best effect on the rheological and electrical properties, and these ICAs can be suitable for many electronic and package processes, so to optimize it as functionalizer.

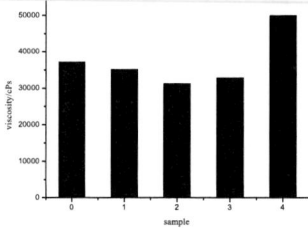

Figure 2. Viscosity of ICAs filled with Silver flakes: 0-untreated, 1-treated with 3-aminopropyl trimethoxy silane, 2-treated with adipic acid, 3-treated with (3-mercaptopropyl) trimethoxy silane, 4-treated with thioglycolic acid.

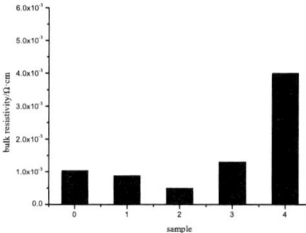

Figure 3. Bulk resistivity of cured ICA filled with Silver flakes treated and untreated (with different functionalizers)

The weight ratio of adipic acid in silver flakes was also optimized. Figure 4 and Figure 5 show the changes in viscosity and bulk resistivity with the weight percentage of adipic acid. It can be seen that when adipic acid was 0.5wt% of silver flakes, the ICA showed the lowest viscosity and bulk resistivity, which indicates the rheological and electrical properties were the best, so 0.5wt% of silver flakes is chosen as functionalizer level. The results show that the properties of ICAs were best when the adipic acid was 0.5wt% of Ag flakes.

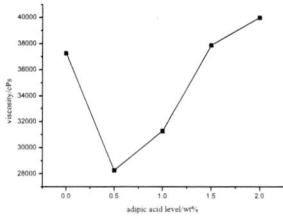

Figure 4. Viscosity changes with the weight ratio of adipic acid

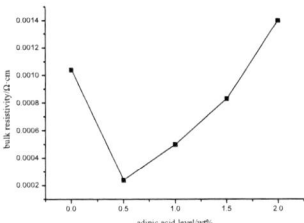

Figure 5. Bulk resistivity of cured ICA filled with Silver flakes treated with adipic acid at different weight ratios

To get a clear profile of whether or not the functionalizers replaced the lubricant on silver flakes surface, silver flakes treated before and after were measured by FTIR. The results are shown in Figure 6. The two characteristic positions of organic acid were 3450 cm^{-1} and 1600 cm^{-1}. After functionalization, there was stronger absorption at 3450 cm^{-1} and 1600 cm^{-1}, and a new position appeared at 700 cm^{-1} except for thioglycolic acid. This indicates that the functionalizers replaced the lubricant on silver flakes surface, but thioglycolic acid did not.

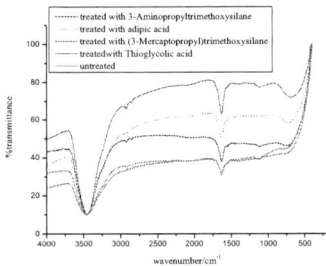

Figure 6. FT-IR of Silver flakes before and after treatment

Conclusion

With the use of low molecular surface functionalizers, chemisorptions took place through the formation of a bond between silver surface and an adsorbed molecule. This improved the dispersion of silver flakes in organic resin and therefore improved the rheological and electrical properties of conductive adhesives. ICAs displayed the best properties when the adipic acid was 0.5wt% of silver flakes. The contact resistance shifts of ICAs on a Sn surface with and without functionalization during 85 C/85% RH aging will be studied in the future.

Acknowledgments

This work was supported by the STC Torch program, contract no: 0903H195300, EU programs "Thema-CNT" and Nanopack. This work was also carried out within the Sustainable Production Initiative and the Production Area of Advance at Chalmers. This support is gratefully acknowledged. The authors are also grateful for the support of the Swedish National Science Foundation under the project "Nanointerconnect" (621-2007-4660) and the Vinnova Program on "Designade Material" through the contract No. 2009-03230.

References

1. *IPC roadmap*: A guide for assembly of lead-free electronics 4th draft. Sanders road, northbrool, 2000-2215.
2. 2nd EU Lead-free Soldering Technology Roadmap and Frame-work for an International Lead-free Soldering Roadmap. Soldertec at Tin Technology, 2003.2
3. J. C. Jsilvert, P.J.M. Beric and G.F.C.M. Lijten, "Electrically Conductive Adhesives: A Prospective Alternative for SMD Soldering" *IEEE Transactions on Components, Packsilvering, and Manufacturing Technology*, Part B, 18(2): pp. 292-298, 1995.
4. C.P. Wong, D. Lu, S. Vona, and Q.K. Tong, "A Fundamental Study of Electrically Conductive Adhesives," *Proceedings of the 1st IEEE International Symposium on Polymeric Electronics Packsilvering*, (Norrkoping, Sweden), pp. 80-85, 1997.
5. J. Liu, "An overview of advances of conductive adhesive joining technology in electronics applications," *Mater. Technol.*, vol. 10 (1995), pp. 247-252.
6. Yi Li, Kyoung-Sik Moon, and C P Wong, "Electrical property improvement of electrically conductive adhesives through in-situ replacement by short-chain difunctional acids," *IEEE Transactions on Components and Packsilvering Technologies*, VOL. 29, No. 1, March 2006, pp. 173-178.
7. Masahiro Inoue, Hiroaki Muta, Takuji Maekawa, Shinsuke Yamanaka, Katsuaki Suganuma. Temperature dependence of electrical and thermal conductivities of an epoxy-based isotropic conductive adhesive. *Journal of Electronic Materials*, 2008, 37 (4): 462–468.
8. Masahiro Inoue, Johan Liu. Electrical and thermal properties of electrically conductive adhesives using a heat-resistant epoxy binder. *ESTC 2008, 2nd Electronics Systemintegration Technology Conference*, Greenwich. UK, 2008, pp. 1147–1152.
9. Chune Fu, Si Chen, Par Berggren, Qiong Fan, Wenhui Du, Balan Ganesh and Johan Liu, "Optimization of stiffness for isotropic conductive adhesives," *IEEE advanced packing materials*, (2010), pp29-33.
10. Zhang Zhikun, Jiang Sijia, Liu Johan, Inoue Masahiro, "Development of high temperature stable isotropic conductive adhesive," *Proceedings of the 2008 International Conference on Electronics Packaging Technology & High Density Packaging*, July 28/31,2008, Shanghai, China, CD Version, paperE3-04.

CHAPTER 7

PACKAGING AND ASSEMBLY

Microstructural Evolution of Sn3.0Ag0.5Cu3.0Bi0.05Cr/Cu Solder Joints During Thermal Aging and Its Effects on Mechanical Properties

F. Lin[a], W.Z. Bi[a], G.K. Ju[a], and X.C. Wei[a, b]

[a] School of Materials Science and Engineering, Shanghai University, Shanghai 200072, China
[b] Key State Lab for New Displays and System Integration (Chinese Ministry of Education), Shanghai 200072, China

The evolution of IMCs in SACBC/Cu solder joint and its tensile property after isothermal aging at 150 □ for 0, 168 and 500 h were investigated. The hollow blocky phase containing Cr was observed in the solder matrix of the joint after soldering, which was not found after aging. The needle-like Ag_3Sn transformed to be short rod-like or ellipsoidal during aging. Bright and sphericized Bi particles precipitate at the interface and suppress overgrowth of IMCs layer, resulting in a relative stable tensile strength.

Introduction

Solder plays a crucial role in the assembly and interconnection of electronic products. As a joint material, solder provides electronic, thermal and mechanical continuity [1]. Primarily, it should wet the substrate and provide good adhesion. A thin, continuous and uniform intermetallic compounds (IMCs) layer formed at the interface between the solder and substrate is an essential requirement for good bonding. With the inevitable trend to implement lead-free soldering due to environmental and health concerns and achieve higher circuit board component densities, severe requirements for mechanical reliability of solder joints need to be met. Researches [2, 3] indicated that the cause of solder joint failure is mainly due to the overgrowth of IMCs which have inherent brittle nature and the tendency to generate structural defects during the storage or service period. Therefore, the science of interfacial reactions for lead-free alloys needs thorough studies, and the core issue is the formation and growth of IMCs layer between the solder and the substrate as well as microstructural evolution at the interface.

Due to its good mechanical properties, adequate wetting characteristics as well as the comparable melting temperature, Sn-3.0Ag-0.5Cu (SAC) solder alloy has been one of the most important candidates replacing lead-tin solder. However, SAC alloy has some adverse properties, such as higher melting point, poor wettability and overgrowth of IMCs. Investigations [4-7] have suggested that the doping atoms in SAC could reduce the atomic activity of Cu and Sn, and thus delay the process of inter-diffusion and the IMCs formation or growth. Li et al. [8] and J. Zhao et al. [9] have reported that adding appropriate Bi in Sn-Ag-Cu solder alloy system could refine the grain size of the IMCs and inhibit the excessive IMCs growth in solder joints.

Previous study [10] has found that the trace of Cr addition observably reduced the growth rate of IMCs layer of Sn-3.0Ag-0.3Cu-0.05Cr (SACC)/Cu joint, whereas the melting point of alloy slightly rose. Then a Sn-3.0Ag-0.5Cu-3Bi-0.05Cr (SACBC) solder alloy with a better melting range of 210-217 □ was developed. The behaviors of interfacial IMCs for SACBC/Cu and SAC/Cu joints during aging at 150 □ for 0, 24, 168, 500 and 1000 hours respectively have been comparatively investigated [11]. It is believed that the microstructural evolution significantly influences the reliability of solder joints. Therefore, to provide useful information for reliability evaluations of solder joints, the present work focuses on the evolution of IMCs and discusses effects of the alloying elements on the evolution and the mechanical properties.

Experimental procedure

The SAC solder used in this study was the commercial solder alloy. The SACBC solder was prepared independently. Alloys used for smelting are pure Sn (99.95%), Ag (99.99%), Cu (99.99%), Bi (99.99%) and Cr (99.99%). The master alloy Sn2.0Cr (in wt.%) was prepared utilizing the vacuum high-frequency induction furnace for 30 min at ~1500□ in high vacuum condition of 1.33×10^{-5} Pa. Then the exact content of Cr was measured by chemical analysis. Afterwards, according to the designed composition, Sn2.0Cr, Ag, Cu, Bi and Sn were mixed together and sealed into the vacuum quartz tubesealed into the vacuum quartz tube, which was put into the resistance furnace for 2 h at 1100□. Lastly, the alloy was reheated for 1 h at 300□ and cooled down in the air.

Solder joints were prepared using the Simulation Preparation Device of Solder Joint. For the substrate, pure copper bars with both ends polished to remove the surface oxide layer were used. After fluxing with Kester®Tacky flux, a piece of solder with a diameter of 6 mm and a thickness of 2 mm was emplaced into the thin clearance of the mating surfaces of two copper samples [12]. Subsequently, the whole setting was heated in a temperature-controlled furnace set at 255±5 □ keeping soldering for 5 min, and then cooled in air. The samples were gently pushed towards each other to obtain a good joint. Fig.1 is a prepared joint.

Fig.1. prepared Cu/Solder/Cu sample

According to the high temperature storage life standard JESD22-A103C by JREDEC (Joint Electron Device Engineering Council), the specimens were annealed at 150 □ for 0, 24, 168, 500 and 1000 h in an oven. For each condition, five samples were prepared, which were used for tensile testing first and then were used to study the interfacial microstructure. Tensile testing was carried out at room temperature at a strain rate of 6.7×10^{-4} s^{-1} by a CMT5305 tensile tester. Afterwards, the specimens were cut by electric discharge machining and cold mounted in epoxy. To analyze the appearance and evolution of the IMCs, they were mechanically ground, polished and etched with a 94%C$_2$H$_5$OH+4%HNO$_3$+2%HCl (in vol.%) solution. The microstructure of joint was observed by means of an Apollo 300 Thermal Field Emission scanning electron microscope (SEM), and the composition was examined by energy dispersive spectroscopy (EDS). The presence of Bi in the solder was detected by transmission electron microscopy (TEM, JSM-2010F) with electron diffraction (ED). A plate sample with a thickness of 1 mm was cut from solder ingot, and ground to about 60 μm. Finally, a disc with a diameter of 3 mm was cut and thinned by an ion-miller for TEM.

Results and discussion

Microstructural evolution in solder joints

The microstructures of SACBC/Cu and SAC/Cu joints after thermal aging at 150 □ for different times are shown in Fig. 1. The soldering reaction layer on Cu substrate is the scallop-type η'-Cu_6Sn_5 phase in both joints. In addition to Cu_6Sn_5 and Ag_3Sn IMCs, which were also observed in SAC/Cu joints, some hollow blocky phases were found in the solder matrix of SACBC/Cu joint after soldering, as shown in Fig. 1(a). According to the EDS analysis results, Cr element was found in the phase, whose structure needs further study and confirmation. However, after aging at 150 □ for 168 h and 500 h, these blocky phases haven't been observed. It can be inferred that the phases containing Cr are unstable during thermal aging.

For SACBC/Cu joint, the needle-like Ag_3Sn transformed to be short rod-like or ellipsoidal with round and slippery ends during aging, as Fig. 1(b) shows. It might be beneficial to improving the mechanical properties of the joints. After aging for 500 h or longer, they coarsened (Fig. 1(c)). The main evolution mechanism may be the spheroidization controlled by the interfacial diffusion, which has been put forward by Allen et al. [13]. As can be seen from Fig. 2, the Cu_6Sn_5 and Ag_3Sn strips are often attached to one another. It may be that this has an effect on the onset of coarsening, as Allen et al. suggested.

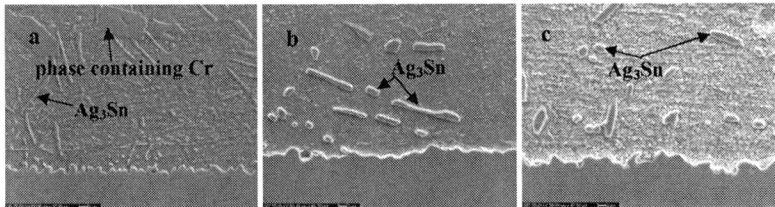

Fig. 1. SEM images of SACBC/Cu joints: (a) as-soldered; (b) after aging for 168 h at 150□; (c) after aging for 500 h at 150□

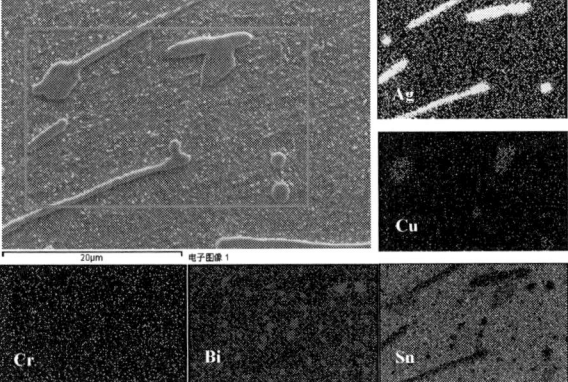

Fig. 2. SEM image and EDX mappings of IMCs in solder matrix of SACBC/Cu joint after aging for 168 h at 150 □.

Evolution of interfacial IMCs

Usually, there are two major stages to form the interfacial IMCs layer of solder/Cu, (a) in the soldering process, the molten solder reacts with Cu substrate to form the IMC layer; (b) after soldering, the layer thickens depending on the solid state diffusion of Cu and Sn atoms. The reason for the fast interfacial reaction in SAC/Cu is that Cu atoms diffuse into solder because of the driving forces from the temperature and concentration gradient. So the activity of Sn and Cu is a decisive factor governing the interfacial diffusion kinetics.

On the contrary, our previous study results showed that the overgrowth of IMC layer was effectively restrained in SACBC/Cu joints, the thickness of which was obviously thinner than that of SAC/Cu, as listed in Table 1 [11]. It is worth noticing that the surface morphology of IMC layer is different in the two solder joints, as shown in Fig. 3. There are a number of bright nano-particles on the surface of IMC in SACBC/Cu joint (Fig. 3(b)), while the surface is quite smooth in SAC/Cu (Fig. 3(a)).

TABLE 1. The average thickness of interfacial IMCs layer in solder joints.

Aging time (h)	Thickness in SAC/Cu (μm)	Thickness in SACBC/Cu (μm)
0	4.56	1.13
24	5.79	1.80
168	7.17	2.95
500	8.63	3.88
1000	11.33	4.72

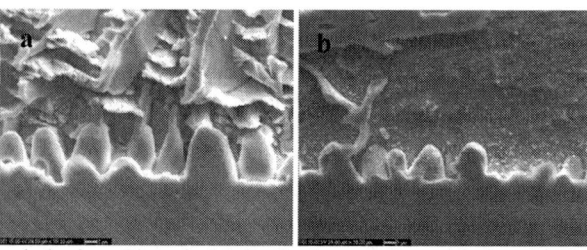

Fig.3. Morphologies of interfacial Cu_6Sn_5 layers after soldering: (a) SAC/Cu; (b) SACBC/Cu

The fine particles may be due to the precipitation of Bi during interfacial reaction process. The solid solubility limit of Bi in Sn at room temperature is about 1% from binary Sn-Bi phase diagram. Furthermore, TEM analysis in Fig. 4(a) shows the sphericized Bi particles precipitating from the Sn matrix in solder alloy. Similarly, more Bi would precipitate near and on the interfacial IMC related to the depletion of Sn close to the IMC layer during aging. Thus, it is indicated that the bright particles near and on the IMC layer are Bi precipitates. The existence of these particles would decrease the interfacial energy and weaken the reaction rate of Sn and Cu, which is the main reason for the slow growth of IMC layer in SACBC/Cu joints. On the other hand, a portion of Bi dissolving in Sn matrix can improve the bonding strength with Cu_6Sn_5 IMC. Fig. 4(b) shows that a thin matrix stick (solid solution of Bi and Sn) can prop up a big Cu_6Sn_5 block.

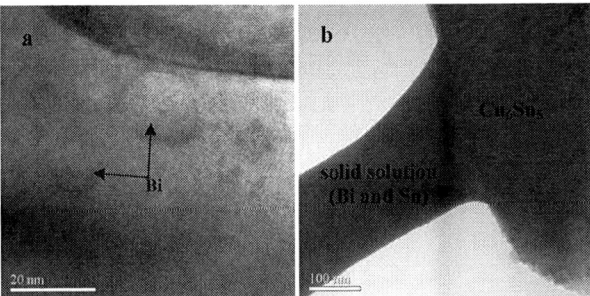

Fig.4. TEM analysis of SACBC solder.

Mechanical properties of Cu/solder/Cu joints

The stress-strain curves of the two joints in Fig. 5 show that the tensile strengths of SACBC/Cu joints are obviously superior to that of SAC/Cu, whether as-soldered or isothermal aged. However, the former has lower elongation than the latter. After aging, the strength of SAC/Cu joint decreases significantly, which results from the overgrowth of interfacial layer and the obvious coarsening of Cu_6Sn_5 and Ag_3Sn IMCs.

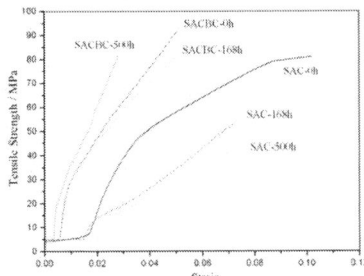

Fig. 5. The Strain-Stress curves of the two joints after thermal aging for 0, 168 and 500 h.

As for SACBC/Cu joint, the relative stable tensile strength with aging time can be attributed to thinner interfacial layer and strengthened Sn matrix with the solution strengthening and precipitating effect of Bi. Zhao et al. [9] also reported this positive impact of Bi precipitation in Sn-3.0Ag-0.5Cu-3.0Bi solder alloy after aging at 120□. But they did not find Bi precipitation in the solder alloy under cast condition and after aged at 150 □ for 100 h. In our work, Bi precipitation is not only found in SACBC solder alloy but also in solder matrix of the joint after aging at 150 □ for 168 h (Fig. 2). On one hand, more Bi precipitating near the interfacial layer acts as obstacles to suppress the further growth rate of IMCs layer. On the other hand, too many Bi accumulate at the interface will result in lower elongation of the joint.

Conclusions

(1) The hollow blocky phase containing Cr was observed in the solder matrix of the joint after soldering, which was not found after aging. The needle-like Ag_3Sn transformed to be short rod-like or ellipsoidal during aging.

(2) There are a number of bright nano-particles which are indicated to be Bi precipitates on the surface of IMCs layer in SACBC/Cu joint. It may suppress overgrowth of IMCs layer of the joint.

(3) It is indicated that Bi precipitating near the interfacial layer acts as obstacles to suppress the further growth of IMCs layer, while too many Bi accumulate at the interface will result in lower elongation of the joint.

Acknowledgments

The present work was carried out with the support of the Foundation for Innovation in Shanghai University (contract No. SHUCX102232), the Special Foundation for Outstanding Young Teachers in Shanghai University (contract No. B.37-0209-09-001), Open Foundation of Key State Lab for New Displays and System Integration (Chinese Ministry of Education) and Open Foundation of Key State Lab for New Materials of tungsten and copper (Jiangxi, No. 2010-WT-02).

References

1. M. Abtew and G. Selvaduray, *Mater. Sci. Eng. R*, **27**(5-6), 95 (2000).
2. H. Matsukia, H. Ibukab and H. Saka, *Sci. Technol. Adv. Mat.*, **3**(3), 261 (2002).
3. E. Saiz, C.W. Hwang, K. Suganuma and A.P. Tomsia, *Acta Mater.*, **51**(11), 3185 (2003).
4. B. Li, Y.W. Shi and Y. P. Lei, *J. Electron. Mater.*, **34**(3), 217 (2005).
5. C.M.T. Law, C.M.L. Wu and D.Q. Yu, *J. Electron. Mater.*, **35**(1), 89 (2006).
6. K.S. Kim, S.H. Huh and K. Suganuma, *Microelectron. Reliab.*, **43**(2), 259 (2003).
7. J. Liu, J. Xu, F. Zhang, F. Yang and X. Zhu, *Chin. J. Rare Met.*, **29**(5), 625 (2005).
8. G. Li and X. Shi, *Trans. Nonferrous Met. Soc. China*, **16**(z1), 739 (2006).
9. J. Zhao, Q. Lin, X. Wang and L. Wang, *J. Alloys Compd.*, **375**(1-2), 196 (2004).
10. G. Su, Y. Han, C. Wang, H. Wang, X. Wei, *Proceedings of the 16th IEEE International Symposium on the Physical and Failure Analysis of Integrated Circuits*, p. 393 (2009).
11. Y. Han, F. Lin, G. Ju, X. Wei, *The 11th International Conference on Electronic Packaging Technology & High Density Packaging*, p. 208 (2010)
12. G. Ju, X. Wei and J. Liu, *Solder.Surf. Mt. Tech.*, **20**(3), 4 (2008).
13. S.L. Allen, M.R. Notis, R.R. Chromik and R.P. Vinci, *J. Mater. Res.*, **19**(5), 1425 (2004).

Study of EMC for Cu Bonding Wire Application

Toshiro Takeda, Hidetoshi Seki, Shingo Itoh , Shin-ichi Zenbutsu

Electronic Device Materials Research Laboratory I, Sumitomo Bakelite Co., Ltd.
20-7 Kiyohara Industrial Park Utsunomiya－City Tochigi Prefecture 321-3231 Japan

> The influence of EMC (Epoxy Molding Compound) and reliability
> with fine Cu wire was studied. It is found that the failure mode was
> corrosion at the intermetallic layer of the wire bonding part by
> failure analysis after HAST (Highly Accelerated Temperature &
> Humidity Stress Test). Al elution and Cl ion was observed at the
> intermetallic layer. EMC with lower Cl ion content and Al inhibitor
> of corrosion shows better HAST property. The selection of frame
> retardant for green EMC is important because some retardants have a
> negative impact on HAST performance.

Introduction

Cu bonding wire (Cu wire) has several advantages over Au bonding wire (Au wire), like better electrical properties, mechanical strength and thermal conductivity. Thick Cu wire has been used for discrete applications already. Recently applications for fine Cu wire have become more and more popular[1].

Experiments

The equipment and materials used for this study are as follows: Wire bonder was an Eagle 60AP from ASM Assembly Technology Ltd. Cu wire is TC-E(4N, 25um) from Sumitomo Metal Mining Co., Ltd.. Package for reliability was SOP16L(7.2x11.5x1.95mmt, Alloy-42 L/F) with TEG(3.5x3.5x0.35mmt, Pad: 95x110um, Al/Ti-TiN-Ti/SiO : 600nm / 10-50-10nm / 1.0um). HAST condition was 130oC, 85%RH and 20V or 140oC, 85%RH and 20V. Judgment criterion of HAST evaluation was the circuit resistance becoming more than 20% compared to the initial one.

Wire bonding parameters adjustment

The structure of the bonding pad is one of the key factors for reliability of fine Cu wire application. Thicker Al pad with Cu layer is preferred [2]. Wire bonding parameters are adjusted before starting HAST assessment because the pad structure for this study is regarded as difficult one from the view points of bonding ability. As the result of bonding parameters adjustment, we've gotten the following values. Bonding diameter : 51um (ave.), 52um(max.) and 51um (min.), Bonding thickness : 14um (ave.), 15um(max.) and 14um (min.), Ball shear strength : 17.0kgf/mm2 (ave.), 18.5kgf/mm2(max.), 16.0kgf/mm2(min.). The cross section of the wire bonding part after bonding parameters is shown in Fig1.

Fig1. Cross section of wire bonding part

EMC sample preparation for HAST study

The four EMC samples based on green (BrSb free) EMC were prepared as shown in table1. Cl ion and IC (Inhibitor of Corrosion) are an important factor of metal circuit corrosion for this case. Different Cl ion content samples were prepared by applying a different purification level of epoxy resin. Low-Cl sample and High-Cl one are obtained from high purity epoxy for a semiconductor application and normal purity epoxy for standard purpose respectively.

Table1. EMC sample comparison for this evaluation

EMC	EMC#1	EMC#2	EMC#3	EMC#4
Epoxy	Low-Cl	Low-Cl	High-Cl	High-Cl
IC	No	Yes	No	Yes
Cl(ppm)	11	13	52	76
pH	4.0	5.7	4.0	5.7

1)IC: Inhibitor of Corrosion
2)Cl, pH: water extraction with PCT 125oC for 20h

The amount of Cl and pH were measured according to the following procedures. Purified water was added to crushed cured EMC and then the water extracted after being in a PCT 125oC for 20hr. Because the high purity epoxy is used in EMC#1 and EMC#2, the amount of Cl is about 1/5 compared to EMC#3 and EMC#4. The pH buffer agent was used with EMC#2 and EMC#4. The corrosion rate of Al is smaller around pH 5.5 and the rate is larger in lower pH and higher pH. The pH of extracted water of EMC#2 and EMC#4 was 5.7. In this pH, Al showes good corrosion resistance. It meets the facts described in former study [3].

HAST result

HAST results for four kinds of EMC are shown in Table2 and Fig2. It was found that the HAST property of Cu wire was inferior to that of Au wire. There was either a superiority or inferiority among the four kinds of EMC, shown in the results with Cu wire. From this point, it is considered that EMC influences the HAST property. HAST was better in the order of EMC#2, EMC#4, EMC#1 and EMC#3 for Cu wire. EMC#3 showed the worst HAST property, because the pH was low and the amount of Cl was high. EMC#4 was inferior to EMC#2, because the amount of Cl was higher though pH was good. EMC#2 showed excellent HAST property for both Cu and Au wire. The circuit resistance became larger only on the anode side. This phenomena also indicates Cl ion (anion) influences the HAST property.

Table2. Results of HAST 130oC, 85%RH, 20V

EMC	Epoxy	IC	MTTF(hr)	
			Au wire	Cu wire
EMC#1	Low-Cl	No	>480	120
EMC#2	Low-Cl	Yes	>480	>480
EMC#3	High-Cl	No	40	20
EMC#4	High-Cl	Yes	>480	300

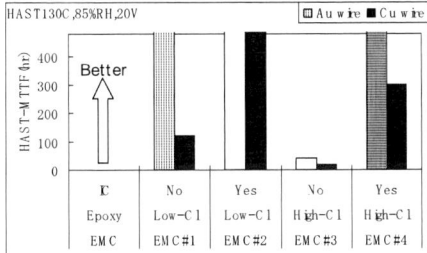

Fig2. HAST MTTF (Mean time to failure) comparison

HAST Failure analysis

Failure analysis in detail was conducted with the sample after HAST 140oC, 85%RH and 20V for 480h. Cross section of wire bonding part is shown in Fig3. A corrosion layer was observed between Cu and $CuAl_2$.

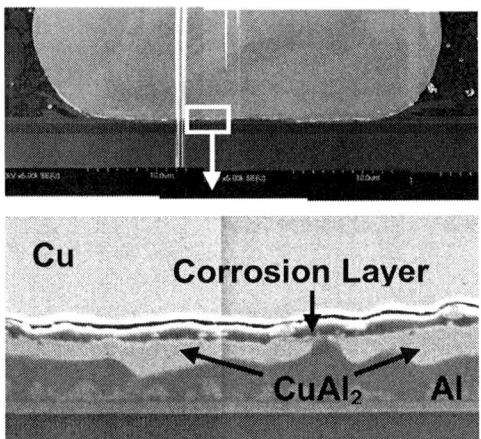

Fig3. Cross section of wire bonding part after HAST; 140oC, 85%RH and 20V for 480h.

TEM analysis was also conducted focusing on a corrosion layer and results are shown in Fig4. Al, O and Cl elements are observed at the corrosion layer. The intermetallic layer of wire bonding part contains Cu_9Al_4 and $CuAl_2$ [4]. $CuAl_2$ was observed but Cu_9Al_4 was not observed by this analysis. It is considered that Galvanic corrosion is the failure mechanism of this HAST property. Eluted Al from Cu_9Al_4 forms a corrosion layer like $Al(OH)_3$ and $AlCl_3$. Cu was not observed at the corrosion layer as the results of reprecipitation because of its ionization tendency characteristics.

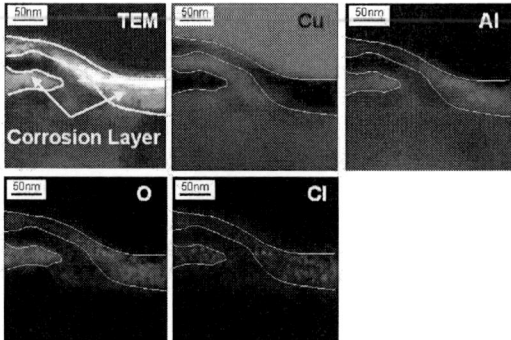

Fig4. Element mapping of corrosion layer by TEM

Study of several kinds of flame retardant

HAST property of EMC with several kinds of metal hydroxide flame retardant were evaluated. ALH(Aluminum hydroxide) and MGH(Magnesium hydroxide) were used for this study as shown in Table3. ALH is very stable and does not affect pH value itself. MGH is alkaline with less stability compared to ALH. EMC with MGH shows higher pH. There is no significant difference in Cl contents among the samples.

Table3. EMC sample comparison with various kinds of metal hydroxide
1)IC: Inhibitor of Corrosion

EMC	EMC#5	EMC#6	EMC#7
Epoxy	Low-Cl		
IC	Yes		
MH	No	Al H8wt%	MGH8wt%
Cl(ppm)	20	19	18
pH	6.0	6.1	7.8

2)MH: Metal Hydroxide

HAST results are shown in Table4 and Fig5. EMC with ALH remains a similar pH as EMC without ALH. EMC#6 shows good HAST property as same as EMC#5. HAST property of EMC#7 is inferior to that of EMC#6 because of its higher pH.

Table4. HAST(130oC,85%RH, 20V) results

EMC	MH	MTTF(hr)
EMC#5	No	>500
EMC#6	Al H8wt%	>500
EMC#7	MGH8wt%	60

1)MH: Metal Hydroxide

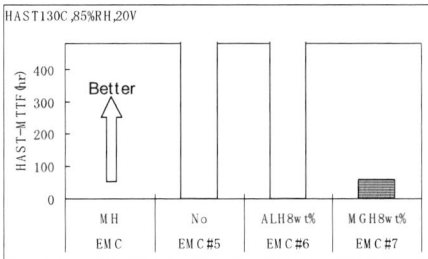

Fig5. HAST MTTF (Mean time to failure) comparison

FEM stress simulation

FEM stress simulation in the 20 pin SOIC package was conducted. Accordingly principal stress at Cu wire bonding connection is compression mode at room temperature. On the other hand, principal stress at Cu wire bonding connection area at high temperature shows rip up mode force. In the case of HTSL, stress generated by EMC may be one of important factor which determines reliability.

Conclusions

1) HAST property of Cu wire is inferior to that of Au wire. Galvanic corrosion of Cu_9Al_4 is the cause of HAST failure with Cu wire.
2) Combining pH buffer type and epoxy with low Cl contents, HAST property with fine Cu wire is greatly improved.
3) Green EMC without flame retardant and using ALH shows good HAST performance. HAST property of EMC with MGH is inferior to that of EMC with ALH because of its high pH.

Acknowledgments

The authors would like to thank Mr. Ryo Togashi from Sumitomo Metal Mining Co., Ltd., Mr. Eiji Murase from Ohkuchi Electronics Co., Ltd and Mr. T. P. Low from Malaysian Electronics Materials SDN.BHD., for providing Cu wire and useful discussion about wire bonding conditions.

References

1. Salim L. Khoury, David J. Burkhard, David P. Galloway, A Comparison of Copper and Gold Wire Bonding on Integrated Circuit Devices, IEEE Trans. on components, hybrids, and manufacturing technology, Vol.13, No.4 (Dec. 1990).
2. KNS Cu Wire Bonding Workshop, Jamin Ling, Enabling Cu Wire Bonding, K&S Copper Conference 2008
3. Materials Science and Technology (vol.19 of series) Ed. by S.Michael, Corrosion and Environmental Degradation - Corrosion of Non-ferrous Alloys- (2) Aluminum-based Alloys, WILEY-VCH (2000/5) pp.113-132
4. M.Drozdov, W.D.Kaplan, G.Gur & Z.Atzmon, In-Depth Microstructural Investigation of Copper Wire-Bonds, K&S Copper Conference 2008

830

Corrosion of Gold and Copper Ball Bonds

C. D. Breach[a], Ng Hun Shen[b], Teck Kheng Lee[b] and R. Holliday[c]

[a]ProMat Consultants, 160 Lentor Loop, #08-05 Tower 6, Singapore 789094
[b]ITE College Central, 20 Yishun Ave 9, Singapore 768892
[c]World Gold Council, 10 Old Bailey, London EC4M 7MG, United Kingdom

> Gold and copper ball bonds were isothermally aged under moist conditions (85°C and 85% relative humidity (RH)) and wet conditions (85°C in DI water with and without NaCl) in an effort to better understand the corrosion mechanisms that operate under moist and wet conditions. The objective of this work is to undertake and report on the initial stages of a research project that aims to compare the performance and assess the performance limits of gold and copper ball bonds.

Introduction

Replacement of gold with copper bonding wire has taken on a prominent position in the microelectronics packaging industry in recent years, largely driven by increasing gold prices [1-6]. There is no doubt that copper, like other materials, has an important role to play as an interconnect material but the expectation that copper is a 'drop in' solution for gold is worrying because there is a trend in the modern electronics packaging industry to forego extensive process and reliability testing, and rapidly implement a gold replacement programme in order to save costs. Copper, like other metals in microelectronics, has benefits and disadvantages and thus far only the benefits are being highlighted, whereas it is in the best interests of design, process and quality engineers to understand the performance limits of gold and copper so that materials selection can be made intelligently, with as much science as possible supporting decisions to replace gold with copper. It's also important to understand the true cost savings with copper wire by as thorough a cost analysis as possible but at present the situation regarding cost is poorly understood [7]. A recent SEMI survey [6] however, shows that the microelectronics packaging industry is beginning to think more critically about the copper versus gold debate. Despite such concerns, in some spheres of the industry there are blatant attempts to only present copper wire as a perfect replacement for gold and quash constructive studies that attempt to show a balanced view of the performance of gold and copper bonding wires.

One area of concern is performance of gold and copper under moist conditions, which is subject to standard test conditions such as PCT, 85°C-85% RH and so forth. According to recent studies, copper is not performing well under moist conditions [3,4], prompting interest in replacing copper wire with Pd-coated Cu wire but from a practical point of view it would be simpler to understand why copper does not perform well under moist conditions and try to find a solution rather than introduce a more complex material like Pd-coated Cu wire. Palladium coated copper wire FABS are harder than copper FABs [8], which causes problems with bond pad damage [9] but on the other hand, one of the main benefits of Pd coated Cu wire, known since introduction into industry in 2002/3, is improved stitch pull strength and better shelf life because of the Pd coating.

Interest in corrosion of copper [10-13] and particularly galvanic corrosion in ball bonds [11, 13] has recently grown due to moisture-induced failures during temperature cycling and PCT. It is usually considered sufficient to pass such tests without further scrutiny and the engineering and materials science that underlies materials performance in these tests is largely neglected. Standard tests are important but in order to understand corrosion processes in ball bonds it is often useful to perform non-standard tests under more extreme conditions to get to the underlying materials science. In line with this more fundamental approach to reliability, this paper presents preliminary results on the materials science of corrosion, failure and reliability of gold and copper ball bonds under dry, moist and wet conditions using a combination of standard (HTS, 85°C-85% RH, PCT) and non-standard (wet conditions with and without controlled ionic contamination).

Experimental Details

2N gold and 3N copper wires were used to make ball bonds of nominal diameter 55μm and height 13μm with an ASM Eagle 60AP. Copper ball bonds were made using an atmosphere of 95%N_2/5%H_2 (forming gas). Bond pad metallization was 1.2μm thick Al-0.5%Cu-1%Si. Ball height and diameter are shown in **Table 1**. Process optimization was performed using statistical optimization with power (P), force (F) and time (t) as parameters. After optimum parameters were determined, a confirmation run was performed. A Dage 5000 series pull/shear tester was used for ball shear testing. Shear velocity was 300μm/s and shear height was 4μm.

TABLE I. Average height and diameter of Au and Cu ball bonds.

Wire	Ball Height (μm)	Ball Diameter (μm)
Au	13.9±0.3	54.8±0.7
Cu	12.9±0.27	56.8±0.5

The following tests were performed:
1. Ageing at 85°C/85%RH.
2. Wet isothermal ageing in DI water at 85°C.
3. Wet isothermal ageing in DI water at 85°C with NaCl.

Neck pull, stitch pull and ball shear testing was performed on devices after ageing under the various conditions.

Results

Isothermal Ageing at 85°C/85% Relative Humidity

Figure 1 shows the effects of moisture on neck pull and stitch pull strength. Consider firstly gold results in Figure 1(a). At 500 hours, average gold neck pull strength deceased slightly and the distribution of pull strengths broadened markedly. The lowest neck pull value at 500 hours was about 60% of the average neck pull strength. Stitch pull strength in the as-bonded condition was initially tightly distributed and after 168 hours average stitch pull strength decreased slightly and the data distribution broadened towards lower strength. After 300 hours average stitch pull strength and data distribution remained more or less constant. Oddly, at 500 hours, average strength remained the same and the data distribution broadened towards high strength values and the data distribution appears almost bimodal.

Copper neck pull strength in Figure 1(b) was initially very tightly distributed and after ageing showed a steady decrease in average strength and broadening of the data distribution towards lower strength. Stitch pull strength was also tightly distributed in the as-bonded condition and after ageing for 168 hours average strength dropped by approximately 15% relative to the as-bonded condition and the data distribution became much broader. Average strength remained constant at 300 hours and increased slightly at 500 hours. The distribution of stitch pull strength remained broad at 300 and 500 hours.

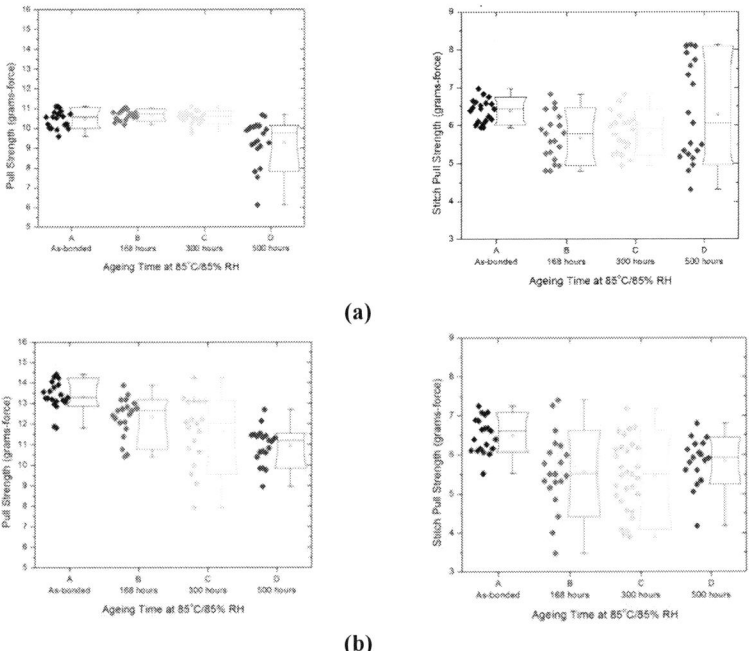

Figure 1. Neck pull strength (LEFT) and stitch pull strength (RIGHT) versus ageing time at 85°C/85%RH. (a) Gold ball bond (b) copper ball bond.

Figure 2 shows shear strengths of gold and copper. Gold shear strength in Figure 2(a) was high and the data tightly distributed in the as-bonded condition and at 168 hours. At 300 hours the average strength reduced slightly, the distribution broadened and some low shear strengths were observed, the lowest value being about 20% less than the average value. At 500 hours average shear strength was similar to the 300 hours average and the distribution somewhat broadened relative to the as-bonded condition with the lowest value being about 10% less than the 500 hours average.

In Figure 2(b) copper had a high average strength and was tightly distributed in the as-bonded condition. After ageing at 168 hours the average strength was similar to the starting condition and the distribution was broader with the lowest shear strength value being approximately 15% lower than the average at 168 hours. The average value at 300 hours decreased slightly but recovered slightly at 500 hours. The major trend in the copper data was broadening of the data distribution, with the lowest values at 300 hours and 500 hours being about 25% and 30% less than the average values for those times respectively.

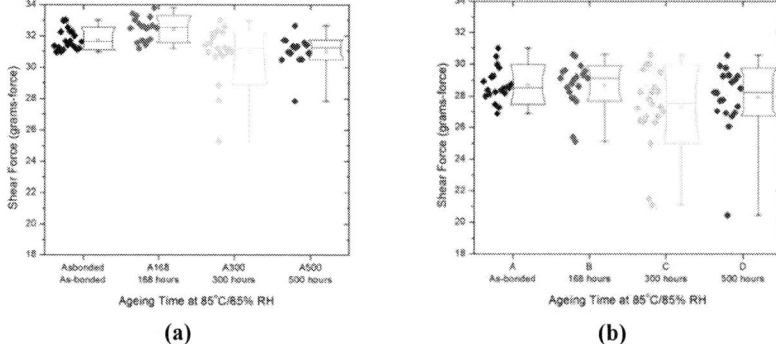

(a)   (b)

Figure 2. Ball shear strength versus ageing time at 85°C/85%RH. (a) Gold ball bond (b) copper ball bond.

Figure 3 shows representative SEM images of sheared gold and copper balls after 300 hours. In Figure 3(a) the shear tool sheared through the gold ball was sheared leaving part of the ball on the bond pad, as expected. At the ball periphery parts of the ball appeared to have sheared away form the intermetallic and had the appearance of a brittle fracture, a morphology that is similar to that of oxidized Au-Al intermetallics. The copper ball in Figure 3(b) showed some regions where aluminium was plastically deformed and others where the surface appeared smooth. The absence of plastic deformation of aluminium in combination with the smooth surface is indicative of probable damage to the interface between the copper ball and the intermetallic. Unfortunately at the time of writing qualitative chemical analysis on the surface of the sheared samples, which would have helped in further analysing the surface, was not available.

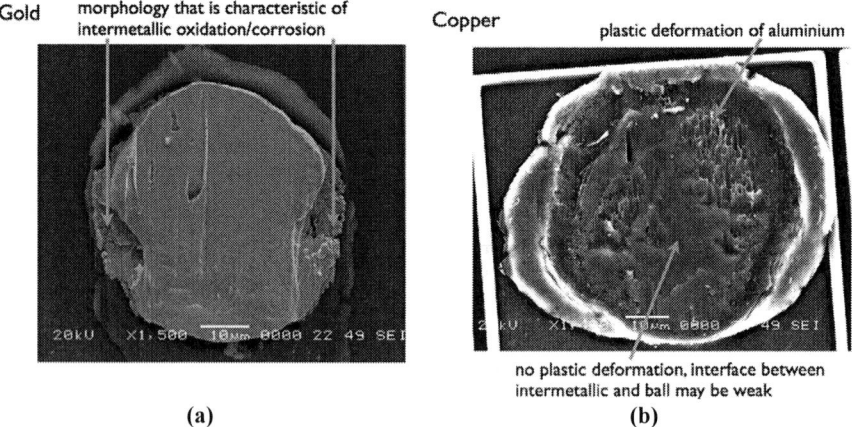

(a)   (b)

Figure 3. SEM images of the die side of sheared ball bonds after 300 hours ageing at 85°C/85%RH. (a) Gold ball bond (b) copper ball bond.

Isothermal Ageing at 85°C in DI Water

Wet ageing is an extreme condition that is not likely to be encountered under normal operating conditions but exposure of devices to wet conditions is expected to accelerate corrosion relative to moist conditions and may give information that can be used to understand intermediate conditions such as 'moist' ageing. Limited shear testing data up to 60 hours in Figure 4(a) shows that gold maintained high shear strength but copper ball bonds showed an increase and then decrease in shear strength, but the balls can still be considered strong. Devices were aged up to 168 hours but gold did not show any degradation or loss of pull or shear strength. After 36 hours, copper ball bonds showed normal shear failures and a typical example is shown in Figure 4(b) where intermetallics clearly have grow very little and failure occurs by plastic deformation in the Al-alloy bond pad.

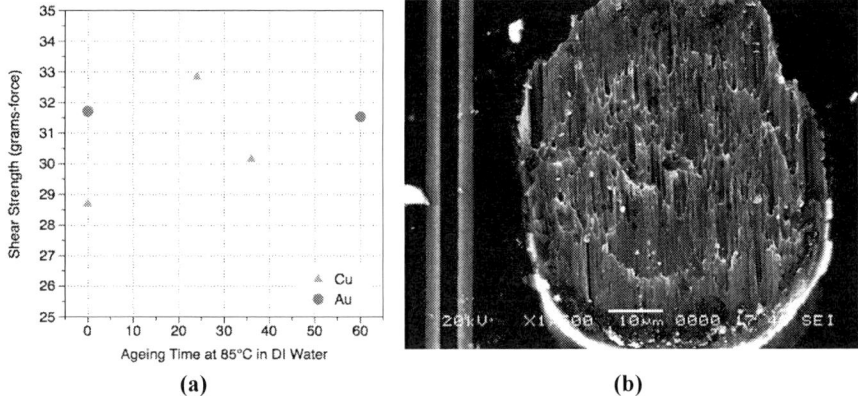

(a)   **(b)**

Figure 4. (a) Graph of shear strength versus ageing time at 85°C (b) SEM image of the bond pad side of Cu ball bonds after 36 hours ageing in de-ionised water for 36 hours and shear testing.

After 168 hours nearly every ball came off the bond pad and various features were observed. Figure 5(a) shows massive cracking of what appears to have been the passivation layer and the Al bond pad appears to have been completely removed with some white residue on the underlying passivation layer. Figure 5(b) shows passivation cracking but with large amounts of a white deposit where the bond pad should be found. The balls were not dissolved, simply detached from the chip.

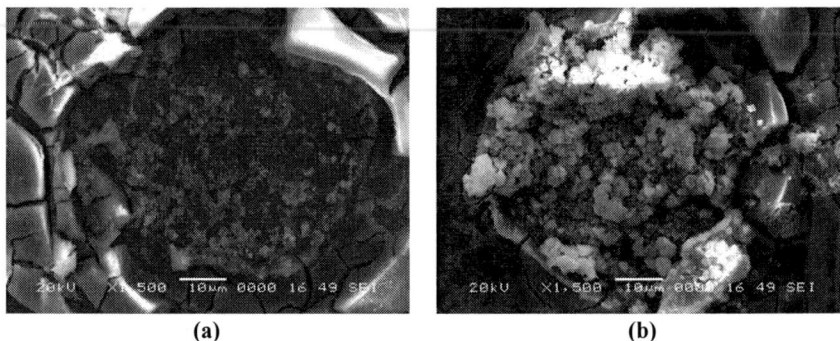

| (a) | (b) |

Figure 5. SEM images of the bond pad side of Cu ball bonds after 168 hours ageing in deionised water for 36 hours and shear testing (a) massive cracking of the passivation layer and appearance of white deposits at the centre of what was the bond pad (b) large white deposits where the ball was bonded.

An EDX analysis of a deposit like that in Figure 5 is given in Table 2. Negligible Al is present and the analysis indicates that the white residue is largely made up of copper and oxygen and is most probably a copper oxide of some form.

TABLE 2. EDX analysis of the white residue in Fig. 5.

Element	Weight %	Atomic %
C	5.37	13.58
O	23.08	43.79
Al	0.91	1.02
Si	2.25	2.43
Cl	17.16	14.69
Cu	51.24	24.48

Isothermal Ageing at 85°C in DI Water and NaCl

NaCl is well known to accelerate corrosion and chemical reactions in microelectronics. Adding a small amount of NaCl to DI water induced very rapid failure of copper ball bonds and as Figure 6(a) shows, a copper ball can be completely removed from the bond pad, leaving the bond pad almost completely intact and appearing to be very planar. EDX analysis of the bond pad confirmed the bond pad to be intact (Table 3). However, part of the bond pad appeared to be removed from the passivation. EDX analysis in Table 4 confirmed that the layer in Figure 6(b) was the Si passivation layer.

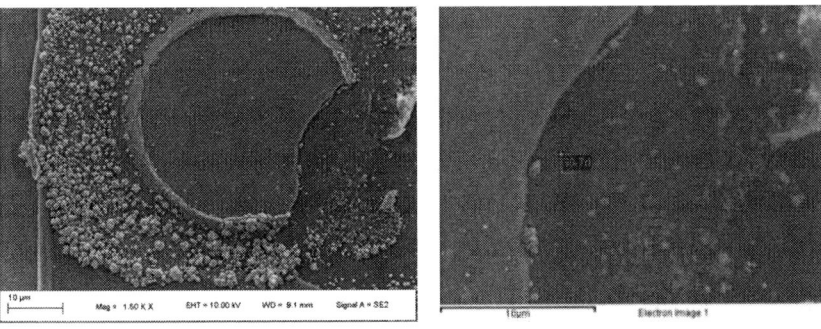

| (a) | (b) |

Figure 6. a) SEM image of the bond pad side of Cu ball bond after 30 minutes ageing in NaCl-DI water solution. (b) area where bond pad has detached from the passivation layer.

TABLE 3. EDX analysis of bond pad in Fig. 6.

Element	Weight %	Atomic %
O	19.31	31.56
Al	56.62	54.87
Si	7.05	6.56
Cu	17.02	7.01

TABLE 4. EDX analysis of passivation in Fig. 6.

Element	Weight %	Atomic %
C	3.65	5.85
O	54.23	65.26
Al	0.52	0.37
Si	41.60	28.52

Discussion

Isothermal Ageing at 85°C/85% Relative Humidity

The results generally show that gold ball bonds are not as susceptible to oxidation/corrosion effects as copper ball bonds. However, the sheared gold ball in Figure 3(a) shows evidence of intermetallic oxidation around the periphery, a conclusion that is drawn based on previous publications that showed this failure mode [14-16]. The Au_4Al based intermetallic (which may contain other elements from the bond pad) can oxidise under dry conditions such that Al forms an oxide film and Au is precipitated. Oxidation and corrosion being essentially surface and interfacial phenomena, it is not surprising that attack of intermetallic typically starts from the periphery of ball bonds and progresses inwards, reducing the bonded area and resulting in lower shear strength. Copper in Figure 3(b) shows areas where Cu-Al intermetallics are well bonded to the copper ball and because the shared chemical bonds between Cu-Al intermetallics and Cu and Cu-Al intermetallics and Al are very strong, plastic deformation of the ball occurs in the weakest phase i.e. the aluminium bond pad. Therefore, a reasonable but currently unproven hypothesis is that in those regions where aluminium plastic deformation occurs the Cu-Al intermetallics are strongly bonded and unaffected by moisture (oxidation or corrosion). Precisely which intermetallics are present at these areas is not known in this study but according to Boettcher et al [13] $CuAl_2$ forms near the Al bond pad and CuAl and Cu_9Al_4 adjacent to the copper ball. Boettcher et al concluded that high reliability copper ball bonds consisted of $CuAl_2$ at the ball periphery and $CuAl_2$ and CuAl at the centre of the balls, whereas lower reliability ball bonds had CuAl as well as $CuAl_2$ at the ball periphery and Cu_9Al_4, $CuAl_2$ and CuAl at the ball centre. Drozdov et al [17] concluded from TEM studies that $CuAl_2$ was the dominant compound in as bonded copper balls but generally the number and type of Cu-Al intermetallic formed seems to vary considerably and it is possible that at some regions there may be multiple layers of intermetallic and at others single layers. In addition intermetallic formation is accelerated at the periphery of ball bonds under the inner chamfer region of the capillary where the highest stresses occur, which may also favour some intermetallics over others. Boettcher et al [13] concluded that moisture had little effect on $CuAl_2$ and corroded or oxidised CuAl and primarily Cu_9Al_4 and that when the latter compounds degraded, an amorphous Al oxide formed over $CuAl_2$ that remained adhered to the bond pad. The Al oxide layer contained Cu precipitates and the microstructure seen in Figure 4 of Boettcher et al [13] resembles the microstructure of Au precipitates and Al oxide in references [14-16] that formed under dry conditions, which is due to selective (internal) oxidation of Al and precipitation of Au. Oxidation of Al in bulk intermetallics [18] and precipitation of Au in thin films of Au-Al intermetallics has been observed [19, 20] and generally occurs in aluminide compounds [21-23]. Apparently the same or a similar process can occur in Cu-Al intermetallics, as Boettcher et al's study seems to show. The smooth regions in Figure 3(b) were unfortunately not analysed and speculatively it is concluded that those regions may be Cu-Al intermetallic that remained adhered to the Al bond pad or a layer of Cu precipitates with Al oxides (as per Boettcher et al). There are then two simple explanations of the nature of the smooth region:

a) Under the assumption that the smooth phase is $CuAl_2$, it would be plausible as per Boettcher et al's study, that if CuAl and/or Cu_9Al_4 compounds were present above $CuAl_2$, the latter would not be attacked but the former may oxidise or corrode and result in weakening of the $CuAl_2$ / CuAl and Cu_9Al_4 interfaces. Figure 7(a) illustrates the mechanism.

b) The smooth region is an amorphous Al oxide with dispersed Cu precipitates as per Boettcher et al's observations. Figure 7(b) illustrates the concept.

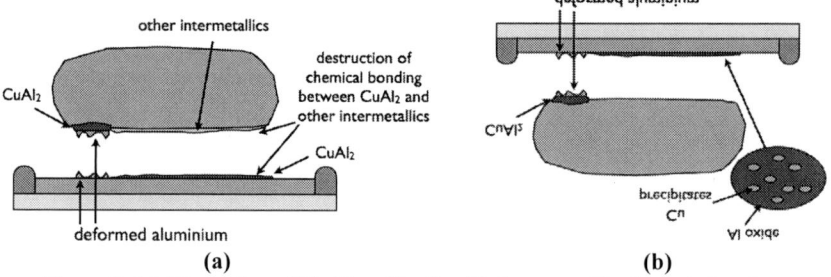

Figure 7. (a) Illustration of CuAl and/or Cu_9Al_4 intermetallic degradation due to oxidation/corrosion and destruction of chemical bond with $CuAl_2$ resulting in separation of ball and bond pad (b) Illustration of intact $CuAl_2$ and internally oxidised other Cu-Al intermetallics. Not to scale.

Confirmation of one or other of these hypotheses requires further analytical work, which is ongoing.

Isothermal Ageing at 85°C in DI Water

The presently limited results show that when immersed directly in water, copper ball bonds on aluminium began to fail grossly after an incubation period of at least 36 hours simply by balls detaching from the bond pads. The mechanism by which failure occurred is not certain, but is believed to be galvanic corrosion i.e. water acts as an electrolyte that completes a circuit between a more noble material (cathode) and a less noble material (anode) with the result that the lower nobility material (the anode) disintegrates and may undergo further reaction to form other substances. Direct immersion in water complicates understanding what is corroding because there are several different materials in contact with water, each of which has an unknown corrosion potential. The residue in Figure 5 is rich in copper but not in aluminium and the aluminium bond pad is no longer present, which suggests that aluminium has dissolved and possibly reacted with hydroxyl ions but has not deposited around the corrosion site. The presence of copper rich residues like those in Figure 5 show that copper has apparently also corroded. Cu-Al intermetallics, whichever composition was initially present, would have been so thin as to be undetectable by routine analysis even before immersion in water and so it is impossible to know if any intermetallics are still present. A key drawback to analysing this situation is that there are no measured corrosion potentials for any of the materials used and therefore only data on materials from other studies, which is not directly representing the actual materials used in this study, can be drawn upon. An example of such data is given in Table 5, which shows that Cu is nobler than Al and can be nobler than $CuAl_2$ with the caveat that the copper purity and composition may possibly change the position of copper relative to $CuAl_2$. Galvanic couples of the materials in Table 5 could be Cu-$CuAl_2$, Cu-Al and $CuAl_2$-Al, where the first material denotes the cathode (non-consumed). Copper would not appear as an anode and would not be expected to corrode unless of course there are other chemical effects and given that copper is corroding, the situation is clearly more complex. The SiN passivation layer was also observed to crack (Figure 5(a)) and may also be involved in some chemical processes.

TABLE 5. Corrosion potentials of various thin film metals and alloys determined in a 0.1M NaCl at pH 6 using the microcell method [24]..

Material	Corrosion Potential φ (mV)
Cu (99.9%)	-232
$CuAl_2$	-665
Al (99.999%)	-823

Isothermal Ageing at 85°C in DI Water with NaCl

There was no apparent chemical damage to the aluminium bond pads after ageing in DI water with NaCl (Figure 6), probably because the time was too short to initiate corrosion. However, while the bond pad was intact, the Cu balls detached from the bond pads very cleanly, which may mean that the solution of NaCl in water attacked and corroded the Cu-Al intermetallic phases. Partial exposure of the SiO_2 passivation layer under the bond pad suggests the salt solution also destroyed the adhesion at the Al-SiO_2 interface. Uno has proposed a mechanism of attack of Cu_9Al_4 by Cl ions from epoxy moulding compounds that is proposed to result in formation of CuAl and cracks [4]. Chlorine ions from the salt could, according to Uno's proposal account for the lifting away of the ball from the bond pad. Uno's proposed reaction scheme is as follows:

$$Cu_9Al_4 + 12Cl^- \rightarrow 4AlCl_3 + 9Cu + 12e^-$$

$$H_2O + \frac{1}{2}O_2 + 2e^- \rightarrow 2OH^- \tag{5}$$

$$AlCl_3 + 3OH^- \rightarrow Al_2O_3 + 3HCl + 3e^-$$

$$Cu_9Al_4 + 5Al \rightarrow 9CuAl$$

However, it is unclear what intermetallics are present in Cu ball bonds on aluminium. Future work aims to thoroughly analyse the corrosion mechanisms and gain a clearer picture of the corrosion mechanism.

Conclusions

Gold and copper ball bonds have been exposed to two conditions at 85°C: one with 85% relative humidity and the other direct immersion in water. The following conclusions are drawn:

a) 85°C/85% RH: Gold ball bonds show minor intermetallic degradation and minor changes in shear strength. Copper intermetallics show signs of gross degradation due to intermetallic oxidation resulting in significant decreases in shear strength.

b) 85°C-direct immersion in DI water: Corrosion products contain high amounts of copper and negligible aluminium and the aluminium bond pad is no longer present, which suggests that both copper and aluminium corrode.

c) 85°C-direct immersion in DI water with NaCl: NaCl accelerates ball lift off while not affecting the aluminium bond pad, probably because of insufficient time for corrosion of aluminium to take place,

These data are preliminary and further detailed analytical work continues in an effort to understand the oxidation and corrosion mechanisms of gold and copper ball bonds on aluminium alloy bond pads.

Acknowledgments

The authors would like to thank the World Gold Council for funding this work that is part of a project to understand the limitations of gold and copper ball bond reliability.

References

1. A. Shah, M. Mayer, Y. N. Zhou, S. J. Hong, J. T. Moon. *IEEE Trans. Elec. Packgg. Manfg.* **32** 176 (2009).
2. S. Kaimori, T. Tonaka, A. Mizoguchi. IEEE Trans. *Adv. Packgg.* **29** 227 (2006).
3. T. Uno, S. Terashima, T. Yamada. *IEEE Proc. ECTC* 1486 (2009).
4. T. Uno. *Microelectronics Reliability.* In Press (2010).
5. C. J. Vath, M. Gunasekaran, M. Ramkumar. *IEEE Proc. 11th EPTC*, 9-11 Dec. 2009 p374.
6. Semiconductor Industry Opinions Concerning the Selection of Bonding Wire Material, *SEMI Market Survey*, January 2010.

7. C. J. Vath, R. Holliday. *Proceedings of CSTIC 2011 Conference*, March 15-17 Shanghai New International Expo Centre, Shanghai China.
8. D. Stephan, F. Wulff, E. Milke. *Proceedings of 12th Electronics Packaging Technology Conference (EPTC)*, Shangri-La Hotel, Singapore 8-11th December 2010 p. 343.
9. O. Yauw, H. Clauberg, K. F. Lee, L. Shen, B. Chylak. *Proceedings of 12th Electronics Packaging Technology Conference (EPTC)*, Shangri-La Hotel, Singapore 8-11th December 2010 p. 467.
10. C. D. Breach, R. Holliday. *Proceedings of 11th International Conference on Electronic Packaging Technology and High Density Packaging (ICEPT)*, Xi'An, China, 16-19th August 2010.
11. C. D. Breach, Ng Hun Shen, Tee Wai Mun, Teck Kheng Lee and R. Holliday. *Proceedings of 34th International Electronics Manufacturing Technology Conference (IEMT)*, Melaka, Malaysia, 30th November-2nd December 2010.
12. C. D. Breach, Ng Hun Shen, Tee Wai Mun, Teck Kheng Lee and R. Holliday. *Proceedings of 12th Electronics Packaging Technology Conference (EPTC)*, Shangri-La Hotel, Singapore 8-11th December 2010 p. 214.
13. T. Boettcher, M. Rother, S. Leidtke, M. Ulrich, M. Bollmann, A. Pinkernelle, D. Gruber, H. J. Funke, M. Kaiser, K. Lee, M. Li, K. Leung, T. Li, M. L. Farrugia, O. O'Halloran, M. Petzold, B. März, R. Klengel. *Proceedings of 12th Electronics Packaging Technology Conference (EPTC)*, Shangri-La Hotel, Singapore 8-11th December 2010 p. 585.
14. F. Wulff, C. W. Tok, C. D. Breach. *Materials Letters* **61(2)** (2007) 452.
15. C. D. Breach, F. Wulff. *Microelectronics & Reliability* **46(12)** (2006) 2112.
16. C. D. Breach, C. W. Tok, F. Wulff, D. Calpito. *J. Mater. Sci.* **31** (2004) 6125.
17. M. Drozdov, V. Gur, Z. Atzmon, W. D. Kaplan. *J. Mater. Sci.* **43** (2008) 6029.
18. C. Xu, C. D. Breach, T. Sritharan, F. Wulff, S. G. Mhaisalkar. *Thin Solid Films* **462-3** (2004) 351.
19. H. Piao, N. S. McIntyre. J. Electron Spectrosc. Relat. Phenom. **119** (2001) 29.
20. H. Piao, N. S. McIntyre, G. Beamson. J. Electron Spectrosc. Relat. Phenom. **125** (2002) 35.
21. G. H. Meier, F. S. Petit. *Materials Science & Engineering* A **153** (1992) 548.
22. H. Schmalzried. *Chemical Kinetics of Solids*. Chapter 9. VCH Publishers (1995)
23. H. J. Grabke. *Materials Science Forum* **149** (1997) 251.
24. N. Birbilis, R. G. Buchheit. *J. Electrochemical Soc.* **152** (2005) B140.

Cost-effective Use of Gold Wire in Semiconductor Packaging

Charles J. Vath, III[a], R. Holliday[b]

[a] ComSol Consulting Pte. Ltd., 369 Holland Road #09-01, Singapore 278640
[b] World Gold Council, 10 Old Bailey, London EC4M 7NG, United Kingdom

Abstract

The semiconductor industry has maintained an average 15% cost down per year over the range of products it offers. Continued cost reduction will be difficult to achieve based on current practices and raw material prices. The need to remain competitive has compelled integrated device manufacturers and subcontractors to look at alternative material types, most notably copper wire. A recent survey of the industry has revealed that there remain very widespread concerns on migrating away from the use of gold. It is important that the cost of using gold wire is minimized without compromising on product reliability and quality. We present a critical review of the wire bonding process focusing on various practices, procedures, and material choices. Such analyses indicate, for example, that a reduction of the FAB ratio in a 32 I/O QFN with 38 wires from 2 to 1.4, would save 1 km per 1,100 strips.

Introduction

The industry has traditionally maintained an average 15% cost reduction per year over the range of products it offers. Part of this reduction comes from packaging where manufacturing efficiencies such as cycle time reduction, supply chain management, and yield improvement allow for continued reductions in the cost per interconnect. In early 2009, this reduction was forecasted to be between 0 to 4.2% (1) over a five-year period, based on package type. Small outline thin (SOT) packages represented the package with no reduction, while Wafer Level Packages (WLP) are expected to show the most reduction as volumes increase and yields continue to improve. As the leaded package families are expected to increase in volume, while die sizes shrink over the same period, the cost reduction targets may be difficult to meet in the future based on current practices and raw material prices.

Wire bonding is still the dominant interconnection technique in semiconductor packaging. Gold wire is the easiest and fastest wire to use in the bonding process and it remains the material of choice for the majority of wire-bonded devices produced today. However, the need to reduce material costs and remain competitive in the current market environment has compelled integrated device manufacturers and subcontractors to look at alternative material types, most notably copper wire.

At face value, copper might appear to be a drop in replacement for gold. However, the results of a recent survey (2), commissioned by The World Gold Council and conducted by SEMI®, the global semiconductor industry association, shows the semiconductor industry has serious reservations about the reliability and yield of copper bonding wire. SEMI® surveyed 46 leading semiconductor companies across the world to determine the extent of copper bonding wire programs in the industry and to identify the key issues and considerations related to decisions in selecting bonding wire material. Issues that were highlighted included: in-service product reliability, process yield, unproven historical performance, the increased costs in having to purchase new equipment, copper's unsuitability for complex wire loop shapes in advance packages, difference in electrical performance and comfort with the established supply base for gold bonding wire. In addition, research over the last two years indicates that there are reliability related issues with the copper/aluminum bond when exposed to tests that utilize high moisture levels (3,4). The performance of such bonds is inferior to the established gold/aluminum bonds. Thus, the semiconductor industry finds itself caught in a dilemma.

In order to minimize the impact of the cost of gold on the overall profit margins of the device assembly companies, a detailed analysis has been performed to show how to use gold in the most cost-effective manner. It is a critical review of the wire bonding process with a focus on the practices and procedures which may contribute to excess wire usage during the normal course of daily operations. It also provides information on how to evaluate current gold consumption by device and how modification of the current bonding procedures can reduce the amount of gold used.

Major Factors Impacting the Cost-effective Use of Gold

There are many variables that have a direct impact on how cost effectively gold is being utilized in package assembly.

- Device interconnect requirements - wire diameter, bonded ball size (reduced free air ball), loop height/shape
- Package design - wire length, loop height, stitch bond position, lead frame/substrate finish
- Capillary design - to facilitate small ball, eliminate second bond issues, materials to extend life time
- Bonding parameters - robust process window
- Bond pad surface condition and underlying material - metallization type and thickness, surface cleanliness

All factors shown in Figure 1 affect the process and, therefore, gold consumption.

Device Interconnect Requirements

Wire Diameter. The device interconnect requirements are those established by the end customer at the time of order placement or contract negotiation. At this time, wire diameter, bond pad pitch, bond pad opening, looping requirements, and wire purity are agreed upon. In addition, quality specifications involving visual inspection and destructive testing such as wire pull, ball shear, stitch pull, and neck pull are determined. Frequencies of testing, sample size and control points are also established.

A key objective is to determine the smallest acceptable wire diameter based on the device performance criteria and physical layout of the chip. This is the first opportunity to manage the cost of the gold wire and care should be taken to obtain consent for the smallest wire diameter the device can use.

Figure 2 shows the relationship between cross-sectional area (DC resistance), the surface area (skin effect for high frequency) and relative cost for various wire diameters. As can seen from the graph the surface area scales with the diameter but the cross-sectional area scales with the square of the ratio of the radii. Thus, a smaller wire will better support high frequency than a DC current. Therefore, it is best to use the smallest wire and finest pitches allowable when taking into account both high frequency and DC electrical requirements (5). This will provide the largest percentage improvement in cost.

Bonded Ball Size. It is on the periphery of the device where most of the bond pads are located. These may be in a single ring or multiple rings depending on the die size and the complexity of the device. The characteristics of the bond pads that define the first bond criteria are the bond pad pitch (BPP) and bond pad opening (BPO). These will determine the maximum diameter of the deformed ball and the maximum wire diameter that can be used. Depending of the customer requirements for ball shear, one should use the smallest ball size necessary to achieve the required shear values.

Figure 3 shows the relationship between the amounts of wire consumed and a given free air ball (FAB) ratio. A wire size of 25.4 µm was used for illustrative purposes. A reduction of 30% in the FAB ratio will result in a decrease of 65% in the amount of wire required to form the smaller FAB. For a device that has 600 bonds in it, this represents a savings of 48 mm of gold wire per unit.

Loop Shape/Height. With multiple rows of bond pads in a staggered configuration, as shown in Figure 4, wire loops will not be coplanar. Many devices require six or more different loop shapes to complete the connections from the device to the substrate. This coupled with multiple wires for power and ground connections (which also require different loop heights), results in a challenge to the engineer to obtain optimized trajectories while minimizing the use of wire. In addition to these constraints, one must also consider the shape with respect to mechanical rigidity. Wires cannot droop or sag or be swayed during molding as these defects will result in diminished performance or device failure.

Figure 5 shows a schematic of a typical loop in a high I/O package. The various waypoints along the loop above the die are used to define the loop shape. The wire span is the straight-line distance between the bond pad and the lead tip. This is the dimension used by many models for wire length. The last kink distance (LK_D) and last kink height (LK_H) also impact the overall rigidity of the loop, wire straightness, and cost. The equations relating all of the geometries to the overall usage such as total wire length (WL_T), wire diameter (d_W), loop height (BLH), wire length for the FAB (WL_{FAB}), FAB diameter (d_B), FAB ratio (R_{BW}), stitch bond length (L_{SB}) are shown below.

$$WL_T = WL_{FAB} + BLH + (L_{WS} + SQRT (LK_H^2 + LK_D^2)) + L_{SB} \qquad [1]$$

Where

$$WL_T = Total\ Wire\ Length \qquad [2]$$

$$FAB_{VOL} = (4/3) \, \pi \, (d_B/2)^3 = (\pi \, (d_W/2)^2) \, WL_{FAB}$$

[3]

$$WL_{FAB} = (4/3)((d_B/2)^3/(d_W/2)^2) = (4/3)(d_B^3/2d_W^2) = (4/3)(((R_{BW}d_W)^3)/2d_W^2)$$

[4]

$$WL_{FAB} = (4/3)(R_{BW}^3 d_W/2) = (2/3)(R_{BW}^3 d_W)$$

[5]

Table 1 provides illustrative numbers for two popular package types using the equations shown above. Different wire diameters were also used for the BGA to illustrate the cost impact. For a reduction of 5.1 microns (22.3%) in the wire diameter there is a 40% reduction in the cost of the gold in the package.

Wire Purity. Decreased wire purity may be an issue with respect to the device performance and it plays a crucial role in providing additional mechanical strength to the wire. Wires of lower purity have higher resistance and hardness. This will affect device performance. Productivity will also be affected due to the loop shape complexity as it takes more control and time to obtain the best shape. But the benefit is that the wires tend to stay in place through subsequent processing. The hardness will also affect the response of the pad to the impact of the ball. But wire bond systems have additional controls to assist in mitigating this impact. Solutions do exist for 2N Pd doped gold wire (6).

Package Design.

Package design is important for the performance of a device, both electrically and environmentally. Examination of the design aspects with respect to wire usage, lead routing and substrate finish highlights areas on which to focus to reduce consumption.

Wire Length. Wire length impacts yield, cost, and performance. The distance that a wire needs to cover to complete a connection should be minimized by design of the substrate or lead frame. The lead frame should be designed such that the gap between the leads and the die attach pad (DAP) are as tight as the etching or stamping process will allow. As lead frame materials become thinner, this gap can be reduced. The DAP should be large enough for the die and any necessary ground bonds. A small die on a large DAP consumes more wire, more time to bond, and runs the risk of wire sweep during molding. To keep costs low for organic substrates, lead routing and lead tip pitches should be optimized for both the substrate manufacturing process and for the wire bond process. Lead lines and spaces should be minimized as much as possible to allow the lead tips to be closer to the die. This reduces wire consumption, increases bonding throughput, and reduces the risks of wire sweep at molding. In a paper presented by Flynn Carson (7), he showed that for a 12 mm X 12 mm die in a BGA package, wire lengths could be reduced by approximately 30% for a reduction of 10 microns (90 μm pitch to 80 μm pitch) in lead pitch.

Loop Height/Shape. Loop heights should be as low as possible to allow for the thinnest package format. Thinned die allow for lower overall loop heights bringing down costs further. Loop shapes should be as triangular as possible. It can be seen from equations [1] how these various parameters influence wire consumption. Lower loop heights decrease *BLH*, and *LK_H*. A more triangular shape will increase *LK_D*, which will also reduce the amount of wire consumed.

Stitch Bond. The second bond should be as close to the lead tip as possible. This will require good lead tip geometry control at etching, stamping, and coining. For organic substrates, features should be sufficiently compensated to allow for the maximum flat tip width. These factors also impact the strength of the second bond and machine performance as irregular lead tips result in weak or lifted stitch bonds and missing FABs. The second bond should lie along a line that denotes the shortest path between the die bond pad and the lead tip. This is only possible when lead tips are flat and well controlled dimensionally. The second bond length should be tightly controlled, as this will impact the strength of the bond and gold consumption.

Lead Frame/Substrate Finish. The last topic to be covered in this section deals with the plated finish of the lead frame or substrate. Not all plated layers are created equal. While the more common-place techniques (XRF, SEM microsection) for measuring the characteristics of the plating may indicate that the plated finishes are the same, they may give very different responses at wire bond causing non-stick or lifted stitches resulting in lost gold and productivity. The morphology of the plated layer has a direct impact on the second bond. Different crystal structures of the nickel plating on pre-plated QFN lead frames with Ni/Pd/Au finishing will allow either smooth bonding or create many occasions for lifted second bonds, open EFO, and other second bond related machine stoppages. All of these process perturbations can contribute to gold loss. Gold plating is also a contributor to second bond problems for the same reasons. L. Crider, et al. reported different gold structures with very different hardnesses even though the thicknesses and roughness measurements were within specification (8). Care should be taken to optimize plating morphology in order to maximize second bond performance.

Coatings. In addition to plating morphology, coatings used to control epoxy bleed out (EBO) on plated lead frames can cause non-stick if not formulated and applied correctly.

Capillary Design

In achieving the most cost effective use of gold wire, it is important to understand how the design of the capillary used in the wire bonder can affect the bonding process. Figure 6 (9) shows the key design features to help those who may not be familiar with the terms used in this section. The size of the free air ball (FAB) is a critical value that must be identified in order to proceed with the design of a capillary. The FAB affects the centering, thickness, shape, and squashed ball diameter. The volume of the FAB must be larger than the sum of all volumes from the other geometries selected for the tip of the capillary.

Mechanical Design to Facilitate Small Ball and Low Loops. Capillary profiles affect the ball bond and the loop control of the wires. The chamfer angle (CA) affects the shape of the squashed ball and how well the ball bond is centered on the wire. It also affects how the ultrasonic energy and stress are transmitted from the transducer to the ball/pad interface during ultrasonic excitation. The strength and intermetallic coverage (IMC) of the resultant bond depends on the proper selection of this value. Figure 7 from a paper by

N. Srikanth et al. (10) shows a model result for the von Mises stress and plastic strain. The chamfer diameter (CD) also affects the ball shape and size as well as centering of the ball prior to squashing of the ball.

Hole Diameter. The hole diameter (H) is selected based on the wire diameter to be used. Ample clearance needs to be allowed to minimize drag on the wire during high-speed motion of the bond head during loop formation. Insufficient clearance will result in broken wires. Excessive clearance will result in inconsistent loop heights and wire straightness.

Tip Diameter. The dimension of the tip diameter (T) is driven by the BPP, BPO and squashed ball diameter. As these dimensions continue to decrease so does the tip diameter (t). Thus, it is even more important to control dimensions and tolerances. Small variations can result in large process variations, as the relationships are non-linear.

Shape. The loop height and bond pad pitch impact the bottleneck height (BNH) and bottleneck angle (BNA). High loops with very fine pitch will result in a capillary design that has a long thin tip. This will lead to failures of the capillary due to breakage at the transition zone. This is another reason for keeping loops as low as possible.

Mechanical Design to Eliminate Second Bond Issues. There are capillary design aspects that impact the second or stitch bond. The face angle (FA) determines the thickness of the material in the stitch bond transition zone. This angle varies depending on the type and hardness of the wire being used as well as the surface on which the second bond is placed. Incorrect selection will result in cut wires or lifted stitch bonds.

Outside Radius. The outside radius (OR) affects the strength of the stitch bond under stitch peel testing. If the radius is too small, the tip acts like a cutting edge and will cut the wire at the stitch/wire transition. If the value is marginal, low stitch peel and broken stitch failures will result. When this occurs, no wire is pulled out to form a tail so that a new ball can be formed. Very often this leads to machine stoppage that requires the operator to rethread the capillary. This could result in a loss of 50 mm or more of wire during the 'rethreading' process.

Tip Finish. Capillary finishing refers to the texture of the tip. A smooth finish has a low build-up rate but also poor coupling between the wire and the tool. A matte finish refers to a roughened surface. This improves the grip of the tool on the wire. However, it also results in faster material build-up on the tool that will lead to a shorter tool life.

Tip Diameter. The tip diameter (T) needs to be considered with respect to second bond as well. Very small T values with thin wires result in smaller stitch bonds with resulting lower stitch peel strengths. The T value should be as large as the BPP and BPO allow.

Materials to Extend Lifetime. As dimensions shrink, the forces per unit area on the capillary increase exponentially. Therefore, material wear-out is a concern. Capillary changes result in lost production time and some loss of gold wire due to the rethreading process. So material selection plays a key roll. Several of the major suppliers have, in development, new materials that are targeted to provide improved strength and lifetimes.

Bonding Process

Process Window. Proper characterization of the process window is necessary in order to maintain a stable production environment and utilize consumables in the most cost-effective manner. All variables in Figure 1 need to be tightly monitored and controlled. This is no small task when one considers how many major and minor variables affect the wire bonding process.

Statistical Process Control. Statistical process control (SPC) techniques with well-tested control limits provide adequate warning before the process is in trouble. Maintaining a very tight standard deviation (σ) for the tests deployed will insure consistent system performance. A drifting or inconsistent σ is a good indicator of capillary wear out, variation in the quality of the material input, or improper machine maintenance. Catching defects where they occur reduces costs significantly based on 'The 1:10:100 Rule' (11).

Bonding Pad Surface Condition. Optimization of the bond pad is often overlooked, as it is typically done without consultation with the engineers in assembly. Large OSATs have rules that govern this critical feature and it is important to pay attention to their recommendations.

Pad Stack. The types and thicknesses of the metal and dielectric layers that constitute a pad 'stack' have a direct impact on the mechanical and chemical stability of the interface between the ball and the pad surface. There is also a direct impact on yield and reliability with respect to these materials. Corrosion, cracking, and cratering may result from the use of a non-optimized process on a non-optimized pad. Other dielectric materials incorporated in the structure may be up to ten times softer than the oxides they replaced, raising serious strength issues for the structure. The methods (vias) used to connect these various metal layers through the dielectric will also impact the mechanical strength of the complete 'stack.' The number and locations of the vias with respect to the target location of the bonded ball have a significant impact on the mechanical robustness of the structure with respect to impact (z-axis displacement) and ultrasonic excitation (in-plane x- or y-axis displacement).

Cleanliness. The cleanliness of the surface of the bonding pad is crucial for a good bond. If the pad is not clean or the surface of the FAB is contaminated, the interface is compromised and the resultant bond will be weak or will not stick at all. The presence of fluorides from wafer fab processes and carbon from upstream assembly processes will result in non-stick or weak bonds. The weak bonds could become field failures compromising product reliability and the image of manufacturer. Lastly, non-stick occurrences require a bond off and new ball then a re-bond. This results in excess wire consumption and loss of productivity.

A Final Word on the 'Hidden' Costs of Copper Use

In evaluating the costs associated with bonding wire selection it is recommended that the total cost (rather than just the basic wire cost) of a copper solution versus a gold solution be considered. There is much more to the story than just the cost per meter of gold versus copper bonding wire.

While it is a challenge to make a comprehensive cost model that is close to any one end-users situation, highlighted below are some of the issues that will help in the development a total solution cost:

- Higher capital outlay for hardware modifications to the wire bond platforms for copper.
- Copper wire bonding is slower due to longer bond times and slower loop motion, resulting in lower units per hour production rates.
- The costs for infrastructure and consumable gases used to protect the free-air ball in copper bonding need to be accounted for during the costing process.
- The costs associated with higher bonding temperatures and bonding forces for copper versus gold.
- Reliability and package testing are more difficult with copper; using normal decapping methods based on nitric or sulphuric acid to remove the thermoset plastic leaving the gold wire intact can't be used with copper. Alternative laser decapping methods are costly and time consuming.
- Copper's inferior corrosion resistance means copper wire has a limited 'floor life' compared to gold. This may result in increased scrapping of material and associated costs.
- …And finally, the significant unknown in costing copper bonding wire technologies is the potential cost of in-service chip failure.

"There is no question copper is a cheaper material but also one that brings new challenges to the bonding engineer. It may be a cost effective process in the eyes of a product manager but when the extra care and attention is required, it is weighed against the existing gold process and then we should ask ourselves, are the net savings worthwhile?"

George G. Harman author of the recent 3rd Edition of the renowned book 'Wire Bonding in Microelectronics' and widely considered to be one of the world's foremost authority on the science of wire bonding (12).

Summary

There are undoubtedly many areas where examination of current practices and processes could lead to a more cost-effective use of gold wire in the assembly of semiconductors. The path most frequently explored is the reduction of the wire diameter, although one needs to be mindful of performance issues as well as operational ones. There are other areas where the per-unit savings may not be as dramatic but the total cost reduction would be significant. In summary, reductions in any of the following variables can reduce costs: wire diameter, free air ball diameter, loop height or substrate lead pitch (shorter wire length).

A reduction in the FAB ratio could save up to 48 mm of wire in high I/O packages. The low I/O devices could see significant improvement overall based on the sheer volume produced. For example, consider a 32 I/O QFN with 38 wires and an initial FAB ratio of 2. A reduction of the FAB ratio to 1.4, would save 3.3 mm per unit, or > 900 mm per strip, or 1 km per 1,100 strips.

Package design is another area where cost-effectiveness can be improved. Reducing wire length through substrate design or lead frame redesign can contribute to improving the cost-effectiveness of using gold. Small changes may not seem like much when considered in isolation. The improvement is significant when the number of traces and units multiplies these individual savings. If one considers a 304 BGA with a wire count of 404, the total wire used could be as high as 1.5 to 1.8 meters. A conservative reduction in wire length due to a redesign could save 10% per unit.

The other factors as shown in Figure 1 directly influence the effective use of gold wire. Bonding parameters, capillary selection and control, bond pad cleanliness - all impact yield and system performance which, in turn, affect gold wire consumption. Ideally, a machine would start bonding with a new spool and capillary. It would not stop until the spool was empty and the capillary had reached its end of life, an engineered coincidence. The only wire not used in producing product for revenue is the tail of the old spool and the short amount required to thread the new capillary.

The above scenario coupled with design changes and parameter optimization will provide for the more cost-effective use of gold wire in the assembly of semiconductor devices of all types.

References

1. The Worldwide IC Packaging Market, 2009 Edition, Electronic Trend Publications
2. WGC/SEMI Survey, January 2010, http://www.semi.org/en/Press/CTR_034053
3. C. D. Breach, Ng Hun Shen, Tee Wai Mun, Teck Kheng, Lee and R. Holliday. Proceedings of 34th International Electronics Manufacturing Technology Conference (IEMT), Melaka, Malaysia, 30th November-2nd December 2010
4. C. D. Breach, Ng Hun Shen, Tee Wai Mun, Teck Kheng Lee and R. Holliday. Proceedings of 12th Electronics Packaging Technology Conference (EPTC), Shangri-La Hotel, Singapore 8-11th December 2010 p. 214.
5. J. Shah, Performance Comparison of Gold vs. Copper Wire Bonding, Semiconductor International, Oct. 1, 2009.
6. J. S. Hwang, B. S. Kumar, J. T. Moon, C. Uhm, Y. N. Kim, M. Sivakumar and S. K. Song, 2N Wire for Ultra Fine Pitch Wire Bonding: Challenges & Solutions, 2008 10th Electronics Packaging Technology Conference, pp. 795-799.
7. Flynn Carson, Package on Package Developments and Trends, 2006 Packaging and Test Workshop, Sept. 10-13, 2006, Napa, California.
8. L. Crider, D. Gerrity, S. Russell, C. Martin, "Conditions for Optimized Deep-Access Gold Ball Bonding Process", IMAPS ATW, Marriott Hotel, San Francisco, CA, July 15, 2010
9. http://www.smallprecisiontools.com/products-and-solutions/chip-bonding-tools/bonding-capillaries/?oid=354&lang=en
10. N. Srikanth, C. T. Lim, M. Kumar, Y. M. Wong, Charles J. Vath III, "Wire Bond Challenges in Low-K Devices", SEMICON Singapore, May 2005
11. http://www.qualityinspection.org/inspecting-quality-earlier-is-better-the-1-10-100-rule/
12. G. Harman, *Wire Bonding in Microelectronics, Third Edition*, p. 76, McGraw-Hill Companies Inc., Columbus, Ohio, (2010)

Tables and Figures for the document

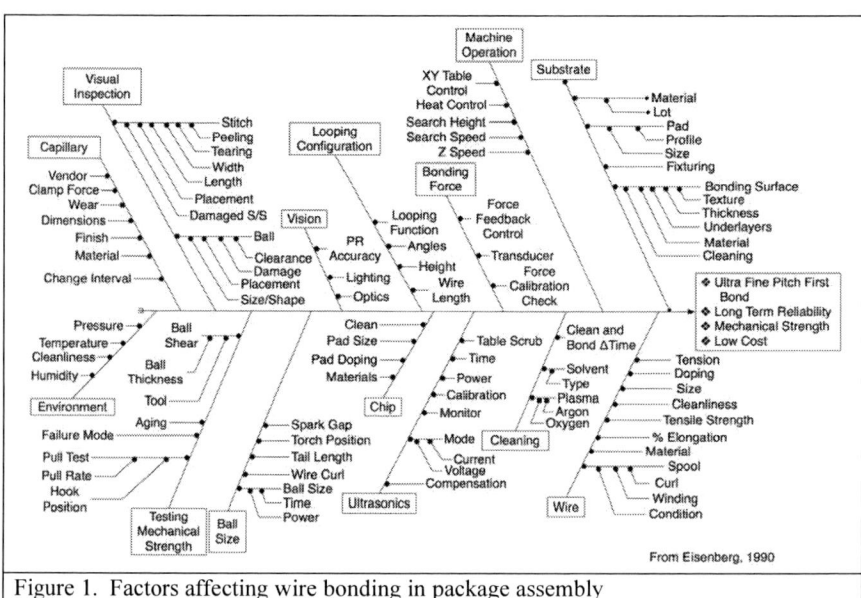

Figure 1. Factors affecting wire bonding in package assembly

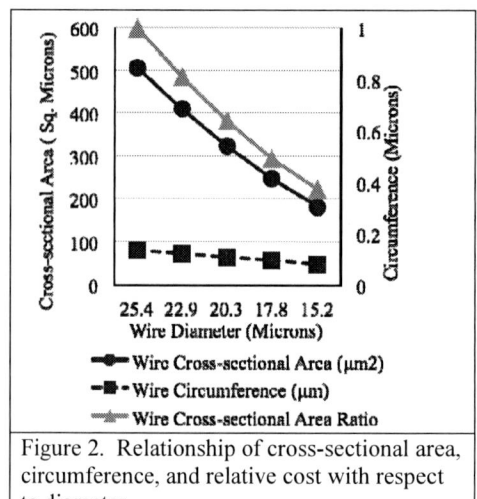

Figure 2. Relationship of cross-sectional area, circumference, and relative cost with respect to diameter

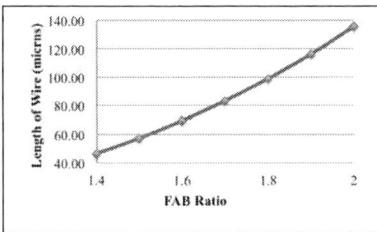

Figure 3. Relationship between FAB ratio and length of wire consumed

Figure 4. Test vehicle with ultra fine pad pitch

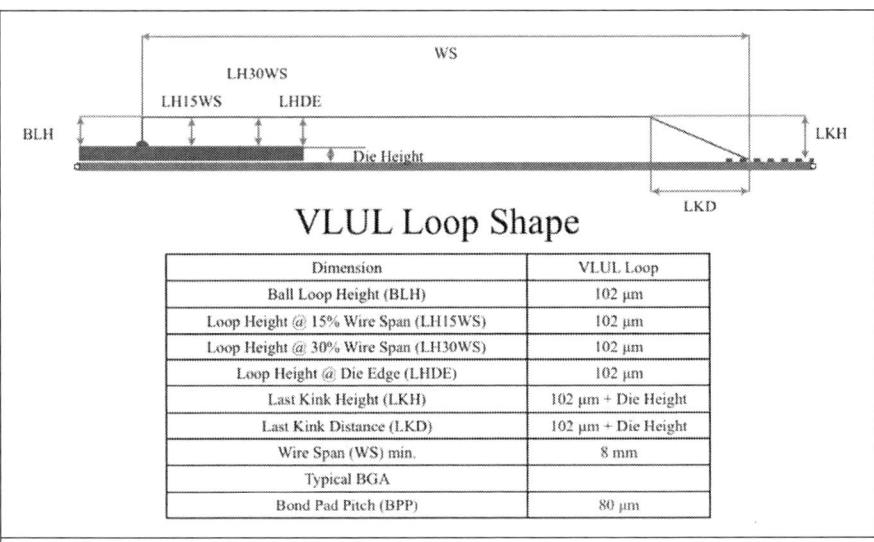

VLUL Loop Shape

Dimension	VLUL Loop
Ball Loop Height (BLH)	102 μm
Loop Height @ 15% Wire Span (LH15WS)	102 μm
Loop Height @ 30% Wire Span (LH30WS)	102 μm
Loop Height @ Die Edge (LHDE)	102 μm
Last Kink Height (LKH)	102 μm + Die Height
Last Kink Distance (LKD)	102 μm + Die Height
Wire Span (WS) min.	8 mm
Typical BGA	
Bond Pad Pitch (BPP)	80 μm

Figure 5. Schematic of a *very low ultra long* loop

Table 1. Calculation of wire length and cost for two popular package types

Package Type	QFN	BGA	BGA	BGA
Pin Count	32	304	304	304
Die Size (μm per side)	2800	12000	12000	12000
Wire Count	38	404	404	404
Wire Count 1/2/3	38	104/160/160	104/160/160	104/160/160
Wire Diameter (μm)	20.3	22.9	20.3	17.8
FAB Ratio	1.6	1.6	1.6	1.6
Wire Span 2D (μm) 1/2/3	750	1500/2500/5500	1500/2500/5500	1500/2500/5500
Loop Shape LKD/LKH 1/2/3	350/350	600/150 150/150 150/150	600/150 150/150 150/150	600/150 150/150 150/150
Loop Height (μm) 1/2/3	150	60/90/90	60/90/90	60/90/90
Die Thickness (μm)	200	75	75	75
Stitch (μm)	51	51	51	51
Wire Length (μm) 1/2/3	1151.41	1678/2759/5759	1678/2759/5759	1678/2759/5759
Total Package Wire Length (μm)	43753	1537283	1537283	1537283
Total Package Wire Volume (cubic μm)	14161086	497550218	497550218	497550218
Total Gold Weight (g)	0.0002733	0.01222	0.009603	0.007383
Gold Value @ US $40.18/g	0.0109843	0.491124	0.385933	0.296729

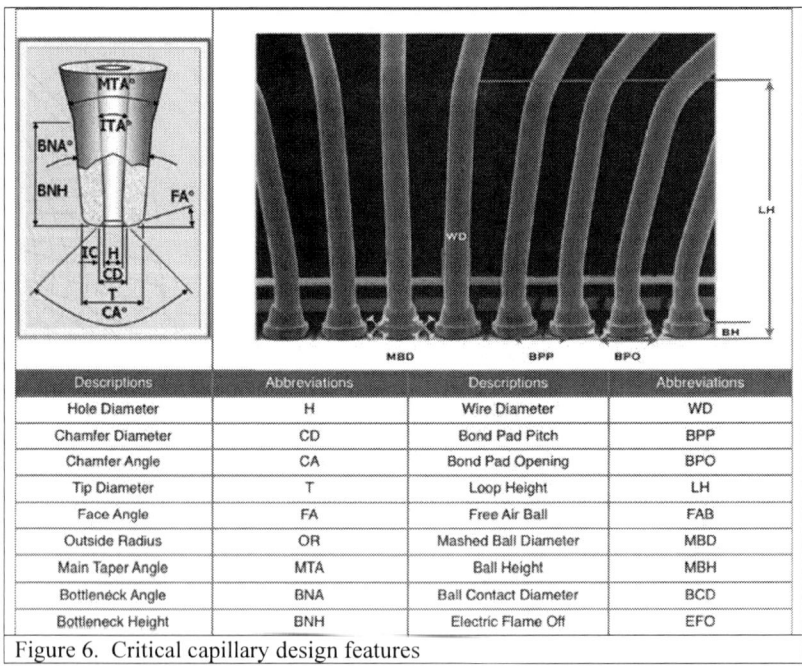

Descriptions	Abbreviations	Descriptions	Abbreviations
Hole Diameter	H	Wire Diameter	WD
Chamfer Diameter	CD	Bond Pad Pitch	BPP
Chamfer Angle	CA	Bond Pad Opening	BPO
Tip Diameter	T	Loop Height	LH
Face Angle	FA	Free Air Ball	FAB
Outside Radius	OR	Mashed Ball Diameter	MBD
Main Taper Angle	MTA	Ball Height	MBH
Bottleneck Angle	BNA	Ball Contact Diameter	BCD
Bottleneck Height	BNH	Electric Flame Off	EFO

Figure 6. Critical capillary design features

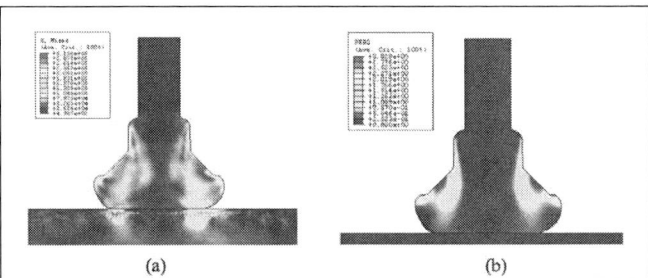

Figure 7. Contour plots showing (a) von Mises stress plot (b) equivalent plastic strain plot under initial contact condition

Copper Wire Bonding in High Volume Manufacturing

Bernd K Appelt[a], Andy Tseng[a], Yi-Shao Lai[b] and Chun-Hsiung Chen[b]

[a] ASE Group, 1255 E Arques Ave, Sunnyvale, CA 94085, USA
[b] ASE Group, Nantze Export Zone, 811 Kaohsiung, Taiwan

Wire bonding is still by far the dominant interconnection technology with about 90% market share. Copper wire bonding has achieved approximately 15% penetration within two years. Considerable amount of fundamental characterization has been performed to understand the nature of Cu wire bonding and even more analysis has been done in manufacturing to ensure reliable packages can be shipped. Here, some of the methodology of successful, high volume manufacturing will be described. Rigorous process optimization and product characterization are key and will be detailed. Supportive data from extended reliability testing demonstrates that 2 x the standard JEDEC life can be achieved. This is in part due to the very slow intermetallic growth rate and in part due to the proper choice of mold compound. The other elements are strict adherence and control of bonding parameters and rigorous clean room management which is also reflected in yields being equivalent to gold wire bonding.

Introduction

Wire bonding has been practiced now well over 40 years and still is the dominant technique for chip to substrate interconnection with a share of approximately 90%. While some have predicted the end of wire bonding due to interconnect density limitations, equipment and wire manufacturers as well as wire bond engineers have been able to advance their technology to < 40 µ bond pad pitches and wire diameters of 0.5 mils when using gold (Au) wires. Copper (Cu) wire bonding also has a long history of more than 20 years but had been limited to high power applications with wire diameters of > 2 mils. Therefore, many of the technical challenges associated with Cu wire like hardness, propensity for oxidation and corrosion were known and mastered. The Cu bonded dice were also designed and built with Cu wire bonding in mind i.e. die pad structure and metal thickness were optimized accordingly.

Fine pitch Cu wire bonding or fine Cu wire bonding had not been seriously considered as long as Au commodity prices were in the low 100s dollar range. Now that commodity prices have surpassed USD 1,400 and appear to stay at these levels (Fig. 1), the continuous drive for cost reduction is demanding its toll from wire bonding. The expected cost reduction is more than can be achieved by reducing Au wire diameters except for the finest diameters of =< 0.6 mils.

The previously mentioned technical challenges have been exacerbated by the advancement of wafer nodes. The development of low dielectric constant (lowK & ELK) wafer dielectrics have resulted in mechanically brittle dice. Every new wafer node is based on lowerK dielectrics and hence ever more fragile dice. This has been difficult already for Au wire bonding and has lead to the development of more robust pad stack structures. Another challenge has been set forth by bonding over active die

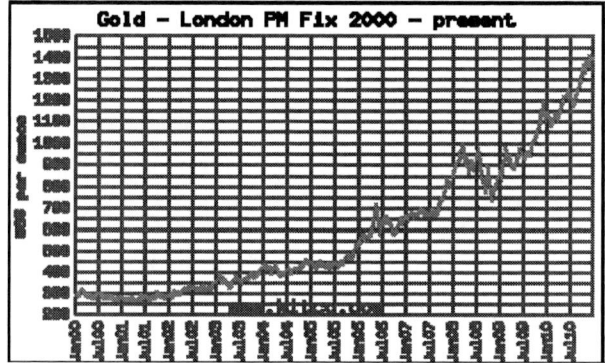

Fig. 1: Gold commodity price evolution over the past ten years

area which also required enhanced pad structures. Some relief was gained from the introduction of Cu metal in the wafers which is more robust than the prior aluminum (Al) die wiring.

To date, all dice are designed and built for eventual Au wire bond assembly. Even Cu wafers have an Al pad finish except for a few products which employ different finishes like nickel/gold or nickel/palladium/gold. Here, the focus will be on dice with Al pads or Al pad finishes. The bonding parameters and reliability performance differ greatly between the different finishes.

Literature Discussion

Cu wire bonding has received great attention only recently, albeit its engineering feasibility studies have been going on for over twenty five years. Key engineering and reliability issues of Cu wire bonding have also been pointed out long ago [1-5]. To overcome Cu oxidation during electronic flame off (EFO) that leads to a free air ball (FAB) of an undesired appearance, forming gas typically composed of 95% N_2 and 5% H_2 was proposed [3] and widely adopted. The H_2 in the forming gas provides additional thermal energy to melt Cu during EFO [6-8], and may convert Cu oxide back to Cu [9]. In response to the increased hardness of the Cu FAB, soft Cu wires with high Cu purity or dopants were sought [10] to soften the wires. However, wire hardness may have little connection with the recrystallized FAB hardness [11]. The key factor may actually be the work hardening that occurs during the actual bond formation. Initial studies to characterize this phenomenon have been performed already [11]. On the other hand, the Al pad can be doped with Si and/or Cu (e.g., [2, 3, 5, 12]) to resist impact and ultrasonic loadings during Cu wire bonding

The most distinguishable feature of Cu wire bonding compared to conventional Au wire bonding, however, is the spotty and small coverage as well as slow growth of Cu-Al IMCs [1, 3, 13 - 22]. Though slow Cu-Al IMC growth is considered an advantage in enhancing reliability of Cu wire bonding [1,3,9,10,12,16,21,23], the strong bondability right upon bonding despite the limited IMC coverage on the Cu-Al interface has attracted intensive investigations; as a result several hypotheses on the bonding mechanism have been proposed [14-16,20,23]. Owing to the ease of oxidation of Cu, long-term durability of Cu wire bonding under varying temperature and humidity conditions have become an essential issue in the industry. However, the reliability data, in particular for fine pitch Cu wire bonding, is so limited and dispersive such that, for instance, whether the ionic impurity in the mold compound leads to the corrosion of Cu wires [4, 8, 24] is still in dispute [1, 15, 25]. Pd-coated Cu wire [8, 26] was developed following the demand of enhancing long-term reliability of Cu wire bonding and shelf life.

A more recent update of Cu wire bonding tooling and general capability has been presented recently [29] and is reflecting the explosive growth of fine pitch Cu wire bonding [30]. In keeping with the Cu wire market penetration, the presentation of Cu wire bond papers at conferences is increasing exponentially as well (see ESTC 2010 and EPTC 2010).

Manufacturing Process Development

For the purpose of this discussion, fine diameter Cu wires refers to wire diameters below 1.2 mils. In fact, the majority of the experiences described here is based on 0.8 mils diameter wire, either 4N Cu wire Maxsoft from Heraeus or 1X Palladium coated Cu wire from Nippon Micrometal. The wire bonders were KnS models Maxum Ultra and Maxum Plus as well as lately Iconn. All wire bonders are equipped with inert gas EFO kits for Cu or CuPd wires. A proprietary capillary design was used and substrates and dice are customer specified with nodes ranging from 180 to 40 nm.

The first step in wire bonding is the free air ball (FAB) formation. For 4N Cu wire forming gas (95% nitrogen and 5% hydrogen) were used as a shroud and for PdCu wire nitrogen was used. The FAB geometry was tuned to yield a spherical FAB without any surface blemishes. This is achieved as usual by optimizing sparc current and duration as well as the gas flow. The spherical ball shape is a good indicator that an 'oxide-free' ball has been formed. This first step is actually one of the easier steps in the process.

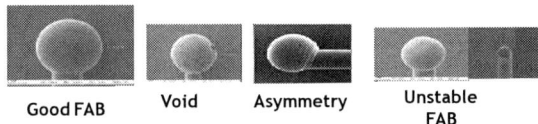

Good FAB   Void   Asymmetry   Unstable FAB

Fig. 2: Examples of good, spherical and unoptimized free air balls.

The second step is the actual bond formation on the die pad. The process of tuning the bonding parameters is essentially the same as for Au wires. Albeit considerably more adjustments are required to ensure that no pad cratering or die cracking occurs due to the more aggressive conditions required to achieve a strong bond. The bond parameter optimization typically follows the standard procedure design of experiment (DOE) as for Au albeit the process window turns out to be considerably smaller than for Au. The boundary conditions are that Al splash, which is usually quite pronounced, must be contained within the bond pad opening (BPO) as shown in Fig. 3. Further, the residual Al thickness has been selected to be 100 nm minimum. It has been shown separately, that this thickness typically survives JEDEC TCT of more than 1000 hrs.

Fig. 3: Examples of Al splash within the bond pad opening.

With the proper bond parameter optimization, dice of any node up to and including 40/45 nm can be bonded successfully without pad cratering or cracking. One great analytical tool employed here is focused ion beam (FIB) microscopy which can provide excellent resolution of pysical structures and grain structures. The time and effort for FIB afford only selective analysis rather than line monitoring. Cross-section examples of dice from different nodes are given in Fig. 4. Typically Cu wafers with an Al cap are more robust in bonding than pure Al wafers. Care must be taken to ensure that the via structures under the pad are not disturbed by excessive force during bonding.

While the exact bonding parameters are dependent on the particular devices and considered proprietary, the bond attributes are open and specified. The wire pull and ball shear strength at time zero are considerable higher than for corresponding Au wires although the AlCu intermetallic compounds are extremely thin. Part of the process optimization is to obtain adequate IMC coverage. During process devlopment this IMC growth is tracked throughout the entire assembly process and at times through reliability testing. In general, the observations in above literature have been confirmed: initial IMC is very thin and difficult to detect and IMC growth is very slow, more than an order of magnitude slower than Au.

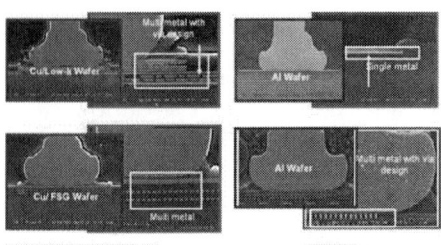

Fig. 4: Cross-sections of dice from different wafer nodes.

Looping to the stitch bond has not presented any difficulties albeit the most aggressive loops are not yet in production. Especially with latest generation bonders and the additional control parameters available, the challenges of very low loop heights and particular shapes should be minimized. The seond bond or stitch bond has not been any challenge to date. No changes in substrate finish has been required either to obtain strong bonds as reflected in the wire pull values. Bond shapes are equivalent to Au. The second bond certainly benefits from an inert gas shroud by minimizing oxidation of the wire even though the temperature is between ambient and substrate temperature. In a high volume maufacturing environment, the shelf / floor life of the wire is not really an issue as the rate of consumption is a fraction thereof. But manufacturing floor management is simpler with a wire of long shelf life like PdCu wire.

The latest advances in bond process development has enabled die stacking with dice gaps of less than 50 um, comparable to Au wires and die over hangs of as much as 1 mm.

A sample is shown if Fig. 5a & b. Like wise reverse bonding for die to die bonding has also been implemented successfully as demonstrated in Fig. 5c.

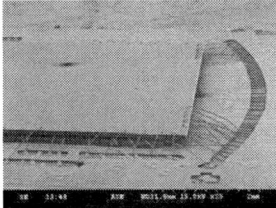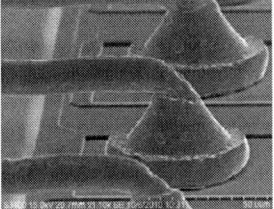

Fig. 5: a- stcked dice with 1 mm overhang, b – low Cu wire loopsof less than 50 μ for dice stacking, c – reverse Cu wire bonding for die to die bonding.

Finally, the package is encapsulated via molding. The mold process and pre-mold plasmas do not require any change other than the usual optimizations of plasma. Concerns have been raised about the reliability of standard mold compounds as do to the propensity of oxidation and corrosion of Cu. Corrosion is a known phenomenon also for Au and the mechanisms are essentially the same for both metals. Extensive efforts have therefore been undertaken to reduce the amount of halogens in the mold compounds. This is an effort that has started long ago for Au and is being continued vigorously for Cu. The mold compound suppliers have an extensive repertoire of actions to minimize the effective amount of halogen ions in the mold compound. Apart from screening resins for intial low halogen content, additives act as ion trappers and buffer the pH as well as mdoify the glass transition temperature. The chemistries involved are of course proprietary.

Lastly, as a general practice of wire bonding and especially for Cu wire bonding, strict manufacturing floor management is key for successful high volume manufacturing. This entails strict clean room management, decidcated tools and operators, as well as strict adherence to hold times between operations to minimize oxidation and surface contamination. Cu wire bonds are reworkable and therefore first pass yields are final yields for bonding. With such a rigorous methodology it is possible to achieve yields equivalent to Au and machine stop rates of less than the traditional Au floor. A reflection of the quality of the Cu wire bond process is a device with 1,478 wires which is running in manufacturing now. Of course the learning can be reapplied to the Au floor and improve its perfromance.

Based on above learning experiences for package types: SO, QFP, QFn and BGA as well as dice from virtually all wafer factories, a rigorous methodology for the evaluation of and qualifcation of new devices has been devised. It is a three step procedure where bondability of a die is established with a base set of parameters. In phase two the bond parameters are optimized and the bond attributes are characterized. In some cases, short loop reliability tests are being performed. Finally in phase three, the actual qualification hardware is being built and tested with the usual JEDEC tests as specified by the application environment. If the new die has some novel attributes like pad stack, metal thickness, etc., additonal characterization loops and tests may be performed to ensure the reliability of the package.

As part of the line quality and reliability monitoring, extended reliability testing is being performed for selected packages and wafer nodes. As shown in Table 1, the standard JEDEC tests like HAST, THT and TCT can achieve at least 2x the usual life times. Depending on the application conditions, these JEDEC data represent multiple life times, especially for many mobile and consumer applications. Above testing is continuing to determine the long term reliability as required by networking and automotive applications.

Table 1: Extended reliability test results

Package Type	Body Size	PCT hrs	TCT cycles	HAST hrs	HTS hrs	THT hrs
QFN	6 x 6		2,000	200	2,000	2,000
QFN	8 x 8		2,000	200	2,000	2,000
QFN	9 x 9	936*	6500*	178*	4500*	
aQFN	11.5 x 11.5	168	500*		500*	
QFP	14 x 20	2016	6000		3500	
QFP	28 x 28	1272	6,000		4,000	
LQP	10 x 10	336	2,000	192	2,000	
LQFP	14 x 14	264	1,500			
LQFP	20 x 20	336	2,000	192	2,000	
LQFP	24 x 24	264*	1,500*	144*	1,500*	
TQFP	14 x 14		2,000	400	2,000	2,000
HQFP	14 x 20	336	1,000*	548*	3,500*	
TFBGA	9 x 9		3,500	144		
TFBGA	12 x 12		6,000*	864	2,000	
LFBGA	16 x 16		1,500	168	1,500	
HSBGA	27 x 27	336	4,000	336	4,000	
* test in progress						

Conclusions

Fine pitch or fine diameter Cu wire bonding has been introduced successfully in high volume manufacturing. Quality and yield has been advanced to levels equal to Au wire bonding. Reliability has been demonstrated to exceed 2x standard JEDEC testing and is continuing. At present, more than 2 billion devices have been shipped from six different factories. More than 1,500 wire bonders are running with Cu wire and it is expected that by the end of the year this number will increase to 4,500.

Acknowledgments

The authors would like the engineering teams under Mike Hung, Louie Huang, Scott Chen and Mike Zhao for their discussions and making the data available for publication.

References

1. 1. Onuki, J. et al., "Investigation of the Reliability of Copper Ball Bonds to Aluminum Electrodes," *IEEE Trans. Comp. Hybr. Manuf. Technol.*, Vol. 12, No. 4 (1987), pp. 550-555.
2. Toyozawa, K. et al., "Development of Copper Wire Bonding Application Technology," *IEEE Trans. Comp. Hybr. Manuf. Technol.*, Vol. 13, No. 4 (1990), pp. 667-672.

3. Khoury, S. L. *et al.*, "A Comparison of Copper and Gold Wire Bonding on Integrated Circuit Devices," *IEEE Trans. Comp. Hybr. Manuf. Technol.*, Vol. 13, No. 4 (1990), pp. 673-681.

4. Caers, J. F. J. M. *et al.*, "Conditions for Reliable Ball-Wedge Copper Wire Bonding," *Proc. 1993 Japan Int. Electron. Manuf. Technol. Symp.*, Kanazawa, Japan, 1993, pp. 312-315.

5. Nguyen, L. T. *et al.*, "Optimization of Copper Wire Bonding on Al-Cu Metallization," *IEEE Trans. Comp. Packag. Manuf. Technol.–Part A*, Vol. 18, No. 2 (1995), pp. 423-429.

6. Tan, J. *et al.*, "Modelling of Free Air Ball for Copper Wire Bonding," *Proc. 6th Electron. Packag. Technol. Conf.*, Singapore, 2004, pp. 711-717.

7. Xu, H. *et al.*, "Effects of Process Parameters on Bondability in Thermosonic Copper Ball Bonding," *Proc. 58th Electron. Comp. Technol. Conf.*, Lake Buena Vista, FL, USA, 2008, pp. 1424-1430.

8. Uno, T. *et al.*, "Surface-enhanced Copper Bonding Wire for LSI," *Proc. 59th Electron. Comp. Technol. Conf.*, San Diego, CA, USA, 2009, pp. 1486-1495.

9. Deley, M. and Levine, L., "Copper Ball Bonding Advances for Leading Edge Packaging," *Proc. Semicon Singapore 2005*, Singapre, 2005.

10. Shah, A. *et al.*, "In Situ Ultrasonic Force Signals During Low-temperature Thermosonic Copper Wire Bonding," *Microelectron. Eng.*, Vol. 85, No. 9 (2008), pp. 1851-1857.

11. Hang, C. J. *et al.*, "Bonding Wire Characterization Using Automatic Deformability Measurement," *Microelectron. Eng.*, Vol. 85, No. 8 (2008), pp. 1795-1803.

12. Kim, H.-J. *et al.*, "Effects of Cu/Al Intermetallic Compound (IMC) on Copper Wire and Aluminum Pad Bondability," *IEEE Trans. Comp. Packag. Technol.*, Vol. 26, No. 2 (2003), pp. 367-374.

13. Murali, S. *et al.*, "An Analysis of Intermetallics Formation of Gold and Copper Ball Bonding on Thermal Aging," *Mater. Res. Bull.*, Vol. 38, No. 4 (2003), pp. 637-646.

14. Murali, S. *et al.*, "Effect of Wire Size on the Formation of Intermetallics and Kirkendall Voids on Thermal Aging of Thermosonic Wire Bonds," *Mater. Lett.*, Vol. 58, No. 25 (2004), pp. 3096-3101.

15. Wulff, F. W. *et al.*, "Characterization of Intermetallic Growth in Copper and Gold Ball Bonds on Aluminum Metallization," *Proc. 6th Electron. Packag. Technol. Conf.*, Singapore, 2004, pp. 348-353.

16. Ratchev, P. *et al.*, "Mechanical Reliability of Au and Cu Wire Bonds to Al, Ni/Au and Ni/Pd/Au Capped Cu Bond Pads," *Microelectron. Reliab.*, Vol. 46, No. 8 (2006), pp. 1315-1325.

17. Murali, S. *et al.*, "An Evaluation of Gold and Copper Wire Bonds on Shear and Pull Testing," *J. Electron. Packag.*, Vol. 128, No. 3 (2006), pp. 192-201.

18. Ibrahim, M. R. *et al.*, "The Challenges of Fine Pitch Copper Wire Bonding in BGA Packages," *Proc. 31st Int. Conf. Electron. Manuf. Technom.*, Kuala Lumpur, Malaysia, 2006, pp. 347-353.

19. Hang, C. J. *et al.*, "Growth Behavior of Cu/Al Intermetallic Compounds and Cracks in Copper Ball Bonds During Isothermal Aging," *Microelectron. Reliab.*, Vol. 48, No. 3 (2008), pp. 416-424.

20. Xu, H. *et al.*, "A Re-examination of the Mechanism of Thermosonic Copper Ball Bonding on Aluminium Metallization Pads," *Scripta Mater.*, Vol. 61, No. 2 (2009), pp. 165-168.

21. Zhang, S. *et al.*, "Characterization of Intermetallic Compound Formation and Copper Diffusion of Copper Wire Bonding," *Proc. 56th Electron. Comp. Technol. Conf.*, San Diego, CA, USA, 2006, pp. 1821-1826.
22. Yeoh, L. S., "Characterization of Intermetallic Growth for Gold Bonding and Copper Bonding on Aluminum Metallization in Power Transistors," *Proc. 9th Electron. Packag. Technol. Conf.*, Singapore, 2007, pp. 731-736.
23. Murali, S. *et al.*, "Effect of Wire Diameter on the Thermosonic Bond Reliability," *Microelectron. Reliab.*, Vol. 46, No. 2-4 (2006), pp. 467-475.
24. Tan, C. W. *et al.*, "Corrosion Study at Cu-Al Interface in Microelectronics Packaging," *Appl. Surf. Sci.*, Vol. 191, No. 1-4 (2002), pp. 67-73.
25. England, L. and Jiang, T., "Reliability of Cu Wire Bonding to Al Metallization," *Proc. 57th Electron. Comp. Technol. Conf.*, Reno, NV, USA, 2007, pp. 1604-1613.
26. Kaimori, S. *et al.*, "The Development of Cu Bonding Wire with Oxidation-resistant Metal Coating," *IEEE Trans. Adv. Packag.*, Vol. 29, No. 2 (2006), pp. 227-231.
27. Murali, S. *et al.*, "Grains, Deformation Substructures, and Slip Bands Observed in Thermosonic Copper Ball Bonding," *Mater. Charact.*, Vol. 50, No. 1 (2003), pp. 39-50.
28. Murali, S. and Srikanth, N., "Acid Decapsulation of Epoxy Molded IC Packages With Copper Wire Bonds," *IEEE Trans. Electron. Packag. Manuf.*, Vol. 29, No. 3 (2006), pp. 179-183.
29. Chylak, R., "Developments in Fine Pitch Copper Wire Bonding Production", *Proc. 11th Electron. Packag. Technol. Conf.*, Singapore, 2009
30. Appelt, B.K. et al., "Fine Pitch Copper Wire Bonding – Why Now?", *Proc. 11th Electron. Packag. Technol. Conf.*, Singapore, 2009

MUF Technology Development for SiP Module

YoungDo Kweon , Job Ha, KiChan Kim, MinSeok Jang, JaeCheon Doh,
ChangBae Lee, DoJae Yoo

Corporate R&D Institute, Samsung Electro-Mechanics Co., LTD, Suwon, Korea

In this study, we developed the molded underfill (MUF) technology for system in package (SiP) module with fine pitch flip chip in RF application, in which two flip chips, LC filter, and additional passive components are integrated side-by-side. This study covered not only MUF reliability performance but also MUF design study focused on minimizing voids between flip chip bumps in the SiP module. The investigation comprises several aspects: A design study that present a printed circuit board (PCB) and epoxy molding compound (EMC) selection approach, air vent design of cavity vacuum molding, and results of void free from Design of Experiment (DOE) of several SiP module layouts.

In the end, Scanning Acoustic Tomography (SAT) result of void, moisture sensitivity test, thermal cycle test and pressure cooker test had also been carried out for reliability evaluation. The test result shows that the optimized SiP module with fine flip pitch has a good reliability performance.

INTRODUCTION

Recently, there is an increasing demand of system in package (SiP) module for movable and reliable wireless solution between two or more networking modules that is connected and transferred with data for ubiquitous convergence system. The electronic package and module are required to make it lighter, thinner, shorter and smaller with lower build cost for easy hand-carry.

So far, molded underfill technology (MUF) to meet these demands has been studied by packaging assembly houses and material suppliers [1-4] as a solution of smaller size with lower assembly cost. It is also an alternative solution to capillary underfill technology that is a mature encapsulation method of flip chip package [5]. Because the capillary underfill process has some limitations. Those are long dispensing process time and hard process control to fill by capillary force, the package size limitation by underfill keep-out area, and the die crack risk of exposure of FC die backside. Even though it is to solve the reliability problem of bumps from higher localized mechanical & thermal strain/stress due to coefficient of thermal expansion (CTE) mismatch between Si chip (2.6 ppm/°C) and organic package laminate substrate (15 ppm/°C) during board level reflow process [6].

In order to solve these problems, a novel flip chip packaging technology has been studied and three types of new concepts are discussed in previous literature [7]. First is the preset underfill (PUF) process. Second is the transfer underfill (TUF) process. Third is an underfill paste (UFP) process which is similar with current underfill process, but the underfill material is developed latent catalyst and phenol cured epoxy resin system. Even though there is the advantage of molded underfill (MUF) to overcome the limitations of capillary underfill, there are still some development items [8] required for SiP miniaturization and low cost high volume production.

In this study, we introduce the methodology required in designing to minimize air-trapped voids in encapsulation inside transfer molded underfill process, and present the test result with reliability performance. Therefore, the study comprises several aspects: A design study that present a printed circuit board (PCB) and epoxy molding compound (EMC) selection approach, air vent design of cavity vacuum molding, and design of experiment (DOE) of several SiP module layouts.

STUDY METHOD AND RESULT

Test vehicle for MUF

The test vehicle used for this study of MUF in the SiP module is shown in Figure 1. and Table I .

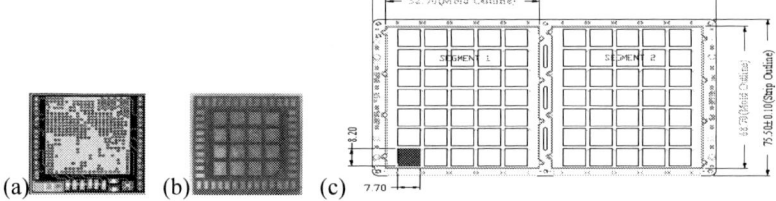

Figure 1. SiP module and Molded outline strip
(a) Top side view (b) Bottom side view, (c) PCB substrate strip outline and mold outline with unit size(a,b)

The dimension of SiP module in this study is 8.2mm*7.7mm*1.13mm with 54 peripheral I/Os, which was sawn from two 52.7mm*68.75mm*0.75mm mold areas of 118.5mm*75.5mm*0.38mm pcb substrate.

TABLE I. **SiP module design specification**

Items	Dimension (mm)
RF/BB IC size, Bump height/Pitch/Count	6.51*5.81*0.41, 0.095 / 0.25/339ea
RF switch size, Bump height/Pitch/Count	0.705*0.705*0.33, 0.085/0.25/8ea
LC filter size	1.6*0.8*0.6
Capacitor size / Count	0603*0.3 / 25ea
Substrate strip size	118.5*75.5*0.38 (6L)
Mold Size/Thickness	52.7*68.7*0.75
LGA Pitch /Count	0.55 / 54ea

Design and DOE for MUF

EMC & PCB material
There are normally two ways to reduce thermal stress on package to reduce module warpage and potential risk of residue stress. One is to reduce the CTE gap among materials of package components. The other one is to use EMC material that has low young's modulus to release thermal stress on package. In addition, to select EMC material for MUF, we considered not only CTE and Young's Modulus but also filler size,

filler content, spiral flow, and gelation time that affect to the molding performance for air void reduction during molding process.

Therefore we selected and developed EMC materials with EMC manufacture and one PCB (as shown in Table II) for final reliability tests referenced on previous literature study [9] [1~5], FEM analysis, and DOE. Table II shows three kind of EMC which is developing by EMC manufacture and one PCB core (CTE $\alpha 1$; 11ppm/°C, CTE $\alpha 2$; 3ppm/°C). We found that all listed materials for MUF can be used for good performance of void free and reliability, but the result depends on the mold process condition and PCB layout pattern.

TABLE II. **Selected EMC and PCB material for MUF**

EMC/PCB	EMCI	EMCII	EMCIII	PCB
Ash Content (%)	88	88	86	
Filler Size (um, Max/AVG)	20/13	30/6	30/7	
Epoxy	MAR4	Biphenyl	Biphenyl	
Spiral Flow (Cm)	150	145	130	
Gel Time (sec)	45	45	45	
Modulus (Gpa, 260°C/RT)	26/0.72	25/0.5	22/0.06	27
CTE (ppm/°C, $\alpha 1$/ $\alpha 2$)	9/37	10/40	10/37	11/3
Tg (°C)	150	135	143	235
Water Absorption (%)	0.14	0.16	0.22	

MUF process & DOE by SiP module layout design

It's important to design vacuum mold tool for molded under-fill to prevent air void trap and mold flash during mold process as shown in Figure 2 that explains mold tool and process.

On the previous literature about flip chip molding process itself [10], the mold tool should be optimized by geometrical means that is creating similar flow resistance over and under the flip chip. It is difficult to control the process and the package design.

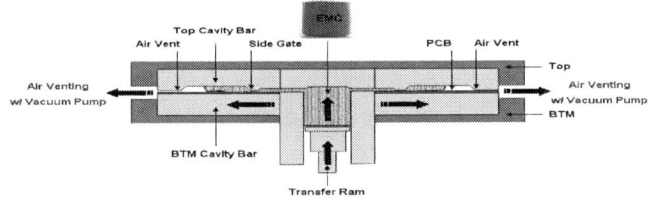

Figure 2. Mold tool & process diagram

For MUF process development, it is required to consider not only the mold process performance such as mold flash and incomplete mold and mold void, but also void occurrence during molding. Important process parameters are mold temperature, clamp force, preheating time and transfer pressure & speed. To determine the range of MUF process parameter, three main process parameters were evaluated on DOE as clamp force is fixed after finding a certain amount of clamping force is needed to prevent flash form from occurring and to prevent substrate damage. Design of Experiment (DOE) about three process parameters for void free process was conducted and the result is shown as Figure 3 that vertical axis is specified for no void (void free, index value 1.0) through SAT(Scanning Acoustic tomography) test, and horizontal axis is process parameter value. (Sample size: 1120units, Adjusted R2 =0.93 in statistic analysis using Minitab 13.).

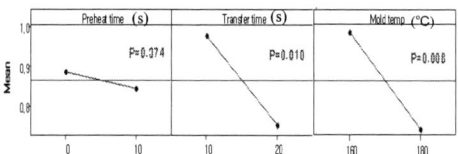

Figure 3. Statistic analysis through Minitab 13

As shown on Figure 3, main factors to yield a complete filling of the mold tool without air void are transfer time and mold temperature that has minus linear relation to void free. Lower mold temperature is recommended because of lower thermal mismatch during molding, curing and cooling; thus a minimized residual stress within assembly. In addition, lower transfer pressure & time is preferable, but it should be optimized to EMC gelation time from transferring of EMC to curing of EMC because of increasing of flow resistivity as time goes on.

In this study, we mainly focused that different substrate layout condition can be affect to void result after molding. We tested four kinds of layout designs to find better flow direction to prevent air void. Figure 4 shows different substrate layout to mold flow.

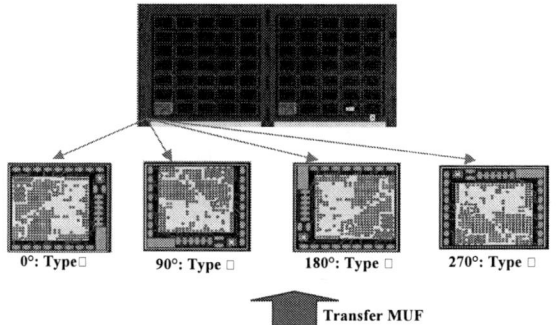

Figure 4. 4 kinds of layout designs to mold flow (Top side view)

Figure 5 shows the SAT test result of four kinds of layout design (different PCB unit rotation) and EMC type to check air void occurrence in the MUF mold. PCB type Ⅱ, Ⅳ and EMC Ⅰ shows best void free performance.

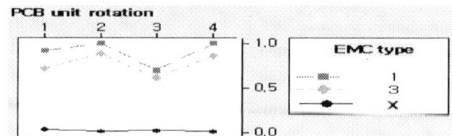

Figure 5. SAT void test result of layout designs

Voids inside EMC are observed after molding processes, especially in the areas around the solder bumps. These voids will be propagated when the package is subjected to reliability tests at higher temperature.

Figure 6 shows the critical air void in EMC after vertical grinding even though it's hard to get a void picture in X-section and this figure was took from another sample. This void can make solder wicking problem and potential moisture pop-con failure that lead to bad reliability.

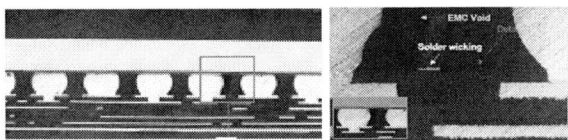

Figure 6. Void failure analysis through X –section

Reliability test result

For reliability assessment of the MUF, in this study, An Acoulab model TITAN SAT operating in the reflection mode with a 110 MHz transducer was used for monitoring the void formation. In addition, the multi-echo inspection allows for accurate detection and detailed detection of trapped voids and any defects. A series of JEDEC reliability tests have been performed on the MUF SiP module.

As shown in Table III, Developed MUF SiP module with PCB type II and EMC type I shows good reliability performance in final reliability test including void check by SAT and no electrical failure before and after all reliability tests.

TABLE III. **Reliability tests results**

Test Items	Condition	Criteria	Result
Void inspection	110MHz Echo-mode	Void failure Criteria	0/373
PreconditioningMSL3	60°C/60%RH, 40hrs, 260°C, Reflow X3	Pass E-Test No die crack, No delamination	0/22
PreconditioningMSL2a	60°C/60%RH, 120hrs, 260°C, Reflow X3	Pass E-Test No die crack, No delamination	0/22
TCT (w/Precon)	-40°C~+110°C, 100cycles,	Pass E-Test No die crack, No delamination	0/10
PCT (w/Precon)	121°C/100%RH, 96hrs	Pass E-Test No die crack, No delamination	0/10

In addition, cross-section optical microscopy is checked too as shown Figure 7 those are the typical cross section images after reliability. No defect between mold compound and underfill has been observed.

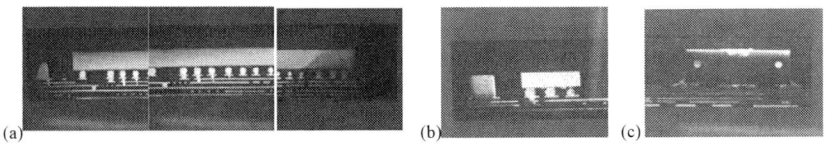

Figure 7. Void analysis through X –section

CONCLUSION

With more integration across IC dies, packages and systematic module through 3D packaging, MUF in SiP module is on its way to play a more and more important role in today and future movable wireless applications. Design & testing in SiP module are very important and it is also required to find the failure mechanism and behavior of no failure. Design is the best stage to systematically implement reliability improvement and monitor the scheme of system. In this study, we discussed designs to improve reliability.

In the MUF in SiP module, it is very important to design PCB layout with passive components and FC bump layout to prevent air voids and minimize it for higher reliability performance. In addition, suitable material property and optimized process parameter are also required. In order to minimize the air voids, four considerations should be developed. First, it is component layout on substrate for geometrical optimization to create similar flow resistance and to reduce obstruction to flow over and under the chip by flux residue and PCB surface condition. Second, it is bump layout design on substrate are important to prevent racing effect by excess material and to have same flow resistance. Third, EMC material property, especially, filler size, spiral flow and gelation time in properties are important to match EMC cure time to fill it in the mold cavity. Forth, vacuum assisted molding with optimized gate & air vent design to create better flow and to prevent air trap by obstructions.

SiP module with the proposed material set can pass MSL level3, MSL L2a, TC and PCT tests and there is no void failure. Furthermore, the proposed process improvement with suggested materials, which use MUF to replace the conventional underfill for SiP module, shows more robust reliability behavior.

References

1. F. Liu, et al., *Characterization of Molded Underfill Material for Flip Chip Ball Grid Array Packages*, Proceedings of 51st. ECTC2001, pp.288~292. (2001).
2. C. Chee, et al., *Underfilling Flip Chip Packages With Transfer Molding Technologies*, Proceedings of 6th. EPTC2004, pp.625~629. (2004).
3. D. Yoo, et al., *Molded Underfill (MUF) Technology Development for SiP Module with Fine Flip Chip*, Proceedings of 43rd. IMAPS2010, (2010).
4. T. Braun, et al., *Improved Reliability of Leadfree Flip Chip Assemblies Using Direct Underfilling by Transfer Molding*, 31st. IEMT2006, pp.27~34. (2006).
5. W. Brown, and R. Ulrich, *Advance Electronic Packaging 2nd edition*, (eds., S. Tewksbury and et al.), IEEE Press. John wiley & sons, Inc., Hoboken, NJ (2006).
6. K. Chen, *Comparing the Impacts of the Capillary and the Molded Underfill Process on the Reliability of the Flip-Chip BGA*, TCAPT, IEEE2008, Vol. 31, No. 3, pp586~591. (2008).
7. H. Usui, et al., *Special Characteristic of Future Flip Chip Underfill Material and the Process*, Proceedings of 50th. ECTC2000, pp.1661~1665. (2000).
8. K. Chai, et al., *Challenge of Vacuum Molded Flip Chip Packaging Technology*, POLYTR2002, 2nd. Int. IEEE Conf, pp.221~224. (2002).
9. C. Chee, and et al., *Lead-free Compatible Underfill Materials For Flip Chip Application*, Proceedings of 52nd. ECTC2002, pp. 417~424. (2002).

10. K. Becker, et al., Advanced Flip Chip Encapsulation: Transfer Molding Process for Simultaneous Underfilling and Post Encapsulation, PoLYTR2001, 1st. Int. Conf. IEEE2001, pp.130~139. (2001).

Multi Beam Grooving and Full Cut Laser Dicing of IC Wafers

Jeroen van Borkulo, Rene Hendriks

ALSI, platinawerf 20G, 6641TL Beuningen Netherlands

Abstract

The traditional blade dicing technology has gone through an impressive evolution keeping up with quality, cost and miniaturization requirements that the semiconductor technology roadmaps introduced and specified. However, since wafer technologies have dropped below 90nm node and low k materials were introduced it became clear that blade dicing evolution is coming to an end and expensive hybrid solutions such as combined laser grooving processes and blade dicing technologies were required to achieve the desired product reliability. Similar situations have been seen with the ongoing trend to thinner wafers that are needed for miniaturization, 3D packaging and IC performance improvements. To achieve sufficient mechanical strength, complex dicing technologies and sequences have been introduced which do not respond to the requirements for current and near future technologies. The paper will discuss how a multi beam process will play a dominant role to achieve a narrow dicing kerf, an extreme small Heat Affected Zone, high die strength values (typical 800-1000 MPa), all while maintaining a high productivity (>12 wafers per hour).

Background

Advanced Laser Separation International (ALSI) was founded in 2001 as a spin off from Philips Semiconductors. ALSI supplies semiconductor laser dicing systems using its proprietary multiple beam technology. ALSI is recognized by the semiconductor industry as a leading laser dicing technology company. ALSI is a market leader for laser dicing applications in RFIC and has a strong position and is a technology leader in the T&D and LED segment. Working with several "early adapter" semiconductor manufacturers to replace conventional mechanical wafer separation technologies, ALSI has achieved a large installed base and considerable experience in laser dicing and related infrastructures.

Challenges in thin Si applications

New IC technology, packaging as well as performance requirements drive the trend to thinner, more sensitive and complex structures and diversify the requirements in the die separation process. MEMS have fragile structures that do not allow dicing processes with high-pressure water cleaning, nor any dust or particle distribution.

With the further shrinking of IC dimensions, low-K material has been widely used to replace the traditional SiO interlayer dielectric (ILD) in order to reduce the interconnect delay. The introduction of low-K material onto silicon has imposed challenges on the dicing saw process. ILD and metal layer peeling and its penetration into the sealing ring of the die during blade dicing are the most common defects. Additionally, the blade costs went up and dicing productivity down for these applications.

The drive to more memory capacity per package volume, to enable the development of portable electronics, requires IC packaging with thinner wafers. 3-D packaging, stacked die, and often with Die Attach Film (DAF). These are common and proven assembly technologies. Further wafer thickness reductions are needed but require new dicing methods to avoid yield loss at the stage where the wafer value is the highest or even worse when a broken die would yield out a package containing 10 or more dies.

The conventional saw dicing technology is preventing the ongoing trend for thin wafer development due to the following main reasons.

- Mechanical forces and vibration cause die crack and chip-outs

- Short lifetime of saw blades due to inability to self-sharpening on the thin substrates
- Difficulty to cutting through the DAF with high dicing speeds
- Low assembly and die picking yield due to smearing of DAF against the die side wall.

As such, major concerns and limitations for the application apply:
- Wafers are extremely sensitive to wafer stress (due to wafer thinning and front end processes)
- Dicing blade selection and exposure to material stacks in the dicing street (polymer, metal, passivation and other test structures)
- Dicing tape material, tackiness
- Dicing speed

Due to these issues and restrictions of conventional saw technology, the thin wafer IC and memory industry are searching for new dicing technologies and have identified laser dicing technology as the new separation technology. This new laser dicing technology has to comply with demanding specifications.
1. A die strength value high enough to prevent die cracks during assembly and operating conditions
2. Compatibility to cutting through DAF, not influencing the DAF properties and resulting in 100% die pick yield
3. Ability to cut through a stack of different passivation materials and test structures (metals, polymers, oxides and nitrides)

Apart from these challenges, the laser dicing also offers additional opportunities:
1. reduction of dicing street width
2. High throughput (>12 wafers/hour)

Laser dicing technologies

Over recent years, several laser-dicing technologies have been developed each having their specific characteristics for a separation process. The main dicing technologies that have become most common are:

Ablation laser dicing

In this process the wafer material is removed by irradiation of laser pulses which locally generate a combination of melt and vapor. The vapor pressure drives the molten material out of the wafer generating an opening also referred to as kerf (see figure 1 below). This technology is predominantly used for dicing through the whole wafer substrate thickness, even with backside metallization although though the use of backside metallization is not common for IC and memory applications. Depending on the application (wafer material, thickness, throughput, and die size), a certain laser type is chosen. Main laser process parameters that determine the interaction of laser light with the wafer material are wavelength, pulse duration and power. The size of the Heat Affected Zone (HAZ), recast and debris depend on these parameters.

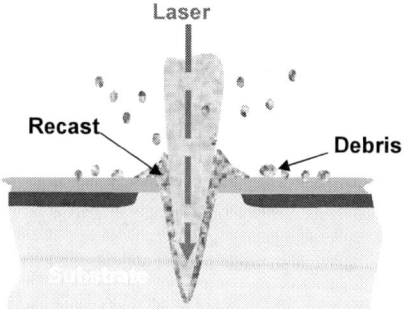

+ Ability to dice through top passivation, Si wafer and DAF
+ High removal rate therefore short dicing time

concerns with common laser technology:

– Reduces die strength
– Rougher side wall roughness
– Possibility of cracks and chip-outs when high power is used

Figure 1. Removal principle for ablation

Subsurface dicing

This separation process focuses laser pulses inside the wafer substrate generating a polycrystalline structure and therefore weakening the material locally (see figure 2 below). After processing the wafer can be expanded, sometimes with the aid of a breaking device, resulting in die singulation. This process works when there is no top passivation and test structures (specifically metal) within the street. These structures would block the laser radiation from going into the wafer material. In addition, any DAF present on the backside of the wafer is still not singulated and would need to be done in an additional step.

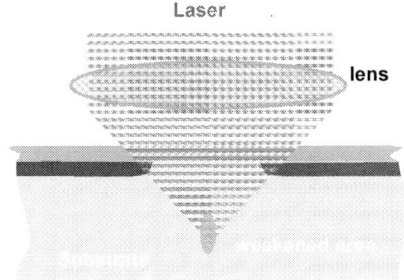

+ "zero" kerf width
+ smooth side wall surface

concerns with sub surface technologies:

– low productivity due to low
  process speed and alignment
– no DAF separation
– not compatible with structures in
  dicing street

Figure 2. Subsurface principle

Hybrid laser technologies

Several hybrid technologies combine laser and the conventional dicing saw or S&B. One typical example is the use of a low power laser to scribe through the top passivation and metal structures within the street, which the mechanical dicing saw has difficulties to cut through. In a subsequent step, the saw is used to cut through the actual Si substrate and possible DAF (see figure 3 below). This process has the advantage of a good interaction of the laser with the passivation and metal layers but still has the negative impact of the dicing saw to the die and the DAF tape (as mentioned above).

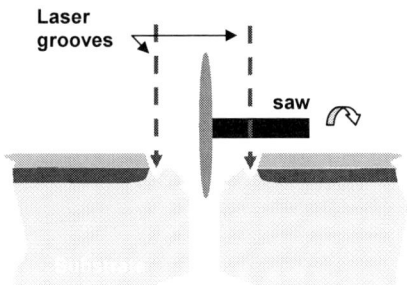

+ removal of passivation out of the
  dicing street
– typical sawing problems remain
– slow due to sawing and several
  process steps

Figure 3. Hybrid principle combination
of laser scribing and saw dicing

Multiple beam technology

When using a laser dicing process to separate the wafer it is always a tradeoff between quality and speed, independent of the above mentioned technology. Nowadays, available industrial lasers can deliver high amounts of power. However, when exerted at such high levels to a thin wafer substrate, the material is not only separated, but may also be damaged severely. To get good quality (no chipping or cracks) with a small or absent HAZ, low laser power levels need to be used. As a result, material removal rates and therefore dicing speed is low and the laser capability with respect to the available power is far from utilized. To solve this issue Philips and ALSI have developed a proprietary multiple laser beam technology. The basic principle is to split the main laser beam up into a plurality of laser beams, each having a low power level and therefore not compromising the overall quality, but as a group of beams keeping the material removal rate and thus the dicing speed high.

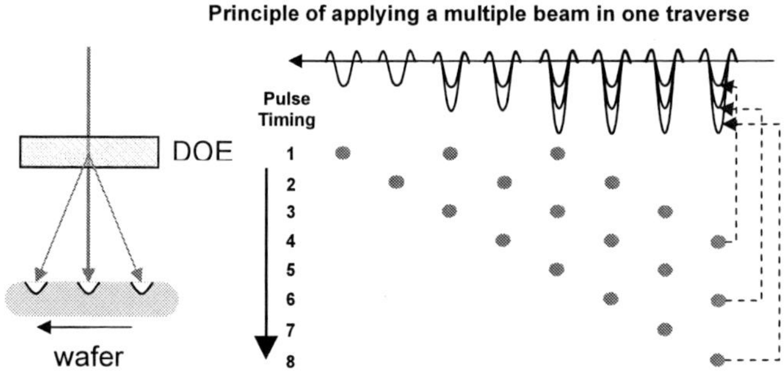

Principle of applying a multiple beam in one traverse

Figure 4. Multiple beam principle

In figure 4, indicated above the principle is demonstrated. The high power laser beam passes through a beam splitting device, in this example a diffractive optical element (DOE). This DOE splits the main laser beam up into a number of beams (in the situation illustrated above three beams) with a certain distance between them. The number of beams generated and the distance between them is depending on the design of the DOE; any number of beams and distance can be generated depending on the design of the DOE.

Using a multiple beam technology allows the dicing process engineer to tune to the optimal power levels for their application (ensuring minimum HAZ and high quality) and at the same time achieve a high throughput.

Using multiple beams gives the advantage of reducing number of traverses, whether either a laser ablation process is the best to use for an application or a subsurface dicing process. A multiple beam technology can be used for any laser dicing process and for each it brings the advantage of speed. When applying the multiple beam laser technology significant advantages can be achieved and concerns for laser dicing are eliminated as demonstrated below in graph 1 and 2 for respectively this Si wafers and for low-K wafers.

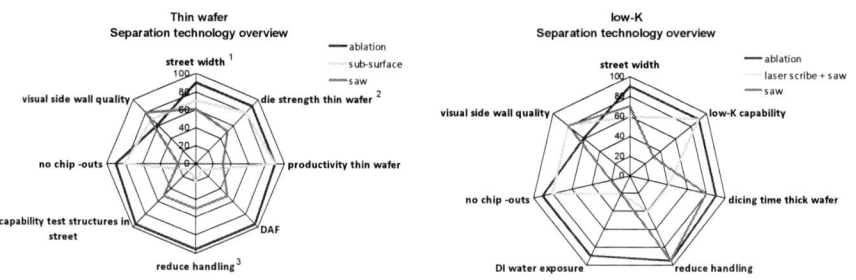

1) sub-surface; aspect ratio 0.4
2) Ablation; achieved via post process or short pulse laser
3) sub-surface; requires retaping, breaking, expansion

Figure 5. Thin wafer separation technology overview

Figure 6. Low-K separation technology overview

Multiple Beam Technology for Low-K Thin Si Wafers

ALSI has developed a full cut laser dicing technology and process for thin wafers mounted on Die Attach Film. Key aspects for thin wafer dicing are die strength and capability to dice Die Attach Film (DAF). The technology is based on ALSI's unique Multi Beam (MB) laser dicing technology. Process evaluations have been performed at several IDMs for memory devices; mechanical die-strength, visual appearance and performed assembly steps, all have demonstrated the excellent capability of this new dicing technology.

Multiple Beam Low-K grooving

ALSI multiple beam technology allows both efficient dicing as well as grooving. In a two step process Low-K wafers can be grooved and subsequently laser diced through or in case of thick wafers (>300um) blade diced. In the first step (indicated below in figure 5) a "low power" opening pass is made at an ultra high speed (300mm/s to 500mm/s). This opening pass is made by a tilted multiple beam. The angle of the multiple beam is recipe driven and therefore allows the opening or groove width of the low-K to be selected depending on wafer type. This multiple beam grooving technology significantly increases productivity by reducing the necessary grooving traverses to just one. In the second step the multiple beams are rotated to be parallel to the dicing direction and the wafer is diced through completely using the standard multibeam process.

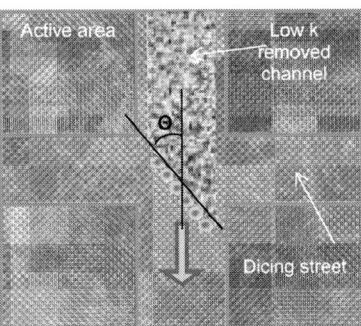

Figure 7. Step1, Low-K grooving using multiple beam under angle

Figure 8. Step2, Dicing through using standard multiple beam

Breakthrough laser dicing process

This newly developed process generates a breakthrough in thin wafer dicing. This process focuses on thin Silicon wafers (<50 µm) including the top passivations and test structures in the dicing street and the wafers are typical mounted on DAF (10-30um thickness). The process dices through both the wafer and the DAF mounted on standard dicing tape and film frames. The new laser dicing concept is a combination of the proprietary beam splitting technology with up to 50 beams and a small focus diameter. Using ALSI's multi beam laser dicing technology there is no need for any post processing for die strength recovery such as dry etching.

The following achievements can be expected:

- Die strength >> 400 MPa
- Dicing width << 24µm
- Through Put Time ≥ 11w/h (300mm wafer, 12mm die size, 50um thick Si on 25um DAF)
- DAF tape tested; Hitachi HF800, HF900 and Nitto EM310. (more types to be tested)
- FOW tape tested; 60um thickness

- Capability to dice Si wafers with different passivations (OxiNitride, Polyimide, SiO2, SiN3) and metal structures (aluminum, Au) in dicing street as well as DAF in same cycle.
- Successful die-picking
- Successful wire-bonding

Figure 7 below demonstrate a good visual appearance of the cut in combination with high die strength without any post processing (graph 3).

The current focus is for wafer thickness ranging between 25-50 um thick wafers. Thicker substrates are possible but this will reduce the throughput time. For now a limited amount of DAF and FOW tapes have been tested and during the coming period this will be extended.

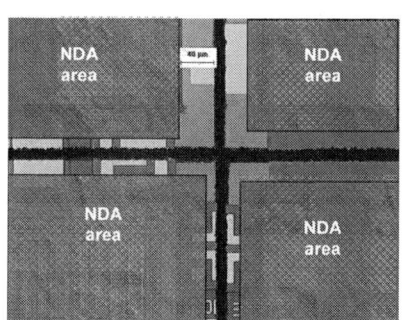

Figure 9. diced through memory wafer

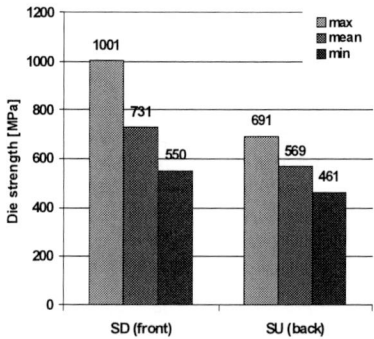

Figure 10. Die strength of a multiple beam laser diced through memory wafer. 50um thick on 30um DAF, measured on 4 point tool.

Conclusion

Memory and IC manufacturers are facing difficult challenges to keep up with the strong demand for packaging. Substrate thicknesses of 50um and less on DAF tape, force them to look to new separation technologies. Laser dicing is providing an excellent solution to meet the strict specifications on the diced die. However, no matter what laser dicing solution is chosen, it will always be a trade off between quality and speed. The multiple beam technology from ALSI provides a solution around this impasse, and maintains both high quality and high speed. Each of the technologies supplied by ALSI; ablation and Low-K grooving, is equipped with this multiple beam technology to ensure better dicing quality and higher yield, in combination with a quick return of investment, utilizing laser dicing technology.

Advanced Bump Structure for Improving the Board Level Characteristics of WLCSP

Chang-Bae Lee, Jongwoo Choi, Jin-Su Kim, Sangsoon Choi, Dojae Yoo,
Seungwook Park and Youngdo Kweon

Corporate R&D Institute, Samsung Electro-Mechanics Co., LTD
314, Maetan3-Dong, Yeongtong-Gu, Suwon, Gyunggi-Do, Korea, 443-743

In this study, to increase solder joint reliability, we developed the epoxy reinforced bump structure surrounding the solder bump joint. The thermal-mechanical reliability of the epoxy reinforced solder bumped wafer level package was carried out during the thermal cycle and drop shock test of the package. The reliability of solder bump joint was evaluated by means of the thermal cycle in the range of -40℃ to 125℃. The experimental results revealed that the thermo-mechanical fatigue properties of the epoxy reinforced bump structure was better than that of the conventional solder bump structure. Epoxy reinforced bump structures improved the WLP characteristic life by at least 2 times. Also, the drop shock reliability of epoxy reinforced bump structure was enhanced by 3 times of drop characteristic life as that of conventional single bump structure. The life time of epoxy reinforced solder bump joint was longer than that of the conventional WLCSP solder joint.

Introduction

Recently the package trend is demanding the smaller package size and the lower impedance electrical path with a short interconnection. The wafer level chip scale packages (WLCSP) is one of them, which has the solution of the package needs above. However, WLCSP technology is still not fully accepted on the large device size that is larger than 5*5mm and this kind of wafer level package(WLP) has still more risk for board level reliability to the failure than conventional QFP, BGA and CSP type components. Therefore, how to optimum the structure and material to secure high reliability of WLCSP is very important (1). Solder joint reliability of traditional WLP is the weakest point in the WLP technology. Especially, the drop shock reliability of solder joints has become a major issue for the electronic industry partly because of the increasing popularity of portable electronics and partly due the transition to lead free solders. One of approach to meet this require, these package usually used underfill process as one of solution to reliability increasing because it has difference in the coefficient of thermal expansion between that of the silicon chip and that of the organic PCB (3 ppm /℃ silicon versus 14~16 ppm/℃ for FR-4) (2). The underfill operation increases the manufacturing cost and reduces the throughput. In addition, reworking an underfilled chip on PCB is very difficult. Therefore, the WLP without underfill gradually evolve into another mainstream in the electronic packaging industry (3).

In this study, the board level solder joint reliability with different bump structure was investigated. The drop test is intended to evaluate and compare drop performance of surface mount electronic components for mobile electronic product applications in an accelerated test environment. Also, another concern in solder joint reliability is the failure due to loading cyclic stress. Therefore, thermal cycling test is therefore an important index of board level solder joint reliability progress. This study focused stand-off height and epoxy reinforced bump structure to enhance the thermal cycle and drop reliability in board level. To increase the stand-off height, we adopt the ball placing process using the epoxy solder paste that contained low melting point solder powder(Sn58Bi: 138°C). The FEM (Finite Element Method) simulation was performed to stress analysis and failure mode by epoxy reinforced structure effect. WLCSP structure and ball type were chosen for board level reliability evaluation.

WLCSP Test Vehicle Design and Procedure

WLCSP Design

Test vehicle must be designed to capture all of the thermo-mechanical and drop failures attributed to accelerated reliability tests. The daisy chain test vehicle is designed to be used for second level interconnect assembly test and continuity verification. Figure 1 shows the daisy chain test vehicle for the current board level thermal cycle and drop reliability test of WLCSP. For the board level interconnection, there are 35 solder balls which are uniformly distributed (shown in Fig.1).

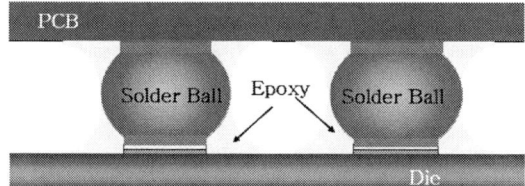 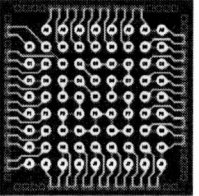

Figure 1. WLCSP test vehicle design and structure

TABLE I. WLCSP specifications

Items	Specifications
Die Size (mm)	3.5*3.5 (7.1*7.1)
Bump Array, I/O	9*9 (18*18), 80 (120)
Bump Diameter / Bump Height (μm)	200 ~ 250/160 ~ 230
Bump Pitch / Pad Size (μm)	300 / 170
UBM structure	TiW/Cu/Ni/SnAg
Die thickness (μm)	350

Thermal Cycling and Drop Shock Test

Board level reliability tests were performed after post reflow and attached WLCSP to PCB test board. As shown in Fig.2, the thermal cycling and drop shock test board was made of FR-4 with a nominal thickness of 1.0 mm. The test board also has ENIG (Electroless Nickel Immersion Gold) surface finish and NSMD (Non-Solder Mask Defined) component attached pads. Board size used for this study is 132*77mm.

Board level thermal cycling test was followed by the JEDEC JESD22-A104, condition G with a temperature range of -40°C to 125°C. During the test, the resistance of daisy chain WLCSP was real-time monitored. Also, JEDEC JESD22-B111 procedure for drop shock is the definitive test method used to evaluate portable electronics. Drop shock test condition for this WLCSP is applied on peak acceleration of 1500G with 0.5msec half sine pulse. The setup of a board level drop tester follows the guidelines of JEDEC standards, having 15 WLCSPs (3*5 matrix) assembled on a test board.

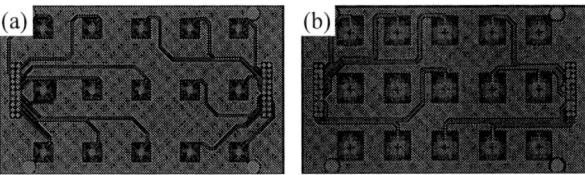

Figure 2. Board level test Board. (a) Die size: 3.5*3.5mm and (b) Die size: 7.1*7.1mm

Wafer Bumping and Board Assembly

The wafer bumping was carried out using 200~220 μm (SAC ball: 200 μm, Cu core ball: 220 μm) spheres assembled on wafer with Ni/SnAg pad finish. Spheres were assembled using an epoxy solder paste that was stencil printed on the wafer. Spheres were placed using a simple alignment assembly setup and reflowed in N_2 atmosphere.

Figure 3 shows the appearance image after bumping. From Fig. 3 (b), it was found that WLP having epoxy reinforcement surrounding the solder bump.

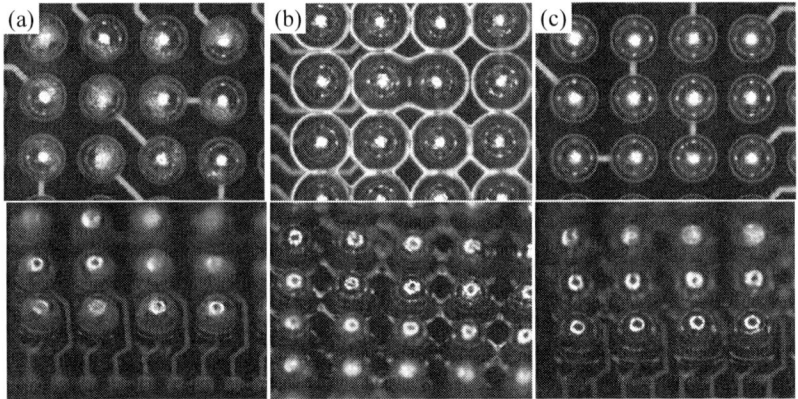

Figure 3. The appearance image after bumping: (a) SAC305 ball+SAC305 solder paste, (b) SAC305 ball+epoxy solder paste and (c) Cu core ball+SAC305 solder paste.

Epoxy solder paste used in this study is a conductive adhesive which is composed of a Sn58Bi low melting point solder and epoxy-based thermosetting resin. Low temperature soldering which are approximately 160 ℃ achieve adequate solder joints and high strength by an adhesive property of thermosetting resin at once, the covered epoxy surrounding the solder bump have high reliability without cleaning. To obtain good characteristic of the adhesive, conduct heating at 160□ for 6 minutes. After bumping process, the bumped die is assembled to FR4 test board with ENIG pad. Assembly process also use screen printing and epoxy solder paste the same as the bumping process.

Figure 4 shows the micrographs of solder bump joint. As shown in Fig. 4 (a) and Fig. 4 (b), it can be clearly seen that the bump stand-off with epoxy collar (about 230 μm) was higher than that of the conventional solder bump (about 160 μm).

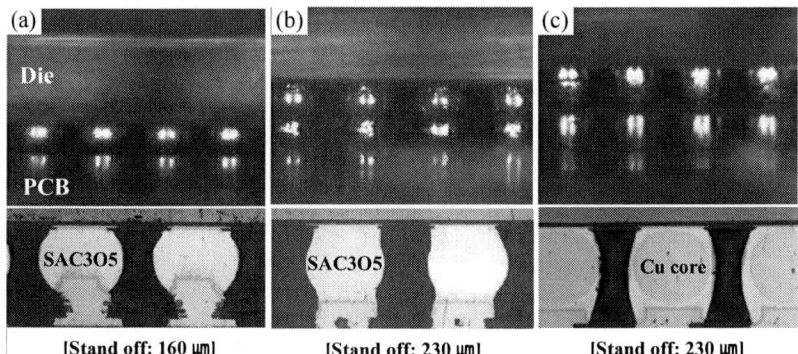

Figure 4. Optical micrographs of side views of solder bump joint: (a) SAC305 ball + SAC305 solder paste, (b) SAC305 ball + epoxy solder paste and (c) Cu core ball + SAC305 solder paste.

Results and Discussion

In this study, to compare the effect of the stand-off height and epoxy supporter, we have designed the three type structures such as conventional single ball, single ball with epoxy collar and Cu core ball.

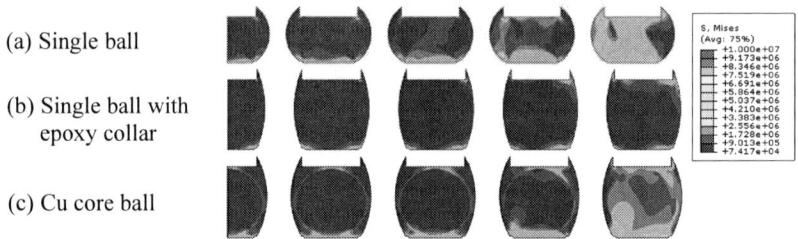

Figure 5. Stress distribution during the thermal cycling test

Figure 5 shows the stress distribution with different WLCSP bump structure during the thermal cycling test. As shown in Fig. 5, the stress of solder joint decreases by the higher stand off and epoxy reinforcement surrounding the solder bump. The Coffin-Manson equation predicts that the thermal fatigue lifetime of a solder joint is proportional to the square of the bump stand off, thus it is well known that a greater bump stand off results in lower solder stress (4-5).

Figure 6 shows the cross sectional SEM micrographs of the solder joint. During soldering, the solder alloy reacts with the pad to form intermetallic compound at the joint interface. From the results of EDX analysis, it was found that only one IMC of Ni-Cu-Sn formed at the interface, no binary IMC (Cu-Sn, Ni-Sn etc) was detected, rather, it was a complex IMC of Ni, Cu, and Sn with a Cu percentage of 2 ~ 6 at %.

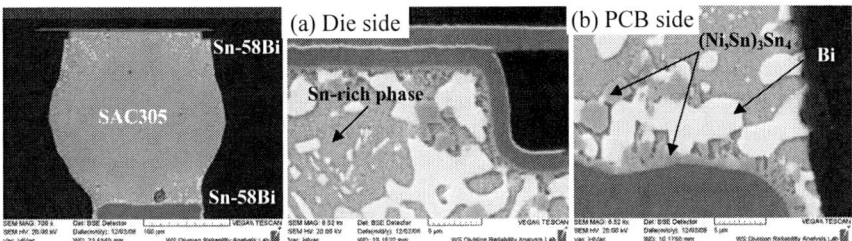

Figure 6. SEM micrographs of WLCSP solder joint with epoxy collar

Table □ and □ show the results of thermal cycling and drop shock test, respectively. The observation of cross-section was performed to understand the failure mechanism occurring, as a result of thermal cycling and drop test, as shown in Fig. 7. The order of thermal cycling performance with different solder composition is SAC305 (3% Ag) > SAC101 (1% Ag). But in the case of drop test, the high Ag solder consistently fail at lower drop cycles than the low Ag solder, this is to be expected due to lower Ag content, it is lower modulus, easier to yield and more ductile (6). So, this is an important factor in selecting solder alloys for high strain rate applications. At lower Ag there is less Ag_3Sn IMC in the bulk solder with concomitant reduction in mechanical strength. Clearly solder with lower Ag content have an advantage in potentially absorbing the effect of high strain rate deformation.

TABLE II. Thermal cycling test summary

Classifications	Die size(mm)	Characteristic life	Failure Mode (Main crack site)
SAC305 (B + SP)	3.5*3.5	887 cycle	RDL Cu
	7.1*7.1	465 cycle	RDL Cu
SAC305 B + ESP	3.5*3.5	1,756 cycle	IMC/solder interface, Bulk solder
	7.1*7.1	959 cycle	IMC/solder interface
SAC101 (B + SP)	3.5*3.5	838 cycle	RDL Cu
SAC101 B + ESP	3.5*3.5	1,465 cycle	IMC/solder interface, Bulk solder
Cu core B + SAC305 SP	3.5*3.5	1,100 cycle	RDL Cu
	7.1*7.1	476 cycle	RDL Cu

B: ball, SP: solder paste, ESP: epoxy solder paste, RDL: redistribution layer

TABLE III. Drop shock test summary

Classifications	Die size(mm)	Characteristic life	Failure Mode (Main crack site)
SAC305 (B + SP)	3.5*3.5	2,994 drop	RDL Cu
	7.1*7.1	923 drop	RDL Cu
SAC305 B + ESP	3.5*3.5	10,887 drop	Bulk solder
	7.1*7.1	1,583 drop	Bulk solder
SAC101 (B + SP)	3.5*3.5	5,805 drop	RDL Cu
SAC101 B + ESP	3.5*3.5	17,641 drop	Bulk solder
Cu core B + SAC305 SP	3.5*3.5	7,344 drop	RDL Cu
	7.1*7.1	815 drop	RDL Cu

B: ball, SP: solder paste, ESP: epoxy solder paste, RDL: redistribution layer

As shown in Table □ and □, the life time of the higher stand off structure (with Cu core or epoxy reinforced joint) was longer than that of the conventional WLCSP solder joint. Especially, the reliability results of epoxy reinforced bump structure were better than that of others (SAC B + SP and Cu core B + SP), due to the role as stress buffer at the critical solder/UBM/die interface. For WLCSP solder joint with epoxy collar, main failure were observed within the solder bulk side or interface and not at the Cu RDL, especially, in case of drop test, all failures occurred inside of solder, as shown in Fig. 7. When the epoxy solder paste was adopted, failure mode changes from vertical RDL crack to ductile failure in solder.

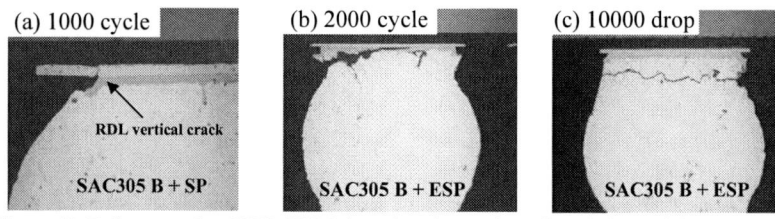

Figure 7. Failure mode of WLCSP joint after thermal cycling (a-b) and drop test (c)

Conclusions

We designed and optimized the new structure of wafer level packaging using the mechanical simulation. Epoxy reinforced bump structure were successfully fabricated with optimized, it will be one of the dominant bumping and board level assembly methods. The results are summarized as follows:

The experimental results revealed that the thermo-mechanical fatigue properties of the epoxy reinforced bump structure was better than that of the conventional solder bump structure. Epoxy reinforced bump structures improved the WLP characteristic life by at least 2 times. It is possible to achieve board level solder joint reliability of more than 1000 cycles under thermal cycles, which is an enhancement of more than 2 times in initial failure life. Also, in the case of drop shock characteristic life, we could get the enhancement of more than 3 times compare with conventional structure. The reliability of large die(5~10mm) without underfill is still in question. But , from the our results, new WLCSP structure with the epoxy reinforced as the stress buffer will have great potential to be applied to electrical devices with a large die.

References

1. S-W. Park et al., *2nd Level Reliability Important on WLCSP*, Proceedings of 43rd. IMAPS2010, (2010).
2. V. Patwardhan et al., *Constrained Collapse Solder Joint Formation for Wafer-Level-Chip-Scale Packages to Achieve Reliability Improvement*, ECTC, (2001).
3. C-M. Liu et al., *Solder Bumps Layout Design and Reliability Enhancement of Wafer Level Packaging*, ICEPT, (2003).
4. P. Garrou, *IEEE TRANSACTIONS ADVANCED PACKAGING*, **23**(2), (2000).
5. H. Yang et al., *Reliability Evaluation of Ultra CSP Packages*, in Proc. SMTA Emerging Technol. Conf., Chandler, AZ, (1998).

6. T-Y. Tee et al., *Design for Board Trace Reliability of WLCSP under Drop Test*, 10th Int. Conf. on Thermal, Mechanical and Multiphysics and Experiments in Micro-Electronics and Micro-Systems, EuroSimE, (2009).

Plasma Cleaning Effect on Automotive Devices

P. Y. Chew[a], and T. A. Aw[a]

[a]Infineon Technologies (M) Sdn Bhd, Batu Berendam, 75450 Melaka, Malaysia

In today's semiconductor market of the growing automotive industry, reliability of the devices becomes a major concern as all these devices require an extremely low stress and high thermal condition. This paper will discuss the critical characteristics that are responsible for the die top corner delamination and focuses mainly on a few key areas such as molding compound, wafer and plasma cleaning process so as to enable the achievable of reliable and robust package for automotive products. The devices are to be characterized with scanning acoustic tomography (SAT) and various package level reliability tests. From the study, it has been found that plasma cleaning process plays an important role in resolving the reliability issue by totally eliminating die top corner delamination, and which has improved tremendously in the life span of the devices and also the reliability of the packages.

Introduction

In automotive industry, reliability of devices becomes a major concern as all these devices require an extremely low stress and high thermal stability. Therefore, the interface between mold compound and silicon chip surface become very critical. This is because delamination between these two interfaces will lead to package cracks resulting subsequently in impacting the functional performance caused by stress damage. Die top corner normally is the affected point due to the low interface strength and focus area of the maximum shear stress. Besides that, R. Ibrahim and et al, did mention that die corner normally is resin rich which means lower filler contents and higher CTE due to turbulent effect during molding. This will directly increase the stress level at die top corner of the package. A few methods can be implemented to improve the adhesion between mold compound and silicon chip. In terms of mold compound, increase adhesion promoter or reduce the stress by increasing filler loading so as to lower down the CTE. In term of chip surface, surface roughening on silicon chip surface through plasma will give better chemical bonding or mechanical interlocking effect through reduction of contamination. All these are covered in the study of this paper and the samples were characterized by scanning electron microscopy (SEM), and scanning acoustic microscopy (SAM).

Experimental Procedures

A. Evaluation on Molding Parameter

As mentioned earlier, it is critical to identify the actual root cause for die top corner delamination by trouble shooting the nearest process which is molding process itself. Firstly, re-optimize the process parameter by including in all the major parameters as per Table I (mold compound A and all other parameters are set to be constant). By setting lower transfer time which will result in faster transfer speed (position remain constant) so as to allow the mold compound to flow in within the effective working window before it start to harden. As for the transfer pressure, it is to allow better compactness especially towards the end flow. For this evaluation, the detectable gate for die top corner delamination is through SAM analysis at zero hours (after PMC) besides ensuring no molding reject issues. Furthermore, the samples are to be subjected to stress test such as thermal cycle (TC) under -55°C to 150°C and pressure cooker test at 96hrs.

Table I: Molding parameter DOE Matrix.

Test Leg	Transfer Time (s)	Transfer Pressure (MPa)	Mold Temperature (°C)
A1	14	11	175
A2	12	11	175
A3	10	11	175
A4	14	11	170
A5	14	11	180
A6	14	13	175
A7	14	15.1	175
A8	14	17.2	175

B. Evaluation on Mold Compound and Plasma Cleaning.

This evaluation is mainly on the approach of either improving from mold compound or silicon chip surface sides. Mold compound plays an important role in improving delamination through
a) increasing the adhesion promoter, or
b) reducing the CTE by increasing the filler loading, or
c) use compound with multi-aromatic (MAR) based resin type which is better in term of adhesion as compared with other resin types.
Plasma cleaning is focusing on removing contamination or roughening effect
in order to have better interlocking between mold compound and silicon chip surface. Therefore, different mold compound types and plasma cleaning as per Table II shown below will be evaluated.

Table II: Mold Compound and plasma cleaning evaluation.

Test Leg	Type
B1	Mold Compound A with OCN type
B2	Mold Compound A with 15% increase of adhesion promoter
B3	Mold compound A with lower CTE by increasing 2% filler loading
B4	Mold compound B with MAR based resin
B5	Plasma cleaning with Ar/H_2 and H_2O_2

Then the samples will be subjected to reliability for stress test for die top corner delamination check at TC, PCT as well as TSK other than zero hours analysis.

Results and Discussions

A. Evaluation on Molding Parameter

The evaluation on molding parameter based on transfer pressure, transfer time and molding temperature had been analyzed based on the mold ability such as wire sweep, short molding or voids other than SAM to check during zero hour and after stress test. The result on the mold ability and reliability with the sample size of 16units are shown as per the table III below.

Table III: Results of molding parameter DOE matrix

Test Leg	Transfer Time (s)	Transfer Pressure (MPa)	Mold Temperature (°C)	Wire Sweep (>10%)	Die Top Corner Delam			
					0hr	PCT 96hrs	TC 500x	HTS 168hrs
A1	14	11	175	0/16	0/16	0/16	3/16	0/16
A2	12	11	175	4/16	0/16	0/16	8/16	0/16
A3	10	11	175	16/16	0/16	0/16	11/16	5/16
A4	14	11	170	0/16	0/16	0/16	5/16	0/16
A5	14	11	180	4/16	1/16	0/16	6/16	0/16
A6	14	13	175	1/16	3/16	0/16	4/16	0/16
A7	14	15.1	175	10/16	1/16	0/16	2/16	0/16
A8	14	17.2	175	16/16	6/16	0/16	10/16	0/16

As an overview, the results had shown that no significant differences on all the test legs observed with die top corner delamination and mostly on TC500x stress test. Further cross section had found that there is a separation between the silicon chip corners with mold compound which is adjacent from the mold gate area (Figure 1). Besides that, by adjusting the transfer time and pressure, there are wire sweep which is more than 10% as compared to test leg A1 (control). With the hypothesis of ineffective working window and not compact therefore transfer pressure, time and temperature are evaluated. However, from the results, it does not show molding parameter influence on die top corner delamination.

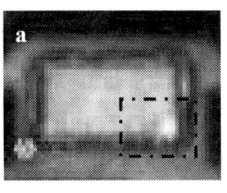

Figure 1: a) SAM scan micrograph with die top corner delamination unit at zero hour a(i) and a(ii) Cross Section with separation between silicon chip and mold compound.

From Figure 1, it shows that there is a separation (delamination) between silicon chips and mold compound which indicates weak adhesion. So in order to improve the adhesion, it has to be done by improving the mold compound adhesion, lowering down the CTE or improving silicon chip surface as per B evaluation.

B. Evaluation on Mold Compound and Plasma Cleaning

This evaluation is to improve the material adhesion strength either on mold compound or silicon chip surface. From mold compound, it is by increasing 15% of the adhesion promoter in order to give higher adhesion and lowering the CTE by increasing the filler loading by around 2%. In term of silicon chip surface, it is by cleaning the surface as there is a suspect presence of contamination. From this evaluation, it is found that with increasing the adhesion promoter and lowering the CTE does not show improvement as compared to the original mold compound. As for plasma cleaning and change of epoxy resin type to multi-aromatic based which show total zero die top corner delamination as per table IV below using test leg A1 parameter.

Table IV: Results of mold compound and plasma cleaning.

Test Leg	Type	Die Top Corner Delamination (unit)			
		0hrs	PCT 96hrs	TC 1000x	HTS1000hrs
B1	Mold Compound A with OCN type	0/16	0/16	3/16	0/16
B2	Mold Compound A with 15% increase of adhesion promoter	63/96	-	-	-
B3	Mold compound A with lower CTE by increasing 2% filler loading	21/96	-	-	-
B4	Mold compound B with MAR based resin	0/16	0/16	0/16	0/16
B5	Plasma cleaning with Ar/H_2 and H_2O_2	0/16	0/16	0/16	0/16

The results show that improvement from mold compound A by increasing the adhesion promoter or lowering the CTE does not shows any improvement as compared to the test leg B1 but die top corner delamination was observed at zero hours after post mold cure (PMC), therefore, no further stress test is necessary for this B2 and B3. This indicates that the compound adhesion is still insufficient between mold compound and silicon chip surface which is the polyimide surface. According to M. Amagai and et al's study, presence of silicon on the chip surface will obstruct the hydrogen chemical bonding between the polyimide surface and the epoxy resin which lead to the decrease of interfacial adhesion as the degree of bonding between the hydrogen of the epoxy resin and the oxygen of the polyimide surface is less. This is further proven and strengthened by the cross section view of the die top corner delamination area as per Figure 2, which shows no presence of polyimide on silicon chip surface.

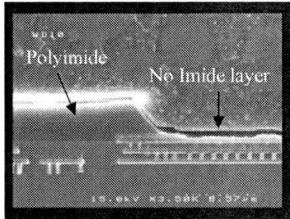

Figure 2: Cross section view shows absence of polyimide at the chip edge with separation line between mold compound and silica chip surface.

Test leg B4 with the multi-aromatic (MAR) based resin had shown a total elimination of die top corner delamination. This is mainly because of the advantages of MAR resin with low moisture absorption and high adhesion properties which lead to better reliability performance. Unfortunately, Tg for this MAR based compound is always below 150°C which is not suitable for this product requirement. But from this test leg, it can be concluded that the adhesion from mold compound is important in order to eliminate of this die top corner delamination.

Plasma cleaning effect is one of the possible methods to remove die top corner delamination as well. Two types of mixture gases Ar/H2 and H2/O2 were also used in this evaluation, both of the results shows significant improvement in the contact angle measurement as compared to non plasma silicon chip surface as per figure 3 below.

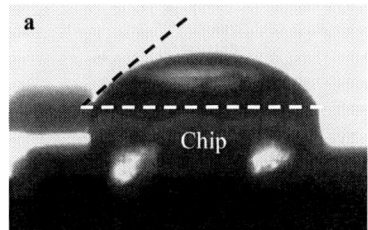
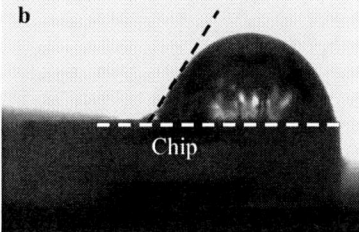

Figure 3: Contact angle measurement. (a) plasma clean (b) non plasma clean

From the contact angle measurement, it indicates that 3(a) is having high surface energy (20deg) which creates a hydrophilic environment on the silicon chip surface while for 3(b) with a hydrophobic (liquid forms spherical shape) surface that resists the formation of hydrogen bonding between mold compound and polyimide. Therefore, with the high surface energy, it helps to eliminate the die top corner delamination and gives good adhesion between mold compound and polyimide even after reliability stress test.

Conclusion

From this study, a good adhesion is defined as the degree of bonding between the hydrogen of the epoxy resin (mold compound) and the oxygen of the polyimide surface. Therefore, die top corner delamination can be eliminated with MAR based type of mold compound which gives much higher adhesion and lower moisture absorption as compared to mold compound A. Besides that, plasma cleaning will also gives good adhesion as it will remove off the contamination on the polyimide surface. With this improved adhesion in automotive devices, it gives a reliable stress level and as such, it helps to prolong the lifespan of the devices as well.

Acknowledgments

Finally, I would like to express my thanks and gratitude to my family members for their encouragement and support in this study. And furthermore, a special thanks to my colleagues who had offered their help and assisted me in this study.

References

1. R. Ibrahim et al, 11th Electronics Packaging Technology Conference, (2009)
2. M. Amagai, Proceedings of the 1993 IEEE IPFA, (1993).
3. M. Amagai, *Microelectronic Reliability*, **40**, (2000).

Packaging Issues for High-Voltage Power Electronic Modules

S. S. Ang, T. Evans, J. Zhou, K. Schirmer, H. Zhang,
B. Rowden, J. C. Balda, and H. A. Mantooth

National Center for Reliable Electrical Power Transmission
University of Arkansas, Fayetteville, Arkansas 72701, USA

This paper addresses interconnect and passivation issues associated with the fabrication of high-voltage power electronic modules. A wire-bondless direct solder attachment hierarchy for interconnection was evaluated using two direct-bond copper (DBC) elements: one acting as a substrate connected to a base plate and the second as an interconnection lead-frame between the power semiconductor devices. Finite element modeling revealed their enhanced performance over wire-bonded modules due to at least an order of magnitude decrease in parasitic inductance and the added benefit of enhanced thermal performance. A two-step passivation/encapsulation method to provide higher breakdown voltages at high temperatures is proposed for future high-voltage power electronic modules.

Introduction

Cost effective high-voltage power semiconductor devices and modules are essential power electronic components in the growing demands of modern power grid systems in terms of improved efficiency, better control of power electronic devices, and integration of renewable energy sources; all essential elements of the so-called "smart grid". The high-voltage requirements in these applications usually require series configuration of the power semiconductor devices in order to increase their voltage ratings. The most common press-pack high-voltage power packages available today are adapted from the traditional "hockey puck" packages. This rigid pressure contact technology, intended for the rugged silicon thyristors, is not optimized for the sensitive microstructures on the surface of modern power semiconductor devices like wide-bandgap silicon carbide devices [1]. As a consequence, a great deal of care is required during module assembly. The issue is further aggravated when the press-pack module size is increased for higher current ratings. There is a significant cost impact on system production cost as a result of these shortcomings [1]. Recent advancements of high-voltage power semiconductor devices such as silicon carbide devices require a different approach in high-voltage power module packaging. Currently silicon carbide (SiC) devices are achieving breakdown voltages above 10 kV and are at the limits of traditional high-voltage packaging solutions. Cree and Powerex have developed high voltage power modules utilizing 10kV SiC MOSFETs in a wire bonded module [2]. However, as these device breakdown voltages increase, new approaches in package design, passivation, and encapsulation become necessary. This paper addresses alternative interconnect and passivation/encapsulation issues associated with the fabrication of high-voltage power modules.

High-Voltage Interconnect Issues

Several interconnect technologies such as wire bonds, solder balls, direct solder, and press pack are used in the packaging of high-voltage power modules. Each solution offers its own unique benefits but it also suffers from limitations. Ultrasonic aluminum wire bonding has become increasingly popular due to the high yield and comparatively low cost [3]. However, the primary limitations of wire bonding are from the inherent parasitic inductances of the bond wires. Additionally, a large package area becomes necessary to facilitate bonding areas on the device and pad area for external connections. Furthermore, wire-bonded devices can only have heat actively removed from one side [3]. However, parasitic inductance of the wire bonds can be reduced and current capacity increased when wires are used in parallel. Solder ball technologies includes flip-chip, under bump metallurgy and solder bumping. An advantage to these approaches can be realized through heat removal from both sides of the device. Conversely, some of the major drawbacks include additional processing steps such as die metallization as well as potential coefficient of thermal expansion (CTE) mismatch between the die and substrate [3]. This can cause additional stress in the solder joints and lead to die cracking. Direct solder interconnection allows for a high power density packages. The structure of such modules usually consists of a thin solder preform on the top and backside of the device that forms a joint with a substrate layer, usually DBC, on either side of the die. Direct solder approaches allow even more heat removal capability from both the top and backsides of the device when compared to solder ball interconnects. This is due primarily to the fact that a thin solder joint is commonly used and the result is an increase in surface area coverage of the solder joint. This also results in improved electrical performance through an increase in current carrying capability and a reduction in parasitic inductance. However, the thin solder joint can lead to increased stress on the device [3]. Press-pack interconnects are typically found in high-voltage applications due to their modular construction that allows for easy series and parallel configurations to increase voltage and current capabilities. Interconnection is typically made through pressure and spring interconnects with the device itself soldered to a substrate or base-plate [3]. Each of these interconnect schemes has its own failure modes that limit the reliability and operating parameters of the module. Typically, die cracking is a potential issue for all the above interconnect technologies. This stems from CTE mismatch at the solder joint between the device and substrate. Wire-bonded modules can fail with fatigue or through lift off of the bond wire from either the device or substrate. Direct solder applications can experience even greater CTE-related stress and strain on the die and substrate. Press-pack modules can fail with spring fatigue or surface wear on the die [4].

The approach developed in this study examines the use of 8kV rated devices with a direct solder attachment hierarchy for interconnection. This involves the use of two direct-bond copper (DBC) elements, one acting as a substrate connected to a base plate and the second as an interconnection lead-frame between the power semiconductor devices as shown in Figure 1 for a two-device power electronic module. This layout was chosen to maximize power density of the module while also achieving the potential for dual sided cooling [5]. The solder joints are 50 microns. Naturally, when operating at full voltage potential, breakdown between the topside and backside of the devices is inevitable due to the small distances between them. In order to eliminate this breakdown potential, studies have been conducted in spin coating the DBC elements with 28 microns of benzocyclobutene (BCB) as the passivation layer.

BCB is chosen as a substrate passivation layer due to its high breakdown strength, which, in some cases, can exceed 300 kV/mm [6]. Since the final thickness after curing of the BCB is less than that of the solder joint, additional underfill can be used to fill in any gaps to relieve mechanical stresses.

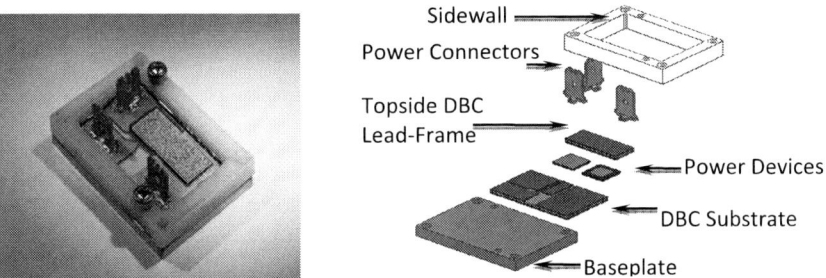

Figure 1. Direct solder attached high-voltage power electronic module.

Finite Element Method (FEM) Simulations

A commercial FEM solver was used to compare the thermal performance of the high-voltage wire-bonded and wire-bondless modules. The ambient temperature was set to $22°C$ with a natural convection coefficient of air of $5W/°C\cdot m^2$ on the top side. In this paper, a practical heat sink equivalent heat transfer coefficient of $10,000\ W/°C\cdot m^2$ is chosen as the value for the bottom substrate boundary condition. All the layer interfaces are assumed to be well contacted. The power dissipation for each device is $7.5W/mm^3$. The simulations used Newton's law rather than Fourier's law when calculating natural convection between the module and the ambient air. Radiation effects were taken into consideration. The simulations were performed under three different configurations: (1) module 1: wire-bonded module with single-sided cooling; (2) module 2: wire-bondless module with a top DBC lead-frame; and (3) module 3: wire-bondless module with a top DBC lead frame and a top heat spreader to promote double-sided cooling. Heat generated by the devices will be dissipated by the contacted layers to the heat sink and ambient air. Steady-state thermal analysis was performed to obtain the temperature distributions of these modules as shown in Figure 2.

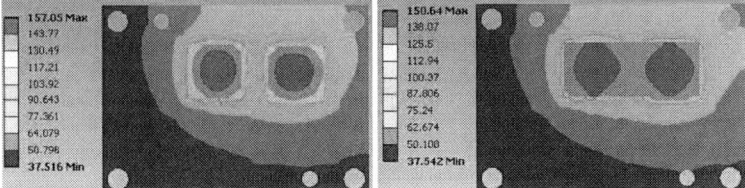

Figure 2 Temperature distributions of modules 1 and 2

In these simulations, the maximum temperature occurs at the junction between the power devices and the DBC base-plate. With the wire-bonded module, its junction temperature reaches $146°C$. For the wire-bondless module, the maximum junction temperature decreases by $5°C$ which shows that with the inclusion of a top DBC lead frame and without forced heat removal, the thermal performance of the module can be improved.

With a top heat spreader to promote double-sided cooling and forced heat removal, a maximum junction temperature of 106°C was obtained. These simulations show that for a wire-bonded module, most of the heat will flow to the substrate, while for modules utilizing direct solder interconnects, double-sided cooling can be implemented to significantly reduce the junction temperature of the devices.

The parasitic inductance associated with interconnections creates significant challenges for packaging design. The FASTHENRY[7] software was used to extract the parasitic inductances of the wire-bonded and wire-bondless modules. Figure 3 shows the simulated results of three different modules as a function of frequency. As can be seen, the parasitic inductance of the wire-bondless lead frame interconnection is in the pH range, which is three orders of magnitude smaller than the wire-bonded modules. The parasitic inductance decreases about 36% when two parallel bond wires were used.

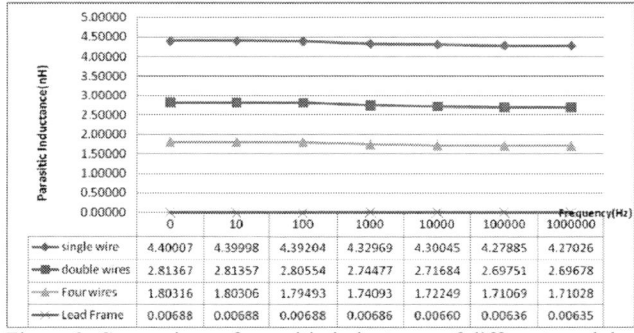

	0	10	100	1000	10000	100000	1000000
single wire	4.40007	4.39998	4.39204	4.32969	4.30045	4.27885	4.27026
double wires	2.81367	2.81357	2.80554	2.74477	2.71684	2.69751	2.69678
Four wires	1.80316	1.80306	1.79493	1.74093	1.72249	1.71069	1.71028
Lead Frame	0.00688	0.00688	0.00688	0.00686	0.00660	0.00636	0.00635

Figure 3 Comparison of parasitic inductance of different modules

High-Voltage Passivation Issues

Impact ionization by electrons is responsible for breakdown in the case of a homogeneous electric field. In inhomogeneous field breakdown, the characteristics of the discharge and its occurrence are significantly affected by the space charge appearing in the dielectric close to the electrode of large curvature [8-9]. As shown in Figure 4, the dielectric strength of an encapsulant for a 10kV module with a 150μm diameter wire bond (height of bond wire from the substrate = 1.8mm, pad distance = 4.73mm) is found to be not less than 45kV/mm. It can be seen that the maximum electric fields occur between the bond wire and the points with the lowest potential on the power substrate.

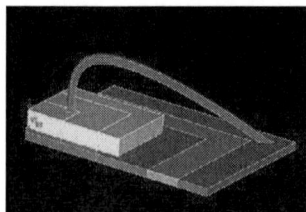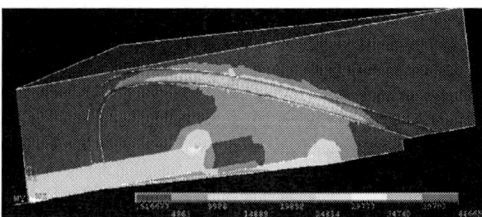

Figure 4. Electric field simulation of a wire bond in a power module

Figures 5(a) and (b) show the leakage current versus temperature at 1kV bias and leakage current versus bias at 150°C for a typical silicone encapsulation material. As can be seen from Figure 5(a), the leakage current is at a minimum at 130°C. This minimum leakage current happens at moderate temperatures rather than at low temperatures because of the decrease of internal stress and crack formation in the binder resin [10-11]. The leakage current increases almost linearly as a function of bias as shown in Figure 5(b).

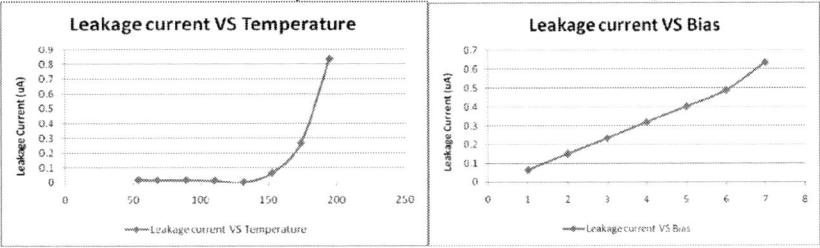

Figure 5. Leakage current (a) versus temperature at 1kV, (b) versus bias at 150°C.

A two-step passivation/encapsulation concept is proposed for the high-voltage power modules above 10kV at 200°C. As shown in Figure 6, a silicone gel or elastomer material is applied after a high-voltage passivation material is applied. As shown earlier, the dielectric strength requirement for this passivation material is 45kV/mm.

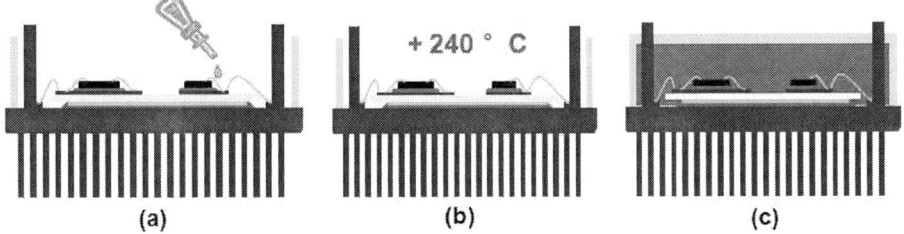

Figure 6. Two-step passivation/encapsulation method for high-voltage power module

A polymer matrix with an inorganic filler using a sol-gel growth technique has been developed as the high-voltage passivation material for the high-voltage power modules. The inorganic filler is used to improve the electrical, mechanical and other properties of polymers as well as to reduce cost. Polyamide imide (PAI), which is known of high thermal resistance and good solubility in polar amide-type solvents, provides a favorable balance between processability and performance. The PAI silica hybrid material can be spin-coated onto the power substrate to a thickness of 40μm prior to the application of an encapsulant.

Conclusions

Interconnect and passivation issues associated with the fabrication of high-voltage power electronic modules were presented. A wire-bondless direct solder attachment hierarchy for interconnection using two direct-bond copper (DBC) elements, one acting as a substrate connected to a base plate and the second as an interconnection lead-frame between the power semiconductor devices was evaluated. Finite element modeling revealed their enhanced performance due to a decrease in parasitic inductances from the interconnects and the added benefit of enhanced thermal performance. In an effort to achieve a functional power module able to accommodate devices with breakdown voltages greater than 10 kV, initial experimentation led to promising results with a BCB substrate and interconnect passivation method. To further increase the reliability and operating performance of future high voltage modules, a two-step passivation/ encapsulation method, using a polyamide imide silica hybrid passivation and a silicone encapsulant, was proposed for the high-voltage power modules.

Acknowledgments

The authors would like to acknowledge the financial support from the National Center for Reliable Electric Power Transmission (NCREPT) and NSF I/UCRC Grid-connected Advanced Power Electronic Systems, and technical contribution of their colleagues at the High-Density Electronics Center (HiDEC), University of Arkansas.

References

1. S. Kaufmann, T. Lang, and R. Chokhawala, *ISPSD* Osaka, Japan (2001).
2. J. Richmond, S. Leslie, B. Hull, M. Das, *IEEE Energy Conversion Congress and Exposition* (2009)
3. J. Calata, G. Lu, and C. Luechinger, *Inter Society Conference on Thermal Phenomena* (2002).
4. P. Hansen and P. McCluskey, *European Conference on Power Electronics and Applications* (2007).
5. Brian Rowden, PhD Dissertation, University of Arkansas, (2010).
6. A. Modafe, N. Ghalichechian, B. Kleber, and R. Ghodssi, *IEEE Trans. Device and Materials Reliability,* **4**(3) 495-508 (2004).
7. M. Kamon, M. J. Tsuk and J. K. White, "FASTHENRY a multi-pole accelerated 3D inductance extraction program", *IEEE Trans. Microwave Theory & Techniques*, **42**(9), 1750 (1994).
8. G. A. Vorob'ev and N. S. Nesmelov, *Electric Breakdown in Solid Dielectrics*, Translated from *Izvestiya Vysshikh Uchebnykh Zavedenii, Fizika*, (1), 90-104 (1979).
9. G. Finis, A. Claudi, and G. Malin, *Trans. Dielectric and Electrical Insulation*, **14**(2), 487-497 (2007).
10. R. Bruetsch, *IEEE International Symposium on Electrical Insulation*, Vancouver, BC, Canada, p. 162-165, June 8-11 (2008).
11. R. A. Bernstorf, *Polymer Compounds Used In High Voltage Insulators"* p. 7-8, Hubbell Power Systems, The Ohio Brass Company (2004)

CHAPTER 8

METROLOGY, RELIABILITY AND TESTING

IDDQ Test Practice in Nanotechnologies

Samuel Ye [a], Clayton Shen [b], Zhe Liu [c], Qin Liyun [d]

[a c d] Availink, Inc., Beijing, China
[b] Availink, Inc., MD, USA

As semiconductor wafer process technology went into nanotech level, I_{DDQ} test faced tremendous challenges than it ever did due to high intrinsic leakage. A case study of I_{DDQ} test in Availink ASICs in 130nm and 65nm process technologies demonstrates that I_{DDQ} test is still feasible. The scheme includes I_{DDQ} test vectors precise selection/generation, I_{DDQ} measurement with high resolution /repeatability, proper statistical post processing algorithm in outlier detection, results verification, etc. As a result, I_{DDQ} test is still considerable in manufacturing test strategy in nanotechnologies era.

1. Introduction

I_{DDQ} testing offers several advantages. Since the supply current can be monitored easily, it provides excellent observability (2). Only a few vectors are usually enough to achieve reasonable high fault coverage. Thus I_{DDQ} testing is quite valuable and an essential testing item in the IC production test in the past. However, deep sub-micron technologies impose difficult challenges for I_{DDQ} testing nowadays and as geometric sizes of transistors of ICs are rapidly scaled down into nanotech level, I_{DDQ} testing becomes less effective. It is mainly because of higher intrinsic leakage of ICs and larger variation in leakage level due to wider process shifting in nanotech level.

This paper presents the cases study of I_{DDQ} testing of ICs fabricated in 130nm and 65nm technologies. Through these cases, we would demonstrate the feasibility of I_{DDQ} testing in current nanotech levels and give the benefits of keeping I_{DDQ} testing in IC's manufacturing test.

2. Background Description

I_{DDQ} testing has been shown to provide several benefits and ensure high reliability of components (3), (4), (5). Since I_{DDQ} testing is defect oriented diagnostic to identify a defective part, it can detect several defect types such as bridging, gate-oxide shorts and breaks (6). It must be used as a supplemental functional testing and reliability testing. For both types, I_{DDQ} adds value by improving test quality at a very low cost.

In nanotech era, since background leakage becomes higher and higher, it has already made fault-free current one or more orders of magnitude higher than the defect current (1). It is much harder to distinguish faulty from fault-free parts in I_{DDQ} testing.

Generally speaking, there are three categories of solutions to extend I_{DDQ} testing usability: (a) technology solutions such as reverse body bias (7), SOI technology (8), etc., (b) design solutions, such as partitioning with built-in I_{DDQ} sensors, multiple threshold transistors, etc., (c) data analysis solutions, which rely on different (statistical) methods of analysis for discriminating faulty chips.

As a startup fabless ASIC design house, technology and design solutions to reduce background leakage are expensive and often limited by various resource limitations. Hence, they're not economically viable to us. So the only choice for us is data analysis solutions to improve I_{DDQ} testing resolution. The solutions are described as, at test application level, taking the leakage into account during the decision making process (1).

The data analysis methods can be generally characterized as using information from intra-die I_{DDQ} variation and information from inter-die for variance reduction. Generally these methods are (a)Current ratios, (b)Delta I_{DDQ}, (c)Clustering techniques, (d)Nearest Neighbor Residual. We adopted current ratio method (9), for I_{DDQ} testing, when we started our initial products in 130nm process technology in 2008. The improvement from original method was that we used direct I_{DDQ} measurements, instead of functional mode comparisons for all I_{DDQ} readings, except minimum current measurement. This was due to measurement accuracy consideration and limited measurement vectors adopted.

With new products being matured gradually, which were fabricated in 65nm process, we worked with foundry to implement stringent stress in I_{DDQ} testing flow with pre- and post- I_{DDQ} measurements and post comparisons/analysis as a way of detecting weak ICs or ICs with latent defects that pose reliability risk (4). Meanwhile, we adopt the combination of current ratios and delta-I_{DDQ} techniques (10) to cater for I_{DDQ} testing of 65nm's products.

3. IDDQ Test Vector Selection

As the use of advanced I_{DDQ} test strategy is current signature based, to ensure a defect is detected, there must be sufficient variation in current across the entire vector set. It is different from conventional I_{DDQ} test that it will still be detected, if all vectors activate the defect. Normally current signature based test vectors would be achieved by having some vectors which activate the defect and some which do not. As a result, a bit difference in vectors requirement.

I_{DDQ} testing vector generation methodologies can be classified into three categories:
(A)Every-vector
By this methodology, the power-supply current is monitored for every vector in a functional or stuck-at fault test set. As a result, it is relatively slow – on the order of 10-100millisecond per measurement: making it impractical in a manufacturing environment
(B)Supplemental
The method bypasses the timing limitation by using a smaller set of I_{DDQ} measurement test vectors to augment the existing test set

(C)Selective

This methodology intelligently chooses a small set of test vectors from the existing sequence of test vectors to measure current

Furthermore, the methodologies are classified as

(a) Ideal: vectors produce a nearly zero quiescent power supply current during testing of a good device.

(b) Non-ideal: vectors produce a small, deterministic quiescent power supply current in a good circuit.

(c) Illegal: an illegal I_{DDQ} vector cannot produce an accurate current component estimate for a good device.

Our EDA tool supports the ideal I_{DDQ} test methodology in supplemental and selective ways for full static, resistive, and some dynamic CMOS circuits. The ideal requirement means we need to reduce the current in I_{DDQ} mode for fault-free device for the sake of good test sensitivity. Several things are done to meet this requirement. Firstly, during ATPG, the tool can perform I_{DDQ} check to ensure there is no internal floating node; pull up or down resistors and bus contention in the design. Secondly, we disable the write, read control line and chip select port for all memories in the chip to further reduce memory power consumption. Lastly, isolate signals from other power down domains.

In the vector generation process, there is a fine tune step to get "best I_{DDQ} pattern", from the considerations of the coverage, fault types, measurement points, as well as total test time consumed. When the chips were in initial debug stage, two or three vector sets would be further verified and compared to determine the final set in use in production. This process also included tens of faulty-free chips characterization for signature-based I_{DDQ} testing to determine min current vector, etc. (9). Generally our I_{DDQ} fault coverage can reach 90% level or higher.

4. IDDQ Measurements, Statistical Analysis and Manufacturing Implementation

Normal leakage, whether high or low, does not impact I_{DDQ} application. The only requirements for proper I_{DDQ} application are current stability at the selected measurement points and repeatable test condition. Maintaining measurement accuracy is important in I_{DDQ} testing and fundamental for repeatability of the measurement. Nowadays, ATE's current meters possess enough resolution to handle advanced I_{DDQ} testing for small current variance among different test vectors. For our test platform, the current meter related to I_{DDQ} measurement supply pins has higher than 19bit resolution. With milliamp range, the meter is able to detect $1 \sim 2$ microampere difference on the basis of higher background leakage activated or deactivated by vectors. On the other hand, repeatability of measurement also plays very important role during I_{DDQ} measurement. At initial test debug stage, proper settling time, noise filtering technique, as well as best sampling numbers should be determined so that the stable and repeatable readings for every test points could be achieved.

All test points data would be collected per device under test (DUT) during the DUT testing and then calculated and analyzed on-site based on linear regression model(9) for final judgment, rejected or passed on ATE. The upper limits were defined based on equation of

$$Tupper = Slope*Min_{meas} + Intercept + Outlier \qquad [1]$$

The lower limits were defined by

$$Tlower = Min_{meas, \, assume} - Outlier \qquad [2]$$

The outliers were confirmed based on post statistical analysis done at preproduction stage which was 130nm products scenario. For the products fabricated in 65nm, the combination of current ratios and delta I_{DDQ} techniques are adopted. Delta I_{DDQ} is defined as the difference between two consecutive I_{DDQ} readings. The reason to have delta I_{DDQ} testing in 65nm products is based on the theoretical assumption: when process parameters increase the background current, all I_{DDQ} tests will observe a similar increase. The delta I_{DDQ}, however, will remain approximately the same for reliable chips and will increase proportionately to the I_{DDQ} current readings. Having a tight delta I_{DDQ} threshold combined with a looser maximum I_{DDQ} threshold (current ratio defined) permits the defects to be detected without discarding reliable dice (12). Physically the determination of the maximum I_{DDQ} threshold is 100% trade-off among the yield loss, outlier's definition, initial rejects lab test confirmation, test escape evaluation, etc... To define delta I_{DDQ} thresholds it is necessary to run post statistical plot for delta distribution in histogram (11). In our cases, the algorithm is built-in into test program to do final judgment as a post process decision.

In 65nm product series, we also added one second voltage stress test, besides normal I_{DDQ} testing, in production flow. It is aimed at further screening out potential early failure parts fabricated in 65nano-tech. The stress condition was recommended by foundry that collaborated with us in migrating products into 65nm technology. By comparing pre-and post- I_{DDQ} readings for same point measurement on same die/chip, we can eliminate suspected weak parts (13), (12). The pass/fail threshold for stress test is purely empirical data dependant. With one fourth to one third amount of test time "increased", one can get back encouraging lower dpm rate in production. It is still a trade-off between quality and overall cost. For oversea customers, such scheme is worthy in mass production mode.

5. Manufacturing Results and Manufacturing Test Strategy

I_{DDQ} testing are essential tests in Availink two generation products in 130nm and 65nm respectively. We found that by comparing with original production lot's data, I_{DDQ} testing could capture failures almost for every field return unit. And I_{DDQ} testing results for these parts could even be helpful for failure analysis in isolating failure cell and capturing either gate-oxide damage or bulk tunneling, due to structure damage, etc. as the evidence to show the chips end-users.

Advanced I_{DDQ} screening methods must be employed to improve current sensing resolution in nanotech products. To cater for leakage current variance from lot to lot, we defined that three different lots' data must be collected by going through post statistical analysis for thresholds characterization and trial run. It is trade-off process for thresholds confirmation. On the other hand, both current ratios or delta I_{DDQ} are all current signature related methodologies for defect detection in better resolution and better immunity on wafer fabrication process drifting. Thus, these methods are good for state-dependent defects while ineffective for passive defects (state-independent) in circuits. The passive defects (e.g. a VDD to ground resistive short) behave the absence of steps in the signature over all vector set. Hence, the signature based methods are not suitable for such defects detecting. In order to detect passive defects, we utilize already stored all measurements in absolute values during I_{DDQ} testing and extract the mean of absolute readings and standard deviations. Based on post statistical calculation built in the tester on-site algorithm we then justify if the mean and standard deviations of absolute values for this chip located at the region of outliers or the area of majority fallen and therefore identify the existence of passive defects. The post statistic model used for analysis is based on lognormal distribution. It is derived from the theoretical assumption that the effective channel length of MOS transistors and threshold voltage follow a normal distribution while the relationship between I_{DDQ} and Leff is exponential (14). This procedure generally would increase the yield loss in production. However, it is still trade-off between quality and cost. If there are sufficient data collected from lots, then the negative impact could be controlled to minimum.

Some simple experimental studies for the effectiveness of our I_{DDQ} testing were in-house done in the past. By simply exchanging I_{DDQ} testing position in the overall test flow, we could verify if the fallouts from I_{DDQ} testing were able to be captured by scan based stuck-at, scan based delay, and functional tests, etc.(6). For example, one of our 65nm design for scan based stuck-at set with 99.5% stuck-at fault coverage; scan based delay testing with 83.5% transition fault coverage, 79.9% SDQM fault coverage, >20% path delay fault coverage, etc.; for functional test set with design verification vectors of >40% coverage; while I_{DDQ} vector set with 91.1% I_{DDQ} related fault coverage; was gone through such experimental tests. Most of fallouts from I_{DDQ} testing could be caught by other tests for the product. Among total fallouts, there is still a small portion which was unable to be captured by all other tests. So they are I_{DDQ}-only-failures. For these I_{DDQ}-only-failures, due to various constraints at start-up fabless design house, we were not able to do further analysis to identify the transistor level defects for further confirmation. However, in 130nm products, one of field return units showed the behavior of I_{DDQ}-only-failures with functional intermittent issue. By means of foundry's physical failure analysis, the gate-oxide damage was finally detected on related power cell as an indirect evidence of I_{DDQ} testing effectiveness in our 130nm's product.

According to our 130nm and 65nm products manufacturing test, I_{DDQ} testing is still usable and value-added in production with refined test methodologies and improved post statistical analysis.

6. Conclusion

Nanotechnologies impose more challenges on I_{DDQ} testing. It causes the total normal leakage of nanotech devices often one or more orders of magnitude higher than the defect current. However, by refined advanced-I_{DDQ}-testing methodologies and combined proper post data statistical analysis, with I_{DDQ} testing vectors precisely selected, we can still extend the usability of I_{DDQ} testing in manufacturing test. As the cases study of Availink 130nm and 65nm products with I_{DDQ} testing implementation in production, this has been practiced and the conventional benefits of I_{DDQ} testing are also visible. As a result, I_{DDQ} testing in nanotech era is still valuable.

In our future 65nm new products, we consider I_{DDT} etc. approaches. I_{DDT} is defined as supply transient current testing, as the counterpart of I_{DDQ}. It is reported that immune to high background leakage, higher coverage and much faster measurement time is with I_{DDT} (15). It might significantly improve defect detections in nanotech era.

7. References

1. Hans Manhaeve, "Current testing for Nanotechnologies: A Demystifying Application Perspective," ats, pp. 456, 14th Asian Test Symposium, 2005
2. Tsuyoshi Shinogi et al., "An Iterative Improvement Method for Generating Compact Tests for Iddq Testing of Bridging Faults," Institute of Electronics, Information and Communication Engineers (IEICE) Trans. On Information and System, Vol. E81-D, No 7, July 1998
3. Steven McEuen, "Iddq Benefits," Proc. Of IEEE VLSI Test Symposium, 1991.
4. Steven McEuen, "Reliability Benefits of Iddq," Journal of Electronic Testing: Theory and Applications, 1992
5. K. Wallquist, "On the Effect of I_{SSQ} Testing in Reducing Early Failure Rate," Intl. Test Conf., 1995
6. Rochit Rajsuman, "Iddq Testing for CMOS VLSI,"proceeding of the IEEE, April 2000
7. Zhanping Chen et al., "Iddq testing for Deep-Sub-Micron ICs: Challenges and Solutions," IEEE Design & Test of Computers, March-April 2002, pp. 24-33.
8. B. Iniguez et al., "Analysis and Future Trends of Iddq Testing for Silicon On Insulator CMOS ICs, "IEEE Intl. Workshop on Defect Based Testing, Los Angeles, CA, 2001, pp. 40-44.
9. Peter Maxwell et al., "Current Ration: A Self-scaling Technique for Production I_{DDQ} Testing" IEEE ITC, Oct. 1999, pp. 738-746
10. Yu Wei P'ng et al., "I_{DDQ} Test Challenges in Nanotechnologies: A Manufacturing Test Strategy", 16th IEEE Asian Test Symposium, 2007, pp. 211
11. C. Thibeault, "Improving Delta-Iddq-based test methods", ITC International Test Conference, 2000, pp. 207-216
12. Theo J. Powell et al., "Delta Iddq for Testing Reliability", VLSI Test Symposium, proceedings, 2000, pp 439-443.
13. Markus Schmid and Hans Manhaeve, "Improving Automotive IC Quality. A Case Study on the Implementation of Advanced I_{DDQ} Strategies Targeting Product Quality Improvement, Burn-in Elimination and Test Cost Reduction", IEEE European Test Symposium, 2006
14. Manoj Sachdev, "Deep Sub-micron Iddq Testing: Issues and Solutions", Proc. Of European Design & Test Conf., 1997
15. Manoj Sachdev, "Current-Based Testing for Deep-Submicron VLSI", IEEE Design & Test of Computers, 2001, pp 76-84

Cost-effective and Accurate Solution for Jitter Performance Test in High-speed Serial Links

Ming Lu

Application Development Center of Verigy, Shanghai, China

Abstract – With the increasing demand for faster data communication, the data rate of communication systems has reached in the gigahertz range and higher. To accurately quantifying the transmission quality of high-speed I/Os with acceptable test time in production becomes one big challenge to ATE test engineers. The major hindrance is to fully measure jitter related specs accurately will lead to excessive cost of test (e.g., capital investment, time-to-market, test time), while BIST loopback testing can not promise a result with a high confidence level. This paper proposes one comprehensive solution for at-speed jitter performance testing on ATE, which includes jitter histogram testing, random jitter / deterministic jitter separation and jitter stardust testing. This solution has been implemented in volume production of a real device and the result shows its high accuracy and cost effectiveness.

I. Introduction

The increasing need for higher bandwidth and the latest achievement in semiconductor technology are continuously driving the integration level of former stand alone high-speed I/Os cells into System-On-a-Chip (SOC). The data rate of embedded I/O cells has already exceeded 5Gbps while the number of parallel embedded I/O cells is constantly increasing. [1]

This is the challenge for manufacturers of leading edge communication devices to apply automated test equipment (ATE) for device characterization and volume production instead of slow and expensive lab equipment. Communication devices demand a high quality level. In the mean time, the competing market is impacting the design cycle time and the time for ramping up the production.

The jitter specifications were thoroughly tested on a bench system during design validation and characterization phases. In a production environment, traditionally, there are two approaches for at-speed testing on ATE. One approach is to perform tests between the transmitter port (TX) and the receiver port (RX) of a device under test (DUT) with a loopback trace on the loadboard. The disadvantage of this method is that the test result is reduced from an originally parametrical characteristic measurement to a simple Pass/Fail decision without any knowledge of jitter parameters. This is definitely not sufficient to guarantee the proper function of high-speed devices. The other approach is to perform tests between the DUT TX/RX ports and tester channels with a simple specification searching to get the peak-to-peak jitter value, or to execute a 6-point functional tests to judge whether it meets the minimum spec of the eye mask. The drawback of this method is that it still can not comply with various specifications and standards as the total jitter budget has to be separated into different components in today's applications. A fast peak-to-peak jitter measurement is ambiguous unless some boundary condition is established. Effects like rising/falling time, random/deterministic jitter are characteristic for each DUT and the customers require all relevant parameters to be measured and documented for every single chip.

This paper focuses on the jitter compliance test of a DUT TX port which includes jitter histogram, RJ/DJ separation and jitter stardust measurements. Jitter tolerance test is not considered in this paper. Section II provides strict definitions of jitter and jitter sub-components. Section III presents a detailed solution for accurate jitter performance test on ATE. In Section IV the measurement results are presented and finally conclusions are drawn in Section V.

II. Jitter Definition and Categories

2.1 Jitter Definition

Jitter is defined as – "Short-term, non-cumulative variations of the significant instants of a digital signal from their ideal positions in time" (International Telegraph and Telephone Consultative Committee). A significant instant for a digital can be defined as the rising or falling edge from a bit transition crossing a voltage threshold level [2].

2.2 Jitter Categorization

Jitter is typically divided into several categories depending on their properties like random jitter and deterministic jitter. These properties sometimes provided clues to the origin of the jitter allowing the designer to more easily find the jitter source in the design. Figure 2.1 shows a typical categorization of jitter.

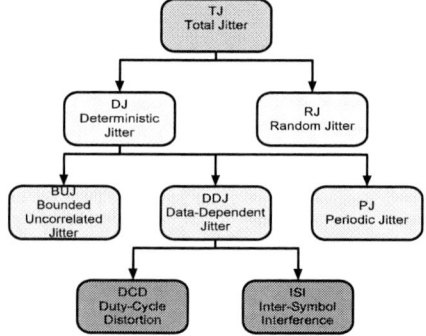

Figure 2.1 Jitter subcomponents

Random jitter (RJ) is defined as being unbounded in the sense that there is always a probability of the jitter value reaching any value. RJ is usually modeled using a Gaussian probability distribution because of the central limit theorem.

Deterministic jitter (DJ) is mainly characterized by being bounded. DJ can be further divided into several subcategories depending on the underlying physical causes: periodic jitter (PJ), bounded uncorrelated jitter (BUJ), data dependent jitter (DDJ) in the form of duty-cycle distortion (DCD) and intersymbol interference (ISI).

Total jitter (TJ) normally is expressed by following equation.

$$TJ = DJ + \alpha(BER)RJ \qquad (E2.1)$$

Where DJ is peak-to-peak value of deterministic jitter and RJ is the standard deviation value of the random jitter. The α (BER) factor is a vlaue that is uesed to compute a "peak-to-peak" value of the random jitter for a given BER. Table 2.1 shows some typical used values (transition densiton = 0.5)

Table 2.1 Values for $\alpha(BER)$	
BER	$\alpha(BER)$
10^{-9}	11.996
10^{-12}	14.069
10^{-16}	16.444

III. Implementation on ATE

3.1 Jitter Histogram Approach on ATE

The jitter histogram approach is for peak-to-peak jitter and rms jitter test which is based on edge transition histogram. Edge transition histogram (number of occurrences vs. jitter range/bins) is calculated by the accumulated edge jitter distribution. The approach of an at-speed test with digital pins of ATE is very similar to the functionality of a BERT. Every bit of the data stream coming from a device transmitter is compared to an expected value (level threshold).

The compare strobes can be shifted in the timing and level threshold axis. In order to acquire an error count curve, the compare strobes have to be moved through the left or right pass/fail transition on the timing axis of the data bits monitoring the number of failing bits with a per pin error counter. And then calculate the edge transition histogram base on the error count curve. There are two points need to be highlighted.

- The total pattern length determines the BER depth of the error count curve. In order to archive stable result, the suitable pattern length should be taken on consideration. (e.g., PRBS23)

- The step width determines the resolution of the error count curve. It is always an tradeoff between test time and test accuracy. The optimal solution is to use two kinds of step width during testing. The bigger one is used for coarse alignment which locates the rising/falling edge position. The smaller one is for accurately getting error count within small timing range around rise/falling edge. This solution will meet both test time and test accuracy requirements.

3.2 Jitter Separation Approach on ATE

For RJ/DJ separation, it is a little complicated on ATE. The following describes the detail procedure.

(1) Coarse Synchronization

Coarse synchronization is to search the position of rising/falling edge quickly with big step width. There are two timing ranges need to be measured, left transition edge and right transition edge of one UI. This synchronization is a Go/No-go approach which will reduce test time to minimum.

(2) Error Count acquirement

After getting rising/falling edge position of one UI (Unit Interval), pin electronics (PE) of ATE will get error count within time range around transition edges. In this phase the step width should be set to very small to meet the high accuracy criteria. The results will be two data arrays which contain right/left error count curves.

(3) BER calculation

From the error count curves which were gotten from step 2, we can calculate the BER curves.

$$BER_{right/left}[i] = \frac{ErrorCount_{right/left}[i]}{TotalCount}$$ (E3.1)

Where TotalCount is the total number of bits in vector stream.

(4) RJ/DJ calculation

After getting BER curves (also called bathtub curves), we transfer them to Q-space for the selected BER range. The transfer equation is shown as below.

$$Q = \sqrt{2}erf^{-1}(1 - \frac{4}{P_B} \cdot BER)$$ (E3.2)

The fitting on the selected range of the Q-space curve will give two straight lines with a certain slope and a certain distance point. The respective slope is determined by the Random Jitter, while the distance of the two slopes is determined by the Deterministic Jitter.

3.3 Jitter Stardust Approach on ATE

Although most standards define the jitter requirements for DUT in terms of RJ and DJ components, for design verification it is important to know the spectral distribution of the jitter generated by the DUT. This information could provide an indication of the root-cause of measured jitter (e.g., crosstalk from a system clock) [2]. We will use an efficient approach on ATE to do this jitter spectrum decomposition (also called Stardust test) which is described in [4]. The main idea is to set the compare strobe timing not at the middle of the day eye as usual for a functional test but at the edge of the data eye. The jitter on the data signal will move the data eye left and right of the strobe point just like modulating. The compare error signal can then be post-processed to obtain the jitter spectrum.

The procedure of jitter stardust test has following steps.

(1) Coarse synchronization and error count acquisition

Same as step (1) and (2) in RJ/DJ separation.

(2) Strobe Edge Relocation

From the error count we can get the mean value of jitter histogram. Then we set the strobe edge position to this mean value by changing spec.

(3) Errormap Acquisition

Execute test vectors the record the errormap sequence. 1 stands for pass and 0 stands for fail.

(4) Stardust Calculation

Then the power density spectrum can be calculated from errormap sequence through FFT arithmetic.

IV. Result Analysis

4.1 Measurement Results

In this section, the measurement results of an experiment DUT board with lossy trace are presented. ATE system is Verigy 93000 with High Speed Extension Card which data rate can reach 12.8Gbps. The experiment data rate is 5Gbps. Pattern is in clock mode, 1M bits.

(1) Jitter Histogram

By using introduced algorithm in section 3.1, loop 20 times. We can see that the measured value of RMS jitter and Peak-to-Peak is very stable. The variation of RMS jitter is less than 0.2 ps, the variation of Peak-to-Peak jitter is only 2 ps.

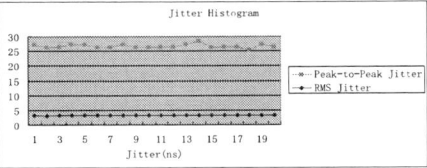

Figure 4.1 Test result of Jitter Histogram

(2) RJ/DJ

By using introduced algorithm in section 3.2, loop 20 times. We can see that the measured value of RJ, DJ is very stable. The variation of RMS jitter is only 0.23 ps, the min-max difference of DJ jitter is about 3 ps.

Table 4.1 Test result of Jitter Separation

Jitter Type	STD (ns)	MEAN (ns)	MAX (ns)	MIN (ns)	DIFF (ns)
RJ	0.236308	4.3706	4.758	3.844	0.914
DJ	0.792116	3.9305	5.624	2.713	2.911
TJ	2.641973	66.25526	70.56208	60.08944	10.47264

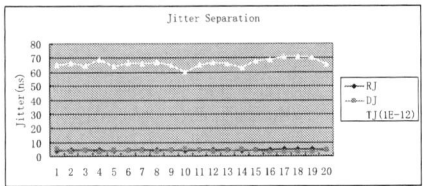

Figure 4.2 Test result of Jitter Separation

(3) Stardust

By using introduced algorithm in section 3.3, we can see that the measured value of jitter frequency. As we added 50MHz, 20ps peak-to-peak sinusoidal jitter on the data, the result shows clearly that 50MHz is the largest bin except DC subcomponent.

Top 10 amplitude elements:

HR1+ frequency[0] = 0.000 Hz amplitude[0] = 0.304

HR1+ frequency[1] = 50001144.409 Hz amplitude[1] = 0.191

HR1+ frequency[2] = 99997520.447 Hz amplitude[2] = 0.000

...

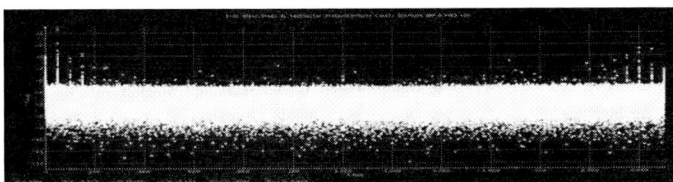

Figure4.3 Jitter Spectrum (5Gbps, 50MHz Sinusoidal Jitter)

4.2 Correlation Discussion

We did an independent correlation with Agilent 86100C Digital Communication Analyzer (DCA). The correlation results show that stardust approach we introduced works well as DCA. For Jitter histogram testing, DCA result is 26.4ps which is within $\pm$ 5ps of measured peak-to-peak value by ATE (23.5ps). For RJ/DJ testing, the value measured by DCA is RJ=1.42ps, DJ=6.7ps. The result from ATE is RJ = 4.37ps, DJ=3.93ps. The result is in reasonable range and causes of value difference are explained in [1]. The main reason depends on jitter separation algorithm difference and intrinsic jitter which tester contains.

4.3 Test Time Results

The cost-effectiveness is one key factor of the suggested jitter solution. Comparing to bench instruments, ATE has an exceptional advantage in parallel testing for multi bins of device or multi devices. In Table 4.2, test time examples are presented. The used test pattern is a clock pattern and there are 2 lanes in parallel testing. If we test more lanes simultaneously, the cost of test will be reduced more.

Table 4.2 Test time example

Test Item	1Msamples	8Msamples
Jitter Histogram	254.519ms	492.043ms
RJ/DJ Separation	402.153ms	785.952ms
Jitter StarDust	3.125sec	3.342sec

V. Conclusion

This paper focuses on how to test jitter on ATE cost-effectively and accurately. Firstly we analyzed the jitter categories and get proper equation for computing. Then based on these requirements, we presented one test solution to cover main jitter features: jitter histogram, RJ/DJ and jitter stardust. Detail methodology and mathematic models were also highlighted. Finally, we showed the actual test result and correlation discussion. The results are promising but we still have lots of work to do in the future, such as more accurate jitter separation algorithm, smarter calculation, hidden data uploading on ATE with shorter test time.

Acknowledgements

I would like to thank Jose Moreira, Hubert Werkmann, Wei-Min Zhang and Callum McCowan for their great support.

Reference

1. G. Hansel, K. Stieglbauer, G. Schulze, and J. Moreira, "Implementation of an Economic Jitter Compliance Test for a Multi-Gigabit Device on ATE", IEEE International Test Conference, 2004

2. Jose Moreira and Hubert Werkmann, "An Engineer's Guide to Automated Testing Of High-Speed Interfaces", page 24, 2010 Artech House.

3. "Understanding Jitter and Wander Measurements and Standards", Agilent Technologies, Inc., Application Note.

4. B.Laquai and R. Plitschka, "Testing High Speed Serial IO Interfaces Based on Spectral Jitter Decomposition", IEC DesignCon 2004, 2004.

Plasma Etching for Failure Analysis of Integrated Circuit Packages

J. Tang[a,b], J. B. J. Schelen[c], and C. I. M. Beenakker[b]

[a] Materials Innovation Institute
[b] Delft Institute of Microsystems and Nanoelectronics (DIMES),
Laboratory of Electronic Components, Technology and Materials (ECTM)
[c] Electronic and Mechanical Support Division (DEMO)
Delft University of Technology, Delft, the Netherlands

Plastic integrated circuit packages with copper wire bonds are decapsulated by a Microwave Induced Plasma system. Improvements on microwave coupling of the system are achieved by frequency tuning and antenna modification. Plasmas with a mixture of O_2 and CF_4 showed a high etching rate around 2 mm^3/min. The role of O_2 and CF_4 in etching molding compound is described. Plastic package with 38 um Cu bond wires and a 2 mm * 3.5 mm die inside is fully decapsulated in 20 minutes. Cu wires remain undamaged after decapsulation proving the efficiency of this method.

I. Introduction

Failure analysis on integrated circuit (IC) packages often requires decapsulation of the package. In plastic IC packages, molding compound is used for encapsulation. Epoxy (20 wt%) and silica filler (60-80 wt%) are the major components. Decapsulation is the step to open an IC package to facilitate the inspection and electrical examination of the die and wire bonds inside. The trend in IC packaging is to use copper as wire bond material due to its economical and performace advantages (1). However, in that case conventional hot nitric acid decapsulation cannot be used because nitric acid quickly reacts with copper and damages the copper wire bond.

Plasma etching is a potential technique for IC package decapsulation with inherent advantages of high selectivity and process control (2). However, with conventional plasma etching systems silica filler in the molding compound appears to be difficult to remove. Additional steps like blowing with high gas flow or chemical etching in HF are required to remove the silica residue. This leads to a painstakingly long time to fully decapsulate an IC package.

Microwave Induced Plasma (MIP) has shown its potential in decapsulation of plastic IC packages (3), (4). By adjusting the composition of the plasma gas the MIP system can easily remove the silica fillers in the molding compound thus the time needed to fully decapsulate a plastic IC package is about ten times shorter than conventional plasma etchers.

In previous experiments we improved the microwave coupling of the system by antenna modification. Power reflection is decreased from 50 % to 15 %. In this paper, microwave coupling is further improved by using a frequency tuning method. Using these improved conditions the influence of O_2 and CF_4 on molding compound removal rate is investigated under less than 5 % reflected power.

II. Experiment Setup

The MIP system consists of a Sairem GMP-180W solid-state microwave generator (f=2450+-20 MHz, P<180 W), a lab made Beenakker type microwave resonant cavity (5), (6) and gas connections. Fig.1 is a schematic of the MIP system. The cavity is designed to resonate in the TM_{010} mode. The electric field strength inside the cavity follows a zero order Bessel function and is at its maximum value in the center and zero near the periphery. The resonant frequency of the cavity is determined by the following equation: $D = 2.405c/\pi f$, where D is the inside diameter of the cavity, f is the resonant frequency and c is the speed of light. For the cavity used in the experiments, D is 93 mm and this results in f to be 2470 MHz. The cavity is intentionally designed to have a larger resonant frequency than 2450 MHz in order to compensate for its decrease when a dielectric material is inserted.

Alumina discharge tubes with 6 mm outside diameter and 1.2 mm inside diameter are used. A K-type thermocouple is used to measure the plasma effluent temperature. Etching depth and diameter are measured with a Veeco Dektak 150 surface profiler.

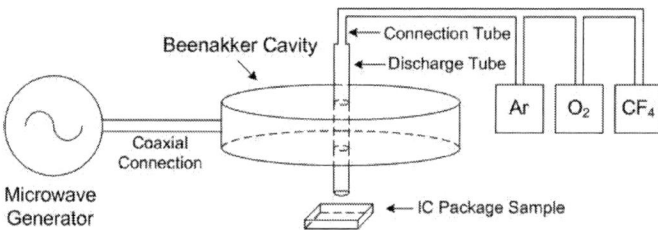

Figure 1. Schematic of the MIP system

III. Improvements on the Microwave System

Resonant cavities with a plasma inside behave differently from empty ones. The plasma inside can be regarded as a dielectric material with losses (7), (8). The equivalent permittivity of the plasma can be written as:

$$\varepsilon_p = \varepsilon_0[1 - \frac{ne^2}{m\varepsilon_0\omega(\omega + j\upsilon)}]$$

where n is the electron density in plasma, ω is the generator frequency, υ is the electron-neutral collision frequency, ε_0 is the permittivity of free space, m is the effective mass of the electron, e is the electron electric charge, j is the imaginary operator. The real part of the permittivity explains the resonant frequency of the cavity to change. The imaginary part of the permittivity is related to the power absorption into the plasma. The plasma characteristic values n and υ determine how the cavity performance changes once a plasma is ignited. In order for the MIP system to have a lowest reflected power, the system should be critically coupled when a plasma is present inside the cavity. This means the system should be designed to be decoupled when there is no plasma present.

In previous experiments we have shown that antenna modification inside the cavity greatly decreases power reflection from 50 % to 15 %. To further decrease the reflected power, frequency tuning by using a variable frequency generator with a 40 MHz tuning range is used. Changes in gas composition cause plasma characteristic to change and in turn influence the coupling. Fig.2 depicts the reflected power versus frequency of the generator and CF_4 flow rate respectively. Input power is kept at 80 W, Ar flow rate is 60 L/h and O_2 flow rate is 4 L/h. The reflected power is given as a percentage of the generator output power. The 3D surface shows the reflected power under different operating conditions. The CF_4 flow is varied form 0 to 1.25 L/h and the frequency from 2430 MHz to 2470 MHz. As we can see from Fig.2 the coupling of the MIP system is sensitive to the frequency of the input microwave power. Frequency tuning with a 40 MHz range proved to be efficient and reduces reflected power from 15 % to less than 5 % in most cases.

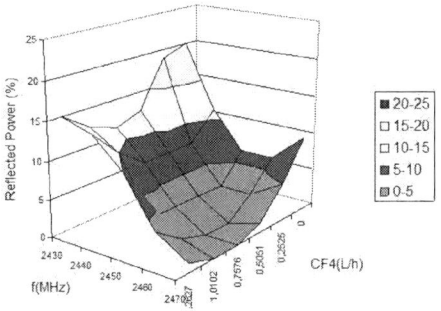

Figure 2. Reflected power versus frequency and CF_4 flow

IV. Etching and Decapsulation Results

Etching of molding compound with the improved MIP system is conducted with $Ar+O_2$, $Ar+CF_4$, and $Ar+O_2+CF_4$ plasmas. In addition we performed decapsulation experiments with plastic packages containing 38 um Cu bonding wires. The reflected power during the experiments is kept below 5 %.

Etching of molding compound using $Ar+O_2$ plasma

Oxygen gas added into the Ar plasma generates atomic oxygen that efficiently reacts with organic materials like photoresist and epoxy. As shown in Fig.3, molding compound exposed to $Ar+O_2$ plasma appears white after a few minutes. This is due to the fact that silica filler is left on the surface after epoxy is etched away. The filler residue does not appear as powder, it is agglomerated instead. Ultrasonic cleaning in water for 5 seconds proved to be efficient in removing the residue. However, the disadvantage of this procedure is that the wires under the residue are often damaged mechanically.

Etching of molding compound using $Ar+CF_4$ plasma

CF_4 gas added into the Ar plasma generates atomic fluorine that reacts with silicon containing materials forming volatile SiF_4. Thus silica filler can be removed by $Ar+CF_4$ plasma. Fig.4 shows the top view on a partially decapsulated package. The etch duration at each position is 4 minutes. Assuming the etching profile to be a spherical cap, the etching rate is calculated to be around 0.15 mm^3/min and is too low for practical applications.

Figure 3. Profile after
Ar+O$_2$ plasma etching

Figure 4. Profile after Ar+CF$_4$ plasma
etching with different CF$_4$ flow

Etching of molding compound using Ar+O$_2$+CF$_4$ plasma

When both O$_2$ and CF$_4$ are added into the Ar plasma, the etching rate dramatically increases. With 80 W input power, 80 L/h Ar, 4 L/h O$_2$, IC package samples are etched for 2 minutes under different CF$_4$ additions. Fig.5 and Fig.6 show the etching profiles. As is shown in Fig.7, the increase of CF$_4$ flow from 0 L/h to 0.25 L/h causes the etching rate to increase sharply from almost 0 mm^3/min to 0.85 mm^3/min. In Fig.8, the addition of CF$_4$ flow from 0.25 L/h to 1.25 L/h causes the etching rate to increase from 0.85 mm^3/min to 2.45 mm^3/min.

The change of curve pattern between low and high CF$_4$ addition can be explained in two ways. The first is when more CF$_4$ is added, the dissociation of CF$_4$ into atomic fluorine comes to a saturation thus the etching rate tends to saturate. The second is the atomic fluorine for the etching reaction with silica filler tends to saturate and causes the tangent of etching rate curve to decrease. The role of CF$_4$ addition is to remove the silica agglomerate during etching. It is only needed that the surface of silica filler is etched by atomic fluorine so that the silica agglomerate structure becomes loose. Then it can be removed by the gas flow of the plasma effluent. White silica powder residue around the sample holder is observed during Ar+O$_2$+CF$_4$ etching, this verifies the second explanation.

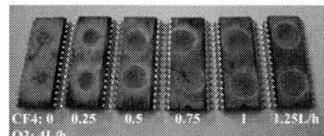

Figure 5. Profile after 2 minutes
plasma etching with CF$_4$ 0-0.25L/h

Figure 6. Profile after 2 minutes
plasma etching with CF$_4$ 0-1.25L/h

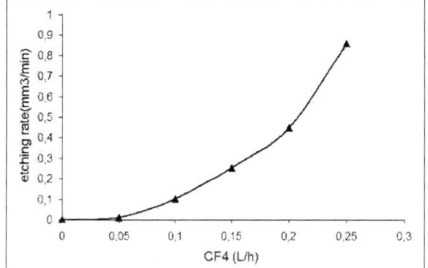

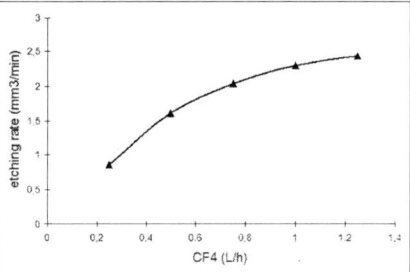

Figure 7. Etching rate versus
CF$_4$ flow rate (0-0.25 L/h)

Figure 8. Etching rate versus
CF$_4$ flow rate (0.25-1.25 L/h)

Decapsulation of plastic IC packages with Ar+O_2+CF_4 plasma

Because of the different structures and materials used in a plastic IC package, it is advisable to perform decapsulation with the MIP system in steps. During the beginning of decapsulation, the top molding compound should be removed as fast as possible. The ratio of O_2 and CF_4 has to be optimized such that the rate of epoxy removal is the same with the rate of silica filler removal. When Cu bond wires and die are reached, recipes should be chosen that give a moderate etching rate, lower effluent temperature, and a higher etching selectivity to reduce possible damage.

In the next experiment a plastic IC package with 38 um Cu bond wires and a 2 mm * 3.5 mm die inside is decapsulated using Ar+O_2+CF_4 plasma under an effluent temperature of around 400 °C. The thickness of molding compound on top of the die is 1 mm. Fig.9 depicts the package fully decapsulated using this method. The volume of molding compound being removed is 25 mm^3 and the etching duration is 20 minutes. All the Cu bond wires and most part of the die appeared to be undamaged as inspected through an optical microscope. Fig.10 shows that the 38 um Cu bond wires are undamaged after MIP decapsulation as is indicated by the smooth surface and the reddish color of the wire. Fig.11 shows damaged copper wire after that we intentionally exposed the wire to extreme conditions of 100 W input power for 15 minutes. The surface of the wire becomes black and rough. Fig.12 shows the patterns on the die after decapsulation. In most places the die appears to be undamaged. However, as Fig.13 shows some damage cannot be excluded. This indicates over etching of the Si_3N_4 passivation layer by the fluorine containing plasma. Improvements on recipe and process control are needed to reduce such kind of damage.

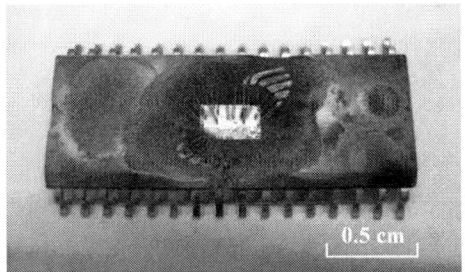

Figure 9. IC package decapsulated by MIP system

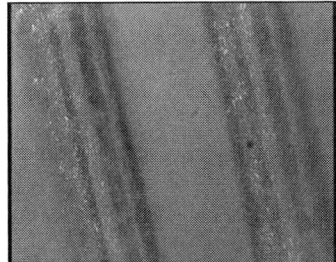

Figure 10. Undamaged Cu bond wires after decapsulation

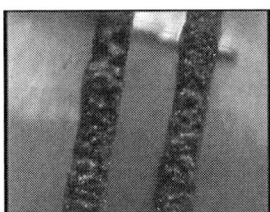

Figure 11. Intentionally damaged Cu wires

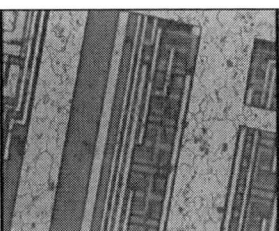

Figure 12. Undamaged patterns on the die

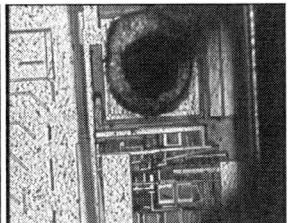

Figure 13. Damaged patterns on the die

V. Conclusions

Microwave coupling of the MIP system is improved by a combination of antenna modification and frequency tuning. Reflected power is decreased from 50 % to 5 %. With only O_2 or CF_4 plasma the etching rate of the molding compound is too slow for practical applications. The possible role of O_2 and CF_4 during etching is explained. With a proper mixture of O_2 and CF_4, the plasma showed sufficiently high etching rate and selectivity. Guidelines for using this MIP system for decapsulation of plastic IC packages are proposed. The high etching selectivity of the plasma ensures all the copper bond wires and most parts on the die to be undamaged after decapsulation.

Acknowledgments

This work is funded by the European Nanoelectronics Initiative Advisory Council (ENIAC) SE2A project. The authors wish to thank Prof. D. C. Schram, H. B. Profijt, Eindhoven University of Technology, for their helpful discussions in the areas of plasma physics. The authors would also like to thank H. Pomp at NXP Semiconductors and DIMES colleagues A. Akhnoukh, A. van den Bogaard, C. C. G. Visser, and R. P. van Viersen for their help on experiments.

References

1. N. Srikanth, S. Murali, Y.M. Wong, Charles J. Vath III, *Thin Solid Films*, Vol 462-463, pp. 339-345, (2004).
2. J. Thomas, J. Baer, P. Westby, K. Mattson, F. Haring, G. Strommen, J. Jacobson, S. S. Ahmad, A. Reinholz, *59th Electronic Components and Technology Conference*, pp. 2011-2015, IEEE Conference Proceedings, (2009).
3. Q. Li, C. I. M. Beenakker, and J. Vath III, *7th International Conference on Electronics Packaging Technology*, IEEE Conference Proceedings, (2006).
4. J. Tang, J. B. J. Schelen, and C. I. M. Beenakker, *11th International Conference on Electronics Packaging Technology & High Density Packaging*, pp. 1034-1038, IEEE Conference Proceedings, (2010).
5. C. I. M. Beenakker, *Spectrochimica Acta*, Vol. 31B, pp. 483-486, (1976).
6. C. I. M. Beenakker, *Spectrochimica Acta*, Vol. 33B, pp. 53-54, (1978).
7. J. L. Shohet and C. Moskowitz, *J. Appl. Phys.*, Vol. 36, No. 5, pp. 1756-1759 (1965).
8. J. Hubert, M. Moisan and A. Ricard, *Spectrochimica Acta*, Vol. 34, pp. 1-10, (1979).

Process Optimization of Contact Module in NOR Flash Using High Resolution e-Beam Inspection

Hsiang-Chou Liao, Che-Lun Hung, Tuung Luoh, Ling-Wu Yang,
Tahone Yang, Kuang-Chao Chen and Chih-Yuan Lu

Technology Development Center, Macronix International Co. Ltd.
No. 16, Li-Hsin Road, Science-Based Industrial Park, Hsinchu, Taiwan, R. O. C.
Phone: +886-35786688 ext. 78173 Fax: +886-3-5789087
ChrisLuoh@mxic.com.tw

The inspection sensitivities and capture rate capabilities of high resolution e-beam inspection system with extracting and retarding mode are evaluated. E-beam with retarding mode inspection demonstrates better performance and reflects the wafer sort yield loss in contact failure items directly. After the contact module process optimization, the yield was improved almost two times above.

Introduction

As device shrink, the contact module in NOR products become more and more tough due to NOR Flash architecture requires one contact per two cells, which consumes the most area of all the flash architecture, and the process window is tight; as compared to NAND Flash sharing the source and the drain of adjacent cell eliminate the need for metal contact and greatly reduces the die size (1). Engineers always struggle with contact module tuning in the beginning of each generation development. Generally, the majority of in-line inspection tool is bright field inspection system which creates an image of microcircuit then inspecting for anomalies. However, such optical images have insufficient resolution to identify the small features, offers insufficient sensitivities to distinguish the electrically significant differences, and have insufficient depth of focus for detection of high aspect ratio defects. Any defect which is not at the surface of inspected sample will be out of focus and therefore undetectable. Electron beam inspection (EBI) system using the voltage contrast (VC) phenomenon is a key technology for root cause analysis of invisible defects. Electron beam system becomes one of critical technologies in advanced semiconductor manufacture, which include conventional scanning electron microscopes (SEMs), focused ion beam microscopes (FIBs) and electron beam (E-beam) defect inspection system. EBI system detect sub-surface defects by measuring the voltage contrast change resulting from the electrical effect of killer defects, i.e., "open" and "short" type defects. The application includes gate leakage, tiny physical defects on the wafer surface or within high aspect ratio contact, via and trench structures, etc (2-3).

Figure 1 demonstrates the relationship between landing energy and electrons yield performance. When landing energy (LE) is under $E_1 < LE < E_2$, more electrons leave surface than that reach surface, where the yield of secondary electron (SE) plus backside scattering electrons (BSE) to primary beam electrons (PE) is greater than 1, causing surface to charge positively. Inspection running conditions under this regime is called positive mode. When $LE > E_2$, less electrons leave surface than that reach surface, causing surface to charge negatively. Inspection running conditions under this regime is called negative mode. In either mode, floating electrical conductors on the wafer under inspection are raised to a potential by pre-charging the surface of the wafer with electrons.

Because they appear in different voltage contrasts, the floating and grounded connectors can be distinguished as demonstrated in Figure 2 (a)(b). In this study, we use different e-beam operation modes to get the NOR Flash contact module optimized monitoring methodologies, and use the focus ion beam (FIB) cross-section images to analyze the failure sites and find out killing defects according to the klarf file position.

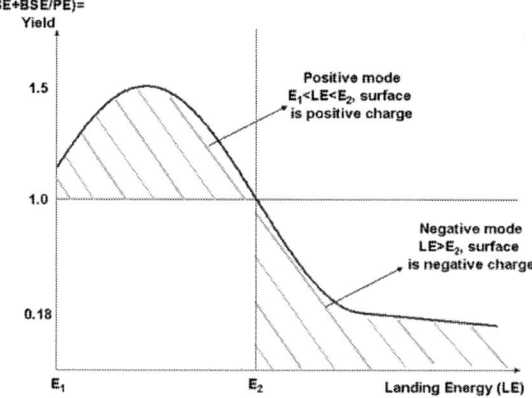

Figure 1. Relationship of LE of PE, SE and BSE yields.

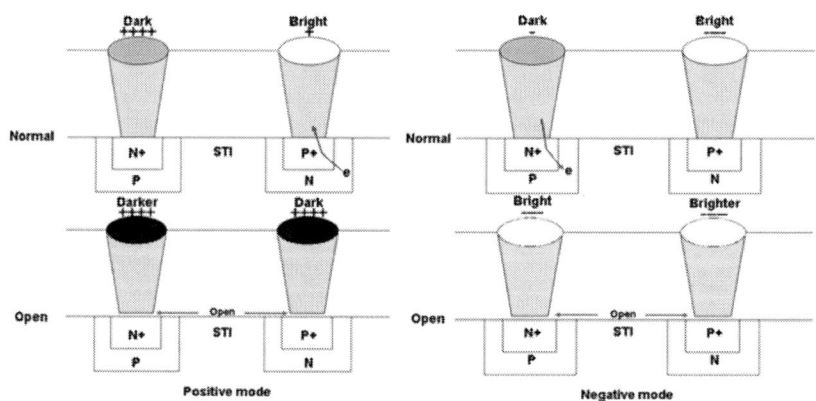

Figure 2. Contact "open" and "short" voltage contrast appears in different operation mode.

Experimental

An advanced e-Beam inspection is applied to monitor blind contact performance right after tungsten chemical mechanical polish process in NOR Flash product. The landing energy of e-beam can be controlled from 250 to 2500eV. Previously, most e-beam inspections of WCMP wafers are using positive mode that can effectively detect P+/N-well open, gate leakage, and N+/P-well leakage when the technology node is 100nm and above. For comparison, e-beam negative mode (also called retarding mode) is also adopted to optimize recipe and compare its performance to positive mode. The defects scan and review monitoring at the others layer are done by KLA2365 bright field inspection system and eDR5210S review system, respectively. Failure analysis are done by FEI DA-300 in-line focus ion beam (FIB) cross-section images and observed by cross-sectional TEM (Philips TECAA1G[2]) for more detail information. And the final yield testing is measured by KALOS tester.

Results and Discussion

e-Beam Positive mode and Negative Mode

The judgment of blind contact defects is to compare the voltage contrast image (patch image) of a die with that of a reference (a neighboring die), one can locate defects in the die. Therefore, the criteria for voltage contrast image are quite important to distinguish the issued defects, are summarized as following paragraph. Firstly, it needs a uniform voltage contrast image refer to the uniform background; secondly, a consistent contrast appearance should be the same under same environment in a circuit design; and finally, a significant contrast difference is needed between the good one and bad one.

In order to get the higher sensitivities and higher blind contact capture rate, two operation modes are optimized to make a comparison with same wafers. Condition A using extractive mode under landing energy 790 eV, the blind contact defect counts are 12485 ea, and Condition B using retarding mode under landing energy 2500 eV with same scan area and same pixel size detected out 35356 ea. The results demonstrate significant differences between two conditions. The reason why is due to condition B use retarding mode with high landing energy have more uniform voltage contrast images and have distinctive voltage contrast differences between the good and bad one, as shown in Figure 3(b). The unwanted variations in the topographic contrast or voltage contrast of an image often exist in extracting mode, especially happen at the products with smaller contact size design rule. Therefore, the retarding mode capture rate of defect of interesting (DOI) is much higher than that of extracting mode.

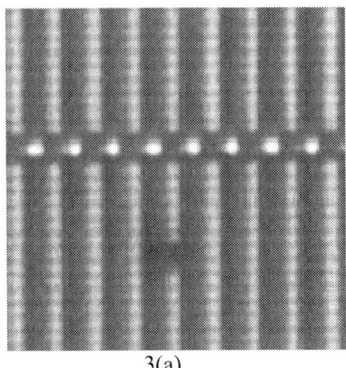
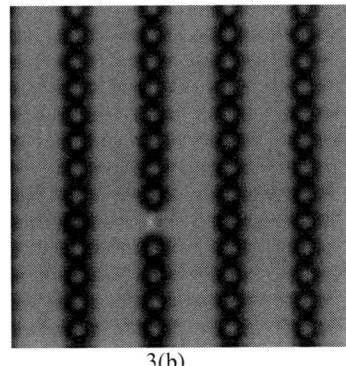

3(a)   3(b)

Figure 3. (a) Condition A: Extracting mode with landing energy 790 eV(b) Condition B: Retarding Mode with landing energy 2500 eV.

Another example demonstrates in Figure 4 that e-beam scan with extracting mode catches less killing defects as compared to retarding mode after post-WCMP in contact module. And the retarding mode gets 100% matching with the wafer sort results contact failure item in wafer #1 and #2. It is proposed that retarding mode of e-beam inspection generates with uniform and distinctive voltage contrast images, makes it locate the targeted defects easily and raises the sensitivities directly. It can distinguish the significant differences between issued wafer (#1) and the process optimized wafer (#2).

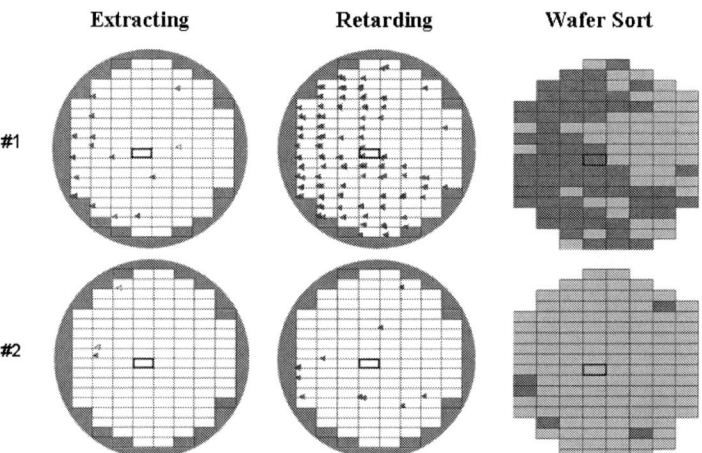

Figure 4. e-beam scan using extractive mode and retarding and match with wafer sort map.

Failure analysis Failure analysis of the killing defects is important to the yield ramp-up. In-line FIB tool is applied for analyzing the failure sits precisely according to the KLARF file position. The marked sites are also analyzed by cross-sectional TEM. Figure 5(a)(b) shows that the failure analysis of single blind voltage contrast (blind contact) and its relative TEM analysis, it indicates that contact hole etch stop at the drain top or bottom sites are the root causes of single blind voltage contrast. Figure 6 (a)(b) FIB analysis demonstrates the same results as TEM results, one failure issue is induced by contact etching stop at the drain bottom site, and the other one is owing to the pre-layer defect block the contact etching and further impeding the Barrier/W-plug fill-in process. After tracing the pre-layer defect performance in post spacer ion implantation by bright field scan/review, lots of residue defects are across the word line or stuffed in the drain bottom sites. As shown in Figure 7 (a)(b), One kind of residue defects is line residue across the word line, the other kind of residue defects is circular residue located in the drain bottom sites. These two kinds of residue defects are the killing defects for etching stop failure.

Yield Ramp-up From the failure analysis of contact module and pre-layer defects, it is verified that the blind voltage contrast defects are induced by etching stop at the residue defects during contact holes patterning. These residue defects are generated by heavy ion implantation on photo resist, and it is hard to remove by normal dry/wet strip process. Increase the CF_4 flow rate of dry strip process after spacer implant will remove the hardened surface residues (the circular and the line residue) on implanted wafers; it can eliminate the blind voltage contrast defects effectively with 86% improvement using retarding mode EBI verification. And it can get almost two times yield improvement with new dry strip recipe as shown in Figure 4 wafer #2.

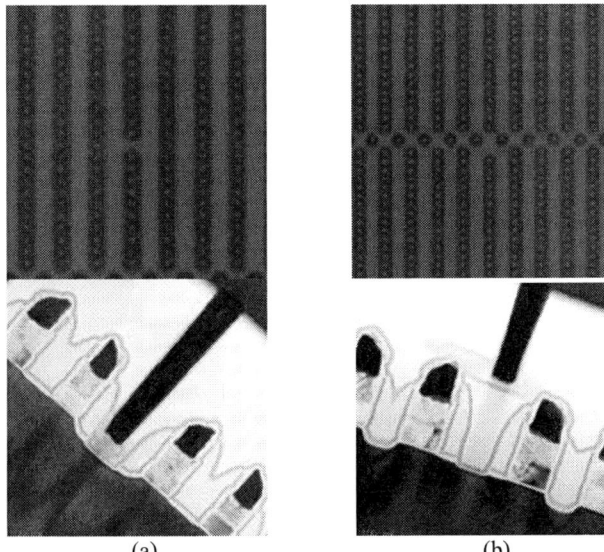

(a)   (b)

Figure 5. (a) Single blind voltage contrast image with its relative TEM failure analysis, drain site bottom defect make the etching stop. (b) Another single blind voltage contrast near source site and its relative TEM failure analysis.

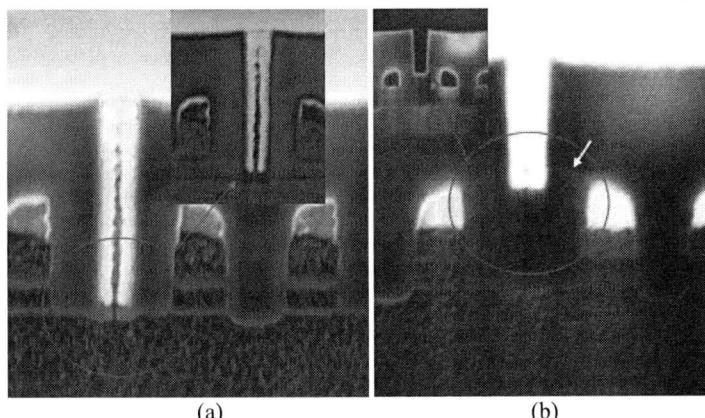

(a)   (b)

Figure 6. FIB failure analysis results: (a) defect located at drain sites bottom and lead to etch stop (b) defect across the word line and lead to etch stop.

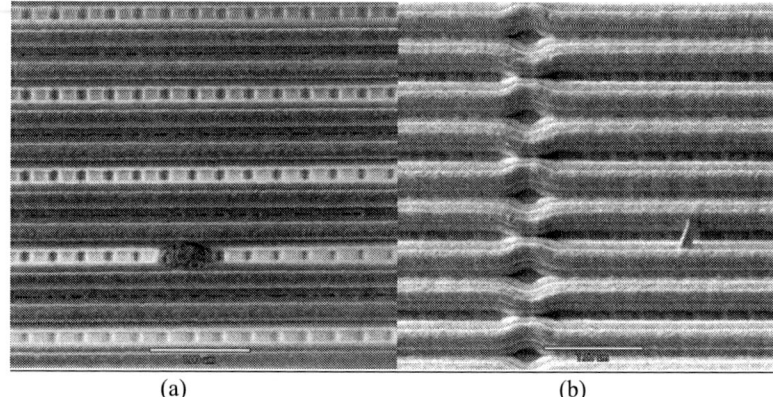

(a)   (b)

Figure 7. Review images at post spacer ion implantation (a) a circular residue located at the drain bottom site. (b) a line residue across the word line.

Summary

In this study, e-beam inspection by retarding mode provide better uniform voltage contact and significant contrast difference to background and references as compared to extracting mode one. Therefore, e-beam retarding mode inspection locates the targeted defects easily and raises the sensitivities directly. Implementing this technique to monitor the sub-surface invisible defects with different voltage contrast is quite efficiency to catch the killing defect of contact module. The defect performance detected by retarding mode e-beam inspection can reflect the wafer sort yield loss directly. With aid of the in-line failure analysis, the defective contact module can be resolved by increasing the CF_4 flow rate of dry strip process after spacer implant, and eliminate the etching stop issue during contact holes opening. It is suggested that e-beam retarding mode inspection methodology should be regarded as routine monitoring of contact module, especially for next generation with tough design rule of contact module.

References

1. R. Micheloni, L. Crippa, and A. Marelli, in *Inside NAND Flash Memories*: *Flash memory architectures,* p.2, Springer Dordrecht Heidelberg London, New York (2010).
2. Oliver D. Patterson, Kevin Wu*, Dan Mocuta, Kourosh Nafisi, *2007 IEEE/SEMI Advanced Semiconductor Manufacturing Conference*, p.48, (2007).
3. A. Miura, M. Ikota, I. Sekihara, H. Miyano, C. Zhaohui, and A. Sugimoto, A.; *2003 IEEE International Symposium on Semiconductor Manufacturing*, p.263, (2003).

ECS Transactions, 34 (1) 925-929 (2011)
10.1149/1.3567693 ©The Electrochemical Society

Verification of Systematic Defects Using e-Beam Defect Review System

Tuung Luoh, Ling-Wu Yang, Tahone Yang, Kuang-Chao Chen and Chih-Yuan Lu

Technology Development Center, Macronix International Co. Ltd.
No. 16, Li-Hsin Road, Science-Based Industrial Park, Hsinchu, Taiwan, R. O. C.
Phone: +886-35786688 ext. 78173 Fax: +886-3-5789087
ChrisLuoh@mxic.com.tw

High resolution e-beam review system with critical point
inspection (CPI) function is implemented to verify the systematic
defects in both array and periphery area of NOR Flash products.
CPI inspection can help us to intercept with unobvious systematic
defects and find the process variation to save the cost of yield loss
from time to time, and it is important for debugging the systematic
defects in irregularly periphery circuits.

Introduction

As technology node of memory devices is approaching 75nm and beyond, the process
window is becoming much narrower and production yield is getting more sensitive to
tiny defects which used to be not. It is a big challenge to find small but yield relevant
defects under a huge amount of nuisance and false defects. Increasing yield loss due to
non-visual defects and process variations requires new approaches in diagnostics and
control. Systematic defect induced yield loss mechanism generally comes from patterning
process variations across the lithographic process window. Sometime, the irregularly
logic or periphery areas makes them more sensitive to have systematic yield loss (1-2).
How to capture the systematic defects in the early stage by scan or review tool becomes a
principal topic of tool development roadmap.

In this study, high resolution e-beam defect review system is adopted to classify the
systematic defects in the early stage both in array and periphery area. Furthermore, the
evolution of systematic defects in different layers also is addressed at this paper.

Experimental

A high resolution e-beam defect review system with critical point inspection (CPI)
function is adopted to verify the systematic defects in the early stage. CPI can monitor
known hot spots and locations where the chip design is less robust to process variation.
CPI inspection sampling rate can select by die or by shot, and it can also execute several
hot spots in a die. The resolution of review system is 1.9nm at 600eV, and the operation
landing energy range is 200~1300 eV, operation beam current range is 50~500pA. CPI
function is implemented on the layers of word line etch and cell etch in NOR Flash
product. The irregularly periphery area of first metal layer is also evaluated by CPI
function.

Results and Discussion

CPI Inspection Applied in Array Area

It is difficult to identify the systematic defects from routine random defects review
(50 to 100 defects per wafer) and classification methodologies. One wafer is processed
with polysilicon line etching scan and the scan map shown in Figure 1(a), there is no
special systematic defect, except one suspected defect with necking poly line at the

925

source contact area after 100 ea random defects review, as shown in Figure 1(b). Therefore, the issued defect position can be regarded as a hot spot; CPI inspection follows the hot spot position to inspect all dice/shots with same position repetitively. In this study, 9 shots with 7X8 dice are selected for CPI inspection. Figure 2(a)(b) show the results of CPI inspection, the map in Figure 2(a) reveals all the inspection points in this inspection. Figure 2 (b) demonstrates the issued systematic defects found by CPI inspection. The systematic defects are always located at the edge of each shot; it indicates that the polysilicon line of source contact region near the photo shot edge area will have necking risk and opening concern; typical polysilicon line with necking shape systematic defect images are shown in Figure 3. After polysilicon line etching inspection, the next inspection stage is cell etching inspection. Same wafer follows the process flow to cell etching inspection, and it is found that the morphology of hot spot positions near photo shot edge change from necking shape to opening one, it will become the major killing defect and lead to yield loss, as shown in Figure 4. In this case, the CPI inspection can help us to intercept with unobvious systematic defects and find the process variation to save the cost of yield loss from time to time. The process variation is due to the KrF lithography tool lens aberrations induce symmetrical behavior through focus. The issue is fixed by changing to ArF or normal KrF lithography tools.

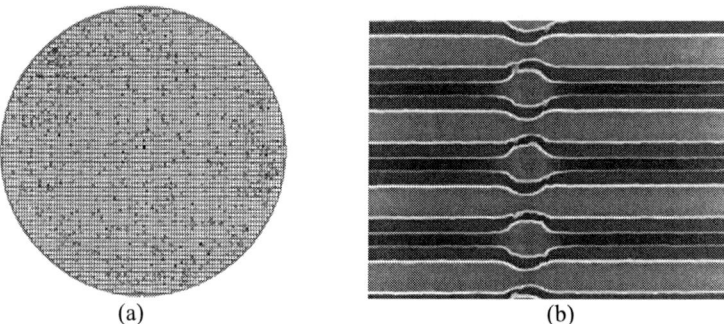

(a) (b)

Figure 2. (a) Scan map of polysilion line etching inspection. (2) One suspected polysilicon line with necking shape is found after 100 ea random review.

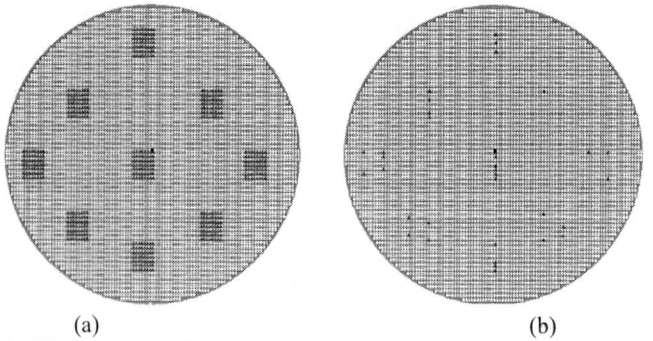

(a) (b)

Figure 3.(a) CPI inspection shot map (b) Necking polysilicon line systematic defects were found near the shop edge.

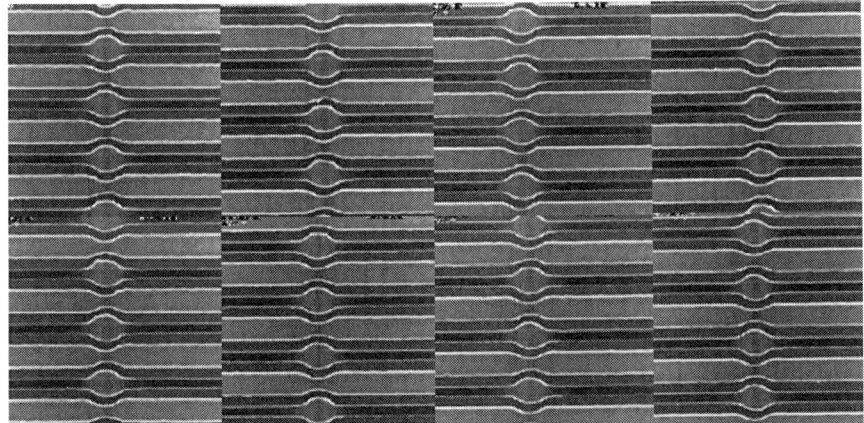

Figure 3 Typical poly 2 necking systematic defect images are inspected and reviewed by CPI inspection.

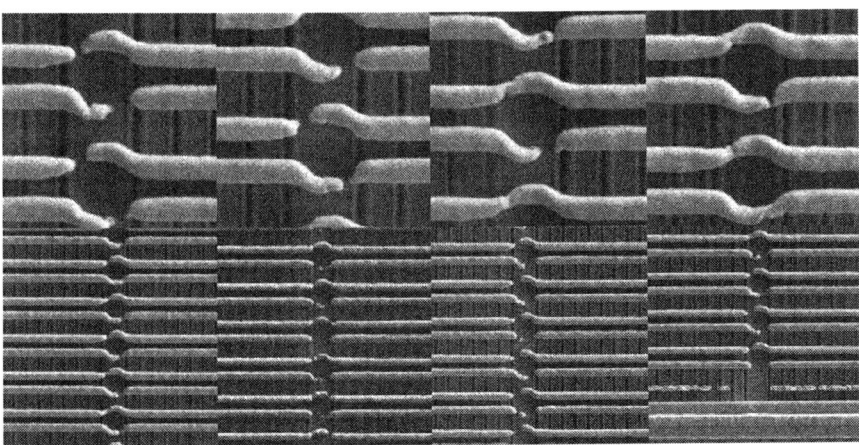

Figure 4. The hot spots near photo shot edge change from necking shape into opening shape after cell etching process.

CPI Inspection Applied in Periphery Area Occasionally, we ignore the defect performance in the periphery area due to the design rule of periphery is looser and less sensitive than the array one. However, many important circuits are located at the periphery, such as the state machine, x-decode, logic circuits, etc. The irregularly logic circuit design makes them more sensitive to have systematic yield loss. Therefore, the CPI inspection methodology is important for periphery circuit debug of systematic defects. Figure 5(a) demonstrates the first metal layer periphery scan results, however, its wafer sort suffer the x-decode failure issue. After implementing the CPI inspection hot spot at x-decode location, local systematic defects are found in wafer upside, as shown in

Figure 5(b). The systematic scum bridge defects are induced by photo developing process, as shown in Figure 6. The scum bridge systematic defects are fixed after optimized the process.

From the above results, it is proposed that the CPI inspection methodology is helpful for the systematic defects verifying and searching.

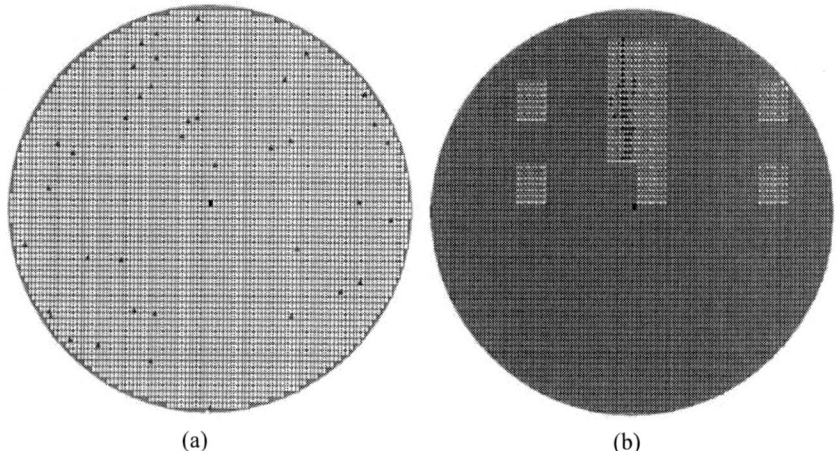

(a)   (b)

Figure 5(a) Periphery scan map of first metal layer (b) Systematic defects are found at x-decode at the wafer upper side.

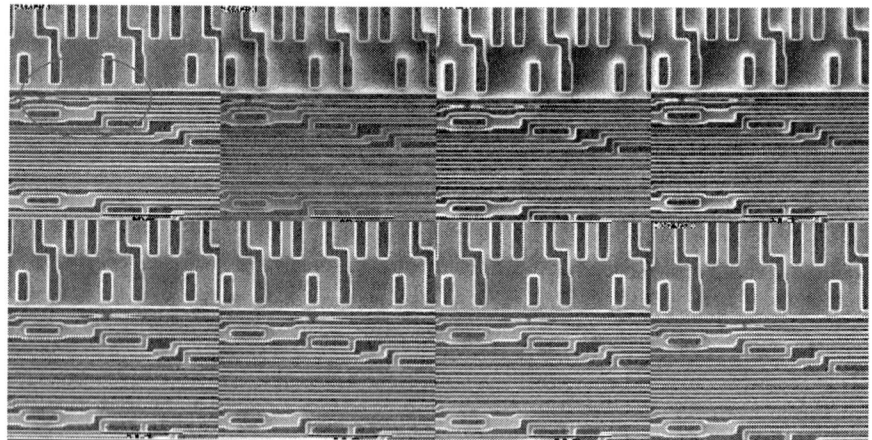

Figure 6. Local systematic scum bridge defects at x-decode area are found. The red circle indicates the scum bridge sites, and located upper side of wafer.

Summary

CPI inspection methodology applied in array area and periphery area is demonstrated in this study. In array area, some of unobvious systematic defects in pre-layer will lead to certain yield loss after CPI verification. In the periphery area, CPI inspection is important for irregularly logic circuit debug. Even there is no pre-scan result; we still can locate hot spots according the wafer sort failure sites, and then using the CPI to inspect the suspected layer and process step. In summary, systematic defects generated from patterning process variations across the lithographic process window; or created by the interactions among the mask and scanner; can be detected by CPI inspection methodology.

References

1. B. Kruseman, A. Majhi, C. Hora, S. Eichenberger, and J. Meirlevede, *International Testing Conference, ITC 2004,* p. 290, (2004).
2. M. E. Lagus, R. Shimshi, and V. Svidenko, *IEEE International Symposium on Semiconductor Manufacturing, ISSM 2005,* p.465, (2005).

ECS Transactions, 34 (1) 931-936 (2011)
10.1149/1.3567694 ©The Electrochemical Society

Determining Coherence Length of X-ray Beam Using Line Grating Structures

Hae-Jeong Lee, Christopher L. Soles and Wen-li Wu

Polymers Division, National Institute of Standards and Technology, Gaithersburg, MD
20899

> Specular x-ray reflectivity has been used successfully to measure
> the cross section of periodic nanopatterns on flat substrates and in
> most of the cases the effective medium approximation is found to
> be adequate for the data analysis. To examine the validity of EMA
> the coherence length of the X-ray reflectometer was measured
> quantitatively by extending the scheme used by Salditt et al. A
> series of linear grating structures with periodicities ranging from
> 300 nm to 16 μm was studies at various azimuthal angles between
> the incident X-ray and the lines. A clear break down of EMA was
> noticed at certain X-ray incident angle depending on the azimuthal
> angle, this provides a useful pathway for determining the coherent
> length of the X-ray reflectometer more precisely.

Introduction

X-ray reflectivity (SXR) is a powerful technique to investigate surfaces and interfaces including their roughness, diffusion across buried layers and thickness of single and multilayer stack by depth profiling the electron density in the direction normal to the surface of a flat sample with a sub-nanometer resolution.[1] We have also shown that SXR is capable of quantifying surface pattern cross section, for example, the cross section of the line gratings fabricated by nanoimprint as well as the molds used to imprint the patterns.[2,3] The efficacy of the SXR is based on the effective medium approximation (EMA) as illustrated in Figure 1. It should be noted that the equivalent concept of EMA has been used in estimating effective refractive index of porous material for ellipsometry or scatterometry.[4,5]
Similarly, when the coherence length of the x-ray source is larger than the lateral dimensions of the line gratings, the density of the lines and spaces (for a grating structure) are averaged together as one "effective" density, thus the EMA. For periodic patterns (such as the gratings studied here), this effective density contains quantitative information about the line-to-space ratio. The opposite extreme would be the limit where the length scales of the surface patterns are larger than the coherence length of the x-rays at which point the patterned surface begins to look physically 'smooth'. In this respect the utility of SXR as a pattern shape metrology is limited to patterns where the periodicity is smaller than the coherence length of the x-ray source. Here coherence length and/or resolution is a property of diffraction instrument and is independent of characteristics of the samples. There are several papers on the so-called 'resolution ellipsoids' for reflectometry with x-rays.[6-9] Mathematical expressions for exact calculations of the ellipsoids are derived for three orthogonal direction.[9] In this manuscript, we experimentally determine the coherence length of x-ray beam in reflectometer to explore the limitations of the SXR technique by analyzing line- gratings with periodicities ranging from approximately 300 nm to 16 μm. [11]

931

Line Gratings Samples and Measurements

A series of parallel line-space gratings with periodicities from 300 nm to 16 μm were prepared utilizing several different fabrication techniques and materials. Gratings with pitches smaller than 1 μm were fabricated by nanoimprint lithography and larger patterns with periodicities on the order of (2, 4, 8, and 16) μm were prepared by optical contact lithography using chrome on quartz mask. The SXR measurements were performed on the X-ray θ-2θ diffractometer with a slit collimation at ambient temperature as described in the previous publication.[2] All measurements were performed in the specular condition where the grazing incidence angle (θ_i) of x-ray beam equals to the reflected angle into the detector (θ_r). The ratio of the reflected (I) to the incident (I_0) beam intensity defines the reflectivity ratio R. The reflectivity collected over a range of incident angles is plotted as a function of log (R) versus the scattering vector q, where $q = (4\pi/\lambda)\sin(\theta_i)$ and λ is the x-ray wavelength.

Figure 1 illustrates a coherence volume associated with the incident x-ray. Axis 3 corresponds to the incident x-ray beam direction and the coherence length is expected to be the greatest in this direction since a 4 bounce crystal monochromator increases the coherence of the x-ray beam primarily in direction 3. The coherence length in direction 1 (ξ_1) and 2 (ξ_2), orthogonal to the incident beam direction will be smaller than direction 3 (ξ_3). These three orthogonal, different coherence lengths define a coherence volume. These coherence lengths are generally classified into two categories; one is transverse or lateral coherence length ($\xi_\perp$) and the other is longitudinal coherence ($\xi_\parallel$). The lateral coherence length includes ξ_1 and ξ_2, which depend on the experimental condition such as effective width of x-ray source (d_s) and distance between sample and x-ray source (R_s) through the relationship, $\xi_t = (\lambda/2) \times (R_s/d_s)$.[12] X-ray source of the diffractometer used is long fine focus tube with focal spot size of 0.4 mm x 12 mm. Given R_s of 350 mm, the estimated values of coherence length in direction 1 and 2 are 2.3 nm and 67 nm, respectively. In the current convention the collimation slits is parallel to direction 1. The longitudinal coherence length of direction 3 (ξ_3) is about approximately 540 nm following the relationship $\xi = (\lambda/2) \times (\lambda/\Delta\lambda)$,[12] where $\Delta\lambda/\lambda$ is the bandwidth defined by monochromator, which is 143.2×10^{-6} for Ge200 crystal.[13]

Based on a simple geometrical reasoning given in Fig. 1 specular reflected beams from a surface pattern can interfere coherently as long as the following conditions are met; $\xi_3 > 2h \times \sin\theta_i$ and $\xi_2 / \sin\theta_i > P_0$ where P_0 is the repeat of the surface pattern in the incident plane. Within these conditions the electron density of a patterned surface is essentially lateral averaged when probed by SXR, i.e. EMA is applicable. Given 540 nm for ξ_3 and for thin films and surface patterns with a thickness or h at 1.5 μm, EMA stays valid as long as the incident angle θ_i is below 10°, a value much beyond the angular ranges used in most SXR measurements which may reach 2° at most. Hence one can ignore the first part of the conditions for the remaining part of this work, i.e. ξ_3 does not enter this study as an experimental parameter.

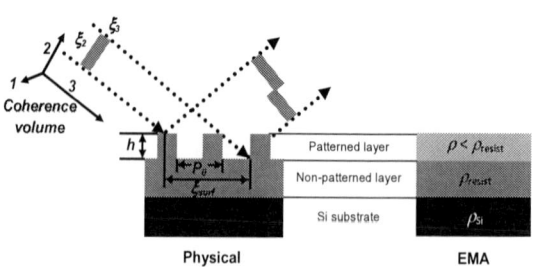

Physical   EMA

Figure 1. Schematics of the incident X-ray beam, the line gratings and the EMA equivalent of the line gratings. The shaded areas denote the coherence volume before and after the specular reflection. The incident beam direction is chosen as axis 3 and the collimation slit has its opening parallel to direction 1.

With ξ_1 being only a few nanometers almost all the SXR measurements were conducted with axis 1 mostly parallel to the line grating, after all, the smallest repeat among all the line gratings is 300nm, a value far larger than ξ_1. With the line gratings available to determine the value of ξ_1 is thus not feasible, by default, ξ_2 remains as the sole subject of study in the rest of this work. The approach to be demonstrated in the rest of this manuscript for determining ξ_2 can be used for ξ_1 determination with samples of appropriate spacing. Even though the value of ξ_2 has been estimated to be 67 nm in the foregoing section, its precise value is difficult to calculate due to the fact that a four-bounce Ge 200 monochromator is located between the X-ray source and the sample. The precise size of the X-ray source seen by the sample with a monochromator between can't be determined readily. In general, we expect the estimated value of 67nm to be a lower bound for ξ_2 . In the above discussion the effect of the optics along the detector side on the coherence length has been ignored completely, including this factor will make the calculation far more complex. Instead, a simple experimental scheme to determine the coherence length, at least semi-quantitatively, will be discussed as follows.

Results and Discussion

The experimental reflectivity data for the different periodicity gratings are shown in figure 2 with the reflection plane (plane 2-3) perpendicular to the line gratings. All the SXR curves display a total of three critical reflection angles or critical wave vector q_c in low q region. The first q_c near 0.017 $Å^{-1}$ corresponds to the lowest density layer, i.e., the patterned layer on the surface of the film. With increasing incident angle two additional critical angles or qc are present, the one near 0.032 $Å^{-1}$ is that of the silicon substrate. The one between that of the patterned surface and the silicon substrate corresponds to that of the residual layer for the nanoimprint samples or the anti-refractive (ARC) underlayer for all line gratings samples with their repeat in micrometers. The oscillations in reflected intensity between critical angles are reminiscent of waveguide mode. The existence of a critical angle below that of the solid under layer or ARC for every sample measured indicates that the value of ξ_2 / $\sin\theta_i$ at q = 0.017 $Å^{-1}$ is greater than 16 µm, the largest repeat in line gratings among all the samples. Since the corresponding value of $\sin\theta_i$ at q_c of 0.017 $Å^{-1}$ is 0.0021, this will put the lower bound of ξ_2 at 33nm; a value less than the estimated one at 67 nm.

In the rest of this work we will discuss our approach to determine ξ_2 via a scheme used by Salditt et al [15], i.e. the effective periodicity (P_{eff}) of a line gratings along the incident X-ray beam can be changed by setting the sample at different azimuthal angle (φ) as depicted in the inset of figure 3(a). SXR results given in Fig. 2 were all obtained at $\varphi= 90°$. The effective periodicity is simply $P_0/\sin\varphi$ and $P_0= 16$ µm for this case. As the azimuthal angle is rotated from 90° towards 0° the effective periodicity increases from 16 µm towards infinity. The SXR curves collected from the 16 µm line gratings as well as the model fittings to the data at several values of φ are given in Figure 3a. Note that the

SXR result at $\varphi = 90°$ can be fitted nicely throughout the entire q range shown in Figure 3a. As the value of φ decreases the range over which the SXR data can be fitted also decreases toward low q region; once φ reaches 25° many additional oscillations appear on the reflectivity curve even near $q=0.01\text{Å}^{-1}$. This is before the first critical q at 0.15 Å^{-1} and the reflectivity did not stay flat at unity, a phenomenon inconsistent with any theoretical predictions in specular reflectivity. This SXR result can no longer be modeled and it marks the complete failure of EMA or the condition $\xi_2 / \sin\theta_i > P_0 / \sin \varphi$. For given values of ξ_2 and P_0 this condition suggest that as φ decreases from 90° the range of θ_i over which the above inequality stays valid has to decrease, i.e. the maximum value of θ_i below which EMA is valid, denoted as θ_{ic}, decreases as the azimuthal angle decrease from 90°.

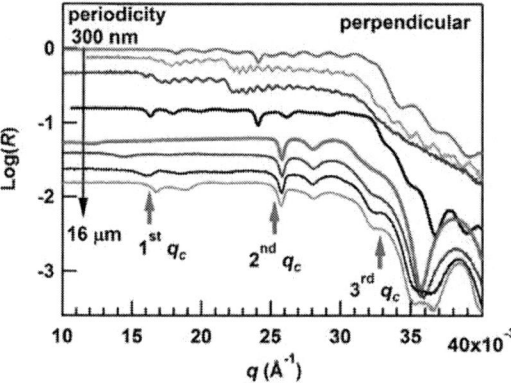

Figure 2. The X-ray reflectivity curves from line gratings with their repeats ranging from 300 nm to 16 μm, data were all collected with the incident X-ray beam perpendicular to the lines on the samples. From top to bottom the periodicity of gratings are 300 nm, 400 nm, 800 nm, 900 nm, 2 μm, 4 μm, 8 μm, and 16 μm. Arrows were added to denote the 1st, 2nd, and 3rd critical wave vectors or q belongs to the patterned layer, non-patterned layer, and silicon substrate respectively. The reflectivity curves have been vertically shifted for clarity.

In order to determine more precisely the value of ξ_2 SXR data from the 16 μm pitch sample at various azimuthal angle φ were fitted with progressively large range of q in order to identify the θ_{ic} position above which the relation $\xi_2 / \sin\theta_i > P_0 / \sin \varphi$ starts to break down. The χ^2 values of the fit were obtained as a function of q range and were given in Figure 3b where the abscissa q range denotes the data was fitted between q=0 to this specific q value. An abrupt increase in χ^2 is evident from all the data fit as the q range increases progressively. The q range position where this abrupt rise in χ^2 occurs is set as q_{ic}. It is noted that q_{ic} decreases as φ decreases from 90° to 40° as expected from the relation $\xi_2 / \sin\theta_i > P_0 / \sin \varphi$. Hereafter q_{ic} is taken as the location where $\xi_2 / \sin\theta_i = P_0 / \sin \varphi$. The values of φ, the calculated P_{eff} or $P_0/\sin\varphi$ and q_{max} are listed in Table 1. The calculated values of ξ_2 used the above rationale are also listed in the last column, they are ranged from 87 to 107 nm with an average value of 96.7 nm. Not surprisingly this value is larger than the estimated value of 67 nm based on the size of the X-ray source and sample-source distance. The corresponding depth profiles in X-ray scattering length density (SLD) expressed in Å^{-2} from fitting SXR data at various φ are given in Figure 3c. As expected, all the SLD profiles are almost identical even they represent the depth profile averaged over different lateral directions. These SLD profiles show clearly three sharp interfaces, a feature consistent with the observation of three critical angles in SXR data. In Figure 3c the silicon substrate is located at a depth of 2000 Å and beyond, the solid anti-reflective coating (ARC) occupies the depth between 1150 to 2000 Å and above it is the line gratings. The SLD value of the line gratings increases from the free surface towards the line gratings/ ARC interface, this indicates that the cross section of the gratings is tapered. A detailed discussion on the use of SXR to measure the cross section of surface nano-patterns can be found elsewhere. [2]

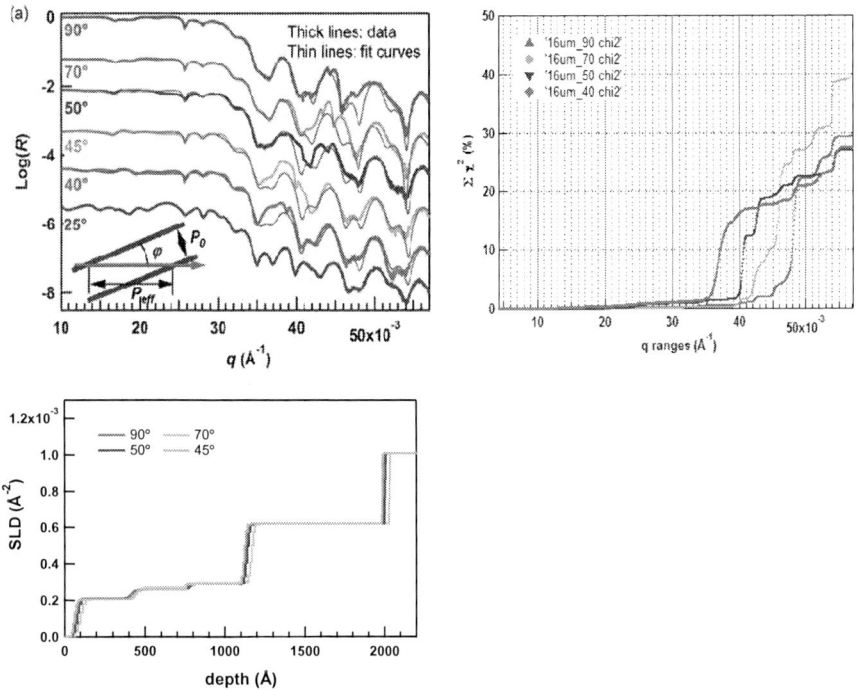

Figure 3. (a) The SXR curves collected from the 16 μm gratings at various azimuthal angle, φ, of the lines with respect to the incident x-ray beam; 0° denotes the incident X-ray beam is parallel to the line direction whereas 90° denotes perpendicular. From the top, φs are (90, 70, 50, 45, 40, and 25)°, respectively. The fits of these SXR results are also presented as colored lines and the experimental data as grey lines. All the fits were performed for the entire q range given in the graph. Inset illustrates the definition of the effective periodicity, P_{eff} as $P_0/sin\varphi$, where P_0 is grating periodicity. (b) χ^2 values of the fit were obtained as a function of q range over which the fitting was performed, i.e. the abscissa q range denotes the data was fitted between q=0 to this specific q value. An abrupt increase in χ^2 is evident from all the data fit as the q range increases progressively. (c) the resultant depth profiles of the 16 μm line gratings in terms of scattering length density from the fits of the data collected at different azimuthal angles.

In summary, we demonstrate that the coherence length of a X-ray diffractometer can be determined from SXR measurements on line gratings with appropriate repeating length P_0. The criterion of $\xi_2 / sin\theta i > P_0 /sin \varphi$ can be experimentally verified readily, conversely, this criterion It is noteworthy that only specular reflection was discussed in this work and the results are not expected to be valid for non-specular conditions.

TABLE I. The calculated effective repeats or periodicities of the 16 μm line grating at different azimuthal angles, the critical incident wave vector qic above which an abrupt increase in c2 of the fit took place. The calculated values of $\xi 2$ based on qic were also given in the last column.

φ (deg.°)	P_{eff} (μm)	q_{ic}(Å^{-1}) EMA breaks	ξ_2 (nm)
90	16.0	0.046	90.2
70	17.0	0.042	87.5
50	20.9	0.040	102
40	24.9	0.035	107

Acknowledgments

The authors acknowledge Jim Wang of the Nano Opto Corporation and Dr. Hyun-Wook Ro at NIST for the imprinted patterns. The authors also thank Dr. Shuhui Kang at NIST to providing helpful advice to prepare micron scale gratings and to Sunil K. Sinha of UCSD for insightful discussions.

References

1. H. J. Lee, C. L. Soles, D. W. Liu, B. J. Bauer, E. K. Lin, W. L. Wu, and A. Grill, J. Appl. Phys. 95, 2355 (2004).
2. H. J. Lee, C. L. Soles, H. W. Ro, R. L. Jones, E. K. Lin, W. L. Wu, and D. R. Hines, Appl. Phys. Lett. 87, 263111 (2005).
3. H. J. Lee, H. W. Ro, C. L. Soles, R. L. Jones, E. K. Lin, W. L. Wu, and D. R. Hines, J. Vac, Sci. Technol. B, 23, 3023 (2005).
4. Z. H. Cen, T. P. Chen, L. Ding, Y. Liu, M. Yang, J. I. Wong, Z. Liu, Y. C. Liu, and S. Fung, Appl. Phys. Lett. 93, 023122 (2008).
5. A. R. Pal, R. L. Bruce, F. Weilnboeck, S. Engelmann, T. Lin, M. S. Kuo, R. Phaneuf, G. S. Oehrlein, J. Appl. Phys. 105, 013311 (2009).
6. R. A. Cowley, Acta Crystallogr. A43, 825 (1987).
7. A. Gibaud, G. Vignaud, and S. K. Sinha, Acta Crystallogr., A49, 642 (1993).
8. M. F. Toney and D. G. Wiesler, Acta Crystallogr. A49, 624 (1993).
9. B. Dorner and A. R. Wildes, Langmuir, 19, 7823 (2003).
10. L. G. Parrat, Phys. Rev. 95, 359 (1954).
11. The data throughout the manuscript are presented along with standard uncertainty ($\pm$) involved in the measurement based on one standard deviation.
12. J. Als-Nielsen and D. McMorrow, Elements of modern x-ray physics, John Wiley & Sons, Inc., New York (2001).J. Doe and R. Hill, *J. Electrochem. Soc.*, **152**, H1902 (2005).
13. A. Authier, Dynamical theory of x-ray diffraction, Oxford University Press Inc., p. 447 (2001).
14. S. K. Sinha, M. Tolan, and A. Gibaud, Phys. Rev. B, 57, 2740 (1998).
15. T. Salditt, H. Rhan, T. H. Metzger, J. Peisl, R. Schuster, and J. P. Kotthaus, Z. Phys. B 96, 227 (1994).
16. M. Tolan, D. Bahr, J. Subenbach, W. Press, F. Brinkop, J. P. kotthaus, Physica B, 198, 55 (1994).E. Gaura and R. M. Newman, *ECS Trans.*, **4**(1), 3 (2006).

TSV/3DIC Profile Metrology Based on Infrared Microscope Image

Jing-Jou Tang[a], Young-Jinn Lay[b,c], Lih-Shyang Chen[b] and Lian-Yong Lin[b]

[a] Dept. of Electronic Engineering, Southern Taiwan Univ., Tainan 710, Taiwan
[b] Dept. of Electrical Engineering, National Cheng Kung Univ., Tainan 701, Taiwan
[c] Dept. of Electrical Engineering, Chung Chou Institute of Tech., Changhua 510, Taiwan

3D IC can provide the advantages of heterogeneous integration through vertical interconnection and high performance without using advanced process. Basically, 3D IC can be implemented based on the technology of TSV (Through Silicon Via; TSV) and micro-bumps. Thus the quality of TSVs is critical to the yield of a 3D IC system. However, due to the nature of miniature and non-easy-probe, it is very difficult to validate the characteristics of topology or electricity of a TSV using traditional probing or metrology methodology during the process.

In this paper, we are going to present a TSV profile metrology system for 3D IC based on the digital image processing (DIP) technique. The profiles of all TSVs in a 3D IC system can be constructed based on the images obtained from IR-microscope or SEM. The results show that our system is helpful to the inspection of TSV duing the 3D IC manufacturing.

Introduction

Over the past 40 years, Moore's law was always true to the fact that the number of transistor was doubled every 18 month. Unfortunately, the risk is becoming higher and higher associated with the design complexity. In 2001, IMEC has proposed their 3D-IC which uses the third dimension (Z-dimension) to provide more routing resource to integrate heterogeneous IC components(1). The key technology used in this technology is so called TSV (Through Silicon Via)(2).

The difference between 3D-IC and 3D package is that the components used in 3D package are discrete, while the 3D-IC will bond the thinned wafer without using traditional wire-bonding for SIP. Basically, 3D-IC can be regarded as a single IC since it can provide more routing resource to increase the integration complexity based on the TSVs. Referring to the Fig. 1, the routing distance and complexity can be dramatically reduced from 2D design to 3D design(3). The data report from IBM shows that the routing distance can be dramatically reduced and the routing resources can be increased 1000 times over the traditional 2D designs(4).

The performance of TSVs is critical to the manufacturing process in 3D IC design. However there is no suitable method to identify and inspect the profile of TSVs in a 3D IC system. In Fig. 2, there are pictures captured from IR-Microoscope at different focuses. It is interesting to see that the diameter of a TSV at bottom (Fig. 2(a)) is smaller than the one on the middle (Fig. 2(b) or the top (Fig. 2(c)). These images validate the fact that

TSVs either are formed by using DRIE or laser drill will be tapered. Unfortunately these images can not be used to inspect the side view of TSVs.

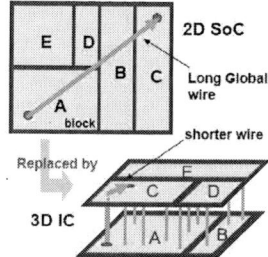

Figure 1. 2D SoC vs. 3D-IC interconnection(5)

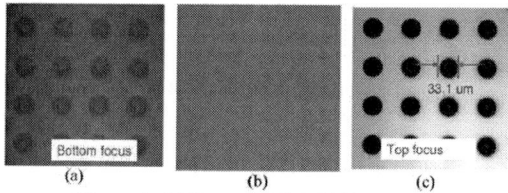

Figure 2. (a) Bottom, (b) Middle, (c) Top

Basically, the DIP technology can be used to construct the 3D profile from a set of 2D images. In the traditional 3D construction algorithm, the threshold value of gray-level should be determined during the reconstruction process. However this method cannot be applied to the construction process for those TSV images captured from IR-microspoce because there may be no deterministic threshold value in a set of available images.

Fig. 3 depicts the results of histogram equalization from those images in Fig. 2 (6). It is clear to find that the gray-level distribution is not uniform in a single picture or among different pictures. Fig. 4 is a simple experiment for the bottom image according to different threshold values. Obviously, there are some unwanted images in every picture no matter what the threshold value it is. Consequently, it is impossible to set a deterministic threshold value for 3D construction of TSV images.

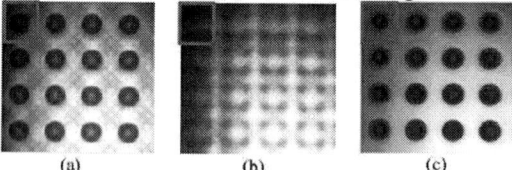

Figure 3. (a) Bottom, (b) Middle, (c) Top

This paper proposes a novel TSV 3D images construction algorithm and system based on the adaptive threshold determination. It is very efficient and helpful to the reconstruction of 3D images for TSV profiles. These images can provide informative data for the process engineers of TSV formation.

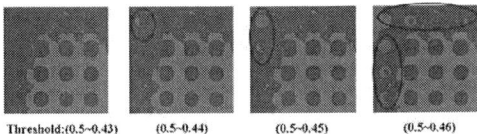

Threshold:(0.5~0.43)   (0.5~0.44)   (0.5~0.45)   (0.5~0.46)

Figure 4. Different threshold range

TSV Images Pre-processing

According to our observations in previous section, it is very difficulty to set a deterministic threshold value for each IR-microscope image. The concept of Region of Interest (ROI) in DIP technique can be employed to deal with this problem. That is to say, we can block out the information outside TSVs before we process the image retrieve of TSVs we intend to inspect. The simplest method is to use the "mask" function for each image. The image data except the TSVs can be filtered and ignored before we start to deal with the construction of TSV profiles. However it is still not clear to draw a region of interest for each TSV in every picture.

Since the diameters of TSVs on top site are larger than the site in the bottom site, we can take the area of top TSV as the ROI. This ROI is obviously can be used as the mask function for the following processing. With the mask function, we can have better image for each TSV in a picture and will be possibly used for construction of TSV profiles.

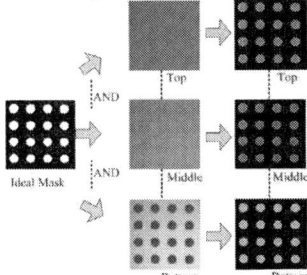

Figure 5. Region of Interest

The mask function determination

In the following section, the mask function will be discussed in detail.

Step 1: Edge Detection (7)

It is clear that the image on the top site is much clear than the ones on the other sites. Thus the contours of the TSVs image can be used to determine the mask function for the following images in different focuses. The method used to determine the contour can be the general edge detection technique in DIP technology. In Fig. 6, a 3x3 mask is used to determine the edge transition. That is to say, we can determine the contour edge from a set of 3x3 matrix scanning throughout the image data. Basically, the edge or contour is located on the place where the gray level transition is large. On the programming point of view, we can set the value as 1 to the place where is determined as the edge while set the value as 0 to the place adjacent to the edge.

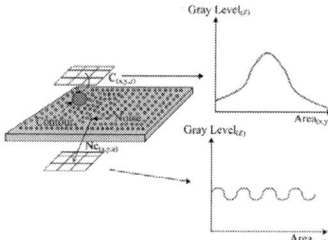

Figure 6. edge detection

Step 2: Median Filter (6):

The contour determination method described above is not robust because there will be some error information due to the content of the images. The contour may be unique. This condition can be avoided by using the median filter function to the data obtained from the above edge detection. Similarly, a 3x3 matrix can be used to scan the data of the image and remove the unwanted noise data. After this processing, a precise contour, i.e., the mask function, can be obtained.

Step 3: Threshold Value Determination

After the processing of edge detection and median filtering, we can have a rough mask contour for the TSVs we are going to process. This rough contour cannot be used as the final contour in the following image process since the gray level on the edge is not sharp enough and will result in the partial volume effect during the threshold value determination in Fig. 7 (8). In order to have a sharp image for the TSVs, another 3x3 matrix can be used to elevate the sharpness on the edge. The pixel of the edge is adjusted based on the minimum two values around this pixel. The final gray level is calculated by averaging these two pixels (i.e., the highest darkness).

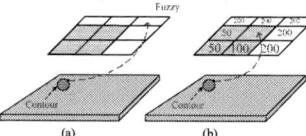

Figure 7. (a) partial volume effect; (b) mean value determination

After completing the above steps, we can have a specific ROI for each TSV. The gray level inside each TSV will be further used to determine the threshold value to the estimation of TSV profile. In the following section, we will present the detail construction algorithm for TSV 3D profile.

TSV 3D Profile Reconstruction

In this section, the algorithm for TSV 3D profile is proposed. Some experimental results are shown and discussed. Here we also illustrate the processing steps.

Step 1: Apply mask function

The mask function developed above is applied to each TSV images at different focuses. The image data outside the mask will be blocked out and set to gray level 255,

Step 2: 3D Image Construction

After the step 1, the area ROI for TSVs in each layer can be obtained. The gray level inside the TSVs can be further binarized based on a suitable threshold value. After this, the gray level for each pixel will be either 0 or 4095 (12bits). Thus a 3D TSV image can be constructed based on the layer (2D) image stacking as shown in Fig. 8.

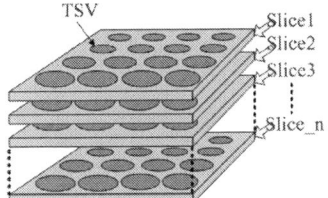

Figure 8. Example of 3D-TSV

Experiment Results

In this section, we will demonstrate and validate our proposed methodology for TSV 3D images reconstruction. The result of the edge detection, median filter, and obtained mask function is shown in Fig. 9(a), Fig. 9(b), and Fig. 9(c), respectively. The image in Fig. 9(a) shows that the contour is not sharp enough to be a mask function. The noise outside the TSV will be removed after the filtering function as shown in Fig. 9(b). However the contrast is still not high enough. The gray level information is then further binarized based on a suitable threshold value.

The images in Fig. 10(a), Fig. 10(b) and Fig. 10(c) are the final TSV images with respect to the top site, middle site, and bottom site. Obviously, the diameter and contour can be determined to reflect the quality of TSV process. Fig. 11 is the final reconstructed 3D TSV profiles images according to the interest of users. In our system, we provide two choices for process engineers to view the TSV profiles in hollow style or in pillar style.

The hollow style can be used to assist the understanding of TSV formation which may be created by using DRIE or laser drill. The pillar style images, on the other hand, can be used to assist the inspection of TSV metallization. By inspection the TSV profiles, either the hollow or pillar style, the yield loss caused by void can be easily located. This information is definitely helpful to the yield analysis of the 3D IC process.

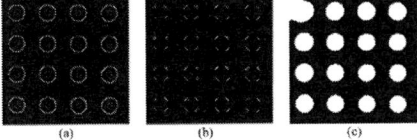

Figure 9. (a)Edge Detection ;(b)Median-Filter ;(c)Mask

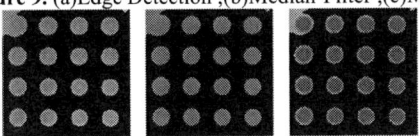

Figure 10. (a)Bottom ;(b)Middle ;(c)Top

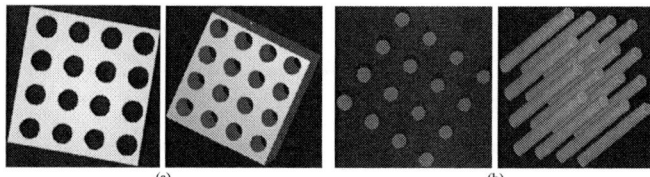

(a)　　　　　　　　　　　(b)

Figure 11. (a) 3D TSV image (hollow);(b) 3D TSV image (pillar)

Conclusion

In this paper, we have presented a TSV profile metrology system for 3D IC. The profiles of all TSVs in a 3D IC system can be constructed based on the images obtained from IR-microscope or SEM. Through our efficient reconstruction algorithm, process engineers can have more information on the performance of TSV formation or filling. For example, the diameter of a TSV at different altitudes of a wafer can be measured and predicted. Also either the profile or void of TSVs can be clearly visualized or located in different angles. Obviously, this information will be helpful to the diagnosis and failure analysis to the 3D IC process such as: wafer thinning, wafer bonding, TSV formation/filling.

Acknowledgments

This work is supported from the projects of NSC 99-2220-E-218 -001 and NSC 98-2220-E-218 -003。

References

1. E. Beyne, "Technologies for very high bandwidth electrical interconnects between next generation VLSI circuits," *Tech. Dig. - Int. Electron Devices Meet.* (2001).
2. K. Navas, V. Rao, L. Samule, S.W. Ho, V. Lee, X.W. Zhang, R. Yang and E. Liao, "Development of 3D silicon module with TSV for system in packaging," in *2008 58th Electronic Components and Technology Conference,* (2008).
3. D. Koen, D. Piet, S. Deniz, B.Beyne, E. Robert and V. Christiaan, "3D interconnect technology for space applications," in *ESA Round Table on Micro/Nano Technologies for Space location.* (2005).
4. "IBM tips TSV 3D chip stacking technique," *Solid State Technology, Tech. Rep.,* (2007).
5. M. J. Wolf , P. Ramm and A. Klumpp, "Thru-Silicon Via Technology: R&D Fraunhofer IZM," *Fraunhofer IZM, Tech. Rep.,* (2008).
6. R. C. Gonzalez and R. E. Woods, *Digital Image Processing,* 3rd Ed., p. 122, Pearson Education, New Jersey (2008).
7. M. A. Ruzon and C. Tomasi, "Color Edge Detection with the Compass Operator," in *Proceedings of the 1999 IEEE Computer Society Conference on Computer Vision and Pattern Recognition,* (1999).
8. D. L. Plam and P-L. Bazin, "Simultaneous"Boundary and Partial Volume Eatimation in Medical Images," in *Proceedings of the 7th International Conference on Medical Image Computing and Computer Assisted Intervention,* (2004).

Endpoint Detection in Plasma Etching using Principal Component Analysis and Expanded Hidden Markov Model

Min-Woo Kim[1], Seung-Gyun Kim[1], ShuKun Zhao[1],
Sang Jeen Hong[2], and Seung-Soo Han[1]

[1]Department of Information and Communication Engineering, Myongji University
Yongin, Kyunggi-Do, 449-728, South Korea
[2]Department of Electronics Engineering, Myongji University Yongin, Kyunggi-Do, 449-728, South Korea

Abstract. In current semiconductor manufacturing, plasma processes such as etch and CVD take the portion at least 40% throughout of integration processes. As the feature size of integrated circuit (IC) devices continuously shrinks, detecting endpoint in low open area plasma etch process becomes more difficult. To solve this problem, a combination of Principal Component Analysis (PCA) and Expanded Hidden Markov model (eHMM) technique is applied to optical emission spectroscopy (OES) signals. Selected patterns are used in PCA, which reduces dimension of the raw data and increases gap between classes. The eHMM is employed to detect endpoint using output of PCA. The eHMM combines the semi-Markov model to enable an arbitrary distribution on the location of the change-point and the segmental HMM to model the configuration in each segment. After modeling using eHMM, real-time OES data were fed to this model to detect endpoint.

1. Introduction

Plasma Etching is an important process in the semiconductor manufacturing. Etching process monitoring influence on silicon wafer is important element in determining the endpoint. When the target layer is removed, the etch process can't be stopped immediately and over-etching is inevitable, because of non-uniformity of the target layers. Over-etching can damage several layers, underlying layers, which can decrease the yield.

One of the most commonly used measurement technique for *in-situ* plasma monitoring and sophisticated endpoint detection is optical emission spectroscopy (OES) (1). Most endpoint detection (EPD) methods using OES focus on identifying a single wavelength that corresponds to a chemical species that shows a transition at the endpoint. As an example, in SiO_2 with CHF_3, carbon combines with oxygen from the wafer to from carbon monoxide (CO) as an etch product. It is known that CO emits light at a wavelength of 482.5 nm, and that this wavelength can be monitored for detecting the endpoint. When the oxide is completely etched, there is no longer a source of oxygen and the CO peak at 482.5 nm decreases, thus signaling an endpoint of etch process (2).

The dry etch process, especially a reactive ion etch (RIE) process, is used for etching thin line patterns in a silicon wafer. In recent years, Conventional EPD method is limited in the etch process using RIE because devices size and open area is smaller than ever before. Typically, differential detection method is usually used to detect endpoint in semiconductor process. By the way, dielectric via or contact etching processes have an open area less than a few percent; however, differential detection method cannot be used for such small open area applications.

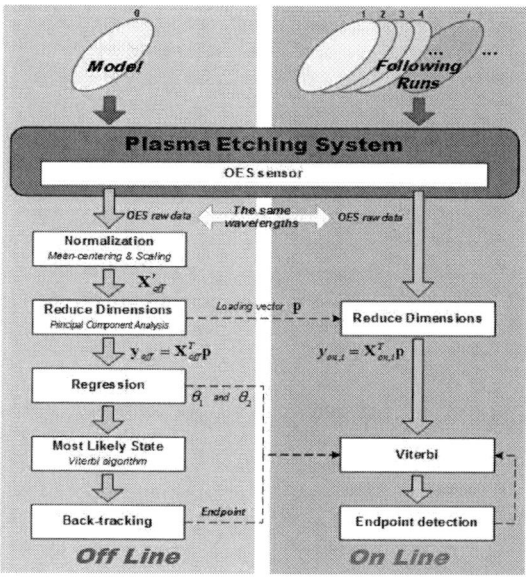

Figure 1. Block diagram of PCA and eHMM in endpoint detection

Figure 1 shows the proposed EPD method using PCA and eHMM. OES data have multi wavelength. The dimension of these data is reduced to one-dimensional data using principal component analysis (PCA). The endpoint is detected using this one-dimensional data by applying eHMM.

2. Methodology

Principal Components Analysis (PCA)

Principal components analysis (PCA) is a famous tool for data compression. PCA decomposes the normal data matrix X as the sum of outer product of vector t_i and p_i plus a residual matrix E:

$$X = t_1 p^T_1 + t_2 p^T_2 + ... + t_k p^T_k + E \qquad [1]$$

where, k must be less than or equal to the smaller dimension of X. The t_i vectors are defined as scores, and contain information on how the samples are related to each other. The p_i vectors are known as loadings containing information on how the variables are related to each other. In the PCA decomposition, the p_i vectors are the eigenvectors of the covariance matrix, i.e. for each p_i:

$$Cov(X)p_i = \lambda_i p_i \qquad [2]$$

where, the λ_i is the eigenvalue corresponding to p_i ; $Cov(X)$ means the covariance matrix of X. Note that for X and each t_i, and p_i pair, there is relations of form(3):

$$t_i = Xp_i \qquad [3]$$

Expanded Hidden Markov Model (eHMM)

In this section, the extended version of segmental semi-Markov model for solving the problem of EPD is suggested. This method detects one time change of the value and resulted in high efficiency in computation. Proposed method has the following specific component (4):

The process is assumed to start at the State 1 and then transition to the State 2 and stays until the end of the process.

- A segmental hidden Markov model is used for modeling the experimental data. It is assume that the regression functions for States 1 and State 2 are linear with values θ_1 and θ_2 plus the additive noise e_t, which is Gaussian with zero mean and unknown variance σ^2.
- Etch process is characterized as semi-Markov, during which a state duration distribution for State 1 is represented as $p_1(t_d)$.

For the problem of EPD, a 2-state segmental semi-Markov model is applied:

- State 1: *1st* segment (from the starting point to the changing point)
- State 2: *2nd* segment (after the changing point, over etch)

The process will start from State 1 and be expected to transit into State 2, with the initial state distribution (π) (2)(5):

$$\pi = \begin{bmatrix} \pi_1 \\ \pi_2 \end{bmatrix} = \begin{bmatrix} 1 \\ 0 \end{bmatrix} \qquad [4]$$

and the transition matrix A is defined by

$$A = \begin{bmatrix} a_{11} & a_{12} \\ a_{21} & a_{11} \end{bmatrix} = \begin{bmatrix} 0 & 1 \\ 0 & 0 \end{bmatrix} \qquad [5]$$

The state duration distribution of State 1 is set by prior knowledge of 'when the change-point will occur'. For example, if determine that the state change occurs approximately at time $\mu_c \pm 20\%$, we can use a truncated normal distribution

$$p(d_1) \propto \begin{cases} \dfrac{1}{\sqrt{2\pi\sigma_c^2}} e^{-\frac{(d_1-\mu_c)^2}{2\sigma_c^2}} & ,\mu_c - 3\sigma_c \leq d_1 \leq \mu_c + 3\sigma_c \\ 0, & otherwise \end{cases} \qquad [6]$$

where, $3\sigma_c = \mu \times 20\%$. As carefree with the duration of state 2, its distribution is set at

$$p(d_2) \propto 1, \qquad for \ d_2 \geq 0 \qquad [7]$$

And then, the linear regression functions of the N states is of form

$$y_t = f_i(t|\theta_i) + e_t, \qquad for \ 1 \leq t \leq N,. \qquad [8]$$

The joint distribution of the model is

$$p(s_1,...,s_t,y_1,...,y_t) = \pi(s_1)\prod_{j=1}^{r-1} p(s_{j+1}|s_j)\prod_{j=1}^{r} [p(d_{s_{n_j}} = n_j - n_{j-1})p(y_{n_{j-1}+1},...,y_{n_j}|s_{n_j})] \qquad [9]$$

Linear Regression Model Linear regression model is defined to accommodate the functional form of the conditional densities $p(y_t|s_t)$ that concern the observed data with the hidden states. In the standard HMM, the observed sequence y_t's rely only on the state sequence s_t, not on the time t. The proposed model allows each state to produce data in the form of a linear regression model (6), i.e.,

$$y_t = f_i(t|\theta_i) + e_t \qquad [10]$$

Where, $f_i(t|\theta_i)$ is a linear regression function with parameters θ_i and e_t is additive independent noise. Gaussian noise often supposed Gaussian, but not inevitably. Accordingly, in the Gaussian noise case we get that $p(y_t|s_t = i)$ is Gaussian with a time-dependent mean $f_i(t|\theta_i)$ and variance σ^2. Note that conditioned on the regression parameters θ_i, the y_t is only depend on the current state s_t, as in the regression framework (2).

Viterbi-Like Algorithm Viterbi-like algorithm is used for calculating the most like state sequence (MLSS) for calculation. For more sensitive EPD, the MLSS $\hat{s} = s_1 \ldots s_T$ for observation sequence $y = y_1 \ldots y_T$ can be represented in equation (7). At each time $t < T$, this algorithm calculates the value $\hat{V}_i^{(t)}$ for each state i, $1 \leq i \leq N$, where $\hat{V}_i^{(t)}$ is defined as

$$\hat{V}_i^{(t)} = \max_s \{p(s|y_1 \ldots y_t)|s = s_1 \ldots s_t, \quad s_t = i, s_{t+1} \neq i \} \qquad [11]$$

In other words, $\hat{V}_i^{(t)}$ is the likelihood chain of the MLSS that ends within State i, and y_t is the last observation of segment i. At time $t = T$, $\hat{p}_i^{(T)}$ is defined as

$$\hat{V}_i^{(t)} = \max_s \{p(s|y_1 \ldots y_T)|s = s_1 \ldots s_t, \quad s_T = i\} \qquad [12]$$

According to the definition, MLSS of the sequence $y_1 \ldots y_T$ could be the state sequence $s_1 \ldots s_T$ with likelihood $\max_i \hat{V}_i^{(t)}$.

The recursive function for calculating $\hat{V}_i^{(t)}$ is

$$\hat{V}_i^{(t)} = \max_{t'} (\max_j [\hat{V}_j^{(t')} A_{ji}] p(d_i = t - t') p(y_{t'+1} \ldots y_t | \theta_i)) \quad for \quad 1 < i < N \qquad [13]$$

When the process reaches the final time, we acquire the maximum $\hat{V}_i^{(t)}$ using equation [13] and subsequently find MLSS using trace back algorithm. In MLSS, changing from State 1 to State 2 is the ultimate endpoint (6).

3. Application for Endpoint Detection

The wavelength to be monitored in OES sensor data is depending on the material. Usually wavelength of CO is used because CO creates the most popular in etch process. Most commonly used wavelength for CO is 482.5 nm, however, CO has multiple wavelengths, as shown in Table 1. In this paper, these multiple wavelengths are used to make eHMM model. The multiple wavelengths are reduced to one-dimensional data using PCA, and is fed to eHMM to make a model. To validate the model using data from the same wavelength, loading vector is applied to convert product values. Viterbi-Like algorithm is applied to the output of the model to detect endpoint.

TABLE I. Several relative wavelengths for the single wavelength method.

Material	Wavelenghts (nm)
CF	240.0, 247.5, 255.8
CF_2	259.5, 271.2, 274.9
CO	219.7, 292.4, 313.9, 349.3, 482.5, 560.9
CO_2	287.7, 290.0
Si	252.0, 252.3, 505.6
SiF	440.2
SiO_2	248.6

4. Result

Figure 2 shows the linear regression model of the raw data using PCA. It can easily be found that the PCA process reduces noise in the raw data and increases the gap at around endpoint. The two slopes are decided to represent two segments in the linear regression model. The larger gap between State1 and State2 is better to detect endpoint. Therefore, creating regression model with one-dimensional data with larger gap is more effective. Theta regression parameters θ_i of each model are 0.36002 and -0.0018, respectively.

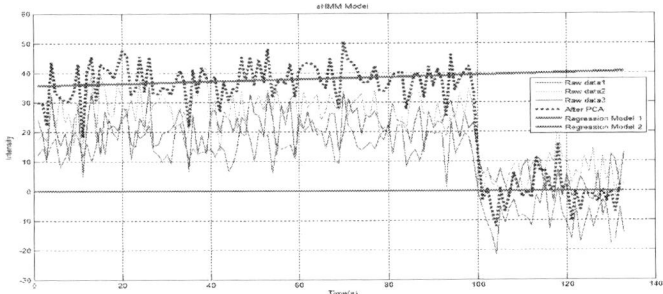

Figure 2. eHMM Model

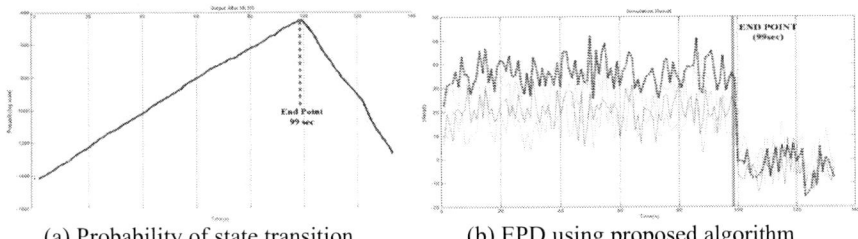

(a) Probability of state transition   (b) EPD using proposed algorithm
Figure 3. EPD results using proposed algorithm

Figure 3 is the results of the proposed algorithm applied to *in-situ* real time data. Figure 3 (a) shows the probability of state transition. The greatest point corresponds to the endpoint. With this algorithm, the only one candidate point is acquired and the endpoint is found correctly. Figure 3 (b) shows the detection of endpoint using eHMM algorithm. This figure verifies that the proposed algorithm accurately detects endpoint. This algorithm is assumed to start from the state 1 until endpoint occurs. At the endpoint, the state is change to the state 2. As state changes, the proposed algorithm detects the correct endpoint.

5. Conclusion

In this paper, a combination algorithm of principle component analysis (PCA) and Expand Hidden Markov model (eHMM) for endpoint detection is proposed. In this algorithm, loading vector is calculated using PCA process and raw data is multiplied to produce one-dimensional data. This data is used by regression model and MLSS algorithm. After this, eHMM and Viterbi-Like algorithm is used to detect the endpoint accurately. Compared to commonly used algorithm such as differential algorithm, the proposed algorithm can detect more accurately endpoint especially in smaller open area ratio in dry etch process. The performance of this algorithm will be compared to other several algorithms and analyze the advantages and disadvantages in the future.

Acknowledgements

This research was supported by Basic Science Research Program through the National Research Foundation of Korea (NRF) funded by the Ministry of Education, Science and Technology (Grant No. 2010-0024021).

References

1. J. Shabushnig P. Demko, Application of optical emission spectroscopy to semiconductor device fabrication, in Amer. Lab., **16**, p. 60–67 (1984).
2. S. Jeon, *et al*, "Endpoint Detection of SiO_2 Plasma Etching Using Expandd Hidden Markov Model". ISNN 2010, Part II, LNCS **6064** (2010).
3. K. Han, "Modified PCA Algorithm for the End point Monitoring of the Small Open Area Plasma Etching Proocess Using the Whole Optical Emission Spectra", International Conference on Control, Automation and Systems (2007).
4. Ge. X., Smyth.P, "Segmental Semi-Markov Models for Change-Point Detection with Applications to Semiconductor Manufacturing", Technical Report UCI-ICS p.00-08 (2000)..
5. Ge. X., Smyth.P, "Hidden Markov Models for Endpoint Detection in Plasma Etch Processes", Technical Report UCI-ICS 01-54 (2001).
6. N. R. Draper and H. Smith, *Applied regression analysis*. John Wiley & Sons, Inc, (1998).
7. Ostendorf, M., Digalakis,V.V., Kimball, O.A,"From HMM's to segment models, a uniedview of stochastic modeling for speech recognition", IEEE Transactions on Speech and Audio Processing, **4**(5):360--378 (1996).

Improvement of In-line SCD metrology on BEOL Copper CMP erosion layers for 65nm technology node logic production application

Clear Rong[a], Zhigao Wang[a], Zhengchao Yin[a], Zhengquan Tan[b], Jolly Zhao*[c]

[a] Semiconductor Manufacturing International Corp, No.18, WenChang Road, Beijing, Economic-Technological Development Area, 100176, P.R.China
[b] KLA-Tencor FaST Division, 1, Technology Dr 3-2028, Milpitas,CA, 95035, USA
[c] KLA-Tencor China, 1109A of Zhaolin mansion, #15, Ronghua Middle road, BDA, Beijing,100176, P.R.China

> Erosion Cu (Copper) interconnect monitor of BEOL (Back End of Line) CMP (Chemical Mechanical Polishing) is a series of key steps for producing high yield Integrated Circuits. For years, ultrasonic was applied to Cu in-line measurement. However, do we have an alternative? In this paper, new solution by the theory of scatterometry was proposed, it is called SCD (Scatterometry CD) metrology. To demonstrate the performance of SCD, multiple layers of Logic 65nm production, from Metal 2 to Metal 5, were developed by SCD and point to point matching with baseline tool were qualified. Furthermore, DOE (Design of Experience) have been employed to verify correlation among SCD, ultrasonic and TEM (Transmission Electron Microscopy). Finally, SCD is proven to be the reliable, high throughput and non-destructive solution of Cu line monitor.

INTRODUCTION

For last several logic nodes and recently DRAM production, copper interconnect replaced the traditional Aluminum to reduce the resistive-capacitive delays and more robust electro-migration. To fabricate the Cu interconnections, the dual damascene technique and chemical mechanical polishing (CMP) process are engaged (2). To maintain the accurate interconnect resistance, it is critical to monitor Cu thicknesses of different metal layers. Since these thicknesses may vary significantly across the wafer and from wafer to wafer, higher throughput and lower CoO (Cost of Ownership) method is preferred.

It is generally believed that ultrasonic can be applied to Cu in-line monitor. Otherwise, TEM (Transmission Electron Microscope) cross sectional electron imaging can be used for pad profile and layer thickness verification. But, ultrasonic has throughput concern and TEM is destructive method not suitable for inline moniroting. So, do we have better choice?

Based on the previous publication (3), SCD was brought to our mind as real-time high volume production control. Compared to traditional technologies, it has significant advantages. Firstly, higher throughput and lower CoO. Secondly, multiply parameters can be measured at the same time. For example, in the cases of these paper, not only Cu thickness but also the MCD (Middle CD), HT (Height), SWA(Side Wall Angle) of BD trapezoid, even bottom thicknesses can be reported by SCD after several seconds measurement. Last but not least, do not need to worried about wafer damage or destructive by SCD.

SCD TECHNIQUE AND MEASUREMENT DETAILS

Most SCD is based on Spectroscopic Ellipsometry (SE) by the analysis of diffracted light. When a light source illuminates a periodic structure which we called grating, light will reflect off the grating. The reflectance properties such as intensity and polarization of the light will depend on the materials, not only the top-most layer but also underlying layers and profile of the grating. It is noted that scatterometry techniques generally require particular measurement targets, typically 50μm×50μm in scribe line. For each application, we try to obtain good fits by optimized film models, layers structure and algorithms, then building libraries based on those fitting and altered grating target parameters. Each measured spectrum is compared to the library, and the library location that gives the best fit is used to determine the answer. Especially, most of these model and recipe developed offline and does not take away the metrology tool time.

The 300mm wafers used in this study and various Cu CMP processes were evaluated, including Metal 2 (M2) to Metal 5(M5). These gratings consist of lines and spaces with fixed pitch 200nm. SCD measurements are done by KLA-Tencor SpectraCD 200 at approximately 70 degrees angle of incident with available 240nm to 780nm wavelength range. Ultrasonic data are obtained by MetaPULSE of Rudolph and TEM cross sections are conducted using Defect Analyzer 300 system of FEI.

STRUCTURE EXPLANATION AND SPECTRA FITTING SAMPLE

Structure Explanation

M2 structure is as figure 1. Structures of M3 to M5 are almost similar, only a little difference of bottom layers. Individually, M3 layout is "F2: Cu(HT180) //F1:TaN (HT6nm) //G1: BD (HT190nm) //B10: BD (HT130nm) //B9: NDC (HT50nm) //B8: BD (HT340nm) //B7: NDC (HT50nm) //B6: BD (HT160nm) //B5: NDC (HT30nm) //B4: ILD HDP(HT360nm) //B3: HTN(HT45nm) // B2: Ni-Silicide(HT7.5nm) //Substrate:Si", F means filled materials, G means grating trapezoid and B means Bottom layers; M4 layout is "Cu(HT120) //F1:TaN (HT6nm) //G1: BD (HT120nm) // B12: BD (HT120nm) //B11: NDC (HT50nm) //B10: BD (HT340nm) //B9: NDC (HT50nm) //B8: BD (HT340nm) //B7: NDC (HT50nm) //B6: BD (HT160nm) //B5: NDC (HT30nm) //B4: ILD HDP(HT350nm) //B3: HTN(HT45nm) // B2: Ni-Silicide(HT7.5nm) //Substrate:Si"; M5 layout is "Cu(HT120) //F1:TaN (HT6nm) //G1: BD (HT120nm) //B14: BD (HT120nm) //B13: NDC (HT50nm) //B12: BD (HT340nm) //B11: NDC (HT50nm) //B10: BD (HT340nm) //B9: NDC (HT50nm) //B8: BD (HT340nm) //B7: NDC (HT50nm) //B6: BD (HT160nm) //B5: NDC (HT30nm) //B4: ILD HDP(HT350nm) //B3: HTN(HT45nm) // B2: Ni-Silicide(HT7.5nm) //Substrate:Si";

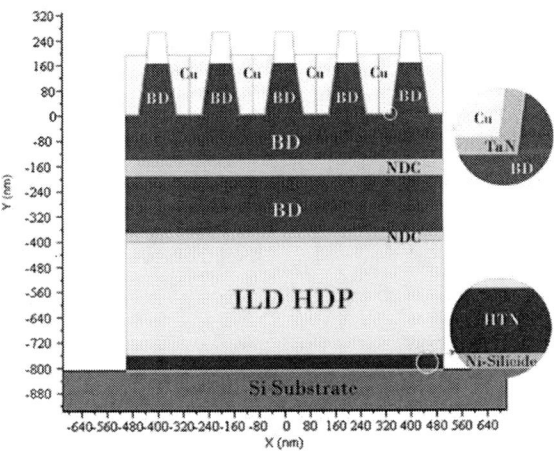

Figure 1. Schematic of M2 Stack

Within complex parameters of metal levels and dielectric layers, after sensitivity and measurement requirement analysis, MCD, SWA, HT of BD grating, Cu thickness are critical and chose to be floated.

Spectra fitting Sample

Figure 2 is the sample of library regression of M2. Red curves are the production spectra and green points show the library modeling fitting points. The nearer they are, the higher GOF (Goodness of Fit) and more confident results we will get. And, top-left picture are the real-time profile report of current measurement location.

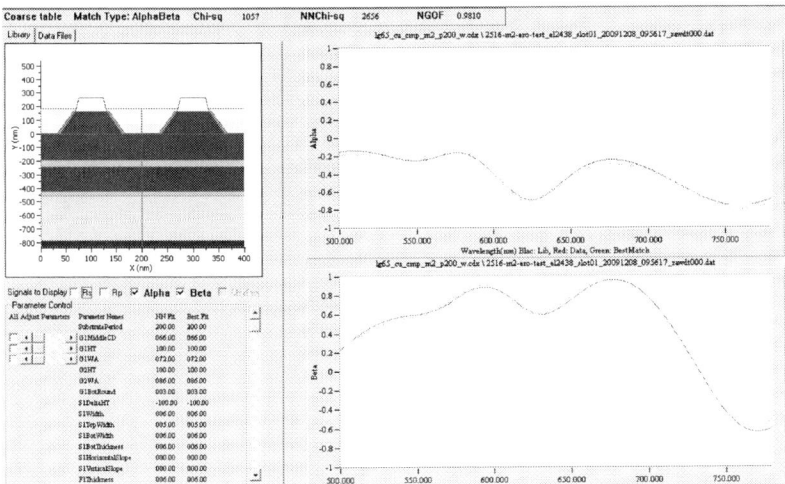

Figure 2. Library offline simulation result snapshot of M2 by SpectraAnalyzer 4.2

Average GOF level of BEOL Cu layers SCD applications is about 0.80, that is lower than most of FEOL GOF. What is the special of BEOL? In contrast to the relatively simple stacks that comprise a FEOL structure, BEOL layers are typically complicated structures with a large number of underlying layers.

CORRELATION TO BASELINE METROLOGY

The common strategy of determining the applicability of the metrology solution is to run parallel measurements with the baseline metrology method. Therefore, in this paper, Cu thicknesses which reported by SCD are compared with the corresponding ultrasonic data for each Cu erosion layers, both production and DOE correlation are studied.

Production Correlation between SCD and Ultrasonic

In terms of point to point correlation of several production lots, as Figure 3 and Figure 4, basically, SCD show the coincident trend with Ultrasonic, including M2 to M5. Even more, for M3 and M4, SCD give more stable and lower range measurement.

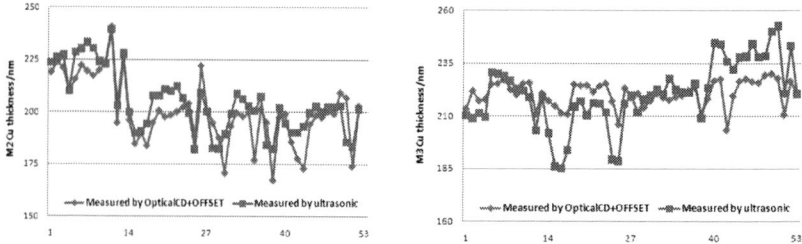

Figure 3. Cu thickness site by site comparison of SCD and ultrasonic. Left is M2 and right is M3.

Shown in figure 4, we collected more data of M4 and M5 to track the libraries performance of such complicated structure of bottom layer. Overall, the matching results are good to be accepted. Y axes range of each figures are determined by process control specification. As follows, if SCD was released to on line measurement and production data were comprised to SPC (Statistical Process Control) chart, Cpk (Complex Process Capability index) will never get worse because new technology metrology system involved. This is the key point for new method adoption.

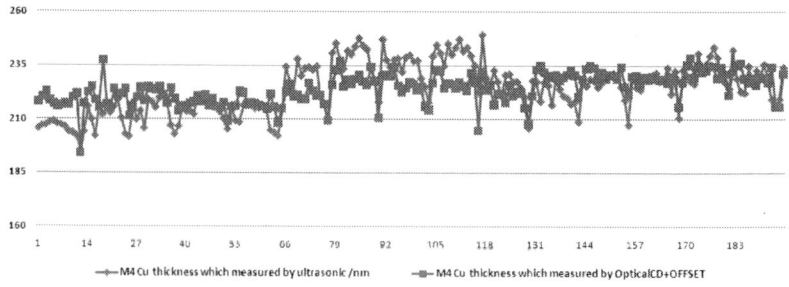

Figure 4. Cu thickness site by site production M4 comparison of SCD and ultrasonic

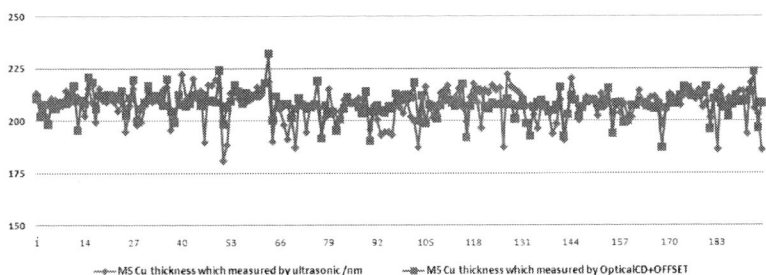

Figure 5. Cu thickness site by site production M5 comparison of SCD and ultrasonic

DOE Correlation among SCD, Ultrasonic and TEM

To demonstrate the enough window coverage of each SCD libraries, three DOE (Design of Experiment) wafers were applied in this paper, the 1st one is the analog of baseline production, the 2nd one maximizes Cu thickness and 3rd one minimizes Cu thickness of possible process window. We gave out a sample as Figure 5 to present correlation R^2 algorithms. What is more, R^2 correlation of SCD, ultrasonic and TEM for each layers are summarized as table 1.

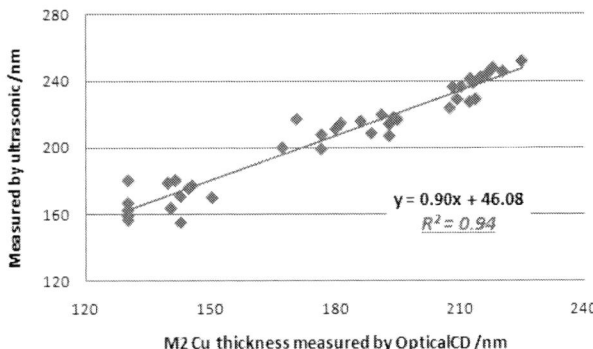

Figure 6. SCD versus Ultrasonic plot for M2 sample to interpret R^2 in below table

It can be observed that there are excellent correlation between SCD, ultrasonic and TEM, most of the R2 are more than 0.90, only except the 0.79 of SCD versus ultrasonic in M3. But, M3 careful examination data of SCD reveals perfect correlation with TEM, R^2 is 0.99. Therefore, compare with SCD vs. TEM, it appears that ultrasonic data shows worse correlation with TEM results in this extraordinary point.

TABLE I. Correlation of SCD with ultrasonic and TEM for M2 to M5 Cu thickness

R^2	M2	M3	M4	M5
w. Ultrasonic	0.95	0.79	0.96	0.98
w. TEM	0.91	0.99	0.98	0.99

SUMMARY AND CONCLUSION

In this paper, we have demonstrated the use of SCD for measuring 65nm technology nodes logic production structures for multiply BEOL Cu CMP erosion layers applications. We compared the correlation performance between traditional technologies. We showed the benefit of SCD as a reliable, high throughput and non-destructive BEOL metrology solution with credible production matching performance. Furthermore, excellent correlation on DOE wafers confirm that SCD can be utilized for multiple Cu CMP layers monitor with complex underlying structures.

Acknowledgments

Give profound appreciation to Xin Zhang and SMIC YE team for their contribution to this project. Furthermore, the author gratefully acknowledges Francis Jen, Bob Dong and Jim Li for the fruitful discussion between us and for their encouragement toward successful outcome of this paper.

References

1. H. Helneder, H. Korner, A. Mitchell, M. Schwerd, and U. Seidel, Microelectronic Engineering, **55, 257–268** (2001).
2. S. Lakshminarayanan, J. Steigerwald, D. T. Price, M. Bourgeois, T. P. Chow, R. J. Gutmann, and S. P. Murarka, IEEE, Electron Device Lett., **vol. 15, no. 8, 307-309** (1994).
3. L. Towidjaja, C. Raymond, and M. Littau, SPIE, Vol. 6152 61521X-1 (2006).

*Contact Author: Jolly.zhao@kla-tencor.com; Tel: 86-10-51062009;

ECS Transactions, 34 (1) 955-960 (2011)
10.1149/1.3567698 ©The Electrochemical Society

Spectral Sensitivity Analysis of OCD based on Muller Matrix Formulism

Shi Yaoming, Zhang Zhensheng, Liu Guoxiang, Liu Zhijun, Xu Yiping

Raintree Sicentific Instrument (Shanghai) Corporation, 68 Huatuo Road, Shanghai, China

The spectral sensitivities of scatterometry based on Mueller matrix formulism at the technology node of 65 nm and beyond were studied. The advantage of Mueller matrix based polarized light scattering technique over conventional ellipsometry technique in measuring patterned wafers was present. RCWA method was used to simulate the 2D gratings scattering. As the analysis shown, Mueller measures produce more signals than SE or SR measurement does and Mueller spectral sensitivities to structure features variation are stronger in most cases.

Introduction

Recently, people start to use Mueller matrix based polarized light scattering experiments to measure the critical dimensions or profile information in every step of lithographical and etch processes in semiconductor industry (1,2,3,4,5,6,7). In this paper, we focus on spectral sensitivity at the technology node of 65 nm and beyond.

In the linear intensity regime, the basic phenomenological equation for polarized light scattering is (8,9,10):

$$S'_a = \sum_{b=1}^{4} M_{a,b}(k,\theta_{inc},\phi_{inc},\theta_{scat},\phi_{scat})S_b \qquad [1]$$

Where S'_a is the 4-component Stokes vector representing the scattered light, $M_{a,b}(k, \theta_{inc}, \phi_{inc}, \theta_{scat}, \phi_{scat})$ is the 4-by-4 Mueller scattering matrix representing the scattering sample, and S_b is the 4-component Stokes vector representing the incident light. The matrix depends on the wave vector length $k=2\pi/\lambda$, the polar and azimuth angles ($\theta_{inc}, \theta_{inc}$) for the incident light, the polar and azimuth angles ($\theta_{scat}, \theta_{scat}$) for the scattered light, and the structure of the scatterer.

For a beam of polarized light traveling towards wafer plane, we can measure its intensity right after a polarizer. If we place an s linear polarizer $P_\perp$, then the intensity we measured will be called I_s. Similarly we use symbol I_p for light intensity after a p linear polarizer P_p, I_{D+} for intensity after a positive diagonal linear polarizer P_{D+} (45 degree from s and p), and I_{D-} for intensity after a negative diagonal linear polarizer P_{D-} (-45 degree from s). If we place a right (left) circular polarizer P_R (P_L), then the intensity we measured will be called I_R (I_L).

The Stokes vector components are related these various intensities via the following relations:

955

$$S_1 = I_{total} = I_p + I_s = I_{D+} + I_{D-} = I_R + I_L \qquad [2]$$
$$S_2 = I_p - I_s, \quad S_3 = I_{D+} - I_{D-}, \qquad S_4 = I_R - I_L \qquad [3,4,5]$$

The advantage of Mueller matrix based polarized light scattering technique over conventional ellipsometry technique in measuring patterned wafer is similar to the advantage of reflectometry technique over unpolarized light scattering technique in measuring the un-patterned wafer. It basically unscrambled the original signals into more primitive signals so that the cancel out effect is greatly reduced. Furthermore, 16 elements of Mueller matrix contained all the information about the scatter that you can extract from the linear polarization scattering. Nothing more can be extracted. If you only measure a few of them or a few linear combinations of them, then you are not making the optimized measurements and calculations.

Using unpolarized light scattering technique, we measure the total scattering light intensity, given the total incident light intensity; namely,

$$I'_{total} = R_{total} I_{total} \qquad [6]$$

By comparing [6] with [1] we found out that:

$$R_{total} = M_{1,1} \qquad [7]$$

So the information in only one element of the Mueller matrix, $M_{1,1}$, is used in deducing the scatter structure. Using refletometry technique, we measure the scattered light intensity in s direction and p direction, given the incident light in s direction or p direction; namely,

$$I'_s = R_s I_s, \qquad I'_p = R_p I_p \qquad [8,9]$$

By comparing [3,4,5] with [1] and [2] we found out that:

$$R_p = \left(M_{1,1} + M_{1,2} + M_{2,1} + M_{2,2}\right)/2 \qquad [10]$$

$$R_s = \left(M_{1,1} - M_{1,2} - M_{2,1} + M_{2,2}\right)/2 \qquad [11]$$

In general two pieces of information (R_p, R_s) is better than one piece of information (R_{total}) when they are used in determining the scatter properties. Using ellipsometry technique, we measure two ratios (α, β) of scattered intensities and it can be shown that (α, β) are related to Mueller matrix elements via:

$$\alpha = \frac{M_{12} + M_{22}\cos(2A) - M_{32}\sin(2A)}{M_{11} + M_{21}\cos(2A) - M_{31}\sin(2A)} \qquad [12]$$

$$\beta = -\frac{M_{13} + M_{23}\cos(2A) - M_{33}\sin(2A)}{M_{11} + M_{21}\cos(2A) - M_{31}\sin(2A)} \qquad [13]$$

Where, A stands for the analyzer angle of the Ellipsometry system.

Using Mueller matrix based technique, we make the M_{11} and the normalization factor and normalize all the other 15 Mueller matrix elements and obtain 15 ratios:

$$m_{ij} = \frac{M_{ij}}{M_{11}} \quad (i, j = 1,2,3,4) \tag{14}$$

In general 15 pieces of information ($m_{a,b}$, $a*b \neq 1$) is better than two piece of information(α, β) when they are used in determining the scatter properties.

Sensitivity Analysis

In this section, several numerical experiments are carried out to quantify the conjectures that are outlined in section I.

We use RCWA method to numerically solve the light scattering problem as accurate as possible. Figure 1 shows the typical Mueller matrix based polarized light scattering measurement setup. Figure 2 shows the cross section of the first sample 2D grating profile in our numerical experiments.

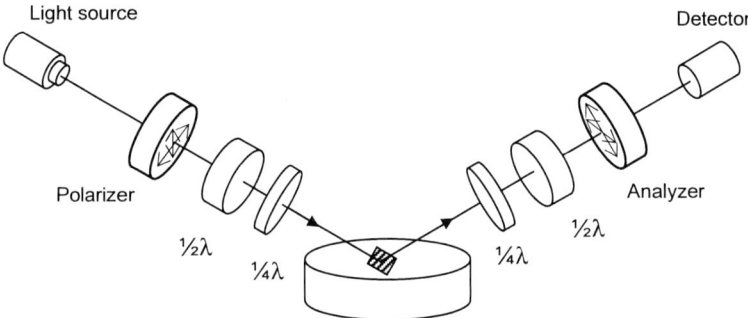

Figure 1. A typical Mueller matrix based polarized light scattering measurement setup with rotating quarter-wave plates and half-wave plates.

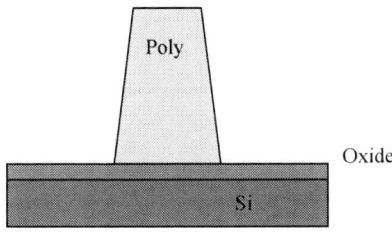

Figure 2. Cross section diagram of the sample.

In Table I below, we provide the incident light information as well as the grating dimensions etc.

TABLE I. Basic parameters of gratings and simulation conditions.

ITEM	Description
Nominal CD	65 nm
Nominal HT	100 nm
Side Wall Angles (SWA)	88°
The thickness of SiO2	2nm
Typical pitches (L:S)	130nm (1:1); 195nm (1:2)
Wavelength (λ) range	250-800 nm, with 5 nm step size
Angle of incidence (AOI)	20°; 69°
Azimuth Angle	0°
Analyzer Angle	20 deg

The 2D grating is placed in x-z plane. Under classical mount, the incident plane is also the x-z plane. The incident light and scattered light are also confined in this plane. The Jones matrix that represents the scatter (or the grating in this case) is diagonal and the corresponding Mueller matrix has the following simplified form:

$$
M = \begin{pmatrix} M_{11} & M_{12} & 0 & 0 \\ M_{12} & M_{11} & 0 & 0 \\ 0 & 0 & M_{33} & M_{34} \\ 0 & 0 & -M_{34} & M_{33} \end{pmatrix}
$$

[15]

We noticed that this Mueller matrix has the helicity property (off-diagonal quadrants are zero) and Perrin symmetry (9). The information we can extract form the unpolarized light scattering is related to Mueller matrix elements via:

$$ R_{total} = M_{11} $$

[16]

The information we can extract from the Spectral Reflectometry are related to Mueller matrix elements via:

$$ R_p = (M_{11} + M_{12})/2, \qquad R_s = (M_{11} - M_{12})/2 $$

[17,18]

The information we can extract from the Spectral Ellipsometry are related to Mueller matrix elements via:

$$ \alpha = \frac{M_{12} + M_{11}\cos(2A)}{M_{11} + M_{12}\cos(2A)}, \qquad \beta = \frac{M_{33}\sin(2A)}{M_{11} + M_{12}\cos(2A)} $$

[19,20]

We will use the MSE to judge if signal (R_{total}) or signals (R_p, R_s) is better in the sensitivity analysis of given profiles appeared in lithography or etching in the manufacturing of Vary Large scale Integrate Circuit (VLIS). We will carry out similar comparison for signal (α, β) against signals ($m_{a,b}$, $a*b \neq 1$).

Here is the definition of MSE for signal B:

$$MSE(B, \Delta SWA)=\left(\frac{1}{N}\sum_{i=1}^{N}\left(B(MCD, HT, SWA, \lambda_i)-B(MCD, HT, SWA+\Delta SWA, \lambda_i)\right)^2\right)^{1/2} \quad [21]$$

Using the spectral with nominal parameters, $B(MCD,HT,SWA,\lambda_i)$, as the base line, we can calculate the MSE for spectral with bias ΔCD, $MSE(B, \Delta CD)$. For the parameters variation shown in Table II., the corresponding spectral variations are shown in Figures 3. We also plotted $MSE(B, n\Delta p)$ (for $p=MCD,HT,SWA$) to show the signal sensitivities with respect to various variable changes in Figures 4.

TABLE II. The variables change and nominal settings.

Variable	Bias Δ	x(nominal)	$\Delta x / x$
BCD	+0.2nm,	65 nm	0.77%
HT	+0.2nm	100 nm	0.5%
SWA	+0.2°	88°	0.23%

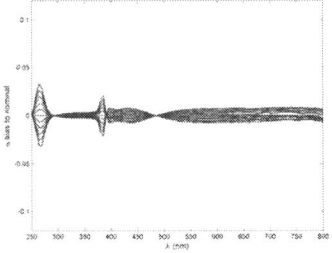 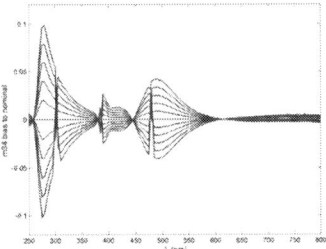

Figure 3. The spectra comparison with CD bias from -1 nm to 1 nm and step of 0.2 nm at Pitch = 195 nm, AOI =69 deg, and phi = 90 deg.

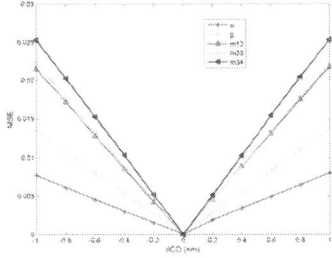 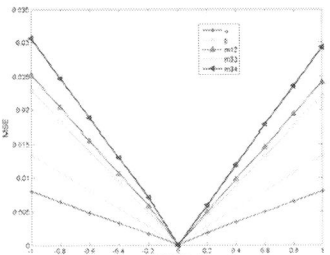

Figure 4. Spectra MSEs change to variables bias at Pitch = 195 nm, AOI =69 deg, and phi = 0 deg. Left: CD bias from -1 nm to 1 nm and step of 0.2 nm ; Right: SWA bias from -1deg to 1deg.

The MSE values between the structure of nominal variable values and that of variable bias are shown in Table III.

TABLE III. MSE data lists at AOI = 69 deg and phi = 0 deg.

Dimension unit: nm
Angle unit: deg

pitch	variable	bias	Rs	Rp	Rtotal	alpha	beta	m12	m33	m34
130	CD	0.2	0.0026	0.0014	0.0016	0.0022	0.0070	0.0044	0.0098	0.0072
	HT	0.2	0.0030	0.0008	0.0016	0.0033	0.0048	0.0065	0.0058	0.0055
	SWA	0.2	0.0028	0.0012	0.0016	0.0024	0.0072	0.0051	0.0100	0.0081
195	CD	0.2	0.0017	0.0010	0.0008	0.0019	0.0029	0.0045	0.0045	0.0051
	HT	0.2	0.0024	0.0006	0.0011	0.0020	0.0020	0.0054	0.0038	0.0044
	SWA	0.2	0.0020	0.0008	0.0009	0.0018	0.0028	0.0050	0.0045	0.0059

Discussion

From the figures and tables in section II, we show that
(1) SR measures produce more signals (R_p, R_s) than unpolarized light scattering measurement does (R_{total});
(2) SR measurements (R_p, R_s) have better sensitivities with respect to minute structural changes than unpolarized light scattering measurement (R_{total}) does.
(3) Mueller measures produce more signals (m_{12}, m_{33}, m_{34}) than SE measurement does (α, β);
(4) Mueller measurements (m_{12}, m_{33}, m_{34}) have better sensitivities with respect to minute structural changes than SE measurement (α, β) does in most cases.

References

1. T. Novikova, A. D. Martino, S. Hatit, and B. Drevillon, *Applied Optics*, **45**, p.3688(2006).
2. M. Foldyna, A. D. Martino, E. Garcia-Caurel, and etc., *Eur. Phys. J. Appl. Phys.*, **42**, p.351 (2008).
3. T. Novikova, A. D. Martino, and etc., *Optics Express*, **15**, p.2033 (2007).
4. R. Silver, T. Germer, R. Attota, B. M. Barnes, B. Bunday,J. Allgair, E. Marx and J. Jun, *Proc. of SPIE*, **6518**, p.65180U-1 (2007).
5. M. Foldyna, A. D. Martino, R. Ossikovski, E. Garcia-Caurel , C. Licitra ,*Optics Communications,* **282**, p.735(2009).
6. Y. N. Kim, J. S. Paek, S. Rabello, S. Lee, J. Hu, Z. Liu, Y. Hao and W. McGahan, *Optics Express*, **17**, p.21336 (2009).
7. M. Foldyna, A. D. Martino, E. Garcia-Caurel, R. Ossikovski, F. Bertin, J. Hazart, K. Postava,and B. Drevillon1, *phys. stat. sol. (a,)* **205**, p.806(2008).
8. Y. Shi and W. M. McClain, *J. Chem. Phys.*, **93**, p.5605(1990).
9. Y. Shi, W. M. McClain, and D. Tian, *J. Chem. Phys.*, **94**, p.4726(1991).
10. W. M. McClain, W. Jeng, B. Pati, Y. Shi, and D. Tian, *Applied Optics*, **33**, p.1230(1994).

A method to determine process capability Cpk and corresponding percentage of non-conforming for non-normally distributed and limited production data

Siyuan Frank Yang
Semiconductor Manufacturing International Corporation
Shanghai, China

For normal distributions, C_{pk} is determined by a popular formula $Cpk = \min\{Cpu, Cpl\} = \min\{\dfrac{(USL - \overline{X})}{3\sigma}, \dfrac{(\overline{X} - LSL)}{3\sigma}\}$. For non-normal distributions, the current practice in determining C_{pk} is the percentile method, i.e. to using 50% percentile (i.e. P(0.5)) to replace the average, (P(0.99865) – P(0.5)) to replace 3σ for C_{pu} and (P(0.5) – P(0.00135)) to replace 3σ for C_{pl}. This seems to be a reasonable modification of C_{pk} formula to accommodate the non-normal distributions and has been accepted and widely used by many practitioners in industries. In order to evaluate process performance, the corresponding percentage of non-conforming (%NC) is a critical quantity and should be determined at the same time when C_{pk} is calculated for non-normal distributions. However, we are unable to use regular percentile method (such as the one popularly used in Excel) to estimate the upper and lower percentiles for limited process data. The same problem also occurs to the %NC when Spec limits are far away from the center location for limited production data. This work provided a practical means to resolve this problem by employing an extrapolation method based on the current percentile method to obtain very high and very low percentiles with limited production data assuming the probability curve is still straight up to lower or upper percentiles or Spec limits. This assumption has been further discussed to assure practicability of its applications.

Introduction

Process capability indices are used to establish the relationship between the actual process performance and the manufacturing specifications. They are practical tools for successful quality evaluation and improvement activities. The first process capability index C_p (when the process mean is on the center of the upper and lower Specification limits) was formalized by Japanese and later was rapidly being adopted in the United States [1] since more than two decades ago.

$$C_p = \frac{USL - LSL}{6\sigma} \quad \dots\dots\dots\dots\dots\dots\dots\dots\dots\dots\dots\dots\dots\dots\dots (1)$$

where USL is the upper specification limit, LSL is the lower specification limit, and σ is the process standard deviation.

The percentage of non-conforming of the products manufactured (%NC) is an important and direct index for the manufacturing quality. When the process is perfectly centered at the center of the specification range and the process is normally distributed, the %NC has the following one-to-one relationship with C_p.

$$\%NC = 2\Phi(-3C_p) \quad \dots\dots\dots\dots\dots\dots\dots (2)$$

where $\Phi(x)$ is the cumulative distribution function of the standard normal distribution. Its numerical value can be easily obtained through formula p=NORMSDIST(x) in Excel (Its inverse function is x=NORMSINV(p) in Excel). When C_p = 1.0, we have %NC=2700ppm (parts per million) and %NC=63ppm when C_p=1.33.

In reality, the process always has certain deviation from the targeted center within the specification range. In order to account the deviation of process mean (μ) from the target value (μ_0), an index C_{pk} was proposed again by Japanese and widely used in practice, which is defined as:

$$C_{pk} = \min\left\{\frac{USL - \mu}{3\sigma}, \frac{\mu - LSL}{3\sigma}\right\} = \min\{C_{pu}, C_{pl}\} \quad \ldots\ldots\ldots\ldots\ldots\ldots (3)$$

or equivalently

$$C_{pk} = \frac{H_w - |\mu - \mu_0|}{3\sigma} = C_p * (1 - k) \ldots\ldots\ldots\ldots\ldots\ldots\ldots\ldots\ldots\ldots (4)$$

where H_w=(USL-LSL)/2 is the half width of the specification interval, μ_0=(USL+LSL)/2 is the mid-point of the specification range, and $k = \dfrac{|\mu - \mu_0|}{H_w}$ is the mean shift relative to the half width of the specification interval.

The relationship between the percentage of non-conforming (%NC) and C_{pk} is also an important aspect in quality evaluation. Under the assumption of normal distribution, it is easy to obtain the equation of %NC expressed in terms of the process mean (μ) and the process standard variation (σ) for given USL and LSL.

$$\%NC = 1 - \Phi(\frac{USL - \mu}{\sigma}) + \Phi(\frac{\mu - LSL}{\sigma}) \quad \ldots\ldots\ldots\ldots\ldots\ldots\ldots\ldots (5)$$

Also it can be easily converted to an equivalent equation for %NC expressed by C_{pk} and C_p as below.

$$\%NC = \Phi(-3C_{pk}) + \Phi[-3C_{pk} * (\frac{2C_p}{C_{pk}} - 1)] \quad \ldots\ldots\ldots\ldots\ldots\ldots\ldots (6)$$

When C_p=2 and process mean has 1.5σ shift, we have C_{pk}=1.5 and the corresponding

$$\%NC = \Phi(-3*1.5) + \Phi[-3*(\frac{2*2}{1.5} - 1)] = \Phi(-4.5) + \Phi(-7.5)$$

$$= NORMSDIST(-4.5) + NORMSDIST(-7.5) = 3.4\,ppm$$

This is the well known six sigma quality level of process performance when C_p=2 and process mean has 1.5σ shift [2].

When there is no process mean shift from the targeted center (μ_0), we have C_{pk}=C_p, therefore Eq. (2) is then easily obtained from Eq. (6).

The C_{pk} formula (Eq.(3) and Eq. (4)) and %NC formula (Eq.(5) and Eq.(6)) are widely used in practice however they assume normal distribution for process data. For non-normal distribution, a widely accepted approach for PCI (Process Capability Indices) computation is to use percentile in probability plot [3]. For Cp, the denominator of Eq. (1) is replaced with (upper 0.135% point-lower0.135% point), i.e.

$$C_p = \frac{USL - LSL}{(upper\,0.135\%\,po\,int - lower\,0.135\%\,po\,int)} = \frac{USL - LSL}{P(0.99865) - P(0.00135)} \ldots\ldots\ldots (7)$$

where $P(0.99865)$ is the 99.865 percentile and $P(0.00135)$ is the 0.135 percentile in the probability plot respectively.

Correspondingly the index Cpk is defined as:

$$C_{pk} = \min\{C_{pu}, C_{pl}\} = \min\{\frac{USL - P(0.50)}{P(0.99865) - P(0.50)}, \frac{P(0.50) - LSL}{P(0.50) - P(0.00135)}\} \dots\dots\dots(8)$$

where P(0.50) is the median in the probability plot.

If we are able to find a good distribution form through modeling [4,5,6] for the process data and obtain a very satisfactory fit, we then are able to obtain accurate measures of these three percentiles used in Eq. (7) and Eq.(8). This involves modeling using alternative probability models, such as the Weibull or Gamma distributions and so on. However there exists a disadvantage of distribution modeling since not all realistic process data can be modeled by existing known distribution forms. In such cases, probability plotting will be a practical approach. However, accurate estimation of the P(0.99865) and P(0.00135) require large sample size such as >1/0.00135=740 data points for empirical percentile method used in some very popular software such as Excel (Percentile[(array),p]). In reality, limited data points less than 740 to calculate PCI is very typical. This paper provides a method of determination of percentile using extrapolation approach for limited process data which cannot be fitted by regular various distribution models.

Determination of Percentiles Using Existing Probability Models to Fit Process Data

A set of Gamma distributed data (150 data points) is used to demonstrate the determination of percentiles using Gamma distribution form to fit and to calculate P(0.99865) and P(0.00135).
This data set shows obvious non-normality since it is not a straight line in the normal probability plot and not symmetric in the histogram as shown in Fig. 1.

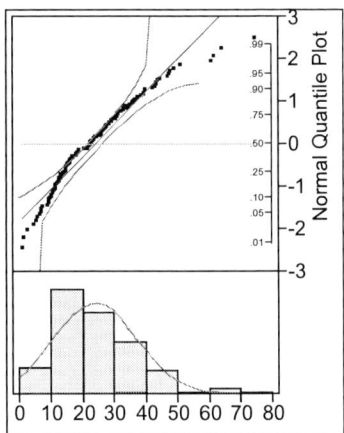

Fig. 1. The normal probability plot and histogram show the non-normality of the data set.

We employ Weibull ++7 [7] to make Gamma probability plot and modeling to fit the shape and scale parameters K and θ of Gamma pdf (probability density function) [8].

$$f(x;\theta,k) = \frac{1}{\Gamma(k)*\theta}\left(\frac{x}{\theta}\right)^{k-1} e^{-x/\theta} \dots\dots\dots\dots\dots\dots\dots (9)$$

Weibull ++7 uses a "reparameterization" with parameter $\mu = \ln(\theta * k)$ to fit the data for the convenience and obtains μ=1.9587 and k=3.2562. We then utilize the model to obtain the upper percentile P(0.99865)=80.0 shown in Fig. 2.

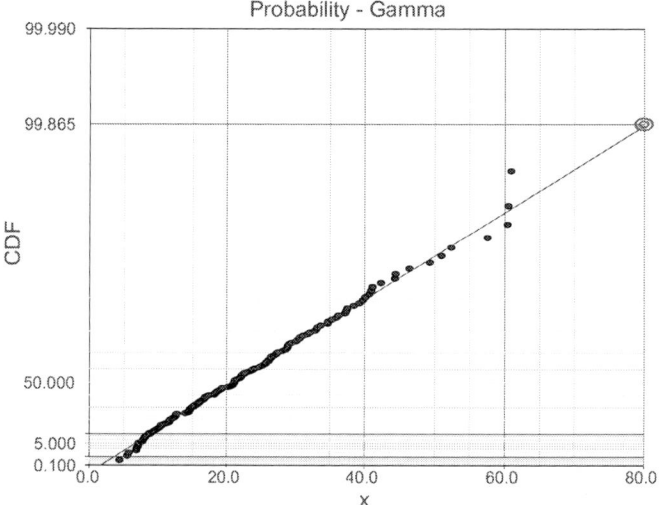

Fig. 2. Probability plot and determination of upper percentile by Weibull ++7.

Determination of Percentiles using Extrapolation Methods for Limited non-Normally Distributed Data

The formula Percentile[(Array),p] in Excel has been widely used to calculate percentiles of p*100% from data array. When we use this formula to calculate P(0.99865)= Percentile[(Array),0.99865] for the above same data set, we obtain P(0.99865)=60.8. This is the same as P(0.99) Percentile[(Array),0.99]. This is simply due to the calculation of P(0.99865) needs more than 740 data points however this data set only has 150.

We developed a technique which determine percentile for limited data with any type distribution form (i.e. distribution free). From the normal probability plot Fig. 3 plotted from the above same 150 data points, the highest percentage among these 150 data points is 99.667% therefore we need do extrapolation to 99.9865%. We assume that the unknown high end portion of the probability plot be linear which is fairly good assumption for most realistic cases. We choose two points (42.339, 93%) and (52.386, 97%) in the probability plot to draw a straight line and extend it to 99.986%. The nature of the linearity enables us to obtain P(0.99865) as

$$P(0.99865) = P(0.93) + k *[P(0.97) - P(0.93)] \quad\quad\quad(10)$$

where the proportion constant k can be obtained from same extrapolation from a standardized normal distribution, i.e.
NORMSINV(0.99865)= NORMSINV(0.93)+k*[NORMSINV(0.97)-NORMSINV(0.93)] and then

$$k = \frac{\text{NORMSINV}(0.99865) - \text{NORMSINV}(0.93)}{\text{NORMSINV}(0.97) - \text{NORMSINV}(0.93)} = \frac{3.000 - 1.476}{1.881 - 1.476} = 3.7634$$

Therefore, using Eq. (10), we can obtain

P(0.99865)=P(0.93)+k*[P(0.97)-P(0.93)]=42.339+3.7634*[52.386-42.339]=80.15

which is very close to the early result of 80.0 obtained through distribution modeling.

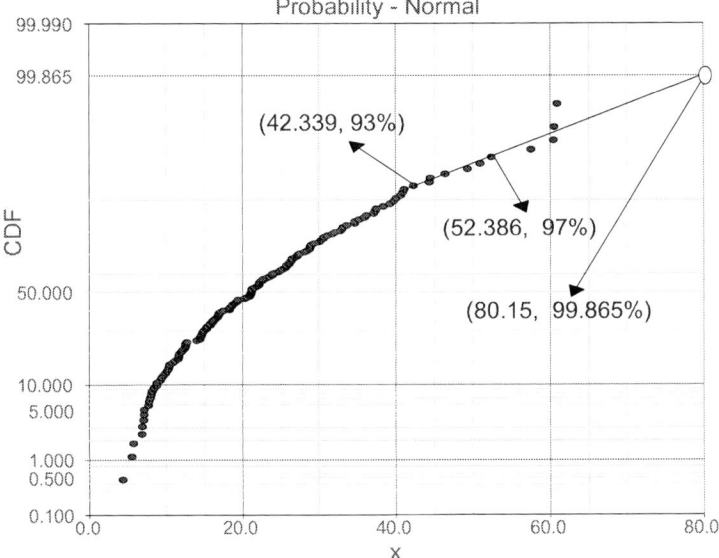

Fig. 3. Extrapolation on normal probability plot

For the lower percentile P(0.00135), we can use the same methodology to do linear extrapolation and obtain the designated percentile like P(0.00135).

For non-normal distribution, the %NC calculations using Eq.(5) and Eq.(6) are no longer valid. We need to employ distribution modeling to achieve this. However, for those distributions which cannot be modeled by regular distribution forms and also for process data with limited sample size, the above technique of determination of percentile using extrapolation method is also very useful here.

Conclusions

Process capability indices (PCI) and percentage of non-conforming (%NC) in production data are important indices to evaluate the production quality performance. The widely used Cpk and %NC calculations require assumption of normally distributed production data. For non-normal distribution, modeling is used to estimate percentiles to calculate Cp and Cpk however sometimes production data cannot be fitted by regular known distribution forms therefore we have to resort to the probability plot method. When the production data cannot be modeled by regular known distribution forms and at the same time the data sample size is limited, the above two techniques are not valid. This paper developed a new technique employing extrapolation in normal probability plot assuming linear behavior at high or low end of the probability plot to have resolved this particular difficult case. This technique has a disadvantage that it is somewhat objective procedure since different analysts might choose two different locations to determine linear extension and hence they will arrive at somewhat different results. However, this can be improved significantly by formal linear regression to obtain the linear equation based on a segment of multiple data points chosen for this purpose.

References

[1] Kane, V.E. Process capability indices, *Journal of Quality Technology*, 18(1), 41-52, 1986.

[2] Stamatis, D.H. Six Sigma and Beyond, Vol. 1: Foundations of Excellent Performance, p.354, St. Lucie Press, Boca Roton, 2002

[3] Pearn, W.L, and Kotz, Samuel, Encyclopedia and Handbook of Process Capability: A Comprehensive Exposition of Quality Control Measures, Series on Quality, Reliability and Engineering Statistics, Vol. 12; World Scientific, New Jersey. 2006

[4] Dudewicz, E.J. and Mishra, S.N. Modern Mathematical Statistics, John Wiley, New York, 1988.

[5] Hahn , Gerald J., Shapiro , Samuel S. Statistical Models in Engineering, Wiley, New York, 1994.

[6] Kotz, S. and Lovelace, C.R. Process Capability Indices in Theory and Practice, Arnold, London, U.K., 1998.

[7] ReliaSoft, A Life Data Analysis Software by ReliaSoft Corporation, ReliaSoft Plaza, 115 Souty Sherworrd Village Drive, Tucson, Arizona, 85710. USA. http://www.ReliaSoft.com , 2005.

[8] Bain, L.J. & Engelhardt, M. Introduction to Probability and Mathematical Statistics, 2nd Ed, Duxbury Thomson Learning, 1992.

High Voltage Device Negative Bias Temperature Instability Improvement with Different Process Conditions

P. C. Sim, S. S. Koo and D. K. Pal

Technology Department, X-FAB Sarawak Sdn. Bhd., 1 Silicon Drive, Sama Jaya Free Industrial Zone, 93350 Kuching, Sarawak, Malaysia

Drastically device dimension shrinkage and rigorous requirement in automotive era puts Negative Bias Temperature Instability (NBTI) at the forefront of reliability issue recently. The PMOS parametric degradation during negative bias high temperature aging can depend on many process variables of the manufacturing flow. A study was carried out to explore the process related dependencies for high voltage PMOS transistor and to increase the device robustness against NBTI stress. In this papers, the process impact on the NBTI degradation were discussed. This investigation work provides methods for significant suppression of the NBTI degradation with silicon rich oxide (SRO) inter layer dielectric (ILD) liner and two-step gate oxidation.

Introduction

The NBTI is a PMOS degradation mechanism due to positive charge build up and donor type interface state generation at the $Si\text{-}SiO_2$ interface under the influence of an applied low negative gate voltage, which leads to a time dependent shift in threshold voltage, channel mobility of the transistors and it induces parasitic capacitances to degrade the device performance. The continuing aggressive MOSFET miniaturization trend without a proportional reduction in supply voltage and stringent automotive requirements have resulted in an increased reliability challenges for modern ICs toward this wear out mechanism. The theoretical background of NBTI is still a hot topic within international literature even though this aging phenomenon has been known for decades [1], [2]. Huge effort has been focused on understanding the degradation mechanisms, models and the effect of the NBTI damage to the device characteristics [3]-[5]. Various improvement techniques have also been reported in [6]-[10] to enhance device NBTI immunity.

The PMOS transistor NBTI performance is influenced by a variety of process mechanisms that cause fixed charges and traps in the oxide as well as at the $Si\text{-}SiO_2$ interface. Thick oxide transistors were found to be more susceptible to NBTI than thin oxide counterparts even though both devices have comparable electric field strength. It was experimentally demonstrated that the degree of process-induced NBTI lifetime degradation in high voltage PMOS device in a $0.35\mu m$ CMOS technology depends on several factors, such as the $Si\text{-}SiO_2$ interface condition, the influence of hydrogen and nitrogen during processing. This paper summarizes the evaluation results of the device NBTI lifetime dependencies on different split conditions during the silicon surface cleaning and "treatment", gate oxidation, and back-end wafer process which show an order of improvement in NBTI lifetime.

Fabrication & Experimental Details

The devices examined in this study were high voltage PMOS transistors fabricated on a p-type silicon substrate in a 0.35μm technology with LOCOS process and 40nm/7nm dual gate oxide growth. NBTI stress was applied with the gate held at a low constant negative bias under elevated temperature of 125°C while the source, drain and body were gounded. The threshold voltage (Vt) was considered as the NBTI test parameter for lifetime extraction throughout our study. A series of experiments were performed to determine the process impact on the NBTI damage to the device.

The evaluated wafers for the study of silicon surface condition influence on the device NBTI performance were split into diluted SC1 cleaning and shorter SC1 dipping time during first gate oxidation pre-cleaning as well as with additional furnace oxide growth and removal to provide a better surface condition before gate oxide formation. Besides, wafers were also processed with and without N_2 annealing during gate oxidation to investigate the impact of nitrogen contents to NBTI degradation. The gate oxidation was also split into three categories: dry long process, dry-wet-dry short process and two steps gate oxide to check the gate oxide growth process effect.

In another case, the impact of different recipes for the ILD liner deposition is investigated with low temperature oxide (LTO) and SRO splits. Systematic sets of experiments have also been carried out to understand influence of hydrogen annealing on NBTI performance. The wafers were separately annealed with pure hydrogen and forming gas (nitrogen + hydrogen). Pure hydrogen sintering is performed at three different positions in the process sequence: post-contact formation, post-first metal layer deposition, and post passivation deposition. In additions, hydrogen gas was added during source-drain rapid thermal annealing (SD RTA) to understand the early device passivation effect.

Results & Discussion

The detailed analysis of the influence of different process on NBTI degradation leads to an understanding of the mechanisms behind the degradation. The direction to improve the device NBTI lifetime is described sequentially.

Pre-Gate Oxidation Preparation

The $Si-SiO_2$ interface is one of the major concerns in device NBTI reliability performance. Silicon surface roughness as a possible cause gate-oxide defect has become a possible contribution to NBTI degradation. The SC1 ($NH_4OH/H_2O_2/H_2O$) cleaning solution has a direct impact on the silicon surface micro-roughness where NH_4OH acts as the etchant of the oxide while H_2O_2 acts as oxidant. The process of etching and oxidizing simultaneously causes silicon surface become rough and this roughness depends on the SC1 mixing ratio (11). In order to have a smooth silicon surface condition before gate oxide growth, reduced SC1 concentration cleaning, reduced SC1 dipping time as well as sacrificial oxide growth and removal for surface treatment have been done during pre-gate oxidation preparation.

The high voltage PMOS transistor NBTI lifetime in Figure 1(a) illustrates that both reducing the concentration of NH_4OH and reducing the cleaning time during SC1 treatment can help to enhance the device NBTI immunity. However, optimizing SC1 mixtures is not a simple task as the tradeoff between surface micro-roughness and particle removal efficiency must be taken into consideration. Figure 1(b) shows that no significant effect has been observed on Vt degradation improvement with the additional sacrificial oxide growth and removal.

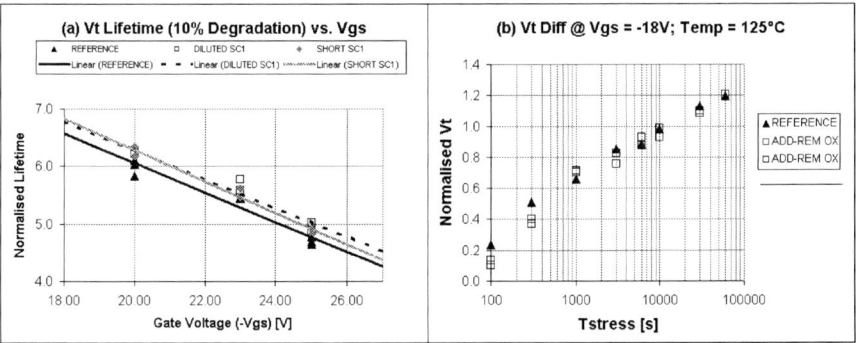

Figure 1. High voltage PMOS NBTI performance for (a) different pre-gate oxidation cleaning solutions and (b) with & without additional oxide growth and removal.

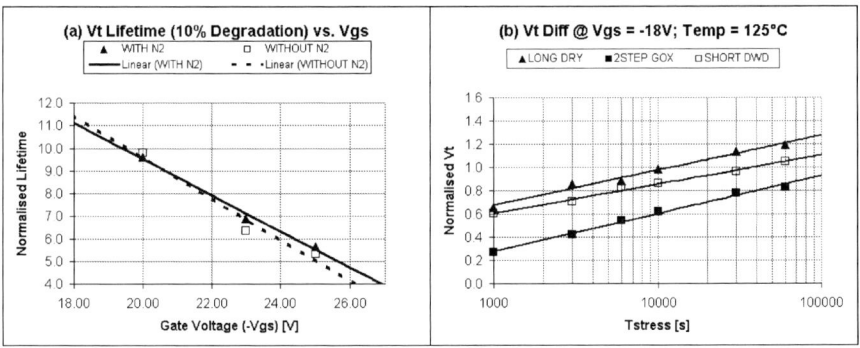

Figure 2. Device NBTI degradation of (a) with & without N_2 post oxidation anneal and (b) different gate oxidation processes.

Post Oxidation Anneal

Post oxidation anneal is a common process to enhance the gate oxide integrity by decreasing any positively charged defects present during gate oxide growth. Several researchers (6)-(8) reported that increasing the nitrogen contents in the oxide yields to a more severe NBTI degradation. Pure thermal oxidation with N_2O annealing shows more significant Vt shift compared to without N_2O. Even though incorporating nitrogen atoms into oxide decreases the total number of dangling Si-H bonds but the nitrided oxide enhances the hole trapping phenomenon which becomes more dominant than interface state generation in contribution to PMOS NBTI degradation.

In our experiment, the introduction of annealing in N_2 ambient did not show significant different in device NBTI performance as shown in Figure 2(a). H. D. Kim et al. (12) suggested an additional formation of interfacial oxide during N_2O annealing process, causing the degradation of oxide property. Thus, without the presence of oxygen during post oxidation anneal, nitrogen contents will not change the interface condition that might degrade the device NBTI performance.

Gate Oxidation

Due to different lattice structures of silicon and gate oxide, an interface is formed, which normally contains dangling bonds. These bonds are later electrically passivated by hydrogen during subsequence high temperature steps. NBTI degradation threatens the device lifetime due to reduced gate oxide thickness and increased interface state density generation at a given stress (13). However, results show that thicker gate oxide device suffered more NBTI failure compared to thin gate oxide device with comparable electric field. Normally, a dry gate oxidation process is long for high voltage devices in order to achieve the desired thick gate oxide. Even though dry gate oxidation can lead to better device NBTI reliability, it is suspected that the longer the time for oxygen to react with silicon during gate oxide grow, the more dangling bonds are formed and interface state density for high voltage PMOS will be much higher during NBTI stress.

In order to reduce the gate oxide growth process time a two step gate oxide split was introduced with deposited oxide toping up the thermal oxide. The two step gate oxide variant has an initial thin thermal growth oxide to ensure good Si-SiO_2 interface for the device followed by furnace oxide deposition to achieve the desired gate oxide thickness. A short oxide growth process also can be achieved with wet oxidation in order to guarantee better gate oxide quality comparing to deposited oxide. The dry-wet-dry short process split has one minute dry oxidation followed by a faster wet process and ending with a dry oxidation. Figure 2(b) shows the Vt degradation of the different splits and it is demonstrated that reducing the long gate oxidation processing time confirms the theory and leads to better device NBTI reliability.

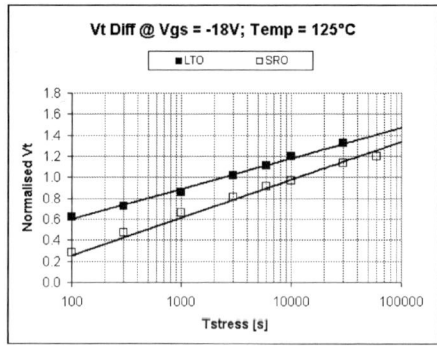

Figure 3. SRO ILD liner shows less Vt shift during NBTI stress.

ILD Liner Deposition

SRO, which is used as ILD bottom layer improves the NBTI performance by almost one order comparing to LTO as can be seen in Figure 3. During device NBTI measurement, the passivated Si-H bonds are broken and interface states are generated within this high electrical field across the gate oxide by means of thermal energy. The SRO liner contains a high concentration of hydrogen that helps to reduce H diffusion from the dangling bond dissociated during NBTI stress and at the same time increases the effectiveness of H to passivate the interface states. Additionally, SRO film can act as a block layer better than LTO film to prevent plasma damage induced by back end of line processing.

Hydrogen Annealing

Hydrogen annealing is often used at the end of the CMOS process in order to passivate the oxide dangling bonds and neutralize the oxide defects. However, metal lines act as diffusion barriers and inhibit sufficient passivation of the interface traps. It is demonstrated in Figure 4(a) that an earlier sintering process before metal routing helps to suppress NBTI Vt shift. However, pure hydrogen and forming gas thermal annealing assessment shows inconclusive result as a post-Metal 1 alloy split has overshadowed the impact. Figure 4(b) shows that by adding hydrogen gas added during SD RTA can provide better NBTI reliability. The PMOS NBTI immunity can be enhanced as the device interface state is effectively passivated.

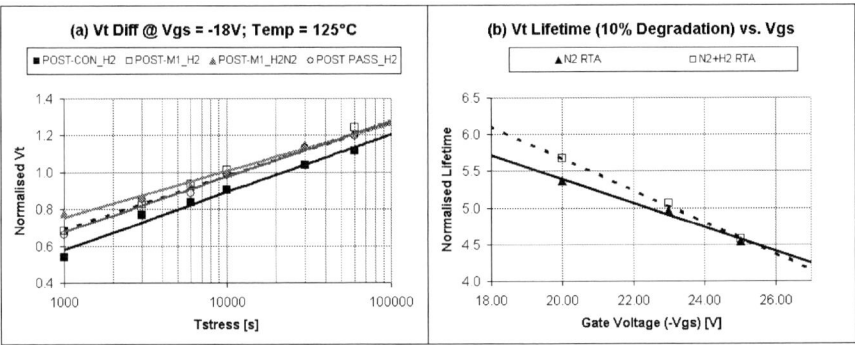

Figure 4. NBTI performance for (a) different sintering splits and (b) with or without hydrogen during SD RTA.

Conclusion

It is a challenge for the high voltage device architects to optimize the trade off between the transistor performance requirements and reliability targets as a more stringent reliability safety margin is required in modern technology. In short, work has revealed that back-end wafer processing particularly ILD liner process and shorten gate growth process time can effectively prevent device NBTI damage and achieved an order of magnitude improvement in device reliability lifetime.

Acknowledgments

The authors would like to thank X-FAB management, Quality Reliability group and Diffusion department for their support, technical advice and contribution to this paper.

References

1. C. Schlünder, "Device Reliability Challenges for Modern Semiconductor Circuit Design – a Review", *Advances in Radio Science*, **7**, 201-211 (2009).
2. B. E. Deal, M. Sklar, A. S. Grove and E. H. Snow, "Characteristics of the Surface-State Charge (Qss) of Thermally Oxidized Silicon", *J. Electrochem. Soc.*, **114**(3), 266-273 (1967).
3. S. Mahapatra and M. A. Alam, "Defect Generation in p-MOSFETs under Negative-Bias Stress: An Experimental Perspective", *IEEE Transactions on Device and Materials Reliability*, **8**(1), 35-46 (2008).
4. T. Nigan, "Pulse-Stress Dependence of NBTI Degradation and Its Impact on Circuits", *Pulse-Stress Dependence of NBTI Degradation and Its Impact on Circuits*, p. 72-78, IEEE Transactions on Device and Materials Reliability, **8**, 1 (2008).
5. V. Huard, M. Denais and C. Parthasarathy, "NBTI degradation: From physical mechanisms to modeling", *Microelectronics Reliability 46*, 1-23 (2006).
6. V. Huard, M. Denais, F. Perrier, N. Revil, C. Parthasarathy, A. Bravaix and E. Vincent, "A thorough Investigation of MOSFETs NBTI Degradation", *Microelectronics Reliability 45*, 83-98 (2005).
7. Y. Mitani, H. Satake and A. Toriumi, "Influence of Nitrogen on Negative Bias Temperature Instability in Ultrathin SiON", *IEEE Transactions on Device and Materials Reliability*, **8**(1), 6-13 (2008).
8. S. Prasad, E. Li and L. Duong, "Process Dependence on Negative Bias Temperature Instability in PMOSFETs", *Electrochemical Society Proceedings*, **2**, 211-216 (2003).
9. T. Ohmi, M. Miyashita, M. Itano, T. Imaoka and I. Kawanabe, "Dependence of Thin-Oxide Films Quality on Surface Microroughness", *IEEE Transactions on Electron Devices*, **39**(3), 537-545 (1992).
10. L. J. Jin, H. P. Kuan, D. Sim and M. Mukhopadhyay, "Influence of Hydrogen Annealing on NBTI Performance", *in Proceedings of International Symposium on Physical and Failure Analysis of Integrated Circuits*, 1-4 (2008).
11. C.Y. Chang and S. M. Sze, *ULSI Technology*, p. 74-79, McGraw-Hill, Singapore (1996).
12. H. D. Kim, S. W. Jeong, M. T. You and Y. Roh, "Effects of Annealing Gas (N_2, N_2O, O_2) on the Characteristics of $ZrSi_xO_y/ZrO_2$ High-k Gate Oxide in MOS Devices", *Thin Solid Films*, **515**, 522–525 (2006).
13. M. Denais, V. Huard, C. Parthasarathy, G. Ribes, F. Perrier, N. Revil and A. Bravaix, "Oxide Field Dependence of Interface Trap Generation During Negative Bias Temperature Instability in PMOS", *Integrated Reliability Workshop Final Report*, 109-112 (2004).

Study The Mixed-mode Delamination of The Epoxy/Cu Interface

Yu Liu[a], Jun Wang[a]

[a] Department of Materials Science, Fudan University, No. 220, Handan Road, Shanghai 200433, China

> The high density packaging, e.g. stacked die packages et al., includes interfaces between different materials. The device reliability is heavily related to the behavior of the interfaces. For instance, the delamination may propagate along dissimilar material interfaces and lead to electrical failure during the package manufacturing, testing processes and even in working state under severe environmental conditions. The interfaces between the epoxy-based composites and copper may be critical due to the large CTE mismatch of the materials. The interface fracture exhibits mixed-mode that combined with mode-I and model-II fractures. In this work, we designed the sandwich samples and a new apparatus to characterize the interface fracture of the epoxy/Cu interface. The sample can be loaded in different angles. The energy release rate and mode mixity were obtained upon the experimental results by the mixed-mode elastic fracture analysis. The method can be applied for other interface characterization in the packaging.

Introduction

The trends of the plastic packaging shift to the higher density, lower cost technologies quickly in recent years. The 3D packaging technologies, such as stacked-die packaging and packaging on packaging (POP), have improved efficiency but introduced complex multilayer structures. The layers are mostly connected by adhesives. The integrity of the interfaces between different materials becomes challenge when the device is subjected to thermal and mechanical loading. For example, the reliability of devices will be heavily dependent on the behaviors of the interfaces during the high temperature assembly processor being used in variant environments [1]. On the other hand, the interfaces between different materials are hard to control and become weak when small defects were introduced during fabrication process. The small defects may propagate under mechanical or thermal loadings. The interface fractures are usually responsible for the failure of the assembled devices [2]. In the high density plastic packages, the CTE miss match between the epoxy based composites and copper is large, the interface between epoxy and copper is critical in the devices.

Interfacial cracking between dissimilar materials is generally in mixed-mode, i.e. both normal and shear stresses take effect near the crack tip [3]. For the elastic materials, Suo and Hathinson [4] derived the energy release rate and stress-intensity factor for the bi-material interface crack under bending and axial loading. Shear stress are lack in the analytical solutions. Qiao and Wang [5] developed a more general analytical solution by including the shear effect. They extend the solution to sandwich adhesive bonded joints [6]. The finite element analysis is applicable numerical method to calculate the mixed-mode interface cracking Li et al. [7]. The shear stress can be included in the analysis. Experimental researches demonstrated that the interface fracture energy is dependent on the mode mixity [8-10]. Evans et al. [11] summarized some of the most extensively used techniques and pointed out the sandwich samples are the most useful for initial investigation of interface fracture energy trends. However, the experimental techniques, such as double cantilever beam, peel test, etc., are not flexible to multiphase angel testing. Wang and Suo [9] proposed the Brazil-nut sandwich sample with center pre-crack to measure the interface fracture energy and mode mixity. The fracture loading phase was controlled by the compression angle along the diameter. Loo et al. [12] also used the Brazil-nut sandwich technique, but the pre-crack was located at the edge of the interface and the tensile loading was applied. Because the shape of the Brazil-nut sandwich is circle, it is an issue when the reformation of interface materials is hard. In this study, we designed sandwich samples with pre-cracks and novel apparatus to measure the interface fracture parameters. The sample mounted on the apparatus can be loaded in 30°, 45°, 60° and 90° angles. Here, the experimental data loaded in 30° was not enough and was left for future work. The experimental results were analyzed by the elastic interface fracture principles and FEA to reveal the energy release rate and the phase angle of the interface crack.

The interface fracture and mode mixity

The sketch of interface cracking in the sandwich is illustrated in Figure 1. The two materials are noted as 1 and 2. Both of them are elastic. Their Young's moduli and Poisson's ratio are E_n, v_n, (n=1, 2). The subscripts 1 and 2 refer to the two materials. The layer of material 2 is in the middle of the structure. A crack is along the interface between the middle layer of material 2 and the top layer.

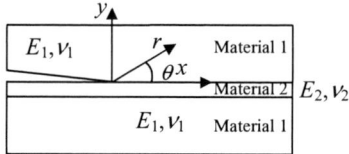

Figure 1. The sketch of interface cracking in sandwich

At the interfaces, the material discontinuity exists. The elastic moduli depend on the biomaterial system. The constrains can be expressed in two non-dimensional parameters, i.e. the Dunders' parameters [13], which are

$$\alpha = \frac{\mu_1(\kappa_2 + 1) - \mu_2(\kappa_1 + 1)}{\mu_1(\kappa_2 + 1) + \mu_2(\kappa_1 + 1)} \qquad \beta = \frac{\mu_1(\kappa_2 - 1) - \mu_2(\kappa_1 - 1)}{\mu_1(\kappa_2 + 1) + \mu_2(\kappa_1 + 1)} \qquad [1]$$

Where, the μ_n are shear moduli; $\kappa_n = 3 - 4v_n$ is defined for plane strain and $\kappa_n = (3 - v_n)/(1 + v_n)$ is for plain stress; $n = 1, 2$ denote the material 1 and 2. The other bi-material constants related to Dunders' parameters are defined as

$$\frac{\bar{E}_1}{\bar{E}_2} = \frac{1 + \alpha}{1 - \alpha}, \qquad \varepsilon = \frac{1}{2\pi} \ln \frac{1 - \beta}{1 + \beta} \qquad [2]$$

Where $\bar{E}_m = E_m$ is the plane stress tensile modulus and $\bar{E}_n = 2\mu_n/(1 - v_n)$ is for the plane strain case; ε is the oscillatory index for the stress filed near crack tip. The stress field near the crack tip is derived by Rice [3]. The singular stress near the crack tip ($\theta = 0$ in Figure 1) is:

$$\sigma_{22} + i\sigma_{12} = \frac{1}{\sqrt{2\pi r}} K r^{i\varepsilon} \qquad [3]$$

Here $K = K_1 + iK_2$ is the complex stress intensity factor; $i = \sqrt{-1}$, r is the distance from the origin along x - aixs. The equation is also the definition of the stress intensity factor K. Following suggestions in [3], the mode mixity Ψ for the interface fracture is defined as

$$Kl^{i\varepsilon} = |K|e^{i\Psi} \tag{4}$$

Note $|l^{i\varepsilon}| = 1$, and $|Kl^{i\varepsilon}| = |K|$. The Ψ is the phase angle of $Kl^{i\varepsilon}$. When $\varepsilon = 0, \alpha = 0$, i.e. cracking in the homogeneous material, the mode mixity is reduced to

$$\Psi = \tan^{-1}\left(\frac{K_{II}}{K_I}\right) \tag{5}$$

Where, K_I and K_{II} are the classical model_I and mode_II stress intensity factors respectively. The relationship between the energy release rate and the stress intensity factor can be written as

$$G = \frac{|K|^2}{E^* \cosh^2(\pi\varepsilon)} \tag{6}$$

Where, $|K|$ is the modulus of complex K, E^* is the average modulus and is defined as

$$\frac{2}{E^*} = \frac{1}{E_1} + \frac{1}{E_2} \tag{7}$$

The J integral far from the crack tip is still valid to calculate the energy release rate [11]. In the general loading case, the finite element analysis can be used to calculate the stress intensities and energy related rate of the interface crack when note the equation [6]. The mode mixity is determined by equation [4] and stress field near the crack tip in equation [3].

Experiments

In order to apply load in different angle, the new loading apparatus was designed. The sample and the loading apparatus are sketched in the Figure 2.

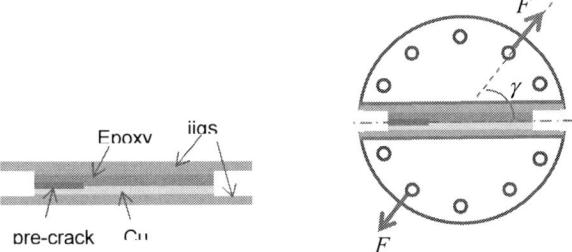

(a)Pre-crack sample   (b)The sample in tension

Figure 2. The sketches of the sample and the loading apparatus

From the above figure, the sample can be loaded in tension in different angle by the holes in the apparatus. We design the loading angles at 30°, 45°, 60° and 90°. The pre-crack were introduced on the edge of the interface. The ratio of samples' width to their thickness was fairly large and thereby the plain strain assumption can be adopted in the analysis. Using the method, the interface between dissimilar materials can be made and tested easily.

Sample preparation

The length, width and thickness of copper plate are 60, 15 and 1.5 millimeters respectively. The surfaces of the copper plates were cleaned in lye at 80°C for 40 min. The lye was made up of tri sodium phosphate and DI water. Then the copper plate was placed at the bottom of the mold before the epoxy liquid was filled. The thickness of the epoxy layer is controlled at about 2 mm. The PI tape was used to form the pre-crack between the copper and the epoxy. The molded sample was put into the oven, in which the temperature was kept at 70℃ for half an hour and raised to 120℃ for additional 30min. After the cure of the epoxy, the sample was mounted on the jigs by adhesion of high strength glue. The pre-crack was ascertained by mechanical method. Finally, the jigs was fixed in the loading apparatus and subject to tension in variant loading angles. The photos of the sample loading at 45° and 90° are shown in Figure 3.

(a) 45°   (b) 90°

Figure 3. Photos of tests at 45° and 90°

When the samples were loaded in different angle, the load-displacement curves were obtained. The further analysis is demonstrated in the follow section.

Results and analysis

When the sample with pre-crack was applied in tension, the sample deformed until the crack began to propagate at the critical loading force and the unloading happened instantly. The load-displacement curve in tension was recorded by computer. In order to get the energy relate rate and the mode mixity, the finite element analysis was carried out. The elastic plane strain was assumed in the analysis. For the steel jigs have much more stiffness than the samples, the jigs were simplified to rigid part in the model. The complete models were also built and computed. The results were similar comparing with presenting results. The models were built in ANSYS according to the dimensions of each sample. The eight-node element was selected and the meshes near crack tip were refined. The middle nodes around the crack tip were shifted to present the singularity of the stress field. The typical meshes of the models are illustrated in Figure 4.

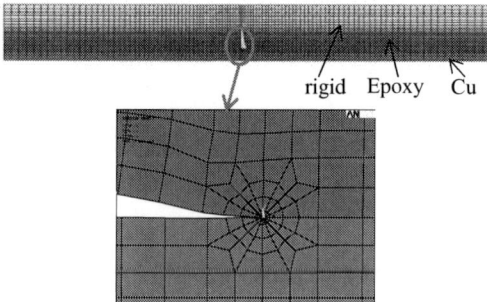

Figure 4. The meshes for interface cracking analysis

In the computation, the Young's modulus of the epoxy was verified by tensile test. All the materials' properties are listed in the **TABLE I**.

TABLE I. The Materials' Elastic Properties

Material	Young's modulus (GPa)	Poisson's ratio
Cu	120	0.35
Epoxy	2.53	0.42

Applying the critical loading forces on each model, the energy release rate was computed and plotted versus loading angle in Figure 5.

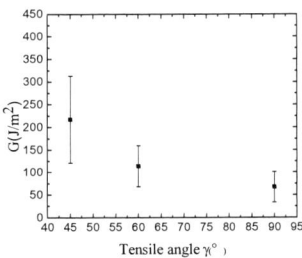

Figure 5. The energy release rate vs. tension angle γ

When the loading force is along different angle, the mode mixity will vary. The stress intensities parallel and normal to the crack surface were analyzed by FEA. Then the mode mixity can be evaluated by the stress intensities from FEA. The mode mixity is plotted with changing of loading angle in Figure 6.

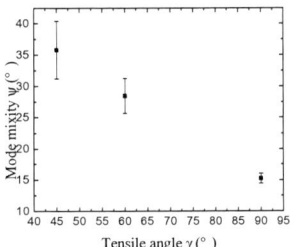

Figure 6. The mode mixity varied with tension angle γ.

According to the curve, the mode mixity is about 15° even in force loading angle is normal to the crack surface. The error increased when the loading angle rise.

For the dissimilar material interface fracture, the energy release rate is not unique but in terms of the phase angle. The energy release rate of epoxy/Cu interfaces is illustrated in the Figure 7.

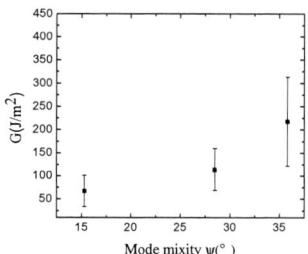

Figure 7. The energy release rate varied with the mode mixity

Concluding remarks

Interface fracture is one of the most popular failure modes in electronic packaging. The energy release rate of interface fracture is function of the mode mixity. This work suggested a valid method to evaluate the critical energy release rate and the mode mixity. During preparing the sample, the dissimilar material can be both in film. It is indicated more application for the interface testing for packaging materials. For the technique can provide flexible loading angle, it is possible to cover large range of mode mixity cases.

Acknowledgments

The work was supported by grants (No. 10972057) from the National Nature Science Foundation of China.

References

1. F. Xuejun, *et al.*, *Electronic Components and Technology Conference*, p. 1054-1066 (2008).
2. J. N. Sweet, in *Integrated Reliability Workshop/1994 Final Report*, p. 30-36 (1994).
3. J. R. Rice, *Journal of Applied Mechanics-Transactions of the Asme*, **55** (1988).
4. Z. G. Suo and J. W. Hutchinson, *International Journal of Fracture*, **43** (1990).
5. P. Z. Qiao and J. L. Wang, *International Journal of Solids and Structures*, **41** (2004).
6. J. L. Wang and C. Zhang, *International Journal of Solids and Structures*, **46** (2009).
7. S. Li, *et al.*, *Journal of the Mechanics and Physics of Solids*, **52** (2004).
8. K. M. Liechti and Y. S. Chai, *Journal of Applied Mechanics-Transactions of the Asme*, **59** (1992).
9. J. S. Wang and Z. Suo, *Acta Metallurgica Et Materialia*, **38** (1990).
10. H. C. Cao and A. G. Evans, *Mechanics of Materials*, **7** (1989).
11. A. G. Evans, *et al.*, *Metallurgical Transactions a-Physical Metallurgy and Materials Science*, **21** (1990).
12. S. Loo, *et al.*, *Journal of Electronic Materials*, **36** (2007).
13. J. Dundurs, *American Society of Mechanical Engineering*, 1969, p. 70-115, New York (1969).

LDMOS Thermal SOA Investigation of a Novel 800V Multiple RESURF with Linear P-top Rings

Aloysius Priartanto Herlambang[a], Gene Sheu[a], Yufeng Guo[b] Hutomo Suryo Wasisto[a]

[a] Computer Science and Information Engineering Department, Asia University, Wufeng, Taichung, Taiwan
[b] School of Electronic Science and Engineering, Nanjing University of Posts and Telecommunications, Nanjing, Jiangsu, 210096, China

In this paper, Technology Computer-Aided Design (TCAD) is used to investigate the thermal characteristic and Thermal Safe Operating Area (T-SOA) of a novel 800V Multiple RESURF LDMOS with linear P-top Rings. Two methods of critical temperature extraction are presented and the agreement between these two methods is proven. The effects of Initial Front Rise time (IFR) and other parameters on the thermal characteristic of the device are inspected. Finally, the simulation result has an agreement with the analytic solution. Therefore, the analytic solution can replace the 3D simulation for large device which is not able to be performed using current available TCAD tools.

Introduction

Safe Operating Area (SOA) is the main concern of high voltage design in IC industry, especially in LDMOS design. Nowadays, the traditional tradeoff between BVoff and Rsp is not sufficient for a novel LDMOS structure. SOA regimes of the LDMOS structure must be investigated in order to ensure the device reliability. IC designer should optimize the device such that the best performance could be achieved from LDMOS triangle design (1).

One of SOA, T-SOA is determined from the turn-on of parasitic BJT due to thermal effect. Stress at medium duration of milliseconds will cause heat generation in the device and increase the lattice temperature. The increment of lattice temperature will affect the thermal stability and lead the device to destruction.

Determining Thermal SOA

Device Structure

The investigation of thermal SOA is performed by utilizing a novel 800V multiple RESURF LDMOS with linear P-top rings that has been optimized in order to obtain the best tradeoff between BVoff and Rsp. (2) Fig. 1.(a) shows the cross section of the device under test.

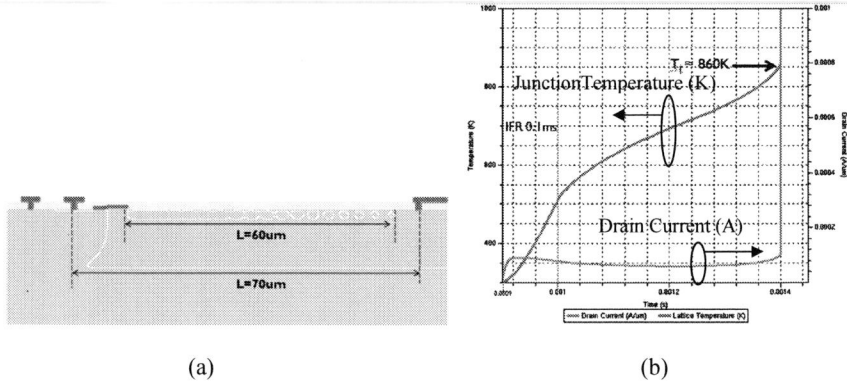

(a)   (b)

Figure 1. (a) Cross section of 800V LDMOS (b) Simulation result

Device Simulation

Thermodynamic and Hydrodynamic methods are used in the Sentaurus device simulation in order to well describe the natural phenomena in the device during the applied stress. Temperature dependent physics models are used to accommodate the device temperature.

Based on those considerations the thermal investigation of the device under test is performed and the result is shown in Fig. 1.(b). As clearly seen that the device temperature is increasing along the stress duration and the current will increase dramatically when the junction temperature is reaching the critical temperature. At this point, the gate will no longer be able to control the current and the device will fail.

Critical Temperature Extraction

Critical temperature is the main parameter which is used to determine the boundaries of thermal SOA. There are two methods to extract critical temperature. First approach is using thermal stability factor to determine the temperature point in which the device starts to become unstable. Time dependent thermal stability factor is calculated using [1] (1).

$$ S = V_{DS} \frac{\Delta I_D(t)/W}{\Delta t} \frac{\Delta t}{\Delta T_J(t)} \frac{2}{K\theta} \sqrt{\frac{D\theta}{\pi}} \frac{1}{HP} \sqrt{t} $$

[1]

where V_{DS} is applied voltage, $I_D(t)$ is drain current, W is device width, $T_J(t)$ is junction temperature, HP is half pinch of the device, $K\theta$ is thermal conductivity of silicon and $D\theta$ is thermal diffusivity. When S is equal to one, the device starts becoming unstable and the gate will no longer be able to control the device current then the stability factor will increase tremendously as shown in Fig. 2.(a). This condition leads the device into thermal destruction known as critical temperature of the junction temperature.

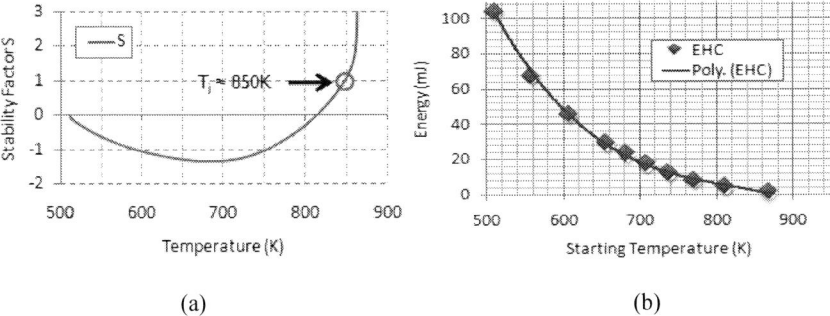

(a)   (b)

Figure 2. (a) S equal to one at approximately 850K (b) Energy handling capability plot shows that the device fail at approximately 860K

Another way is using extrapolation of energy handling capability plot to determine the temperature point in which the device no longer needs any energy to fail.(3) Energy to failure and starting temperature are plotted, then the intersection with x axis (temperature) can be defined as the critical temperature shown in Fig. 2.(b). Those two methods indicate the agreement on the critical temperature extraction results. It is found out that the extrapolation of energy handling capability is not linear as reported previously.

Parameter Effect on Thermal SOA

Some parameters affect the thermal behavior of the device and degrade the Thermal SOA boundary. Therefore, some investigations are established in order to verify the dependence of each parameter.

Drain Voltage Effect

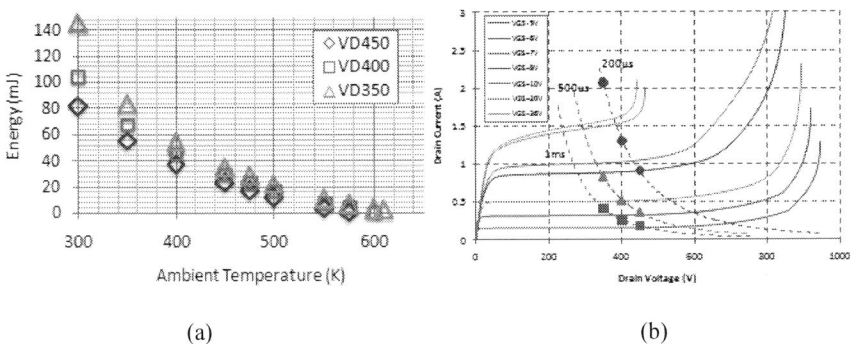

(a)   (b)

Figure 3. (a) Drain voltage effect (b) Thermal SOA boundary at different pulse duration

Drain voltage induces the avalanche generation in the device and will deteriorate the critical temperature as reported by Hower (4). Fig. 3.(a) shows that lower drain voltage is able to withstand the higher ambient operating temperature. Using pulse time normalization method and energy handling capability value, the boundary line of T-SOA with different pulse duration can be estimated as shown in Fig. 3.(b). It is shown that the boundary line will reduce the safe operating area when the pulse duration is longer.

Temperature Coefficient Effect

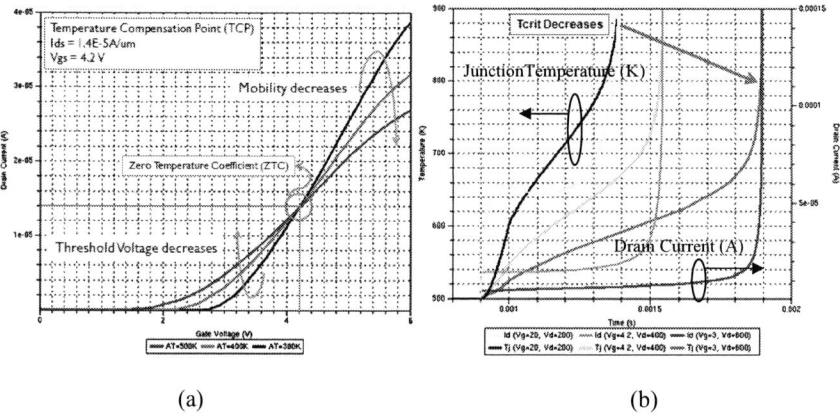

(a)   (b)

Figure 4. (a) Temperature compensation point (b) Critical temperature decrease as effect of gate voltage

Temperature compensation point is a condition where the temperature coefficient is zero. (5) Below the TCP, temperature coefficient is positive and causing threshold voltage degradation and increasing the drain current. Above the TCP, temperature coefficient becomes negative and causes the mobility degradation and decreases the drain current as shown in Fig. 4.(a). Fig. 4.(b) shows that critical temperature changes when different temperature coefficients are applied in the device. Temperature coefficient effect is clearly visible on the drain current characteristic.

Initial Front Rise Time Effect

Initial front Rise time (IFR) is a part of the stress pulse. As the initial part of the stress, IFR has strong influence on device performance under the stress. The device fails faster when the longer IFR is applied to the device as shown in Fig. 5.(a). Junction temperature plot shows that the temperature at starting point (end of IFR) is increasing when the longer IFR is used.

Longer IFR will increase self heating effect on the device and cause higher starting temperature as shown in Fig. 5.(b). It is also shown that the energy to fail is decreasing when the longer IFR is used. This indicates that the device will have lower energy handling capability when the IFR is longer. This condition will lead to deterioration in thermal SOA boundary lines obtained by normalization method.

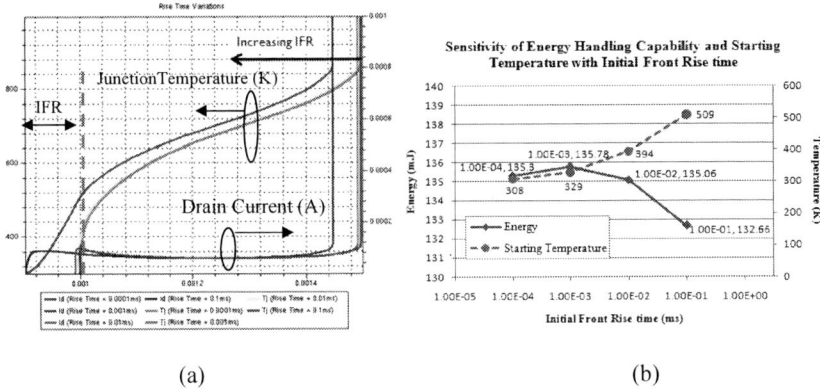

(a)   (b)

Figure 5. (a) IFR effect on time to failure (b) IFR sensitivity

Aspect Ratio Effect

Aspect ratio of the device is investigated in order to study about the heat dissipation effectiveness. 3D simulation is used up to 20μm device width. It is shown in Fig.6.(a) that wider the device, the failure time will reduce and stop at some device width. Even though the failure time is reduced but the critical temperature is not affected at all. Further investigation for large device cannot be performed with current available TCAD tools. 3D simulation needs a lot of mesh nodes in order to keep the accuracy. This condition leads to a penalty of longer computation time and also higher memory requirement. Therefore, these TCAD limitations in 3D simulation become the main issue for device modeling.

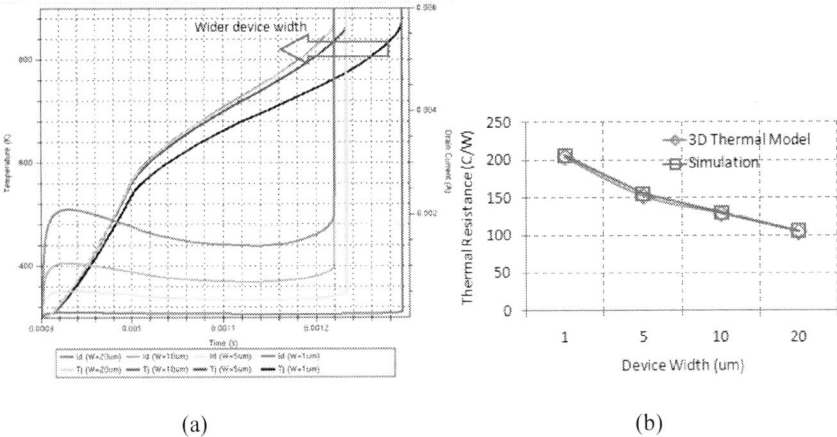

(a)   (b)

Figure 6. (a) Aspect ratio effect (b) Well agreement between analytical and simulation result on thermal resistance value

Fig. 6.(b) shows that the 3D simulation and 2D simulation are in agreement with 3D thermal model that is proposed by Hower (4). The well agreement of thermal resistance value between the value which is calculated using Hower's model and the value which is calculated based on simulation result using Joy and Schlig method (6) is presented. Based on these results, the analytic solution can be used to replace the 3D simulation for device with large area.

Conclusion

Thermal SOA investigation can be performed by using Technology Computer-Aided Design (TCAD) simulation. Drain voltage and temperature coefficient are the parameters that affect the critical temperature. Initial front rise time (IFR) may affect the device performance and deteriorate the thermal SOA boundary. IFR plays the major role to the self heating and device failure and should be considered as a part of pulse duration. 3D simulation is in agreement with 3D thermal model; hence the 3D simulation for large device can be replaced by analytical model.

References

1. P. L. Hower, "Safe operating area - a new frontier in ldmos design," *Proc. ISPSD '02*, Santa Fe, pp. 1-8 (2002)
2. H. S. Wasisto, et al., "A Novel 800V Multiple RESURF LDMOS Utilizing Linear P-top Rings," *Proc. TENCON '10*, Fukuoka, (2010)
3. V. Khemka, V. Parthasarathy, R. Zhu and A. Bose, "Correlation between static and dynamic SOA (energy capability) of RESURF LDMOS devices in smart power technologies," *Proc. ISPSD '02*, Santa Fe, pp. 125-128 (2002)
4. P. Hower, et al., "Avalanche-induced thermal instability in ldmos transistors," *Proc. ISPSD '01*, Osaka, pp 153-156 (2001)
5. M. Denison, et al., "Influence of inhomogeneous current distribution on the thermal SOA of integrated DMOS transistors," *proc. ISPSD '04*, Kitakyushu, pp. 409-412 (2004)
6. R. C. Joy, E. S. Schlig, "Thermal properties of very fast transistors," *IEEE Transactions*, Vol. 17, No. 8, pp.586-594 (1970)

Investigation of Lateral Die Crack Failure at Reliability Test

Yuen Chun SOH[a], Chin Meng Jimmy TAN[a], Xin CHEN[a], Kok Yau CHUA[b], Ruomin Mike DU[a], Yongjie XI[a] and Thiam Huat LIM[b]

[a] Infineon Technologies (Wuxi) Co. Ltd.
[b] Infineon Technologies (Malaysia) Sdn. Bhd.

"Goes Green & Copper", is the strategy to introduce 1) environmentally friendly materials to meet RoHS (Restriction of the use of Hazardous Substances) & WEEE (Waste Electrical & Electronic Equipment) requirements and 2) Cu wires for superior electrical, mechanical & intermetallics properties compared to Au wires. A complete package level qualification was planned to assess the reliability performaces in discrete packages.

This paper will discuss in details on investigation of lateral die crack failure during reliability. Process & materials investigation revealed that interaction of leadframe paddle construction, die attach method, die construction & mold compound factors play important role in aggravating lateral die crack. FEA (finite element analysis) was performed to understand the thermo-mechanical stress distribution within the die & package. Re-qualification was performed & reliability results showed that lateral die crack failure could be eliminated with proper combination of BOM selection.

Introduction

As the electronics industry direction is to strive for environmental aspect, it motivated integrated semiconductor (IC) manufacturers to move towards introducing exclusively with new environmentally friendly materials in their products. IC manufacturers continue to launch new products that meet RoHS & WEEE requirements. Typically, an IC package that is produced using materials with <1000ppm of lead (Pb) and <900ppm of chlorine & bromine (Cl + Br) could fulfill environmental legal demands.

Test Vehicles & Qualification Matrix

Generally for discrete packages like SOT & SOD (Small Outline Transistor & Diode), customers demand for green packages conversion is aggressive. Consequently, backend manufacturing operation continues to drive for green packages & products qualification to enhance business portfolio. Cu wire bonding will be incorporated with green conversion to maximize product & cost performances.

This section will discuss the SOT green package qualification in which BOM consists of Cu wire & halogen free mold compound as part of aggressive conversion plan. Scope of test vehicles & technical information will be shared in Table 1 below.

TABLE I. Test Vehicle Selection & Technical Information.

Description	Control	Qualification
Package Type	SOT223	
Package Size (mm)	6.5 x 3.5 x 1.6	
Device	BDxxx; BCxxx; BSxxx; PZxxx	
Die Size (mm)	Smallest 0.6 x 0.6; Largest 1.0 x 1.6	
Die Thickness (μm)	140	
Die Backside Metallization	AuSn (Eutectic Die Attach)	
Bond Pad Metallization	Al; AlSi; AlSiCu	
Leadframe	Cu Base with Ag Plating	
Bonding Wire (μm)	Au (22, 38, 50)	Cu (22, 43)
Mold Compound	Halogen Compound	Halogen Free Compound
Plating	Pb-free	

Parts were assembled with the identified devices & it's BOM; these units were then submitted into a full package level qualification for reliability robustness assessment. Table 2 showed the reliability test & condition that were being used in this project.

TABLE 2. Reliability Test & Condition.

Description	Condition	Readout
MSL 1 Preconditioning (Precon) with Bosch Profile	• Bake 125°C + Soak 85°C/85%RH + IR 260°C Reflow + TC (-55°C/150°C)	Bake@24hrs + Soak@168hrs + Reflow 3x + TC100x (Bosch Profile)
Pressure Cooker Test (PCT)	121°C; 100% RH; 2 bar	96hrs
High Temperature Reverse Bias (HTRB)	Ta = 150°C; VR = 300V	168hrs; 500hrs; 1000hrs
High Humidity High Temperature Bias (H3TRB)	Ta = 85°C; VR = 85V; 85% RH	168hrs; 500hrs; 1000hrs
High Temperature Storage (HTS)	Ta = 175°C	168hrs; 500hrs; 1000hrs
Intermittent Operating Life (IOL)	Ta = 75°C; Δ Tj = 100°C	168hrs; 500hrs; 1000hrs
Temperature Cycling (TC)	Ta = -55°C to 150°C (condition C)	300x; 500x; 1000x

Reliability Results & Failure Analysis

BSxxx & PZxxx devices observed high failure rate of electrically failed upon completion of full Precon & at early readout of TC@300x. Table 3 showed the Precon & TC@300x failure rate for all the test vehicles. It was unusual to see such high failure rate during Precon & early cycle readout of TC, thus immediate failure analysis (FA) was performed to understand the failure mechanism. FA summary is shown under Table 4, it observed delamination (due to separation between chip & leadframe) at SAM & upon decapsulation similar failure mechanism was observed on both devices with lateral crack through silicon (Si) die.

TABLE 3. Precon & TC@300x Results.

Test Vehicle	Precon	TC@300x
BDxxx	0/240	0/60
BCxxx	0/240	0/60
BSxxx	16/240 (6.7%)	1/60 (1.7%)
PZxxx	38/240 (15.8%)	5/60 (8.3%)

TABLE 4. FA Summary.

No	Description	Findings
1	External Package Inspection	No abnormality.
2	X-Ray	No abnormality.
3	Scanning Acoustic Microscopy (SAM)	Observed delamination.
4	Decapsulation	Observed lateral crack through Si die.

Investigation, FEA & Design of Experiment

Technical gap analysis was performed to check if there is any commonality of BSxxx & PZxxx, details information is shown in Table 5. Strong commonalities pointed at chip size, bond pad metallization & leadframe paddle construction.

TABLE 5. Technical Gap Analysis

Description	BDxxx	BCxxx	BSxxx	PZxxx
Die Size (mm)	1.0 x 1.6	0.63 x 0.63	0.7 x 0.7	0.6 x 0.7
Bond Pad Metallization	3.0µm AlSi	1.4µm AlSiCu	1.4µm Al	
Leadframe Paddle Construction	Type A 1.6 x 1.9mm		Type B 2.65 x 3.00mm	
Bonding Wire	43µm Cu			22µm Cu
Mold Compound	Halogen Free			

Investigations were carried out on remaining unstressed qualification samples. Zero hour FA showed no abnormalities, neither delamination nor lateral die crack. This indicated that the failure mechanism happened at reliability test. Bosch profile Precon (more stringent than normal Precon) steps analysis revealed that lateral die crack happened at TC100x. The failure mechanism was aggravated by moisture & thermo-mechanical related stress. Table 6 showed the investigation summary of unstressed samples & Precon steps.

TABLE 6. Unstressed & Precon Steps Investigation.

Item	Description		Findings
1	Unstressed units.	No abnormality.	
2	IR reflow 3x only.	No abnormality.	
3	TCC 100x only.	No abnormality.	
4	IR reflow 3x + TCC 100x	No abnormality.	
5	Bake@24hrs +Soak@168hrs + Reflow 3x	No abnormality.	
6	Bake@24hrs + Soak@168hrs + Reflow 3x + TC100x (Bosch Profile)	Observed lateral crack.	

Simple FEA modeling was planned, to understand the stress distribution on the die meanwhile also trying to narrow down the root causes area. Simulations were performed to verify whether die size or compound type are the important factors of causing lateral crack at Bosch profile Precon. Table 7 & 8 showed the FEA results, which pointed to bigger die size & halogen compound have higher stress on the die. It indicated that bigger die size & halogen compound will exhibit more severe lateral crack. However, all the FEA results have to be validated by actual experiments data.

TABLE 7. FEA Simulation 1: Different Die on Type B Leadframe + Halogen Free Compound.

Die Size A (0.6 x 0.7) mm^2 Max principal stress = 2.71 x 10^9 (dyne/cm^2)	Die Size B (2.0 x 2.0) mm^2 Max principal stress = 3.04 x 10^9 (dyne/cm^2)

TABLE 8. FEA Simulation 2: Different Compound on Type B Leadframe + Die Size (0.6 x 0.7) mm^2.

Halogen Free Compound	Halogen Compound
Max principal stress = 2.36 x 10^8 (Pa)	Max principal stress = 2.76 x 10^8 (Pa)

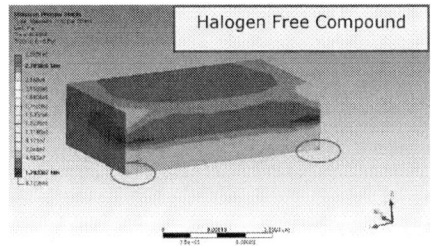 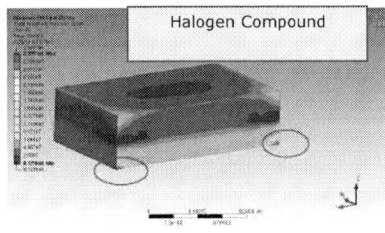

With the information that gathered from technical gap analysis & FEA modeling, design of experiment (DOE) was planned to identify the important main effect or interactions that would exhibit lateral die crack at Bosch profile Precon. Full factorial DOE with 3 factors of die size, leadframe & compound type was performed. Table 9 showed the details DOE run. Further analysis of DOE results such as main effect plot, interaction plot & response optimizer were done to confirm the significant factor & verify if there is any interaction effects among these factors. Apparently, actual DOE results didn't correlate with the earlier FEA modeling. No lateral crack was observed on runs with halogen compound & bigger die size. Run 6 observed lateral similar lateral die crack as previous qualification lot. Table 10 showed the DOE analysis results.

TABLE 9. DOE Runs

Run	Die Size	Leadframe	Compound	Lateral Crack (0 = No; 10 = Yes)
1	1.0 x 1.6	Type B	Halogen Free	0
2	1.0 x 1.6	Type A	Halogen Free	0
3	0.6 x 0.7	Type A	Halogen	0
4	0.6 x 0.7	Type A	Halogen Free	0
5	1.0 x 1.6	Type A	Halogen	0
6	0.6 x 0.7	Type B	Halogen Free	10
7	1.0 x 1.6	Type B	Halogen	0
8	0.6 x 0.7	Type B	Halogen	0

TABLE 10. DOE Analysis Results.

No	Analysis	Comments
1	Main Effect Plot 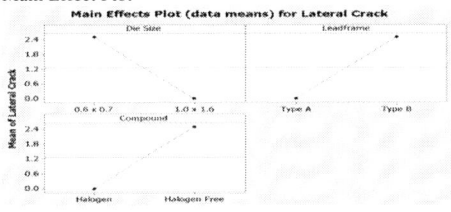	• All the factors are equally significant.

2   Interaction Plot

Interaction Plot (data means) for Lateral Crack

- Strong interaction among all factors.

3   Response Optimizer

- Lateral die crack will only occur (d = 0) if there are 3 ways interaction:
  - 0.6 x 0.7 Die Size
  - Type B Leadframe
  - Halogen Free
- 2 ways interaction will NOT cause lateral die crack to occur.

Conclusion

Based on the above investigations & studies, it was concluded that proper combination of BOM selection are critical & have huge impact on products quality & reliability performances. It was also learnt that FEA modeling is not always correlate with actual experiments results. FEA modeling will be a quick reference but have to be validated. Re-qualification was performed on new BOM combination; in this case Type B leadframe was excluded. All lots passed reliability test without lateral die crack.

Acknowledgments

The authors would like to thank Ong Tiam Sen & Ding Junwei for managerial related support. Lee Lik Wei & Zhang Zhenting for failure analysis related support. Sim Kar Yi & Ma Linda for reliability stress related support.

References

1. The FUTURE of TECHNOLOGY, Infineon's way to "Green Products".
2. Selection Guide "General Purpose Diodes, Transistors & Small Signal MOSFETs.
3. MINITAB website, www.minitab.com
4. ANSYS website, www.ansys.com

10.1149/1.3567704 ©The Electrochemical Society

Study on the Reliability of Nano-structured Polymer-Metal Composite for Thermal Interface Material

Lei Zhang[a], Xin Luo[a,b], Xiuzhen Lu[a] and Johan Liu[a,b]

[a] Key Laboratory of Advanced Display and System Applications, Ministry of Education & SMIT Center, School of Mechatronics Engineering and Automation, Shanghai University, P.O.B. 282, Shanghai 200072, China
[b] SMIT Center & Bionano Systems Laboratory, Department of Microtechnology and Nanoscience, Chalmers University of Technology, SE-412 96 Gothenburg, Sweden
Email: xzlu@shu.edu.cn

The need for higher performance electronic products is posing an urgent challenge for us to improve thermal management techniques in order to maintain reliability of systems and devices. Nano-TIM, a new type of thermal interface material, was developed to improve the heat dissipation of electronic devices. In this paper, the reliability of Nano-TIM was studied. The samples were subjected to thermal and humidity tests with the condition of 40 ℃, 93%RH for 56 days. Pull tests were used to investigate the shear strength of the samples with Nano-TIM coalesced between two PCBs with Sn coating. The shear strength did not change very much after the thermal and humidity tests. Scanning Electron Microscopy (SEM) analysis techniques were used to determine the morphology of the shear fracture section after pull tests and observe the structure of the cross section of Nano-TIM coalesced between two PCBs with Sn coating.

Introduction

With the dramatic increase in demand for higher performance electronic products, fast heat dissipation is the key issue in thermal management for electronics packaging [1]. According to International Technology Roadmap for Semiconductor (ITRS), the power density for a single chip microprocessor will continue to climb even though the high-performance microprocessor is expected to stay relatively flat [2]. It is an urgent challenge for us to come up with assembly technology solution to effectively and efficiently transport the increased heat generated by the microprocessors away from the silicon. Failing to do so will cause the heat to accumulate and cause deterioration in the semiconductor behavior, fractures, delamination, melting, creep, corrosion, electromigration and even burning of packaging materials. Finally it will lead to thermal runaway catastrophic failure. [3]

The roughness of the chip surface and heat sink creates a gap between them that increases the thermal resistivity of the devices. Thermal interface material is used to fill the gap and improve the thermal dissipation of the devices [4][5]. Thus, thermal interface material is one of the key factors in thermal dissipation. However, the low thermal conductivity (2.3~6 $Wm^{-1}K^{-1}$) and different thicknesses of traditional TIM mean that it cannot meet the urgent need. *Liu et al.* have already developed a new class of Nanostructured Polymer-Metal Composite Thermal Interface Material (Nano-TIM) using the electrospinning process. The thermal resistance of the Nano-TIM is 8.5 Kmm^2W^{-1} at bond line thicknesses of approximately 70 μm, corresponding to an effective thermal conductivity of 8 $Wm^{-1}K^{-1}$[6]. The present work aims to study the reliability of this Nano-TIM.

In order to analyze the mechanical properties, pull tests were used to investigate the shear strength of the samples which were made using Nano-TIM coalesced between two PCBs with Sn coating. The samples were divided into several groups for the pull tests. Compared to one group without aging, the other groups were subjected to thermal and humidity tests with the condition of 40 °C, 93 % RH for different days. SEM analysis techniques were used not only to determine the morphology of the shear fracture section after pull tests, but also to observe the structure of the cross section of Nano-TIM coalesced between two PCBs with Sn coating.

Experimental procedure

Nano-structured Polymer-Metal Composite Thermal Interface Material (Nano-TIM) was used in this paper. The composite system consists of polymer fibers with an isotropic three-dimensional network which were manufactured using a technology known as electrospinning and an eutectic In/Bi/Sn alloy filler with the melting point of 60 °C infiltrated into the highly porous elastomeric polymer film. The effective thermal conductivity of it is 8 $Wm^{-1}K^{-1}$.

The samples used for pull tests were manufactured using Nano-TIM coalesced between two PCBs with Sn coating under a pressure of 100 Psi. All samples were subjected to reflow at 85 °C for 5 minutes in air. Figure 1 shows the schematic of the samples. The size of the PCB is 50 mm X 5 mm X 1.5 mm, while the area of Sn-coating is 5 mm X 5 mm.

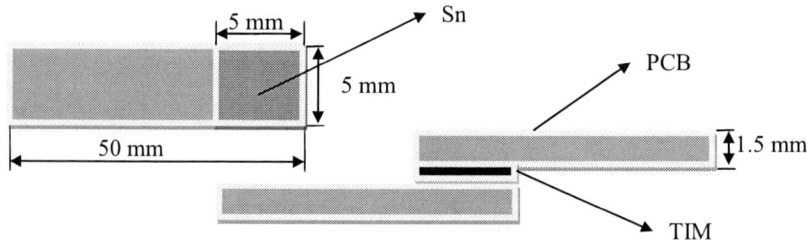

Figure 1: Schematic of samples for pull test.

These samples were divided into 6 groups. According to the Basic Environmental Testing Procedures for Electric and Electronic Products Test Ca: Damp Heat-Steady State, the 6 groups were subjected to thermal and humidity tests with the condition of 40 °C, 93 % Relative Humidity (RH) for 0 days, 2 days, 4 days, 10 days, 21 days, 56 days, respectively.

Shear tests were then carried out on all samples using a shear tester XLD with a shear speed of 2 mm/s. Finally, SEM analysis techniques were used to determine the morphology of the shear fracture section after pull tests and give a cross-section of the samples.

Results and Discussion

The results of pull test for all samples are shown in Figure 2. The results did not show much difference for the shear strength of the samples with different treatment. The shear strength decreased a little at first after thermal and humidity test for 2 and 4 days and then increased a little after thermal and humidity tests for long time. It is indicated that the thermal and humidity test at 40 °C, 93 % RH has little influence on the shear strength of the samples.

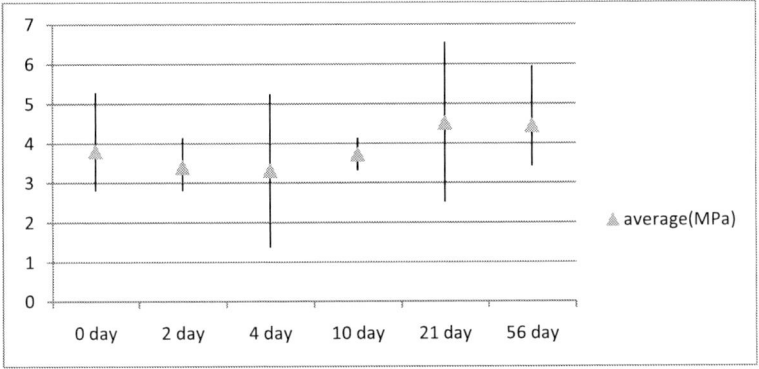

Figure 2: Results of shear strength of all samples after thermal and humidity tests

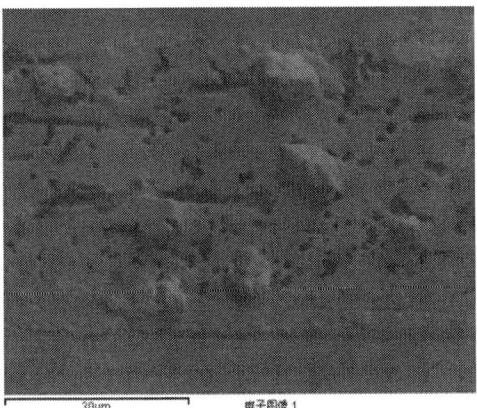

Figure 3: Cross section of TIM between two PCBs

Figure 3 shows that the alloy filled the polymer fiber uniformly. The PCBs were bonded perfectly by the TIM. No cracks or gaps are found on the interface between the PCBs and TIM or in the TIM.

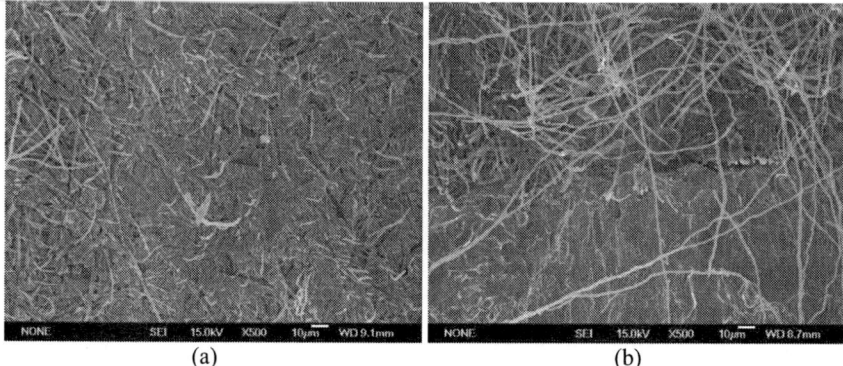

(a)   (b)

Figure 4: (a) The surface after pull test with shear strength of 6.54 MPa; (b) The surface after pull test with shear strength of 5.26 MPa.

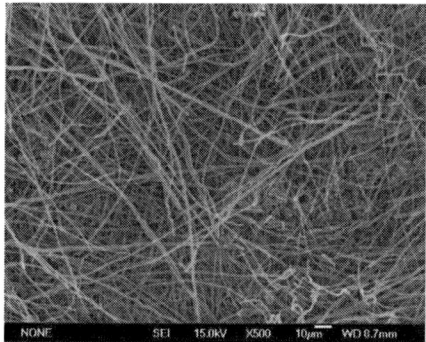

Figure 5: The surface after pull test with shear strength of 2.52 MPa.

According to the results of the shear strength and the SEM pictures, when the sample reflowed under enough pressure, the alloys flowed out, which will greatly affect the reliability of the samples. Figure 4(a) shows that there is almost no overflow of this sample. The alloy filled the fibers completely. The fracture surface shown in Figure 4(b) demonstrated that some overflow had occurred in the samples. No alloy existed on the fracture surface shown in Figure 5. The shear strength of the three samples is 6.54Mpa, 5.26Mpa and 2.52MPa respectively. The shear strength will drop dramatically with the level of overflow increasing. This indicates that the alloy contributes much more to the adhesive strength of the samples.

Conclusion

Due to low thermal conductivity of traditional TIM, Nano-TIM is becoming more and more important for thermal manage today and in the future. The reliability of a new type TIM named Nano-TIM was studied through the measurement of shear strength and SEM observation of the microstucture. The TIM is not influenced very much by the thermal and humidity condition, since the shear strength just changes a little after thermal and humidity tests at 40 ℃, 93 % RH. The shear strength of the nano-TIM was influenced by the filling level of the alloy in the fiber very much. If the serious overflow of the alloys happened during the reflowed, the shear strength will drop.

Acknowledgments

This work was supported by the STC Torch program, contract no: 0903H195300, EU programs "Thema-CNT" and Nanopack. This work was also carried out within the Sustainable Production Initiative and the Production Area of Advance at Chalmers. This support is gratefully acknowledged. The authors are also grateful for the support of the Swedish National Science Foundation under the project "Nanointerconnect" (621-2007-4660) and the Vinnova Program on "Designade Material" through the contract No. 2009-03230.

References

1   E. Pop, S. Sinha and K. E. Goodson, *Heat Generation and Transport in Nanometer-Scale Transistors*, Proceeding IEEE, Vol. 94, No. 8, p. 1587-1601 (2006).
2   ITRS (International Technology Roadmap for Semiconductor), Assembly & Packaging (2007).
3   Johan Liu, Michael Olugbenga Olorunyomi, Xiuzhen Lu, Wen Xuan Wang, Tomas Aronsson and Dongkai Shangguan, *New Nano-Thermal Interface Material for Heat Removal in Electronics Packaging*, pp.1, Electronics Systemintegration Technology Conference Dresden, German(2006).
4   D.D.L. Chung, *advances in Thermal Interface Materials,* Advancing Microelectronics, No.33, p. 8 (2006).
5   R. Prasher, *Thermal Interface Materials: Historical Perspective, Status, and Future Directions*, Proceedings of the IEEE, Vol. 94, No. 8, p. 1571 (2006).
6   Björn Carlberg, Teng Wang, Yifeng Fu, Johan Liu and Dongkai Shangguan, *Nanostructured Polymer-Metal Composite for Thermal Interface Material Applications,* 58th Electronic Components and Technology Conference, p. 191 – 197, Lake Buena Vista, FL(2008).

ECS Transactions, 34 (1) 997-1002 (2011)
10.1149/1.3567705 ©The Electrochemical Society

Failure Mechanism and Testing of PCB Pad Cratering

Dongji Xie[a], Miao Cai[b], Boyi Wu[b], David Geiger[a], Dongkai Shangguan[a] and Ivan Martin[c]

[a] Flextronics International USA 847 Gibraltar Drive, Milpitas, CA 95035, USA
[b]Flextronics Manufacturing Zhuhai No.168, ZhuFeng Road, Xin Qing
Science&Technology Industrial Park, Zhuhai, China
[c]Ivan Martin, Flextronics, 21 Richardson Side Rd, Kanata, Canada

Pad cratering of printed circuit board (PCB) is becoming a prevailing issue encountered in the PCB assemblies which is accelerated when switching to leadfree process. These units with pad cratering may not fail during functional testing but raise potential failure in the field. This paper uses both experimental and finite element analysis (FEA) approaches to understand the pad strength and pad stresses. Extensive mechanical testing by pin pull tests were performed on PCB materials. Cohesive elements are employed to simulate the bonding at the interfaces of pad, laminate and fibers. The results from FEA show that the laminate cracking can be successfully simulated. The design variables such as trace geometry and solder fillet impact are also critical for pad cratering.

Introduction

Pad cratering is defined as a separation of the pad from the PCB resin/weave composite or within the composite immediately adjacent to the pad. Pad cratering in the PCB (printed circuit board) is a failure which occurs in the prepreg underneath the pad. This is one of the major failure modes encountered in electronic assemblies detected recently especially for leadfree and halogen free processes. The units with pad cratering may not fail immediately but fail during rework later in the field. It triggers failures only if the crack propagates into a copper trace or conduction pad and makes the circuit open or intermittent. It can have a significant impact on both production yield as well as long term reliability.

Pad cratering has been widely seen in drop and shock tests as one of the major failure modes, giving an impression that pad cratering only occurred in drop and shock tests or other mechanical tests. In reality, pad cracking could actually occur under thermal stress or thermomechanical stress, especially for large BGAs (body size >30mm).

Pad cratering starts from cracking of the laminate. The crack may initiate from the intersection of the solder, copper pad and laminate, as this is a stress concentration point for crack initiation [1]. The crack may propagate from the interface of the epoxy to the glass bundle under certain stresses.   Depending on the thickness of the resin in the prepreg, the glass bundle may be buried in a distance of 5 to 50 microns from the pad. The interface usually acts as a path for crack propagation. The crack may then propagate through the pad and create a total pad cratering.   The propagation of the crack through the glass reinforcement and along the glass-resin interface may take a significant amount of time and stress to become a complete failure depending on the stress conditions.

997

In order to characterize the pad cratering performance, three levels of tests can be performed: PCB materials, soldered PCB (pad-solder level) and system level. The PCB material is associated with the raw materials (such as epoxy resin, filler and fiber), and the layer structure and fabrication process. The test methods at the PCB level are mostly defined by IPC TM650 in various tests including modulus, flexural strength, peel strength, fracture toughness, etc. Modulus and flexural strength are basic mechanical properties of the laminate. They are dependent on the resin content and curing system. Peel strength test is to test the copper clad adhesion to the laminate. The fracture toughness is not a standard IPC test yet, so it is unavailable in most materials data sheets.

The pad-solder level test includes ball shear [2, 3], ball pull [3] and pin pull [3,4], which are being drafted in a new IPC standard (IPC 9708) [4]. Those tests are used to test the adhesion of the pad to the solder and pad to the laminate as well as the strength of the bulk prepreg layer. The ball shear test and ball pull are more convenient to execute. However, the failure mode is not consistent. In most cases, it may fail in the solder and the failure by pad cratering may only account for 50% on average depending on the pad condition and test set-up. The pin pull test [3] is performed by soldering a pin to the pad and can be done hot (up to 180C) or cold (at the room temperature). This method is attractive because it can detect very high percentage of failures of laminate cracking (>90%).

The system level tests are non-standard tests and performed at the product level, which includes PCB, component, solder joints, and box build. The BGA pads could perform much differently under different loading conditions even if the PCB materials and design are the same. The test method at this level is limited because of complexity. The test methods used most widely include cross sectioning dye and pry after thermal cycle stressing, and/or drop and vibration tests. These tests are more representative of the field usage and hereby more realistic, but at the same time more complex.

Strength of FR4 and Modeling

Flexural strength is widely used in the evaluation of the PCB materials subjected to a bending stress. According to IPC TM-650 2.4.4 [5], the bending strength is a maximum stress of the sample at the breaking point using Equation 1.

$$\sigma_b = \frac{3PL}{2wt^2}$$

(1)

where, σ_b is the flexural strength; P is the bending force; L is the length span of the sample; w is the width; and t is the board thickness. The flexural strength for epoxy is usually above 500MPa at room temperature for FR4 cured with either Dicy or phenolic novolac[6]. This flexural strength represents the performance for the whole layer. The flexural strength is lower if only prepreg layer is used. Fig. 1 shows the setup for the bending test for PCB. Here only a layer of prepreg materials (resin content 46%, 7628x1) with a thickness of 0.1mm is used. The bending test results show that the flexural strength of the prepreg is about 270MPa at 50% failure rate according to Equ. 1.

This is equivalent to half of the regular FR4 materials (the flexural strength from the data sheet shows 474MPa for this material).

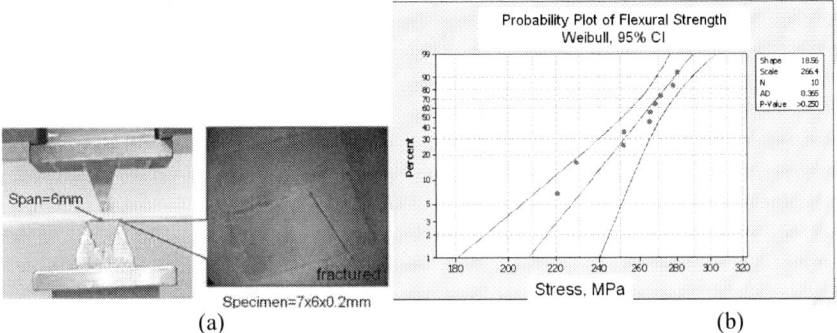

(a)   (b)

Fig. 1 Flexural test on prepreg materials using three. (a) bending fixture and a specimen after bending test and (b) Weibull plot of flexural strength of the prepreg.

Pad Strength Test by Pin Pull

As suggested by IPC 9708 for the pin pull test by soldering a pin to the BGA pad, it may fail at the laminate (prepreg), solder and pad to laminate interface. To understand the failure mechanism, an FEA model is constructed as shown in Fig. 2. In Fig. 2, a copper pin is soldered on the BGA pad. Several layers underneath the pad are considered. The adhesive layer accounts for delamination or peeling of the pad. A cohesive layer is placed between the prepreg and the glass fiber layer which could be the dominant failure during the pin pull test. All the other layers are combined into bulk FR4. These layers are underneath the pad with a thickness between 5 to 10 um. The total thickness of the board is about 1.6mm. The pad size is about 0.43mm. The materials properties used in this simulation can be found in Ref 7. Only cohesive element is used for the cohesive layer with the damage stress at 69MPa [1,8].

FEA results for the pad without traces attached are shown in Fig. 2b. Fig. 2b shows the variation of the maximum principal stress in different layers vs. pull force. It is noted from Fig. 2b that the unit failed at a pull force of 9.9N at the cohesive layer. At that stress level, the stress at the prepreg level is the highest which is close to 225MPa. This is close to the failure level indicating that there is a possibility that the bulk prepreg has failed. The solder could also have a higher stress and risk of failure. This shows that the solder volume and soldering quality are still critical to preventing the solder fracture. When comparing the adhesive layer and the cohesive layer, the stress at the adhesive layer is normally higher, suggesting that the pad peeling or lift may occur earlier than the cohesive layer if both have the same damage strength. As shown in previous study [1], the peel strength of the copper pad can be estimated by Equation 2 as follows:

$$F_p = K * \sigma_L * 3.14 * d \qquad (2)$$

where, F_p is the peel force at a pad size of d, and σ_L is the peel strength for laminate materials, K is a scale factor to accommodate the pulling effect. It shows that F_p is equivalent to the pad strength by pin pull if K=5. Usually, the pad peeling is not a common failure mode for a virgin PCB board [1~3], because the failure probability of laminate cracking is 90%. This shows that K>>5. The stress at the bulk PCB is always the lowest indicating no risk of the failure in the bulk PCB. The cohesive layer could fail earlier than the other layers which could be the initiation point of the pad cratering.

To characterize the pad cratering performance of the BGA board, a standard leadfree test board is tested by soldering copper pins on several pad locations. To model the impact of the traces, a copper trace with three types of pads, Pad O, T and V. Pad O has no trace and Pad T has one narrow trace (0.1mm) which is a typical signal pin. Pad V is the pad attached to a via with a trace width of 0.4mm. All of those pads are non solder mask defined. In this experiment, the PCB boards were reflowed at a peak temperature of 260C.

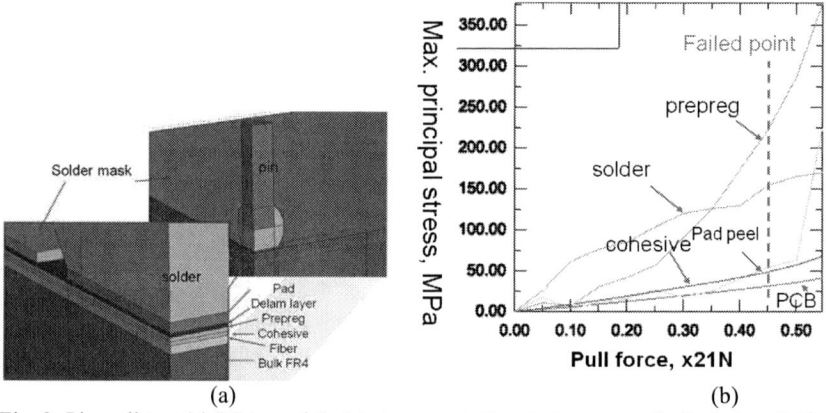

(a)   (b)

Fig. 2 Pin pull test (a) FEA model; (b) stress variation during pin pull. Pad size=0.43mm, non solder mask defined pad.

The trace width does impact the pad pull force as shown in Fig. 3. Fig. 3 demonstrated that Pad T and V have a higher strength compared to that of Pad O which is attributed to an enlarged perimeter of pads. Compared to Pad O and T, the pull force of Pad V is clearly higher as the pad width is close to the pad diameter so that the total area for pulling is significantly larger. A comparison of FEA and experimental results for various pad types is summarized in Fig. 3. For the experimental, a total of 30 pads per type are tested and the minimum and maximum data are illustrated in Fig. 3. All of them failed due to pad cratering. As noted from Fig. 3, the FEA results agree well with the average pull force for different pad conditions. Interestingly, the FEA prediction without solder fillet is close to the average value of the experimental results while the predicted value for pads with solder fillet is close to the maximum of pad pull readings. This suggests that solder volume and quality impact on pad cratering.

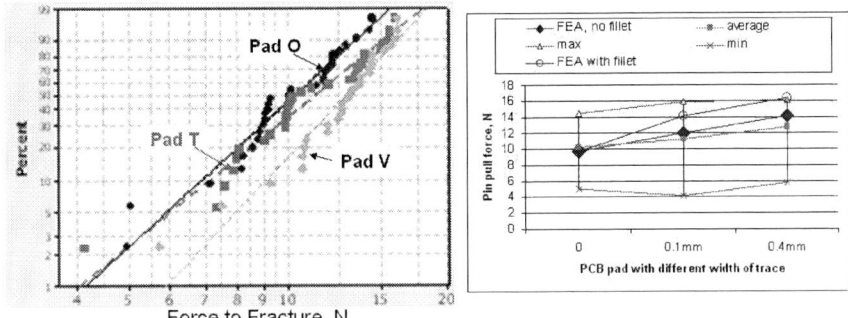

Fig. 3 Pin Pull Forces in (a) Weibull plots and (b) impact of pad width.

Failure mode during Pin Pull Test

To get a valid pin pull test, pad cratering has to be achieved, i.e., the units shall fail mostly inside the laminate or prepreg. The pin pull test failure in this work is illustrated in Fig. 4. The failure rate by pad has achieved 95% in this study. As shown in Fig. 4, a standalone pad failed at the interface between the epoxy and glass bundle, i.e. a cohesive type of failure. Similarly, the same failure mode has been achieved for the pad with a signal trace or via.

Summary

Pin pull test and FEA analysis were performed on the BGA pad with and without traces. The results show that cohesive failure mode is dominant and with the strength predictable by FEA. The pin pull strength can be enhanced by the soldering traces for non solder mask defined pad. The flexural strength test shows that a lower strength could be one of the reasons that the prepreg layer has a higher probability to fail. However, the cohesive failure could occur earlier than cracking in the prepreg. To characterize the pad cratering, several tests can be used at the PCB materials level, board level, and system level. At the materials and bare PCB level, a flexural strength test on the prepreg and pin pull test on the BGA pad are recommended. The qualification test shall be performed in the system level [7] and use actual PCBA and simulates multiple reflows and board handling as well as stresses in the field.

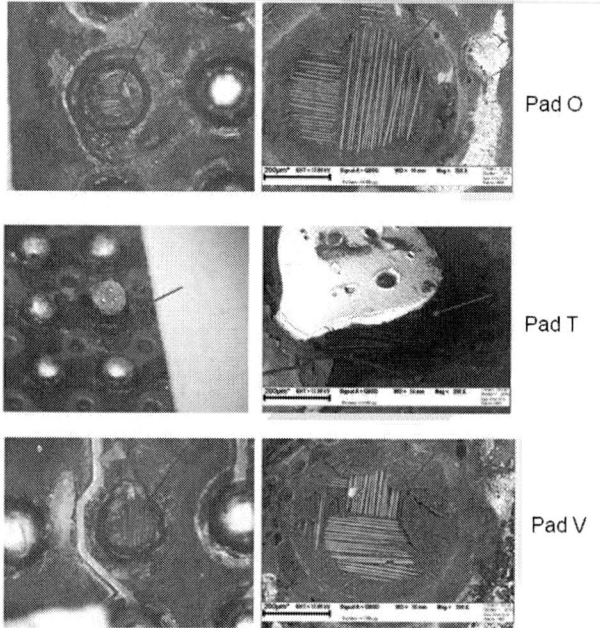

Pad O

Pad T

Pad V

Fig. 4 Typical failure modes of pad cratering in various pads. Two sets of pictures are shown: optical on the left and SEM on the right. It shows clearly the glass bundle is exposed and the failure is inside the prepreg layer.

References

1. D. Xie, D. Shangguan and H. Kroener, APEX 2010, Las Vegas, NA.
2. D. Xie, C. Chin, K. Ang, D. Lau, and D. Shangguan, Proceedings of Electronic Components and Technology Conference, 2008
3. M. Ahmad, J. Burlingame, and C. Guirguis, Apex 2008.
4. IPC 9708: Test Methods for Characterization of PCB Pad Cratering, 2009.
5. IPC TM650-2.4.4: Flexural strength of laminate (at ambient temperature).
6. FR-370HR Laminate data sheet, Isola Group.
7. D. Xie, D. Geiger, D. Shangguan, M. Cai, B. Wu, B. Hu, H. Liu and I. Martin, Proceedings of Electronic Components and Technology Conference, 2009
8. B. Roggeman and J. Li, UNOVIS AREA ARRAY CONSORTIUM 2007.

CHAPTER 9

EMERGING SEMICONDUCTOR
TECHNOLOGIES

FPGA Design with Double-Gate Carbon Nanotube Transistors

M. H. Ben Jamaa[a], P.-E. Gaillardon[a], S. Frégonèse[b], M. De Marchi[c],
G. De Micheli[c], T. Zimmer[b], I. O'Connor[d], and F. Clermidy[a]

[a] Commissariat à l'Energie Atomique (LETI), Minatec Campus, 38054 Grenoble, France
[b] Laboratoire IMS, CNRS-UMR 5218, Université Bordeaux 1, 33405 Talence, France
[c] Ecole Polytechnique Fédérale de Lausanne (EPFL), 1015 Lausanne, Switzerland
[d] Institut des Nanotechnologies de Lyon (INL), Site ECL, 69134 Ecully, France

> *Double-gate carbon nanotube field effect transistors (DG-CNTFETs)* are novel devices showing an interesting property allowing to control the p- or n-type behavior during the device operation. This opens up the opportunity for novel design paradigms. Based on a compact physical model of these devices, we demonstrate the benefit of designing *field-programmable gate arrays (FPGAs)* using fine-grain DG-CNTFET logic blocs rather than traditional look-up tables and coarse-grain DG-CNTFET logic blocs. In particular, we show a reduction by 13% to 48% on average in terms of delay of FPGA benchmarks.

Introduction

The scaling down of complementary metal-oxide-semiconductor (CMOS) technology has led to the emergence of novel post-CMOS devices, such as *carbon nanotubes (CNTs)*. One of the challenges of using CNT technology for building transistors is the chemical doping. Using undoped CNTs is possible, but it results in an ambipolar behavior of the *carbon nanotube field effect transistors (CNTFETs)*, meaning that undoped CNTFETs conduct under both positive and negative gate bias. This issue is addressed using a second gate, which controls whether the device operates as p- or a n-type [1]. The polarity of *double-gate* (or *dual-gate*) *CNTFETs (DG-CNTFETs)* can be selected during operation time.

This property offers the opportunity to design logic gates with reconfigurable devices, leading to more logic functions drawn on the same silicon area. We leverage this property by constructing a full reconfigurable logic system that is reminiscent of a *field-programmable gate array (FPGA)*. Reconfigurable circuits are gaining interest because of their low technology cost, fast design time and enhanced fault-tolerance compared to *application-specific integrated circuits (ASICs)* [2]. However, FPGAs necessitate a large number of configuration memories and have a higher cost in terms of area and power consumption compared to ASICs. Our approach to address these issues is to implement FPGAs using reconfigurable DG-CNTFET logic gates instead of *look-up tables (LUTs)* as *basic logic elements (BLEs)*. Our BLEs require less configuration memory, and their design is area efficient.

In order to assess the benefits of the proposed approach, we first build a family of basic logic elements with DG-CNTFETs, which we characterize using a compact model.

This work was funded by the French National Research Agency under the program ANR-08-SEGI-012 "NANOGRAIN".

We then evaluate the FPGA performance on a large application benchmark for different architecture scenarios. The same design flow is used with LUT-based FPGAs. We demonstrate that the proposed approach with fine-grain DG-CNTFET gates is 48% faster than coarse grains and 13% faster than LUTs. FPGAs with fine-grain DG-CNTFET gates have a cost in terms of area with respect to LUTs (10%), while they are 45% smaller than coarse-grain DG-CNTFET gates.

The paper is organized as follows. We first survey previous works dealing with DG-CNTFET technology. Then we introduce a compact physical model for the considered devices. Subsequently, we introduce the design of reconfigurable DG-CNTFET FPGAs. Finally, we benchmark the presented approach and compare it with the standard LUT-based implementations.

Background and Related Work

Following more than one decade of research on CNTFETs, several issues have been addressed and some challenges have been identified. In this section, we first survey the state-of-the-art of CNTFET technology, before we focus on the physics of DG-CNTFETs.

Depending on the used materials and doping profile, different types of CNTFETs are reported in literature: the major distinction is between MOSFET-type CNT and Schottky-Barrier-type [3]. The first family is characterized by an intrinsic CNT channel and highly doped drain/source regions forming an ohmic contact to the metal. For the second family, the whole CNT along the channel and the drain/source regions is intrinsic. These devices have a Schottky barrier and are ambipolar, *i.e.*, they conduct both electrons and holes, showing a superposition of n- and p-type behaviors. The Schottky barrier thickness is modulated by the fringing gate field at the CNT-to-metal contact; allowing the polarity of the device to be set electrically [1].

Whereas the uncontrollable ambipolar behavior, is undesirable, the ability to select the CNTFET polarity (p- or n-type) in-field by controlling the fringing gate field suggests the innovation of using a second gate, the *polarity gate*, to control the electrical field at the CNT-to-metal junctions and to set the device polarity [1]. The physics of the considered dual-gate device is illustrated using its energy-band diagram. By setting the polarity gate to a high value, the drain and source regions let electrons e⁻ pass and block holes h⁺. Then, the device operates as a n-type transistor. The opposite happens when the polarity gate is set to a low value, and then the device operates as a p-type transistor.

Modeling Double-Gate Carbon Nanotube Field Effect Transistors

In order to assess the benefits of the considered devices at the circuit architecture level, we first need to simulate the electrical behavior of the building blocks of the circuits. This requires an accurate physical model, which, in turn, depends on the underlying technology. Therefore, we first start by explaining the technological assumptions. Then, we introduce the physical compact model.

Fabrication Process

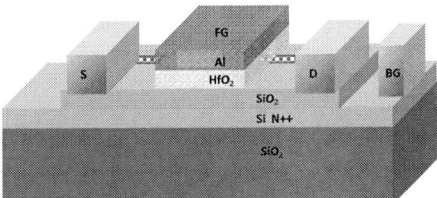

Figure 1: DG-CNTFET device showing the source (S), drain (D), front (FG) and back (BG) gates

At present there is no standard CNTFET process. All developed devices have been produced experimentally. In order to use a realistic and CMOS-compatible process flow, we suggest the following process steps (Figure 1). We start with a SOI wafer, on top of which intrinsic CNTs are deposited or transferred. Then, the gate oxide (HfO$_2$) and the metal (Al) of the top gate are deposited. Following these steps, the top gate is patterned. Then, the active area and the back gate are defined by SiO$_2$ and Si etch respectively. Subsequently, the metal is sputtered onto the contacts to drain, source and to both gates.

Physical Device Modeling

The physical model described in [4] is related to the structure in Figure 1, which is made of three different regions: source access, inner part (underneath the front gate) and drain access. In this structure, four energy barriers appear in the device: at the metal to source (or drain) access junction, two SB-like barriers appear, while at the source (or drain) access to the inner part junction, the barrier is more conventional and is of a pn-junction shape. Depending on the work function difference between the metal contact and the CNT, carriers at the metal-CNT interface encounter different barrier heights: Carriers with energies above the Schottky barrier height reach the channel by thermionic emission. On the other hand, carriers with energies below the Schottky barrier height have a probability to reach the channel according to a transmission function describing the tunnel effect which can be calculated from WKB approximation.

To overcome the complexity of WKB expression for compact modelling, an approximation based on works from [5] is applied. This effective barrier height model is described in [6]. The electron (hole) current is calculated through the Landauer equation, by integrating over energy from the dominating barrier to infinite. The dominating barrier position depends on the applied bias. In fact, the electron current can be limited by three barriers: (i) the Schottky barrier from source, (ii) the one from drain and (iii) the conduction (valence) band of the inner part.

The analytical expression of the drain current is given in [6]. In this model several other features are included. On the one hand, the band-to-band tunnelling has been developed for MOSFET-like CNTFET in [7] and has been validated through NEGF simulation. On the other hand, charges have been modeled according to the ballistic assumption and the analytical expression of charge in each region is given in [1]. The potential calculation inside the device is given in [4].

Fine-Grain Reconfigurable Architecture

The dual-gate feature of the considered devices enables the possibility to leverage the in-field polarity control. The in-field control means the ability to control the device polarity during the operation of the system, following the design and fabrication steps. This opportunity enables a reconfiguration of the circuit at a very fine grain, which is ultimately the device-level configuration. Today's most used reconfigurable circuits are FPGAs. These regular circuits formed by several reconfigurable logic blocks called *configurable logic blocks (CLBs)*, in addition to other logic modules and reconfigurable interconnects [8]. Every CLB consists of a set of N *basic logic elements (BLEs)*. A BLE is a K-input *look-up-table (LUT)*, whose output can be possibly routed to any other LUT input through a latch. Every CLB has I inputs coming from other CLB outputs.

A standard FPGA architecture is depicted in Figure 2, showing CLBs, connected to the routing lines through *connection blocks (CBs)*. The routing blocks are connected through *switch boxes (SBs)*. We focus on the BLE design and we optimize it in the following. The standard BLE design is based on LUTs (see Figure 3(a)). The K-input LUT is a set of 2^K *static random access memory (SRAM)* cells. In our approach, the novelty is threefold. We first replace the K-input LUT by a K-input logic gate designed with DG-CNTFETs, which can be reconfigured with both gate signals. Secondly, we suggest using the input signals not only to make the calculation, but also to perform the configuration by providing the power supply signals as additional inputs, that we multiplex with the initial ones. Finally, we allow the permutation of the power supply of the logic cell, since the *pull-up (PU)* and *pull-down (PD)* networks of a DG-CNTFET can be designed with the same size [9]. The obtained novel reconfigurable cell is depicted in Figure 3(b).

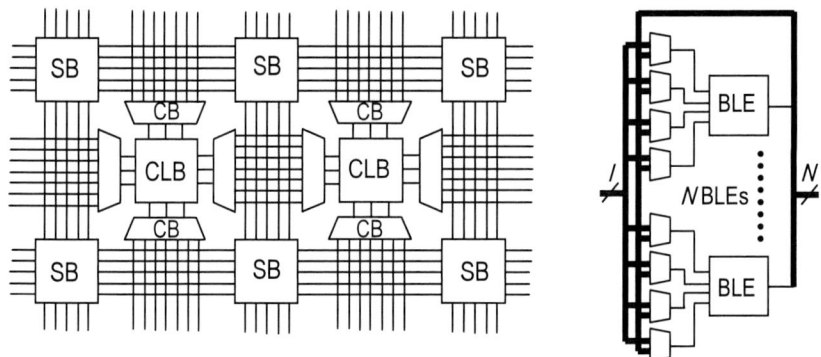

Figure 2: FPGA organization from [8]: Island-style FPGA (left), zoom-in into a CLB (right)

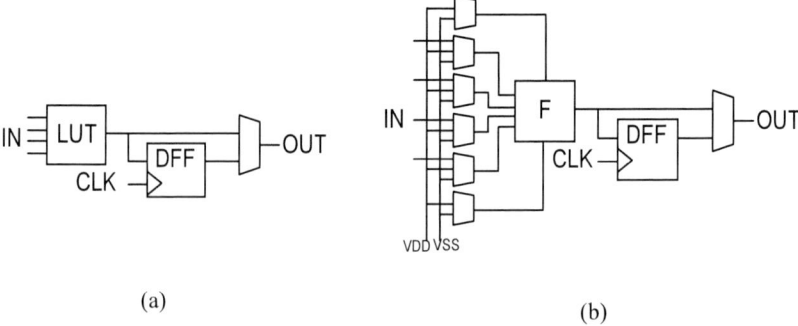

(a)   (b)

Figure 3: (a) LUT-based BLE, (b) BLE based on a reconfigurable DG-CNTFET gate

Simulation Results

We used a set of logic circuits taken from the MCNC and ISCAS'89 benchmarks [10]. We defined a gate library corresponding to the reconfigurable 4-input logic gate (Figure 4), which we used in order to synthesize the benchmark using the ABC tool [11]. It has been reported that gates similar to the used one have a high degree of reconfigurability [9]. The gate library was characterized using the presented DG-CNTFET compact model. We then performed the technology mapping with a library of 4-input LUTs (K=4) using ABC as well. Subsequently, we performed the logic packing of the mapped circuit into CLBs with (N,I) set to (10,22) then (1,4) using T-VPACK [12]. Finally, the placement and routing were carried out using VPR [12].

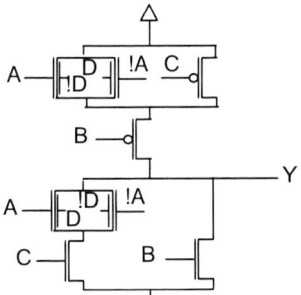

Figure 4: Reconfigurable 4-input DG-CNTFET gate

We compared the geometric average of the synthesis results over the 17 benchmarks for 3 scenarios S1 to S3: reconfigurable gate and LUT implementation at a fine granularity (N,I) = (1,4) for S1 and S2 respectively and reconfigurable gate implementation with a coarse granularity (N,I) = (10,22) for S3. We first note that the reconfigurable gate implements only 17 functions, compared to the 2^{16} functions implemented by the LUT. However, the logic gate efficiently implements most of the functions required for logic synthesis.

TABLE I. Implemented architectural scenarios for FPGA synthesis

Scenario	N	I	CLB area (norm.)	Intra-CLB delay (ps)	Inter-CLB delay (ps)
S1	1	4	2419	46.9	25.1
S2	1	4	2560	50.4	25.4
S3	10	22	17167	199.8	423.3

Table I shows the normalized CLB area (to unit transistors area including contacts), the delay between neighboring CLBs and the delay with the CLBs for the three scenarios simulated with the compact DG-CNTFET model. We notice that the fine-grain architectures naturally provide a smaller CLB area and a faster delay within CLBs because of the lighter input multiplexing, and a faster delay between CLBs because of the lower load of CLB in- and outputs. On the other hand, gate-based CLBs are slightly slower than the LUT-based CLBs because of the more compact BLE design, and they have a smaller area because of the smaller number of required SRAM cells, thanks to the multiplexing of reconfiguration and logic on the same input signals.

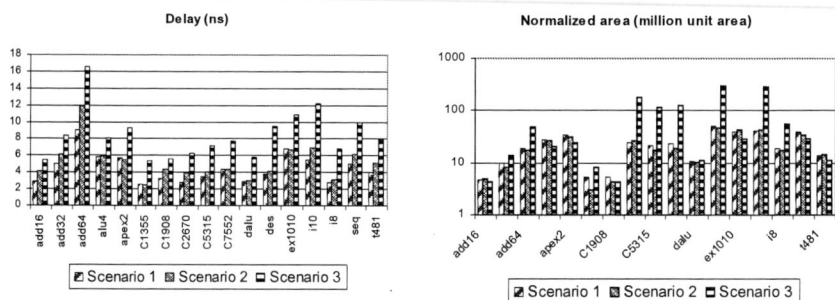

Figure 5: Delay and area comparison between the considered scenarios for different circuits

Figure 5 illustrates the delay and normalized area of the benchmark in the considered architectural scenarios. On average, the fine-grain logic implementation is 48% faster than the coarse-grain logic implementation, and it is 13% faster than the LUT implementation. The fine-grain implementation has therefore the highest performance. This comes, however, with a cost in terms of area. Because of the larger number of required CLBs, namely for pure routing, the fine-grain logic-based circuits are on average 10% larger than the LUT-based ones. On the other hand, they are 45% smaller in size than the coarse-grain logic-based counterparts because of the less inter-CLB routing resources.

Conclusions

In this paper, we introduced a novel family of DG-CNTFETs, which shows the unique property of in-field controllability. Based on a compact physical model of the considered devices, we characterized logic gates, which have been specifically chosen because of their efficient reconfigurability. We enhanced the FPGA architecture with those properties and we mapped a benchmark of logic functions in different architectural scenarios. The approach demonstrated its efficiency with fine-grain architectures, showing a delay reduction up to 48% with respect to coarse-grain architectures and 13% with respect to state-of-the-art LUT-based architectures. While the fine-grain architecture has a cost in terms of area of 10% with respect to the LUT implementation, it is 45% smaller is size than the coarse-size architecture.

References

1 Y. Lin et al, *IEEE Trans. on Nanotechnology*, **4**, 481 (2005).
2 M. H. Ben Jamaa et al., *Proc. of DATE Conference* (2009).
3 I. O'Connor et al., *IEEE Trans. Circuits and Systems I*, **54**, 2365 (2007).
4 M. Najari et al, *IEEE Trans. on Electron Devices*, **58** (1), 206 (2011).
5 J. Knoch et J. Appenzeller, *Physica Status Solidi (A) Applications and Materials*, **205**, 679 (2008).
6 S. Frégonèse et al, *IEEE Trans. on Electron Devices*, **58** (1), 206 (2011).
7 S. Frégonèse et al, *IEEE Trans. on Electron Devices*, **56**, 2224 (2009).
8 V. Betz et al., *"Architecture and CAD for Deep-Submicron FPGAs"*, Kluwer Academic Publishers, New York (1999).
9 M. De Marchi et al., *Proc. of International Symposium on Nanoscale Architectures*, 65 (2010)
10 BLIF circuit benchmarks, http://cadlab.cs.ucla.edu/~kirill/
11 ABC: Berkeley logic synthesis tool, http://www.eecs.berkeley.edu/~alanmi/abc/
12 Versatile packing, placement and routing tool for FPGA, http://www.eecg.utoronto.ca/vpr/

ECS Transactions, 34 (1) 1011-1016 (2011)
10.1149/1.3567707 ©The Electrochemical Society

Three-Dimensional (3D) Integration Technology

T. Ohba

Institute of Engineering Innovation (IEI)
The Univ. of Tokyo
2-11-16 Yayoi, Bunkyo-ku, Tokyo, Japan

Three-dimensional (3D) integration and bumpless TSV (Through-Silicon-Via) technologies beyond post-scaling have been described. Since the extreme scaling is limited by physical and economic sense, 3D will be used concurrently with tow-dimensional legacy process. Because vertical wiring in WOW can be connected directly to the upper and lower substrates by self-alignment, bumps are not necessary when TSV interconnects are used. The low aspect ratio of TSVs allows a higher process margin and throughput in etching and metal filling. Multiple TSVs enable die-to-die connections independently, which improves the total yield in wafer-scale stacking. Stacking at the wafer level drastically increases the processing throughput, and bumpless multi-TSVs provide a yield equivalent to or greater than 2D scaling beyond 22-nm nodes.

Introduction

Two-dimensional (2D) scaling based on planar technology that started in the 1960s and has led the semiconductor industry so far no longer makes economic sense in many situations. The industry is facing a major turning point in how to realize the next generation of large-scale integration (Figure 1). In this context, many three-dimensional

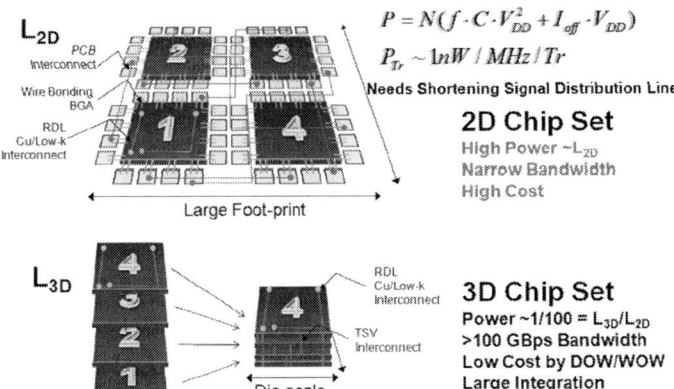

Figure 1. A comparison with wiring length and layout for 2D and 3D chip sets. Miniaturizing layout using 3D provides low power consumption, higher bandwidth, and large integration.

1011

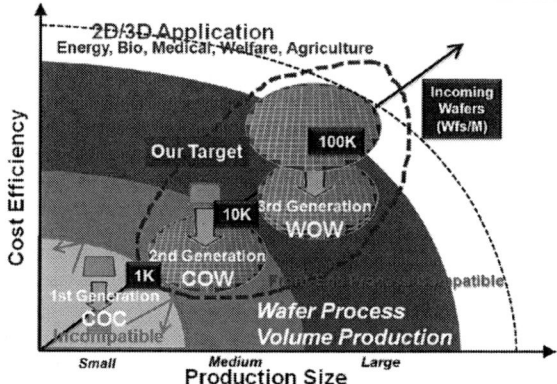

Figure 2. Cost efficiency as a function of production size in the three-dimensional device manufacturing. COC obviously differs from COW and WOW in that it is not wafer processing and thus incompatible for process facilities in the front-end-of-line. COW and then WOW will be used for volume production in the next 3D manufacturing.

(3D) approaches have been proposed (1)-(5), and recent attention has focused on productivity and the costs involved in volume production (6)(7). The key features in bumpless WOW (Wafer-on-Wafer) technology are the fabrication of three-dimensional structures in which any number of thinned-down 300-mm wafers can be stacked back-to-front and self-aligning multi-TSV interconnects without bumps (8)-(10). WOW is the third generation, succeeding COC (Chip-on-Chip) and COW (Chip-on-Wafer) technologies, and is applicable to bumpless wafer-based COW (Figure 2). Stacking at the wafer level drastically increases the processing throughput, and bumpless multi-TSVs provide a yield equivalent to or greater than 2D scaling beyond 22-nm nodes (11)(12). Also, since it is compatible with existing manufacturing facilities in wafer processing, transistors through three-dimensional structures can be designed in a continuous manufacturing line.

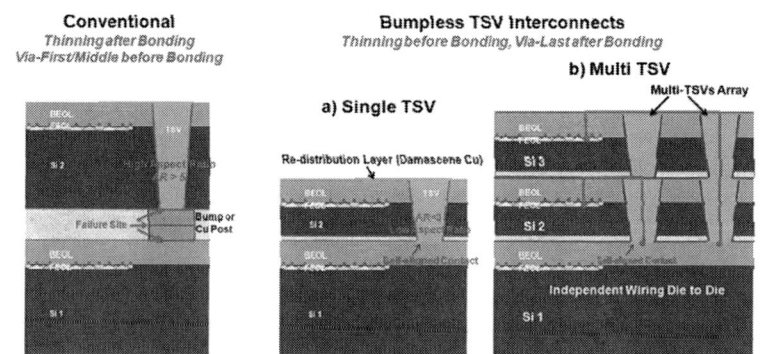

Figure 3. A comparison of conventional TSV using bump electrodes and TSV interconnects without bump in the WOW technology.

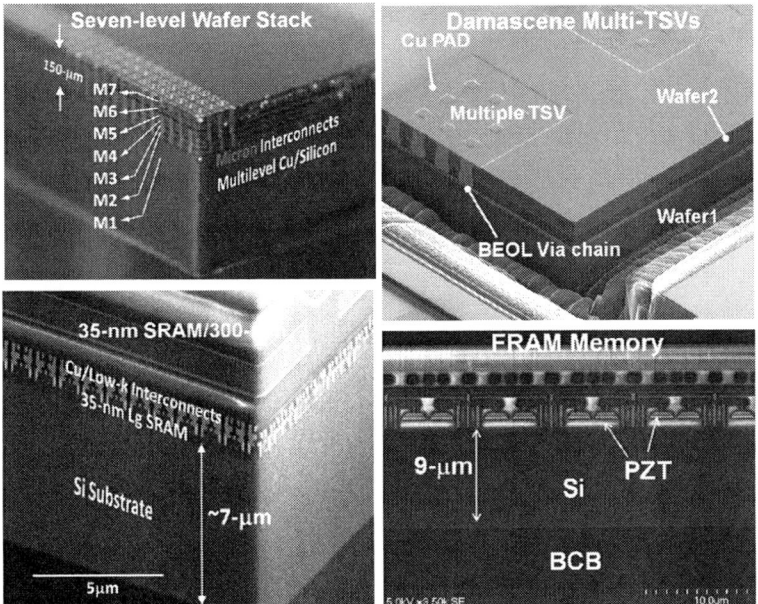

Figure 4. SEM pictures of seventh-level wafer stack with single TSV, Damascene multilevel TSVs, and ultra thinned Si substrates using 35-nm SRAM logic and FRAM memory.

Bumpless TSV and WOW Applications

Bumpless TSV Process

WOW is classified into two types, according to the stacking method: *Thinning before Bonding* and *Via-Last after Bonding*. Figure 3 shows a comparison of conventional TSV with and without bump in this study. The development of WOW has proceeded through four modules, classified along the process flow. The modules include a thinning module for thinning the wafer substrates in which devices are implemented, a stacking module for bonding and stacking the wafers, a TSV interconnects module for forming Cu interconnects embedded in upper and lower wafers with TSVs, and a packaging module for singulating the stacked wafers. Damascene interconnects form a so-called redistribution layer and also serve as a counter electrode for the subsequent stacked wafer (13).

Electrical Characteristics

Because wafer thinning to 10-μm or less does not affect the device characteristics (13)-(15), thinned wafers can be stacked one on another without causing any problems (Figure 4). For instance, a 300-mm wafer in which 35-nm gate length logic devices are implemented can be thinned to 7-μm without degrading device performance, i.e. current density, threshold voltage, and lekage current. Figure 5 shows cumulative failure distribution of Cu interconnects via chain resistance with and without Cu-TSV. Contact

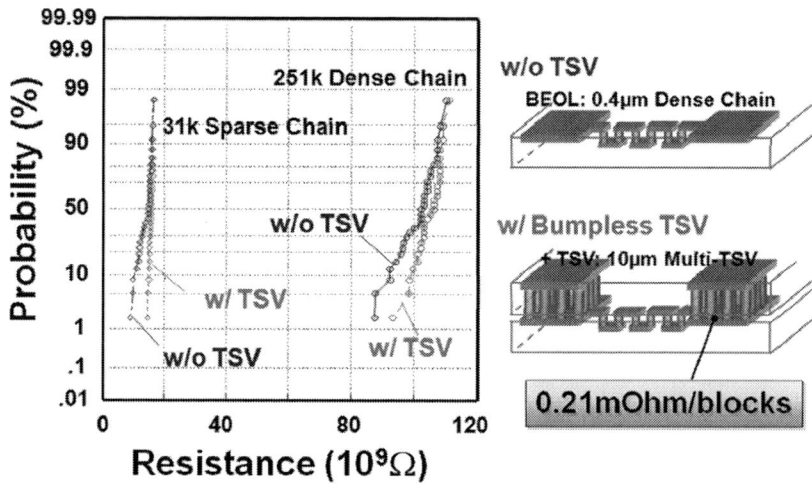

Figure 5. Via chain resistance cumulative failure distribution of Cu interconnects with and without Cu-TSVs compared with via chain density. Schematic diagram of test structure for electrical measurement with and without TSV formed on 65-nm Cu interconnects.

resistance of Cu-TSV to Cu-BEOL is estimated at 0.21-mΩ. There is no open failure and significant change in resistance distribution with TSV. Leakage current characteristic between multiple TSVs blocks was as low as 2.3 x 10^{-11} A at 4.0 V. This suggests that mutiple TSV is suited electrically for 3D stacking.

Concurrent Manufacturing and Prospect

Since cost reduction requires the adoption of advanced lithography technology, advanced lithography and peripheral support facilities account for one-third to one-fourth of the total cost of a manufacturing line. In short, while useful for reducing chip cost, scaling is very burdensome in terms of capital investment. Large-scale investment has so far been made considering two-to-three generations ahead. This is based on the empirical rule that profits are made several generations after investment for reasons of the tradeoffs between products sales and facility depreciation (16).

According to this empirical rule, the investment in 22-nm technology needs to be made in consideration of its applicability to 10–15 nm technologies. However, the price of extreme ultra violet (EUV: λ = 13.5-nm) lithography machines is more than twice that of ArF immersion lithography machines, and their current throughput is around one-tenth or less. When converted into the processing capacity of a current large-scale fabrication facility in accordance with this system performance, an investment of approximately 2000 million USD is required for EUV technology. Considering that the past lifelong sales for each generation were approximately 10 times the corresponding business investment, the expected market size necessary for this investment is 5 to 10 billion USD. Based on the 300 billion USD semiconductor market, this expected market size is not realistic. In short, this is the limit of scaling in light of the economics of the industry, and it is difficult to find a scenario of victory at present.

In combination with three-dimensional stacking for such problems associated with scaling, a roadmap towards high-density integration backed up by production costs can be made. Keeping the wafer shape as-is for stacking ensures compatibility with manufacturing facilities in front-end processing and helps utilize the process technology nurtured in wafer processing. If the processes up to three-dimensional stacking can be handled as units in the manufacturing line, the throughput will be one-hundredth of that in stacking starting with chips. Therefore, future semiconductor manufacturing is expected to advance with a roadmap in which the number of stacked wafers, the wafer thickness, and the number of TSV interconnects serve as indices (Figure 6).

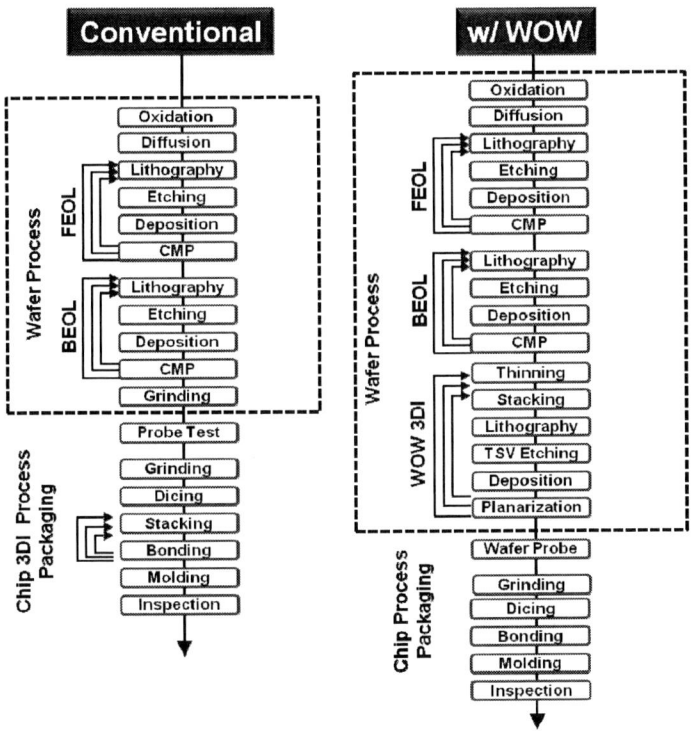

Figure 6. A comparison of the conventional 3D manufacturing line and combination of WOW wafer processing. In case of conventional line, 3D integration is carried out using chips in which singulation process for out-coming wafers are needed after front-end processing. The WOW process enables 3D integration and wafer processing continuously from front-end and the throughput will be one-hundredth of that in stacking starting with chips.

Conclusions

Concurrent three-dimensional integration and bumpless WOW technology for three-dimensional wafer-scale stacking has been described. Because vertical wiring in WOW can be connected directly to the upper and lower substrates by self-alignment, bumps are not necessary when TSV interconnects are used. The low aspect ratio of TSVs allows a higher process margin and throughput in etching and metal filling. Multiple TSVs enable die-to-die connections independently, which improves the total yield in wafer-scale stacking. Thus, high-density integration, such as Terabit capacity with manufactured wafer stacking will become more flexible, and manufacturing costs will be lowered compared with extreme scaling.

Acknowledgements

This study was carried out based on the three-dimensional integration development program by the WOW alliance of the University of Tokyo. The author thanks the more than 20 alliance members and WOW Research Center Ltd. for their cooperation.

References

1. J.F. Gibbons et al., EDL-**1** (6), 117 (1980).
2. S. Kawamura et al., EDL-**4** (10), 366 (1983).
3. S-M. Jung et al., IEDM Tech. Dig., 265 (2004).
4. J.A. Burns et al., IEEE Trans. Elect. Dev. **53**, 2507 (2006).
5. F. Liu et al., IEDM Tech. Digest, 599 (2008).
6. T. Ohba et al., Microelectronic Eng., Elsevier, **87**, 485 (2010).
7. T. Ohba, IEEE ICSICT, 70 (2010).
8. N. Maeda et al., AMC 2008, MRS, 501 (2009).
9. H. Kitada et al., IEEE Proc. of IITC, 107 (2009).
10. K. Fujimoto et al., Solid-State Sensors, Actuators and Microsystems Conference, 1877 (2009).
11. T. Ohba, IEEE Proc. of IITC, 12.1 (2010).
12. T. Ohba et al., IEICE Trans. and Electronics, J**93**-C (11), 464 (2010).
13. H. Kitada et al., IEEE Proc. of IITC, 9.6 (2010).
14. N. Maeda et al., IEEE VLSI Symp. 150 (2010).
15. Y. S. Kim et al., IEDM Tech. Dig. 365 (2009).
16. T. Ohba, Pioneer, ITRI/EOL, Taiwan, Nov., 51 (2010).

ECS Transactions, 34 (1) 1017-1022 (2011)
10.1149/1.3567708 ©The Electrochemical Society

Electrical Quality of III-V/Oxide Interfaces: Good Enough for MOSFET Devices?

G. Brammertz[a], A. Alian[a], H. C. Lin[a], L. Nyns[a], S. Sioncke[a], C. Merckling[a], W.-E Wang[a], M. Caymax[a], T. Y. Hoffmann[a]

[a] imec, Kapeldreef 75, 3001 Leuven, Belgium

We will present the defect density at $In_{0.53}Ga_{0.47}As$ and InP interfaces with ALD Al_2O_3 derived by use of the conductance method and from simulation of low frequency CV-curves. Consequences of the interface state distribution for MOS transistor device operation will be highlighted through 1-dimensional electrostatic simulations. The simulation results will be compared as much as possible to different state-of-the-art transistor results presented in literature.

Introduction

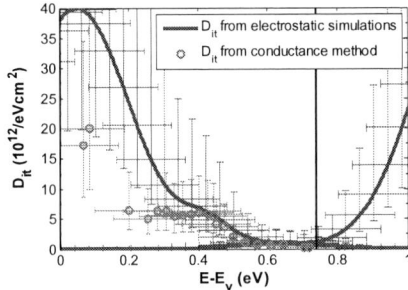

Figure 1. Interface state density at the $In_{0.53}Ga_{0.47}As$-Al_2O_3 interface as a function of energy in the bandgap as measured with admittance spectroscopy at various temperatures (red circles, for details see (7)) and as derived from electrostatic simulations of low-frequency CV-curves (blue line, for details see (8)). Zero on the energy axis corresponds to the $In_{0.53}Ga_{0.47}As$ valence band maximum, the $In_{0.53}Ga_{0.47}As$ conduction band minimum is represented by the vertical black line. Interface states in resonance with the conduction band (at energies inside the conduction band) can be derived from the electrostatic simulations.

High electron mobility materials like III-V semiconductors are currently being studied because of their potential to improve Metal-Oxide-Semiconductor (MOS) transistor performance at scaled drive voltages (close to 0.5 V), as compared to more traditional Si MOS transistors at similar drive voltages (1). Transistors based on $In_{0.53}Ga_{0.47}As$ channels have until now shown promising device properties and are therefore being widely studied in the community (2-6). As a possible interfacial passivation layer, InP has been used in MOS transistor devices and much increased drive currents have been measured for otherwise same device geometry and processing (3). In the following, we will present the interface state density distribution of $In_{0.53}Ga_{0.47}As$ and InP interfaces with atomic layer deposited (ALD) high-k dielectrics such as Al_2O_3, HfO_2, $LaAlO_3$ and $GdAlO_3$ as

1017

measured by admittance spectroscopy and as derived from low frequency CV-simulations. We will use these interface state density distributions in order to derive the surface potential movement of III-V MOS transistor devices and to verify the effect of the interface state density on the sub-threshold slope (SS) of MOS transistors.

Discussion

Figure 1 shows the interface state density of $In_{0.53}Ga_{0.47}As/Al_2O_3$ interfaces as derived from admittance spectroscopy measurements (7) and from electrostatic simulations of low frequency CV-measurements (8). The 10 nm Al_2O_3 high-k dielectric presents a k-value of 9 and was deposited by ALD at 300°C with trimethylaluminum (TMA) and H_2O precursors. Prior to ALD deposition the $In_{0.53}Ga_{0.47}As$ surface was cleaned with a 5 min 26% $(NH_4)_2S$ clean followed by 5 min DI water rinse.

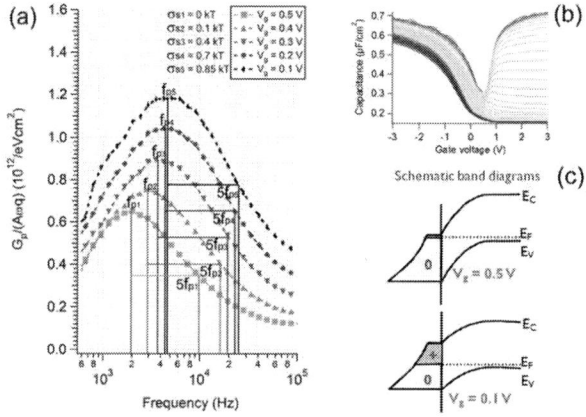

Figure 2. Normalized experimental conductance $G_p/A\omega q$ of an $In_{0.53}Ga_{0.47}As$-Al_2O_3 p-MOS capacitor as a function of measurement frequency for five different bias voltages going from depletion (V_g=0.1 V) to the onset of inversion (V_g = 0.5 eV) (a). The multi-frequency (100 Hz - 1MHz) CV-curves corresponding to the conductance measurements are shown in the upper right hand side corner (b). Schematic band diagrams for V_g = 0.1 V and V_g = 0.5 V with schematic drawings of the density of donor-like interface state charges are shown as well (c).

The agreement between the interface state density derived from admittance spectroscopy and the one derived from the electrostatic simulations is quite good within the estimated errors of the methods. It can be seen that mainly close to the conduction band minimum the interface state density is low. As this is the energy range in which the surface potential moves for nMOS device operation, to first order it can be estimated that the surface potential movement will not be too much hindered by this interface state density. The density of interface states inside the conduction band can also be derived from the electrostatic simulations. With the conductance method these defect states cannot be assessed, as they are in resonance with the conduction band and present time constants in excess of 10 MHz, which is the usual maximum frequency accessible for AC-conductance measurements. Also important to know is that the interface state density inside the whole InGaAs bandgap is donor-like, even for the defect states in the upper half of the bandgap. Figure 2 shows the experimental normalized conductance $G_p/A\omega q$ as a function of measurement frequency measured for five different bias voltages on a p-type $In_{0.53}Ga_{0.47}As$-Al_2O_3 MOS capacitor. As can be seen on the left hand side figure, the full width at half maximum (FWHM) of the conductance peaks decreases as the Fermi level approaches the conduction band minimum. The FWHM of the conductance peak can be directly linked to the band bending fluctuations, which are generally assumed to increase as the amount of interface state charge increases. From figure 2(a) it can be seen that the FWHM of the conductance peaks decreases as the surface Fermi level moves up inside the bandgap. This would therefore imply that the interface states neutralize as the Fermi level moves upwards in energy, which is only the case for donor-like interface states. When the surface Fermi level is close to the conduction band minimum, the extra-broadening due to band bending fluctuations σ_s is close to zero, which would imply that at this energy the total charge at the semiconductor/oxide interface is close to zero. The Fermi level stabilization energy of $In_{0.53}Ga_{0.47}As$ interfaces is therefore close to the conduction band minimum energy. No large flat-band voltage shifts of n-type $In_{0.53}Ga_{0.47}As$ MOS devices are therefore expected as well as a free flat-band voltage modulation with gate metal.

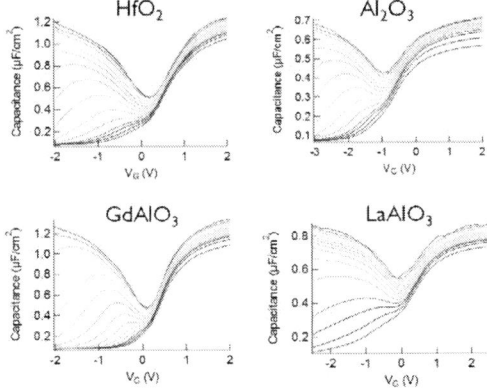

Figure 3. AC-measurements of four n-type $In_{0.53}Ga_{0.47}As$ MOS capacitors with different 10 nm ALD-deposited high-k dielectrics. 15 different frequencies varying logarithmically from 100 Hz to 1 MHz are shown.

Presently, in all measured cases, the $In_{0.53}Ga_{0.47}As$/high-k oxide interface state density does not vary fundamentally when the nature or the processing condition of the high-k dielectric is varied. Although some minor differences can be seen between different high-k dielectrics, the overall interface state distribution is in all investigated cases similar to figure 1. To illustrate this statement, figure 3 shows CV-curves of n-type $In_{0.53}Ga_{0.47}As$ MOS capacitors with four different 10 nm thick ALD-deposited high-k dielectrics. Prior to ALD deposition all the $In_{0.53}Ga_{0.47}As$ surfaces received a 5 min 1:10 $HCl:H_2O$ clean followed by 5 min DI water rinse. Although the k-values of the oxides differ depending on the material, the overall CV-curve shape is similar in all cases, which implies an interface state distribution that is quite similar in all four cases.

The InP/high-k oxide interface seems to present quite similar interface state properties with low interface state density close to the conduction band minimum and increasing donor-like interface state density towards the valence band maximum side. Figure 4 shows the interface state density derived from room-temperature and low temperature CV-measurements on n-type InP MOS capacitors (9).

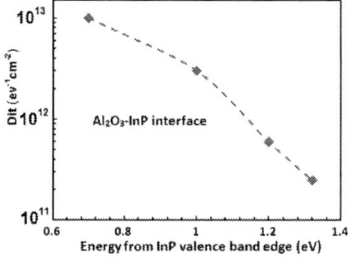

Figure 4. Interface state distribution at the InP-Al$_2$O$_3$ interface as derived from variable temperature admittance spectroscopy measurements (for details see (9)).

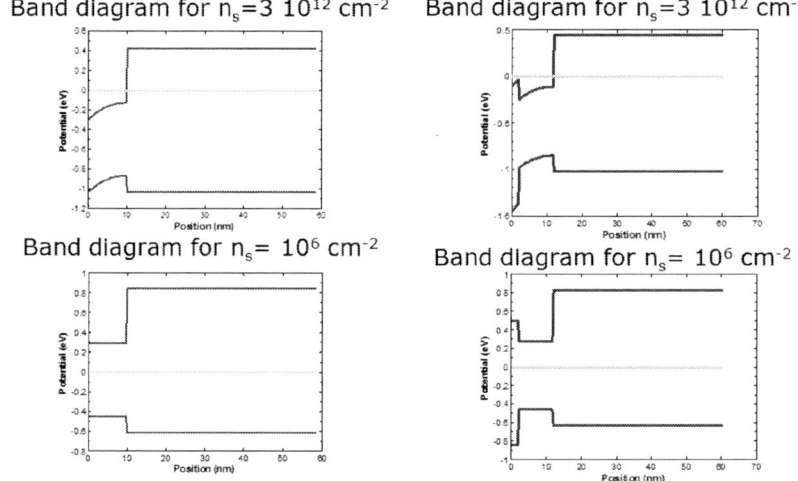

Figure 5. Band diagrams of III-V quantum well MOSFET devices. The buffer layer consists of undoped In$_{0.52}$Al$_{0.48}$As, whereas the channel layer consists of 10 nm undoped In$_{0.53}$Ga$_{0.47}$As. The left hand side figures correspond to the situation with the high-k dielectric (not shown in the figure) deposited straight on the channel layer, whereas the right hand side figures correspond to the situation with a thin InP layer in between the channel and the high-k dielectric. The top figures show the band structures when the devices are in the ON-state, with $3 \ 10^{12}$ cm^{-2} electrons in the channel, whereas the bottom two figures show the band structures when the devices are in the OFF state, with 10^6 cm^{-2} electrons in the channel.

Knowing the interface state distributions of In$_{0.53}$Ga$_{0.47}$As and InP interfaces with high-k dielectrics, we can easily simulate the effect that the interface state charges have on the SS of the devices. We therefore simulate the Poisson equation for a typical quantum-well MOSHEMT structure consisting of a thick undoped $In_{0.52}Al_{0.48}As$ buffer layer covered by a 10 nm thick undoped $In_{0.53}Ga_{0.47}As$ channel layer. We will investigate the case where the high-k dielectric is deposited directly on top of the $In_{0.53}Ga_{0.47}As$ channel layer as well as the case where a 2 nm thick InP cap layer is deposited on top of the $In_{0.53}Ga_{0.47}As$ channel, such that the interface with the high-k dielectric is on top of the InP cap layer. Figure 5 shows the band diagrams of the two previously described quantum-well MOSFET devices. From the band diagrams we can see that in the case of the devices without InP cap layer, the surface Fermi level moves from about 0.3 eV below the $In_{0.53}Ga_{0.47}As$ conduction band minimum to about 0.3 eV above the $In_{0.53}Ga_{0.47}As$ conduction band minimum, when the device switches from the OFF-state to the ON-state. In the case of the device with the InP cap, the surface Fermi level moves from about 0.5 eV below the InP conduction band minimum to about 0.1 eV above the InP conduction band minimum, when the device switches from the OFF-state to the ON-state. These are therefore the regions of interest for our interface state density analysis.

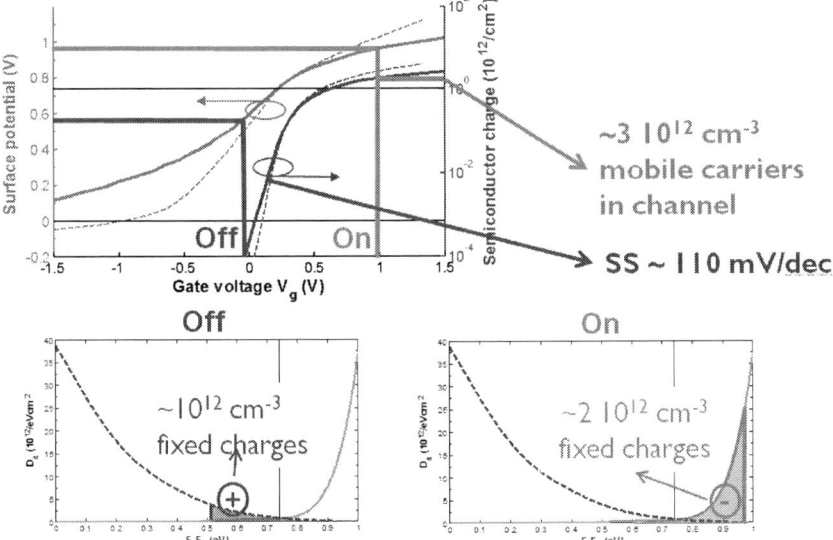

Figure 6. Surface potential and channel electron charge as a function of gate bias voltage for a device with a 4 nm EOT oxide deposited directly on the InGaAs channel. The gate metal work function was chosen as being 4.7 eV. The two lower curves show the amount of charged interface states in the ON- and OFF-state of the transistor. Calculations correspond well to experimental data in (10).

Using the interface state density from figures 1 and 4, we can calculate the amount of charge at the semiconductor-oxide interface due to interface states in the ON- and OFF-state of the devices, as well as the effect that these charges have on the SS of the devices. Figure 6 shows the results of such a simple first order analysis for the case of a device with a 4 nm EOT gate dielectric deposited directly on top of the channel layer. Such a device corresponds to the device presented in ref (10). The same analysis applied to a device with a 1 nm EOT gate dielectric deposited on top of a 2 nm thick InP cap layer is shown in figure 7. Such a device corresponds to the device presented in reference (11). In both cases it can be seen that the effect of the interface states on the SS of the devices is relatively limited, though not completely absent, as measured experimentally on the devices. In order to reach close to 70 mV/dec SS, a further reduction of the interface state density at the $In_{0.53}Ga_{0.47}As$ and InP interfaces with high-k dielectrics is desirable.

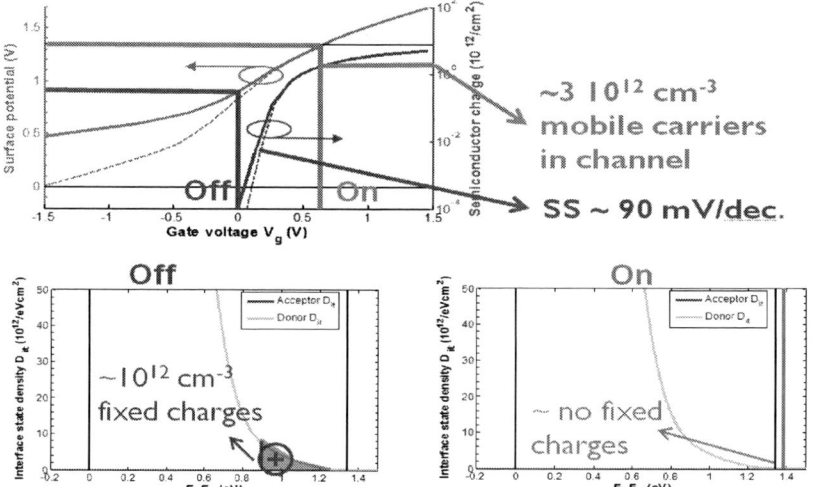

Figure 7. Surface potential and channel electron charge as a function of gate bias voltage for a device with a 1 nm EOT oxide deposited on a 2 nm InP cap layer on the InGaAs channel. The gate metal work function was chosen as being 4.9 eV. The two lower curves show the amount of charged interface states in the ON- and OFF-state of the transistor. Calculations correspond well to experimental data in (11).

References

1. G. Dewey, M. K. Hudait, K. Lee, R. Pillarisetty, W. Rachmady, M. Radosavljevic, T. Rakshit, R. Chau, *IEEE Electron Dev. Lett.*, **29**, 1094 (2009).
2. N. Goel et al., *IEDM Tech. Dig.*, 363 (2008).
3. H. Zhao, Y.T. Chen, J. H. Yum, Y. Wang,1 N. Goel, J.C. Lee, *Appl. Phys. Lett.*, **94**, 193502 (2009).
4. H. D. Trinh, E.Y. Chang, et al., *Appl. Phys. Lett.,* **97**, 042903 (2010).
5. H. C. Lin, W. E. Wang, G. Brammertz, M. Meuris, M. Heyns, *Microelectronic Eng.* **86**, 1554 (2009).
6. R. J. W. Hill, R. Droopad, et al. *Electronics Lett.* **44**, 498 (2008) and **44**, 1283 (2008).
7. G. Brammertz, H.C. Lin, K. Martens, D. Mercier, C. Merckling, J. Penaud, C. Adelmann, S. Sioncke, W.E. Wang, M. Caymax, M. Meuris, M. Heyns, *ECS Trans.*, **16**, 507 (2008).
8. G. Brammertz, H.C. Lin, M. Caymax, M. Meuris, M. Heyns, M. Passlack, *Appl. Phys. Lett.*, **95**, 202109 (2009).
9. H.C. Lin, G. Brammertz, S. Sioncke, L. Nyns, A. Alian, W.E. Wang, M. Heyns, M. Caymax, T.Y. Hoffmann, *ECS Trans.*, this volume.
10. A. Alian, C. Merckling, G. Brammertz, M. Meuris, K. De Meyer, M. Heyns, as discussed at the 2010 IEEE SISC, San Diego, CA. (2010).
11. M. Radosavljevic, B. Chu-Kung, et al., *IEEE IEDM Tech. Digest*, 319 (2009).

Low Temperature Bonding with Thin Wafers for 3D Integration

T. Matthias[a], B. Kim[b], P. Kettner[a], M. Wimplinger[a], and P. Lindner[a]

[a] EV Group, DI Erich Thallner Strasse 1, 4782 St. Florian/Inn, Austria
[b] EV Group Inc., 7700 S. River Parkway, Tempe, AZ 85284, USA

The ITRS roadmap for high-density TSV interconnects specifies maximum layer thicknesses of 5-15 μm in 2013 with a sub-micron layer-to-layer alignment accuracy. For 3D chip stacks with multiple layers it is not only necessary to handle and process such thin layers, but later on these thin layers have to be stacked and bonded. When it comes to wafer bonding for high-density TSV integration, most efforts have been given to developing Cu-Cu thermo-compression bonding and direct oxide bonding. Stacking of thin wafers can be performed either after the thin wafer is debonded or while the thin wafer is still bonded onto a carrier wafer. Bonding of the thin wafer while still mounted to the carrier wafer allows comfortable and safe wafer handling, but adds some complexity to the wafer bonding process.

3D interconnects with TSVs provide a large number of technical benefits in terms of the electrical performance e.g. higher I/O bandwidth, shorter interconnect length, lower power consumption and better signal-to-noise ratio. The shorter interconnect length allows to focus on one device function per layer e.g. memory on one chip and logic on another layer.

In the past high performance devices had to be designed and manufactured with the System-on-chip (SoC) approach. A SoC provides very good performance, but this comes at the cost of a very high design complexity. Chip stacking with TSVs enables devices with the highest performance within the System-in-Package (SiP) family. Compared to SoCs, which require with each device to "re-invent the wheel" and start all over, with SiPs it is possible to follow a true modular design approach. Each company can focus on their device core competence e.g. an ASIC or MEMS and add standardized components like memory or logic controller from other suppliers.

Vertical interconnects enable a smaller form factor not only laterally, but also vertically. Stacking of dies allows working with smaller dies, which increases the wafer yield. Even more important is that stacking of dies enables heterogeneous integration. Silicon dies can be bonded to dies made of GaAs or InP.

3D Integration Schemes: Chip-to-Wafer vs. Wafer-to-Wafer

There are many different integration and manufacturing schemes for 3D interconnects. An important distinction within the various integration schemes is based on wafer or die level processing: chip-to-chip (C2C), chip-to-wafer (C2W) and wafer-to-wafer (W2W).

C2C is mainly been used for high performance, high margin devices. For lower margin devices like consumer electronics C2C is not very suitable due to the single die processing. Of course W2W integration allows wafer-level processing after stacking. W2W integration gives the highest throughput and the highest alignment accuracy. But

W2W integration requires that the dies have the exact same size and it has the inherent risk that a defective die is bonded to a good die, thereby destroying the whole stack.

C2W is a hybrid process and combines the single die placement with the feasibility of wafer-level processing after die placement. With C2W integration it is possible to stack dies of different size. Especially for heterogeneous integration C2W is the method of choice as currently only Silicon devices are manufactured on 300mm wafers, whereas all other semiconductor materials are being manufactured on smaller wafer sizes. In addition C2W enables to test every die prior to stacking, which allows true "Known Good Die" (KGD) manufacturing.

A new process flow, the advanced chip-to-wafer (AC2W) bonding, allows significant throughput improvements by splitting the bond process into two sub processes. The alignment and temporary pre-bonding is performed on a pick-and-place machine, whereas the permanent bonding of the dies is performed as a batch process in a dedicated bond chamber (Figure 1).

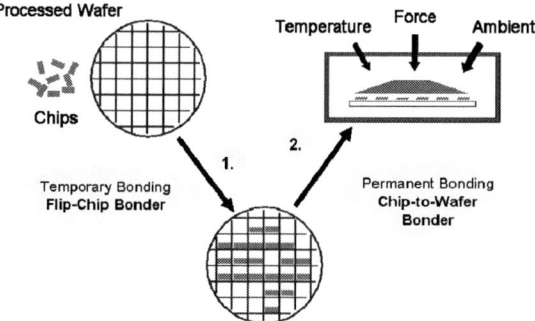

Figure 1. Schematic process flow of Advanced Chip-to-Wafer (AC2W) bonding; AC2W enables enhanced throughput and true known good die (KGD) manufacturing as only the known good dies are placed.

Chip stacking with thin wafers

Thin wafer handling and processing is necessary for nearly all 3D integration schemes. During the past three years, significant progress has been made in bonding and processing of thin wafers; merits and demerits of various bonding processes suitable for TSV integration have been discussed [1], the impact of wafer thinning and packaging on the transistor device performance has been analyzed [2], and complete TSV integration schemes with a temporary carrier have been qualified [3]. Bonding of the thin wafer while still mounted to the carrier wafer allows comfortable and safe wafer handling, but adds some complexity to the wafer bonding process. Fig. 2 illustrates example process flows for first wafer manufacturing and subsequent wafer-to-wafer and chip-to-wafer stacking. Most temporary bond adhesives are organic and cannot withstand temperatures higher than typically 250 °C. Therefore, low temperature is a crucial factor for wafer bonding especially when the wafer is mounted to a carrier.

For chip-to-chip integration it has been assumed that 25μm is minimal thickness where dies can be handled without support; thinner dies have to be diced together with the carrier wafer and have to be stacked while the carrier is still attached. Figure 3 shows the 2 different process flows for C2W stacking.

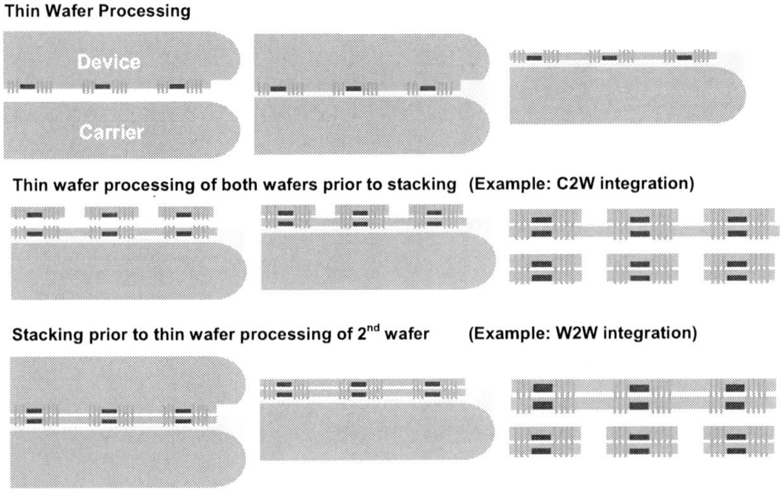

Thin Wafer Processing

Thin wafer processing of both wafers prior to stacking (Example: C2W integration)

Stacking prior to thin wafer processing of 2nd wafer   (Example: W2W integration)

Figure 2. Example process flows for thin wafer manufacturing and subsequent wafer-to-wafer and chip-to-wafer stacking. For many applications e.g. memory stack on interposer all the layers incl. the first wafer have to be thin with I/Os on both sides. The dark rectangles show the frontside (transistor side) of the wafers. Shown are the process flows for face-to-back (F2B) integration. The process flow in the middle shows thin wafer processing and debonding prior to chip stacking. This process flow can be applied for C2W and W2W. The process flow on the bottom shows stacking prior to thinning and backside processing of the 2nd wafer, which is mainly applied for W2W integration.

Permanent Bonding

Cu-Cu bonding looks most preferred for TSV integration because Cu-Cu joints can provide much lower electrical resistivity, higher density and higher EM resistance than any other joints. High process temperature is one of the major bottlenecks of Cu-Cu direct bonding because it can negatively influence wafer alignment, device reliability and manufacturing yield. Our evaluation showed that post-bond thermal annealing is a good way to improve bonds qualities. When thermo-compression bonding was performed at a relatively lower temperature and reduced process time (e.g., 300 °C for 30 min in this experiment), post-bond annealing at moderate temperatures (250-300 °C for an hour) drastically improves the interfacial adhesion energy (from 2.8 J/m2 to 8.9-12.2 J/m2) and reduces the interfacial seam voids as illustrated in Fig. 4. This study is still ongoing to further reduce both bonding and annealing temperatures down to 200-250 °C from the first achievement (250-300 °C).

In a recent study a team from SEMATECH analyzed Cu-Cu thermo-compression bonding on 300mm wafers. The bond quality was evaluated for blanket wafer bonding as well as for patterned wafer bonding. The void density was less than 0,1% for all measured wafers [6]. In the same study the post bond alignment accuracy for a fully integrated 300mm Cu-Cu thermo-compression process with the EVG Gemini wafer bonding system was evaluated. 1-Sigma alignment accuracy of 0.34µm could be verified, which already meets the ITRS roadmap up until 2012 [6,7].   A different study on non-thermo-compression Cu-Cu direct bonding by Radu et al. reported alignment accuracy of <0.2 μm across the whole wafer with an EVG SmartView® NT system[9].

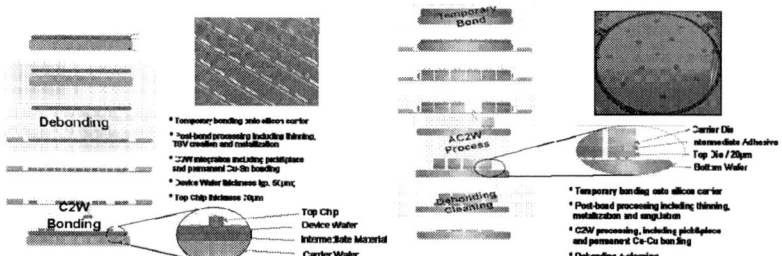

Figure 3. 3D Chip-to-wafer integration: The thickness of the top die determines whether debonding can be performed prior to die singulation and stacking (left side) or whether the die is singulated and permanently bonded prior to debonding (right). chip stacking with Cu-Sn Transient Liquid Phase (TLP) bonding. For ultra-thin dies below 25μm the dies are singulated while they are still on the carrier. The carrier die gives mechanical stability during permanent bonding. Finally the carrier dies are debonded and the wafer stack is cleaned.

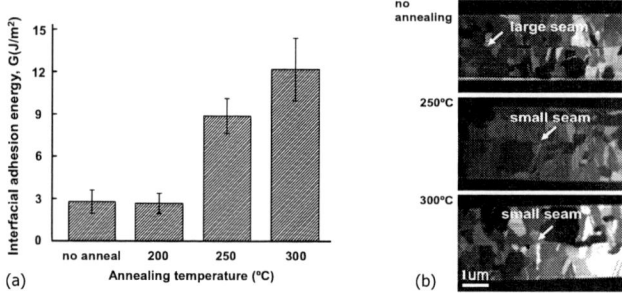

Figure 4. Effect of post-bond annealing temperature on bond properties; (a) interfacial adhesion energies and (b) sample microstructures.

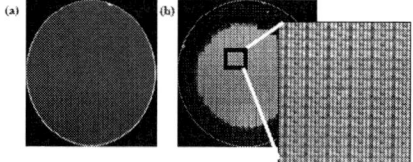

Figure 5: Scanning Acoustic Microscope (SAM) images of Cu-Cu thermo-compression bonded 300mm wafers: a) blanket wafers, b) patterned wafers (the dark ring is a metrology artifact from SAM due to water seeping in through recessed pattern structure) [6]

Fusion wafer bonding is a 2-step process consisting of room temperature pre-bonding and annealing at elevated temperature. The classical annealing schemes, which were developed for SOI wafer manufacturing, require annealing temperatures in the range of 800-1100°C. A novel surface pre-processing step, LowTemp® plasma activation, enables to modify the wafer surface in such a way that the annealing temperatures can be reduced to 200-400°C. Therefore LowTemp® plasma activation enables the usage of fusion wafer bonding for 3D integration. Fig. 5 illustrates the variation of bond strength (expressed as surface energy) with thermal annealing time at 200 and 300 °C. In a standard fusion bonding process using high temperature annealing the full bond strength (in the range of substrate bulk fracture strength, ~2.5 J/m2) is reached after few hours of annealing at temperatures higher than 900 °C. As shown in Fig. 3, annealing process of two-four hours at 200 °C or only one hour at 300 °C results in maximum bond strength. Fusion wafer bonding brings several advantages:

Highest alignment accuracy. Due to bonding at room temperature the misalignment contribution based on thermal expansion of the wafers is eliminated completely, which enables very good post bond alignment accuracy. Sub-micron post bond alignment accuracy has been reported by multiple authors [6,8,9]

Highest throughput. Fusion wafer bonding has a higher than metal-metal thermo-compression bonding as it is a room temperature process. The subsequent annealing can be performed as batch process in a furnace or oven.

Inspection capability after pre-bonding prior to final annealing. After the room temperature pre-bonding step the bond strength is sufficiently high to enable inspection of bond quality and alignment accuracy. In case of misalignment or bond quality problems e.g. voids, the wafer pair can be separated and reworked. This concept of inspection and, if necessary, reworking prior to final annealing has been used in SOI wafer manufacturing since many years.

Low Cost-of-ownership. The combined effects of in-situ bonding in the aligner module, highest throughput, increased yield due to the ability to rework and reduced capital costs results in low cost-of-ownership for manufacturing schemes based on fusion wafer bonding.

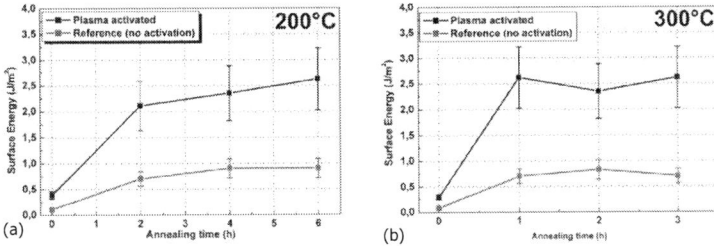

Figure 5. Surface energy vs. thermal annealing time at different annealing temperatures; (a) 200 °C and (b) 300 °C [10].

Conclusions

Chip stacking can be implemented with Chip-to-wafer or wafer-to-wafer integration. We have been developing low temperature bonding (Cu-Cu and SiO2) for stacking thin wafers on the carrier wafer, where the target temperatures for whole processes such as bonding and post-bond annealing are less than 250 °C which is close to the maximum stability temperature of most temporary adhesives. We lowered the bonding temperature to 300 °C with annealing temperatures of 250-300 °C for Cu-Cu bonding (further study is ongoing to reduce temperatures down to 250 °C), reduced post-bond annealing temperatures down to 200 °C for direct oxide bonding, and developed temporary bonding and debonding solutions with various benefits including high temperature stability up to 250 °C.

References

1. B. Kim, et al., *Advances in Wafer Bonding Techniques Enabling Vertical Integration*, Proceedings of IMAPS/SEMI Topical Workshop on Advanced Interconnect Technologies, July 14, San Francisco, CA (2010).
2. D. Perry, et al., *Impact of Thinning and Packaging on a Deep Sub-micron CMOS Product*, Poster Presentation at DATE 09, April 20-24, Nice, France (2009).
3. J. Charbonnier, et al., *Integration of a Temporary Carrier in a TSV Process Flow*, Proceedings of ECTC 2009, May 26-29, San Diego, CA (2009).
4. Holger Hübner et al., *Face-to-face chip integration with full metal interface*, Proc. of Advanced Metallization Conference 2002 , San Diego (2002).
5. B. Kim et al., *Effects of bonding process parameters on the interfacial properties of Cu-Cu direct bonds for TSV integration*, in Proc. International Wafer Level Packaging Conference, October 2009, San Jose, USA (2009)
6. W. H. The, et al., *Recent Advances in Submicron Alignment 300 mm Copper-Copper Thermocompressive Face-to-Face Wafer-to-Wafer Bonding and Integrated Infrared, High-Speed FIB Metrology*, Proceedings of IEEE-IITC (2010).
7. ITRS roadmap 2007
8. G. Gaudin et al., *Low temperature direct wafer to wafer bonding for 3D integration*, Proc. IEEE 3D-IC Conference, November 16-18, Munich, DE (2010)
9. I. Radu et al., *Recent developments of Cu-Cu non-thermo compression bonding for wafer-to-wafer 3D stacking*, Proc. IEEE 3D-IC Conference, November 16-18, Munich, DE (2010)
10. Viorel Dragoi et al., Mater. Res. Soc. Symp. Proc. Vol. 872 (2005), J7.1.1

Vertical LED with Diamond-like Carbon Interface for High-Power Illumination

James C. Sung, Kevin Kan, Michael Sung
SinoDiamond LED, Hai-An, Jiangsu, China

Abstract

Most blue light LED chips are made by growing GaN epitaxy on a sapphire substrate. Because sapphire is an insulator, the two electrodes on a conventional LED die must lie on the same side. To improve upon existing technology, several companies are developing vertical stacked LED designs by coating P-type GaN with a reflector (e.g. Ag) that is soldered (e.g. via Au-Sn) to an electrode as the substrate. These techniques have their limitations. In this report, we introduce a methodology to producing a high-powered LED with a thin-film DLC (diamond like carbon) interface that can effectively bridge the semiconductor GaN and metallic substrate. DLC can not only moderate the thermal mismatch, but also to enhance the heat spreading since DLC has a thermal conductivity (475 W/mC) that is significantly higher than even copper. In addition, the metal substrate of the LED can optionally be replaced by a diamond-metal (Ni or Cu) composite that further minimizes the CTE mismatch and boost heat spreading efficacy. Such a DLC LED design can sustain a high drive current so that reduced number of enhanced dies can be used in place of conventional chips for small form-factor general illumination applications.

Introduction

In the mainstream manufacturing process for blue light LED wafers, GaN is hetero-epitaxially deposited on an insulating sapphire substrate (single-crystal corundum or Al_2O_3). In 2010, Taiwan has the largest production capacity while Japan with highest sales volume. However Cree in the U.S is the most profitable LED manufacturer, growing epitaxial GaN on SiC substrate to produce a Vertical LED design. The manufacturing process can reclaim the expensive SiC substrate by laser lift-off methods, which also has the additional benefits of enhancing lighting efficiency and reducing brightness attenuation.

Ranking	Company	Millions (USD)	Ranking	Supplier	Millions (Grain)
1	Nichia	900	1	Epistar	28800
2	Osram Opto	460	2	Samsung LED	18500
3	Samsung LED	450	3	LG Innotek	17500
4	Epistar	400	4	Sanan	16700
5	Lumileds	370	5	Cree	15000
6	Cree	370	6	Nichia	13500
7	Everlight	300	7	Hung	12400
8	Stanley	250	8	Osram Opto	10400
9	Lite-on	220	9	Showa Denko	10200
10	Citizen	200	10	Forepi	10000
				Other	96700
Reference: LEDInside				Total	249700
			Reference: TMS Research, Semi		

Figure 1: Top 10 companies with the largest LED Product Output Value in the World (2009, 2010)

In recent years, Taiwan has also started developing vertical LED manufacturing capability. Companies such as the Taiwan Plant of the U.S. Silicon Valley company SemiLED applies conductive metal (specifically copper alloy) as the anode electrode as a differentiating design from Cree's silicon carbide substrate. However, the thermal expansion coefficient of copper is much higher than that of GaN. This mismatch in thermal expansion coefficient results in very high surface stress that is exacerbated by the high current density. The GaN layer is a hexagonal wurtzite structure with piezoelectric effect that can disrupt the current distribution due to interface stress thus incurred. Moreover, the expansion of metal substrate will split GaN lattice with very low tensile strength to create defects inside the lattice, which results in points for hot spots to develop, resulting in a vicious cycle that decreases light efficiency.

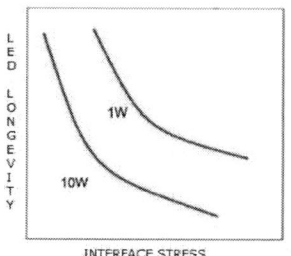

Figure 2: LED lifetime will reduce from increasing interface stress and also higher current density.

In view of these design limitations, this paper describes a new design of the vertical LED which utilizes diamond-like carbon (DLC) films as the interface of semiconductor and metal substrate which not only reduces the stress between two different materials but also provides a high thermal conductivity path to remove the waste heat generated.

It is estimated that in 2010 global production of about 10 million of GaN LED wafer (2-inch) with a total value of USD 2 billion, only 5% which utilizes a vertical design. The design improvements of using a DLC-based vertical design can potentially be a disruptive technology that can significantly increase the luminous efficiency of an LED by a large factor and rapidly gain marketshare to replace conventional LED dies if successful.

LED Lighting the World

Started from 2010, a large percentage of LED lighting is used in the TV and tablet PC backlighting market, with a growing market for indoor and outdoor general illumination lighting. Regular indoor lighting (Incandescent Lamp), fluorescent (Fluorescent Lamp), LCD Cathode (Cold Cathode Fluorescent Lamp) backlight are being rapidly replaced by LED lighting.

The growth market for LED application are in high brightness white lighting products, most of which are based on blue LED excitation of white yellow phosphors. Mass production is based on sapphire substrate, with GaN deposition by MOCVD. Currently Taiwan is the leading production base in the world. However, the mainland China government provides RMB 10 Million subsidies for a single set of MOCVD equipment so China is likely to surpass Taiwan in manufacturing capacity in the next year. The main Taiwanese LED manufacturers (Epistar, Formosa, G-photonic) have set up plants in Mainland China. In Korea, Samsung has purchased 200 MOCVD machines for 2010. Therefore, LED production capacity made by MOCVD in Taiwan may be in surplus for the year 2011, therefore a decrease in production capacity is expected. Thus, it is imperative that Taiwan based manufacturers develop high-end products such as improved efficiency vertical LED technology to survive from the severe price competition from China.

Company	MOCVD QTY
Epistar	225
Huga Optotech	75
Tekcore	50
Nanya Photonic	27
Epistar Group	375
Samsung LED	250
LG Innotech	230
Chime Lighting	90
Lextar (AUO)	70
In-House Camp	640
Reference: Company data, Yunta Research estimates	

Figure 3 : The production capacity of company with MOCVD Equipment (2010)

Conventional LED chip design

In a conventional LED design, the sapphire substrate is electrically insulated so electrodes have to be placed on the same side of the chip, resulting in a horizontal design whereby current flow forms a bending path from one electrode to the other. Therefore the entire light emitting area can not be used efficiently to further enhance the light emission efficiency. Moreover, the hot spots thus incurred by current accumulation will cause additional defects in the lattice that will further reduce the brightness of LED. In order to extend the lifespan of LEDs, the input current must be reduced to (for example 350 mA) so that the light emission efficiency per unit area is lowered. A few conventional commercially-available LED chip designs are as follows:

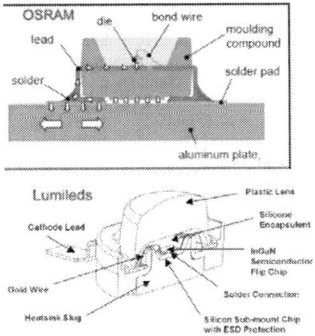

Figure 4 : Conventional LED designs

The current LED efficiency problems cannot be fixed by packaging design. The key to a solution is to find an alternative electrode design to efficiently conduct the current so that the brightness of the LED can be increased significantly. One way to improve LED performance is to change the current flow directionality in the vertical direction from the standpoint of the substrate. By using a vertical design, the LED chip requires less area (since the epilayer is utilized fully), thus we can reduce the production cost by increasing the amount of LEDs that can be created using a single wafer. The brightness of a vertical LED will be increased with the extension of light-emitting area. Therefore, a larger area vertical LED will create more brightness than a few horizontal LEDs of the same size.

LED Vertical LED Manufacturing

The basic concept of the vertical LED design is to place one conductive substrate such as CuW soldered on the P-type face of the GaN epilayer and then focus laser energy (e.g. KR Wavelength at 248nm) onto the opposite side of the transparent sapphire substrate so that the sapphire will lift-off fro the N-type GaN.

After removing the sapphire substrate, the GaN negative electrode can be formed by doping. Its conductivity and transparency are positive (e.g. Mg dopant) so that no ITO coating is required as a current spreading layer.

Interface Stress Problem

Before laser lift-off, a reflective metal coating (such as silver) is needed on the P-type side of the die with a conductive substrate. If we apply copper alloy as the substrate, the thermal expansion coefficient is far more than that of GaN, which makes it very difficult to properly adhere the two surfaces together and also generate a great deal of interface surface stress. In operation, the current will travel along with minimum resistance to increase the temperature of the stressed portions of the die to further stress the GaN lattice. Due to frequent switching, the GaN lattice will be extended repeatedly to form miss row and/or dislocation defects, with the ultimate result of reduced LED brightness. If we insert layers of ceramics such as Tic and DLC between metal substrate and semiconductor chip, the stress can be reduced a great deal. An additional benefit is the thermal conductivity of DLC is significantly higher than copper. Thus DLC is a very good material system for vertical LED designs to dissipate the waste heat in an efficient manner.

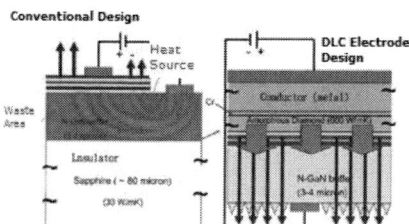

Figure 5: Horizontal LED current will result in the high temperature in partial areas of the die layer (left picture). In contrast, with a vertical LED the internal current flow is very smooth (right picture). Also, the light-emitting area is fully utilized to increase the brightness.

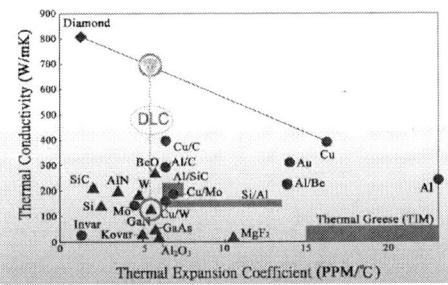

Figure 6: Comparison on thermal conductivity and thermal expansion coefficient of various materials. DLC not only has high thermal conductivity but also an adjustable thermal expansion coefficient which can be perfectly matched to GaN in order to remove internal film stress.

The heat dissipation effect of DLC

In the vertical LED manufacturing process, the surface stress can be relieved by compression of GaN to semiconductor materials such as Si, but the thermal conductivity of such a material is quite low. If the heat is allowed to accumulate too long, defects caused inside the chip will reduce the light emission efficiency and hence the lifetime and reliability of the die. By applying DLC on GaN, we can not only increase thermal conductivity but also reduce interface stress. The improved heat dissipation effect is shown below:

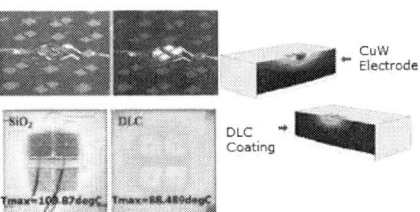

Figure 7 : DLC coating chip can quickly eliminate hot spots to avoid the attenuation of LED brightness. Actual

DLC-coated Si submounts can reduce surface temperature by over 15 C over non-coated samples (left). Simulation of DLC-electrode over CuW shows significantly improved thermal spreading and transport (right)

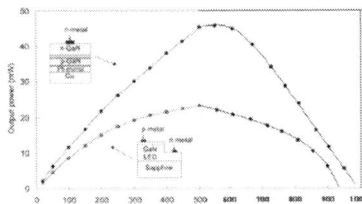

Figure 8 : Luminous performance comparison between Vertical Led and Horizontal LED. Under high current

density, Vertical LED with copper substrate as heat sink will have faster light decay while DLC interfaced Vertical LED results in significantly better performance.

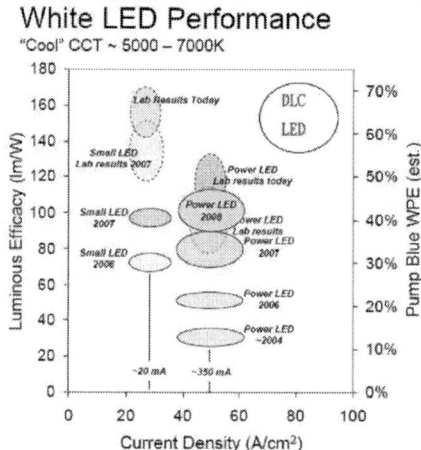

Figure 9: *Because of the non-linear current path, the luminous efficiency of conventional LED dies under high current density will be reduced, whereas a vertical LED die can sustain higher current density without reducing the luminous efficiency. A DLC-based vertical LED can sustain much higher current densities.*

DLC coating does not require soldering

The DLC thin-film coating can also save on the cost of soldering and increase yield rate of LED manufacturing. The alternative to a thin-film coating is to chemically braze or solder the electrode to the GaN layer. To do this at the wafer level is a very difficult technical challenge to do reliably, so adhesion quality suffer, reducing the yield of the product. DLC being a deposition process, is a low-cost process that can adhere to the GaN surface in a highly reliable way, suitable for high-volume, high-yield manufacturing.

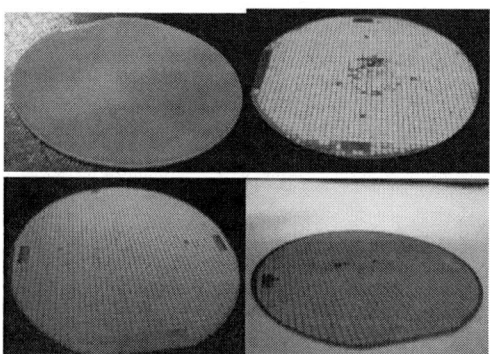

Figure 10 : Sample LED wafers showing the yield issue in attaching the electrode (such as CuW). After laser lift-off, many dies on from the wafer are damaged due to bad adhesive strength. The bottom right picture showed recovered sapphire substrate.

Alumina Abrasives for Sapphire Substrate Polishing

David Merricks
Ferro Electronic Materials
1789 Transelco Drive
Penn Yan, NY 14527, USA
merricksd@ferro.com
phone: +1-315-227-5345

Abstract

The development of Light Emitting Diode (LED) technology over the past thirty years, has led to an exponential increase in device efficiency and light output, such that LEDs are now becoming the light source of choice for virtually all applications. High brightness LEDs (HB LEDs) are driving market growth of around 20% year on year, for applications such as LCD TV backlighting and general illumination, where LED lamps are replacing standard incandescent and fluorescent lighting.

Sapphire (alpha aluminum oxide) is used as the substrate of choice for around 90% of LED chips, with 2" and 4" wafer diameters being the most commonly used. Some leading companies are transitioning to 6" wafer production next year, with 8" wafer processing on the horizon. It is important to produce defect-free sapphire surfaces, in order to efficiently grow the III-V semiconductor stack, responsible for generating the desired wavelengths of light and this is done by CMP. Silica slurries are mostly used as the final polish, but these have very low removal rates, with polishing times of up to a few hours. Silica slurries also exhibit instability at elevated temperatures generated during long polishing and temperature control is critical.

As the industry moves to larger diameter wafers, it will become more important to realize increased throughput and therefore, a faster polishing slurry would have a big advantage. Ferro has developed alumina-based slurries for sapphire polishing, which give faster removal rates and greater stability, whilst producing the required surface quality.

1. Future Requirements of LED Industry

Light emitting diode (LED) usage is now at record levels in mobile applications (cell phones, PDA's, digital cameras, etc) and for signage and display applications. LEDs are fast being adopted into emerging applications, such as backlighting units for computers and TVs, as well as automotive lighting. The biggest driver is sapphire for High Brightness (HB) LEDs for the general illumination market, as these gradually replace incandescent and fluorescent bulbs due to their durability, energy efficiency, enhanced aesthetics and low voltage operation.

The HB-LED revenues for 2010 were around $8.2B, reflecting a 53% growth and by 2014 the market is expected to pass $20B. Substantial growth in HB-LED production capacity will be required and regarding this, many IC fabs and foundries (e.g. TSMC, Samsung, UMC, Micron) are now entering into LED manufacturing, in order to utilize their redundant 150mm and 200mm fabs. These fabs are already experts at high volume, low cost manufacturing, which could have some interesting developments in shaping the future LED industry. Larger chips for greater power and efficiency are required for HB LEDs, which will mean larger diameter wafer processes being required. These processes will demand high throughput, requiring faster process steps at each stage of the LED chip manufacturing process.

The drivers to larger diameter wafers are; cost reduction (due to more efficient use of wafer real estate and lower production costs per chip, through economy-of-scale); improved LED lamp performance (ability to efficiently produce larger size chips to increase lumens output, and thus simplify packaging); growth for high power LED applications (commercial, automotive, and residential lighting applications and large display backlight units); large diameter LED processes and equipment are developing rapidly (for both front-end GaN epi-deposition and back-end wafer processing, testing and packaging).

2. Solid State Lighting Roadmap 2010

The SSL Roadmap[1] is now an annual publication from the US Department of Energy and is an attempt to standardize and integrate the various parts of the LED (and OLED) value chain. The main goals are to reduce costs, improve consistency and promote US-based manufacturing.

Part of the roadmap deals with substrates and for both hetero-epitaxial (e.g. sapphire, silicon carbide) and homo-epitaxial (e.g. bulk gallium nitride) approaches, improvements in substrate quality (surface finish, defect density, flatness, etc) and product consistency are required, in order to meet the demands of high volume manufacturing. For GaN substrates, cost must also be dramatically reduced in order to become a viable option for LED manufacturing.

Figure 1 below shows the 2010 substrate roadmap.

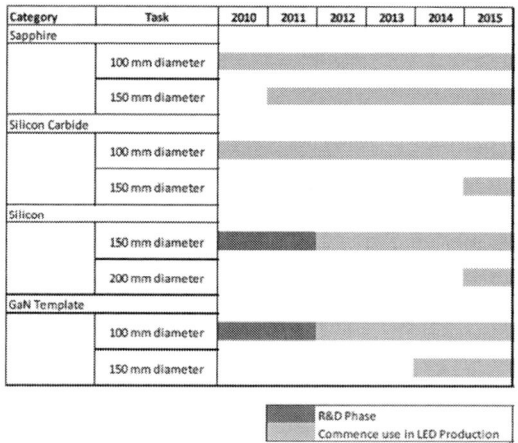

Figure 1: SSL Substrate Roadmap 2010: The roadmap includes two paths, with hetero-epitaxial substrates toward the top and homo-epitaxial substrates toward the bottom

Sapphire is currently the substrate of choice in approximately 90% of LED chips and, according to the SSL Roadmap, sapphire will continue to be the dominant substrate for high power GaN-based LEDs, especially for larger diameter wafers.

3. Sapphire Substrate Polishing

The sapphire substrate market exceeded $200M in 2010, despite the economic downturn, with c-plane sapphire for GaN-based LEDs dominating the market. Revenue increased dramatically under the combined effect of price and volume increase. For the first time, less than 50% of the processed surface area was on 2-inch wafers, as leading companies move to processing larger diameter wafers (mainly 4- and 6-inch).

Sapphire is used as a substrate for growing the III-V semiconductor layer and there are stringent requirements for producing the flat, defect-free surfaces, following production of the rough cut wafers, which lead to efficient performance. Specifications vary from one sapphire manufacturer to another, but surface roughness (Ra) requirements are typically <0.2nm.

Most sapphire substrate manufacturers use similar processes which are variations of a three-step process; a lapping step to reduce the rough topography produced after wafer slicing; a polish step to improve surface roughness (usually with slurry recirculation); a final polish step to eliminate defects (with no recirculation). Diamond or silicon carbide is normally used for the lapping step, with colloidal silica slurry used for the polish steps. Although the use of silica produces acceptable surface roughness on 2" wafers, the polishing time is too long, affecting throughput and also both polish and slurry stability. These effects will become more pronounced as the industry moves to larger diameter wafers. However, silica slurries tend to be cheap and the long polishing times are currently tolerated.

A small number of companies use alumina slurry, which they believe has several advantages over silica slurries, the main one being increased removal rate which increases throughput. It is also possible to use alumina slurry as the sole slurry following wafer slicing, simplifying the whole process.

4. Slurries for Sapphire Polishing

Many leading sapphire manufacturers are now planning to evaluate alumina-based slurries for the larger diameter wafer sizes. Ferro has developed alumina-based slurries building on earlier work from Corning[2, 3] which showed that alumina slurries performed better than other abrasives, with respect to removal rate and surface quality. They found that application of polishing in an environment where the abrasive chemically reacts with the surface, can eliminate subsurface damage and significantly increase surface quality. It is this chemical-mechanical mechanism, rather than solely mechanical polishing from abrasives such as diamond and silica, that leads to excellent surface quality.

Using the Ferro slurry Tizox[TM] 8102, they showed that alpha-alumina slurries led to the best performance and surface quality when polishing c-plane sapphire substrates. The reasons for this are threefold;

1. Hydration layer formation is increased – during polishing, a hydration layer is continuously formed on the base sapphire substrate, which is softer than the base layer and it is the formation of this layer that facilitates material removal and leads to high surface quality.

2. Surface quality is improved – alpha-alumina has the same hardness as base sapphire and so it is extremely difficult for the abrasive to scratch the sapphire bulk material.
3. Removal rate is increased - alpha-alumina abrasive experiences surface hydration similar to base sapphire and a chemical-mechanical reaction between both the base sapphire and abrasive hydration layers promotes accelerated material removal. When the surfaces of the abrasive and base sapphire are brought together during polishing pressure and shear, mutual adhesion occurs and further shear allows the particle to 'tear away' the bonded hydrated layers, promoting further removal by the leading edge of the particle.

Using the above results as a baseline, Ferro has further optimized its sapphire polishing slurries, resulting in the SRS-1652 product. Typical particle characteristics of SRS-1652 are shown in figure 2.

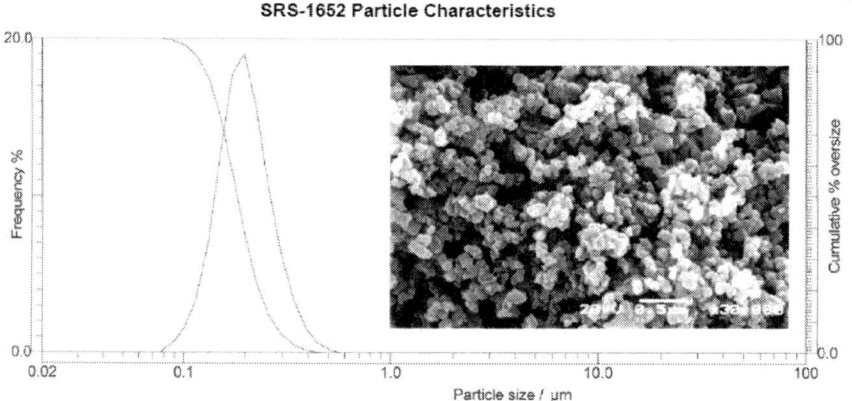

Figure 2: SRS-1652 Alumina Particle Characteristics

Removal rates can be optimized by variation of particle size and formulation chemistry and a surface roughness (Ra) of <0.2nm can be achieved. Proof-of-concept performance on 200mm sapphire wafers has also been demonstrated on an Applied Materials Mirra CMP tool and a typical removal rate curve is shown below in figure 3.

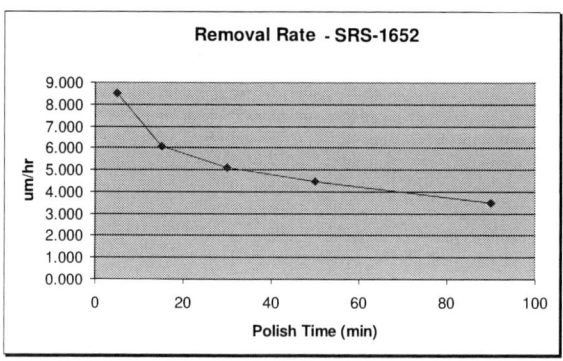

Figure 3: Removal Rate curve for SRS-1652 on 200mm Sapphire Wafers

In comparative studies vs several different silica slurries, SRS-1652 has been shown to give faster removal, equivalent surface quality and also better stability during recirculation. The temperature of silica slurries has to be controlled very tightly during use to prevent gelation, whereas alumina slurries are more robust. SRS-1652 also showed better cleanability. Figure 4 shows the Prestonian behavior of both alumina and silica slurries.

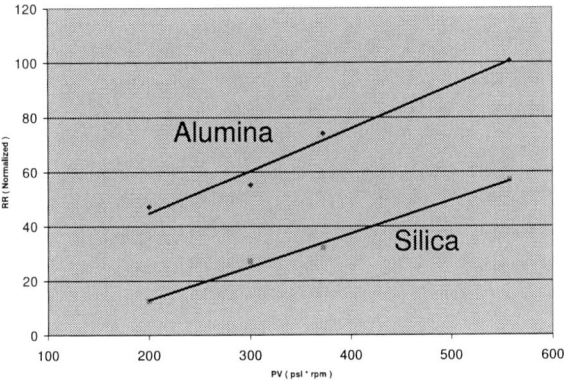

Figure 4: Removal rate (normalized) vs Pressure x Velocity (PV) for SRS-1652 and a silica slurry

5. Conclusions

Sapphire substrates, in one form or another, will continue to be the dominant substrate for LED chips for many years. Usage is expected to increase as sapphire suppliers transition to growing 6" and 8" wafers and also as 150mm and 200mm IC fabs are converted to LED chip manufacturing. As the LED industry moves to larger diameter wafers, more efficient, faster processes will be required. Customer evaluations have shown that process time can be cut in half by using an alumina slurry to replace a silica slurry. It has been demonstrated that not only can an alumina slurry can replace the silica slurry, but it can also replace the diamond slurry and some customers use only alumina slurry after wafer slicing.

Ferro has developed an alumina-based slurry, SRS-1652, which has been shown to give superior performance compared to other commercially available slurries.

6. References

1. Solid State Lighting Research and Development: Manufacturing Roadmap, US Department of Energy, July 2010
2. Honglin Zhu, Dale E.Niesz, Victor A.Greenhut, Robert Sabia, J.Mater.Res., 2005, 20, p504
3. Honglin Zhu, Luiz A.Tessaroto, Dale E.Niesz, Victor A.Greenhut, Robert Sabia, Maynard Smith, Applied Surface Science, 2004, 236, p120

ECS Transactions, 34 (1) 1041-1046 (2011)
10.1149/1.3567712 ©The Electrochemical Society

Experimental and modeling on atomic layer deposition Al₂O₃/n-InAs metal-oxide-semiconductor capacitors with various surface treatments

H. D. Trinh[a], E. Y. Chang[a], G. Brammertz[b], C. Y. Lu[a], H. Q. Nguyen[a], B. T. Tran[a]

[a]*Department of Materials Science and Engineering (MSE), National Chiao Tung University (NCTU), 1001 University Road, Hsinchu 300, Taiwan.*

[b]*Interuniversiy Microelectronics Center (IMEC vzw), Kapeldreef 75, B-3001, Belgium*

Ex-situ sulfide and HCl wet chemical treatments in conjunction with *in-situ* trimethyl aluminum (TMA) pretreatment were performed before the deposition of Al₂O₃ on n-InAs surfaces. X-ray photoelectron spectroscopy analyses show a significant reduction of InAs native oxides after different treatments. Capacitance-voltage (C-V) characterization of Al₂O₃/n-InAs structures shows that the frequency dispersion in accumulation regime is small (<0.75% per decade) and does not seem to be affected significantly by the different surface treatments, whereas the latter improves depletion and inversion behaviors of the nMOS capacitors. The interface trap density profiles extracted from simulation show mainly donor-like interface states inside the InAs bandgap and in the lower part of the conduction band. The donor-like traps inside the InAs bandgap and in the lower part of the conduction band were significantly reduced by using wet chemical plus TMA treatments, in agreement with C-V characteristics.

Introduction

High electron mobility and high-low field drift velocity GaAs, InGaAs and InAs compounds have been considered as potential candidates to replace Si as a channel material for future complementary metal-oxide-semiconductor (CMOS) technology nodes. High-k/GaAs and InGaAs structures therefore, have been attracted much attention in recent studies in order to enhance the performance of metal-oxide-semiconductor (MOS) transistor devices (1,2). Compared to numerous studies of high k/GaAs and InGaAs structures, the study of the high k/InAs structure is still relatively unexplored (3-11). Besides the interest in the inversion mode MOS field effect transistor (MOSFET), the application of InAs as a channel for MOS high electron mobility transistor (MOSHEMT) devices is also very promising. Recently, among a few reports about high-k/InAs structures, studies about the reduction of InAs native oxides and simulation of narrow band gap high-k/InAs MOS capacitors (MOSCAP) have been presented (5,6,8,9,11). Some experimental studies have been reported as well (3,4,7,11) but the influence of surface treatments on the capacitance-voltage (C-V) behavior of high k/InAs MOSCAP structure have not been investigated in details yet. In this work, we study the electrical characteristics of atomic layer deposition (ALD) Al₂O₃/n-InAs with various surface treatments, including sulfide and HCl treatment in conjunction with an *in situ* trimethyl aluminum (TMA) pretreatment (5,12,13). Experimental results and C-V simulations (14) are combined to investigate the electrical properties of Al₂O₃/InAs

1041

MOSCAP structures. Effects of surface treatments on the C-V behaviors in accumulation, depletion and inversion regimes are discussed.

Experiments

The wafers used in this work consist of ~ 2 to 5 x 10^{17} cm^{-3} Si-doped n-type 5nm InAs - 3nm $In_{0.7}Ga_{0.3}As$ - 100nm $In_{0.53}Ga_{0.47}As$ multilayer stacks grown by molecular beam epitaxy (MBE) on n-type InP substrates. The wafers were degreased in acetone and isopropanol at room temperature (RT) before surface treatments. HCl treatment was employed by dipping samples in HCl (38 %) : H_2O (1 : 10) solution for 1 min followed by rinsing in deionized (DI) water. For sulfide treatment, the sample first underwent HCl treatment, and was then dipped in the $(NH_4)_2S$ (20 %) : H_2O (1 : 3) solution for 20 min followed by rinsing in DI water. After that, control sample (native-oxide- covered InAs surface without treatment), HCl-treated sample, and sulfide-treated sample were loaded into the ALD chamber. In the ALD chamber, ten pulses of TMA/N_2 were employed for *in situ* TMA self-cleaning (5,12,13) before the deposition of Al_2O_3. Samples were kept at 300°C for both TMA treatment and oxide deposition processes. The number of ALD Al_2O_3 cycles for MOSCAPs fabrication and XPS measurement were 180 CYC (~ 18 nm) and 15 CYC (~ 1.5 nm) respectively. Samples used for MOSCAPs fabrication were annealed at 400°C in N_2 for 30 s. After that, Ti/Pt/Au gate metal and Au/Ge/Ni/Au back side ohmic contact were deposited, followed by annealing at 400°C in N_2 for 30 s.

Result and discussions

XPS analysis was performed using a commercial Microlab 350 system equipped with an Al K_α source, in an ultra high vacuum (UHV) chamber (~10^{-9} Torr) with 60°-take off angle. Fig. 1 shows the In $3d_{5/2}$ and As 3d XPS spectra of the InAs native-oxide-covered surface, the Al_2O_3/HCl plus TMA treated InAs interface, and the Al_2O_3/sulfide plus TMA treated InAs interface.

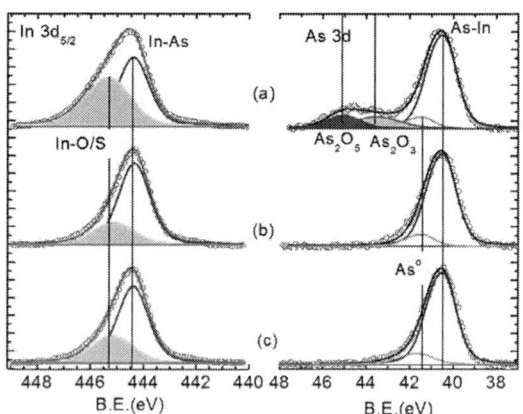

Figure 1. The In $3d_{5/2}$ and As 3d XPS spectra of (a) native-oxide-covered InAs surface; (b) 1.5 nm ALD Al_2O_3/HCl+TMA treated InAs interface; (c) 1.5 nm ALD Al_2O_3/sulfide+TMA treated InAs interface. Native oxides including In_2O_3, As_2O_3, and As_2O_5 were significantly reduced after the use of surface treatments

For both surface treatments, As-related oxides were removed to below the XPS detection level (Fig. 1, As 3d spectra). In $3d_{5/2}$ spectra indicate a significant reduction of

In-related oxides after surface treatment. The sample with sulfide plus TMA treatment exhibits larger In-O/S signal sample due to the contribution of In-S bonds.

Figure 2 shows the multi-frequency C-V responses and quasi-static C-V (QSCV) curves of 18 nm Al_2O_3/n-InAs MOSCAP samples. The multi-frequency C-V measurement was employed using an HP4284A LCR meter, and QSCV curves were acquired using an Agilent 4156C analyzer, with an integration time of 2 s. This integration time was confirmed long enough to get full thermal equilibrium in InAs MOSCAP, longer integration time of 5 s was also performed but the QSCV curve did not change. In the accumulation regime, the multi-frequency responses do not show the obvious difference in frequency dispersion between samples. As shown in Figs. 2(a)-2(c), the values of frequency dispersion of samples are small, in the range of 0.65-0.75% per decade. These low frequency dispersions including the control sample indicate that surface treatments do not seem to affect significantly the C-V responses in the accumulation regime.

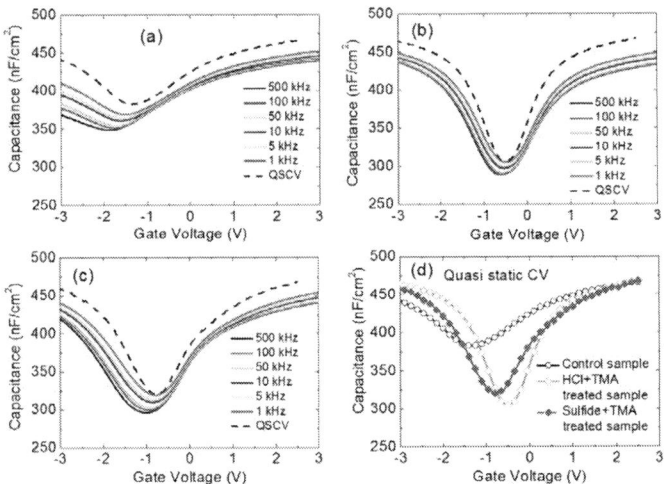

Figure 2. Multi-frequency C-V responses (solid lines) and QSCV curves (dashed lines) in (a) control sample, (b) HCl plus TMA sample, and (c) sulfide plus TMA treated sample of 18 nm ALD Al_2O_3/InAs MOSCAPs; (d) QSCV curves of all three samples, for comparison.

Low frequency-like C-V behavior is observed for all samples in the whole range of measured frequencies. This behavior originates from the short minority carrier response time (τ_R) in very low band gap, high intrinsic density materials as InAs. As shown in Figs. 2(a) and 2(d), the control sample exhibits high value of depletion capacitance (C_{dep}) in depletion regime which reveals large values of semiconductor capacitance (C_S) and/or interface trap capacitance (C_{it}). Large frequency dispersion in inversion regime in this sample as shown in Fig. 2(a) also implies high contribution of interface traps. In chemicals plus TMA treated samples, nice C-V curves with small frequency dispersion in inversion regime are observed [Figs. 2(b)-2(c)]. In Fig. 2(d), smaller C_{dep} of these two samples compared to the control sample indicates that the contribution of C_{it} and/or C_S is reduced.

Out of the two chemical plus TMA treated samples, the HCl plus TMA sample exhibits better electrical characteristics as compared to the sulfide plus TMA treated sample, with smaller frequency dispersion in inversion regime and smaller stretch out [Figs. 2(b)-2(d)]. This result seems contradictory to most of reports on high k/ GaAs (InGaAs) but it is consistent with the report on HfO_2/InAs structure (4).

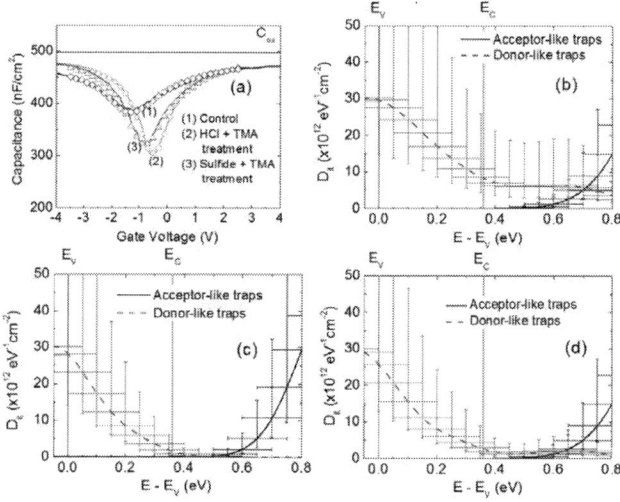

Figure 3. (a) Experimental data (symbols), simulated C-V curves (solid lines) of ALD 18 nm Al_2O_3/n-InAs MOSCAP samples with various surface treatments. Interface state density profiles of all three samples, extracted from simulation, are shown as well: (b) control sample, (c) HCl plus TMA treated sample, (d) sulfide plus TMA treated sample.

Low frequency CV-simulations were performed by full numerical solution of the Poisson equation taking into account the complete multilayer structure as well as the interface states at the InAs/high-k oxide interface, similar to approach in ref. 14. The interface state density (D_{it}) at the InAs/high-k interface was varied, until good fit to experimental data was obtained. All experimental QSCV curves (symbols) were well fitted by the simulations (solid lines). D_{it} profiles of samples extracted from simulation are shown in Figs. 3(b)-3(d), where the estimated error bars of the extracted D_{it} are shown as well. Errors were estimated and taken into account due to the following reasons: (i) error on metal work function, (ii) charge quantization effects and non-parabolicity in the conduction band which were not included in the simulation and (iii) uncertainty on absolute value of oxide capacitance, C_{ox}. The derived D_{it} profiles present a U-shape with minimum in D_{it} profile located around the conduction band minimum (E_C) for all samples. The interface state density shows strong similarities with the $In_{0.53}Ga_{0.47}As/Al_2O_3$ D_{it} profile (14). It can be clearly seen that the two different surface treatments significantly reduce the donor-like traps over the full energy region as compared to the control sample

[Figs. 3(b)-3(d)]. The D_{it} of the sulfide treated plus TMA sample shows slightly higher values of donor-like traps as compared to the HCl plus TMA treated sample at, in agreement with the comparison of C-V characteristics between these two samples.

Conclusions

We have examined the effect of surface treatments on the physical and electrical properties of Al_2O_3/n-InAs structures. The effect of interface states on accumulation capacitance behavior is small and does not depend on the surface treatments. In contrast, the C-V characteristics of Al_2O_3/n-InAs in depletion and inversion region were significantly improved by surface treatments. The D_{it} profiles extracted from simulation shows a significant reduction of donor-like traps after surface treatments in complete InAs bandgap, as well as in the lower part of conduction band. Results also revealed that HCl plus TMA treatment has stronger effect on the reduction of donor-like traps than sulfide plus TMA treatment.

Acknowledgments

This work was supported by Taiwan National Science Council (under contracts Nos. NSC 99-2120-M-009-005- and NSC 98-2923-E-009-002-MY3). The authors would like to thank to ITRC, Taiwan, for equipment support.

References

1. Y. Q. Wu, W. K. Wang, O. Koybasi, D. N. Zakharov, E. A. Stach, S. Nakahara, J. Hwang, and P. D. Ye, *IEEE Electron Device Lett.,* **30**, 700-702 (2009).
2. T. D. Lin, H. C. Chiu, P. Chang, L. T. Tung, C. P. Chen, M. Hong, J. Kwo, W. Tsai, and Y. C. Wang, *Appl. Phys. Lett.*, **93**, 033516 (2008).
3. H.-S. Kim, I. Ok, M. Zhang, F. Zhu, S. Park, J. Yum, H. Zhao, J. C. Lee, P. Majhi, N. Goel, W. Tsai, C. K. Gaspe, and M. B. Santos, *Appl. Phys. Lett.,* **93**, 062111 (2008).
4. D. Wheeler, L. E. Wernersson, L. Fröberg, C. Thelander, A. Mikkelsen, K. J. Weststrate, A. Sonnet, E. M. Vogel, and A. Seabaugh, *Microelectron. Eng.,* **86**, 1561-1563 (2009).
5. A. P. Kirk, M. Milojevic, J. Kim, and R. M. Wallace, *Appl. Phys. Lett.,* **96**, 202905 (2010).
6. R. Timm, A. Fian, M. Hjort, C. Thelander, E. Lind, J. N. Andersen, L. E. Wernersson, and A. Mikkelsen, *Appl. Phys. Lett.,* **97**, 132904 (2010).
7. Y.-C. Wu, E. Y. Chang, Y.-C. Lin, C.-C. Kei, M. K. Hudait, M. Radosavljevic, Y.-Y. Wong, C.-T. Chang, J.-C. Huang, and S.-H. Tang, *Solid-State Electron.,* **54**, 37-41 (2010).
8. A. Lubow, S. Ismail-Beigi, and T. P. Ma, *Appl. Phys. Lett.*, **96**, 122105 (2010).
9. E. Lind, Y.-M. Niquet, H. Mera, and L.-E. Wernersson, *Appl. Phys. Lett.*, **96**, 233507 (2010).
10. N. Li, E. S. Harmon, J. Hyland, D. B. Salzman, T. P. Ma, Y. Xuan, and P. D. Ye, *Appl. Phys. Lett.,* **92,** 143507 (2008).
11. H.-D. Trinh, E. Y. Chang, Y.-Y. Wong, C.-Y. Yu, C.-Y. Chang, Y.-C. Lin, H.-Q. Nguyen, and B.-T. Tran, *Jpn. J. Appl. Phys.,* **49**, 111201 (2010).
12. M. Milojevic, C. L. Hinkle, F. S. Aguirre-Tostado, H. C. Kim, E. M. Vogel, J. Kim, and R. M. Wallace, *Appl. Phys. Lett.,* **93**, 252905 (2008).

13. H. D. Trinh, E. Y. Chang, P. W. Wu, Y. Y. Wong, C. T. Chang, Y. F. Hsieh, C. C. Yu, H. Q. Nguyen, Y. C. Lin, K. L. Lin, and M. K. Hudait, *Appl. Phys. Lett.,* **97**, 042903 (2010).
14. G. Brammertz, H. C. Lin, M. Caymax, M. Meuris, M. Heyns, and M. Passlack, *Appl. Phys. Lett.,* **95,** 202109 (2009).

ECS Transactions, 34 (1) 1047-1052 (2011)
10.1149/1.3567713 ©The Electrochemical Society

Effects of Surface Pretreatments on p-GaN/GZO Contact by rf magnetron sputter

W. J. Wang[a], X. F. Li[b], J. S. Zhang[a], J. H. Zhang[a,b*]

[a] School of Mechatronics Engineering and Automation, Shanghai University, Shanghai, P.R. China
[b] Key Laboratory of Advanced Display and System Applications of Ministry of Education, Shanghai University, Shanghai, P.R. China

The Effects of surface pretreatments of p-GaN using H_2SO_4, HF, KOH solution on the optical and electrical properties of GZO/p-GaN contacts have been investigated using a rf magnetron sputter deposition. The current-voltage characteristics of GZO/p-GaN contact is slightly improved using KOH solution treatment. Contact barrier of GZO/p-GaN can be considerably reduced on annealing at 530 °C for 15 min in N_2 ambient and the ohmic contact is almost achieved for with KOH-treated sample. the light luminance of 86.7 cd/m^2 at current of 20 mA is obtained for GZO electrodes of LEDs with KOH pretreatment.

Introduction

As the new generation of solid lighting source the GaN based LEDs have achieved the blooming applications in the urban lighting, transportation illumination and backlit due to its green and environment protection[1]. At present, the ITO thin film is the main material for the transparent electrodes in the high power LEDs. Many researches have been exploring the new materials to substitute ITO. Zinc oxide is a potential selection since it has some merits of the wide band gap, low resistively, high visible light transmittance, non-toxicity and son on. ZnO is a rich resource and friendly to the environment with a simple preparation technology. Comparing with other elements doped in ZnO, GZO (ZnO:Ga) has a higher transmittance, smaller lattice distortion and easier doping technology. Ohmic contact is one of the key characteristics to estimate the electrical prosperities of the deposited thin films. The ohmic contact resistance is lower and the deposition quality of the thin film and the forward voltage of the lighting devices are better. It also affects the luminous efficiency, the junction temperature, the device stability and the operation life[2]. The technology bottlenecks are how to obtain the good performance of the GZO thin film and the low ohmic contact resistance between the GZO thin film and the p-GaN substrate. Various

---
* Corresponding author.
E-mail address: jhzhang@staff.shu.edu.cn

methods have been introduced to develop the p-type GaN-based LEDs with low-resistance, such as the large work function, proper surface treatment, super lattice and strained layers, hydrogen extraction and interlayer[3]. In general, the surface oxide is easy to formation on the semiconductor because of the oxidation in the air. Surface oxide often takes a role as the barrier to retard the carrier transporting from the metal side to the semiconductor side. Thus, it is essential for the oxide removal to get the better ohmic contact[4-9]. The surface treatment has usually been performed before the deposition of the alloy compound. The dilute HCl solution is the most common technique for the surface pretreatment[10-15].

This paper studied the relationship between the surface pretreatments and the lighting performance of LED chips by the analysis of the ohmic contact of the p-GaN /GZO interface and the change of the luminance intensity.

Experiments

All the GaN substrates had been cleaned in the five steps, which were: 1) the isopropanol cleaning by the ultrasonic energy for 10 min, 2) the acetone cleaning by the ultrasonic energy for 10 min, 3) the chemical reagents soaking at 60 °C for 30 min, 4) the deionized water cleaning by the ultrasonic energy for 10 min, 5) the drying by Nitrogen.

In the study, the experiments executed the following steps. First, one of the chemical reagents, which was the sulfuric acid (25%), the hydrofluoric acid (25%) or the potassium hydroxide (9% wt), could remove the oxide layer of the GaN substrate. Before GZO deposition by the magnetron sputtering the p-GaN deposited on the GaN substrate to form a diffusion barrier. After deposition the GZO film grew on the p-GaN substrate and the specimens endured the annealing treatment in oven. Some of the specimens were tested for the analysis of the ohmic contact characteristics. Other specimens underwent the lithography, ion etching and annealing treatment for the study of the chip lighting performance.

JPGF-700B magnetron sputtering system was the equipment for the GZO film deposition. The applied sputtering parameters in the experiments were the sputtering power for 400 w, the substrate temperature for 300 °C, the sputtering atmosphere for Argon with 16 sccm, the sputtering time for 1 h. The GZO films with different substrate pretreatments were about 250 nm in thickness. The patterns were defined by standard photolithographic technique. The specimens had been endured the baking at 100 °C for 90 s and at 110 °C for 120 s before and after the photolithography process, respectively. The etching equipment was the ion-beam lithography system (LKJ-1C-150I type) and the etching time was 40 min. RTP-300 rapid annealing oven raised the specimen temperature responding to the setting positive slope. The surface morphologies of the GZO films were examined by Atomic Force Microscopy (Nanomavi SPA-400 SPM type). X-ray Diffraction (DLMAX-2200 type) were used to characterize the samples. The LED devices were measured with a Hewlett–Packard

4140B source measure unit and a PR-650 luminance color meter.

Results and Discussions

Fig.1(a) shows the AFM topographic image as-deposited GZO films. It can be seen that uniform grain size exists in sample. Root mean square(RMS) of GZO film were approximately 2 nm according to AFM. The XRD pattern of typical as-deposited GZO film is shown in Fig. 1(b), the diffraction peaks related to quartzite ZnO structure are identified, no peaks pertinent to other crystalline structures were observed exhibiting a highly (002) preferential orientation.

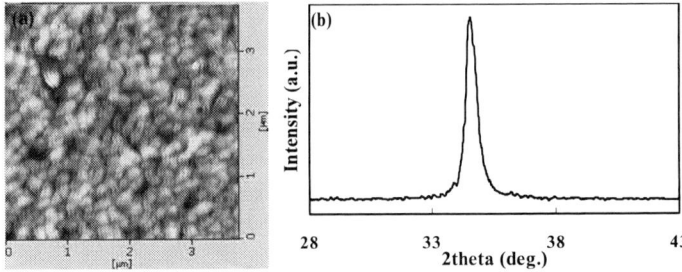

Fig.1 (a) AFM of the GZO films (b) XRD pattern of the GZO films

the current-voltage (I-V) curves of the GZO films pretreated by different solutions have been plotted in Fig.2. The I-V curves of GZO without the annealing process is showed in Fig.2a. The specimen pretreated by HF has the similar I-V curve with the non-pretreatment specimen. The KOH and H_2SO_4 pretreatments have significant effects on the electrical properties to the specimens under the same conditions. Specially, the KOH pretreatment improves the current highest comparing with other pretreatments while the GZO film has been loaded the same positive voltage. The currents of the specimens with KOH or H_2SO_4 pretreatments have the similar values in the regions of the negative voltage. It is true KOH is the best of the four solutions for the p-GaN surface pretreatment in the experiments. After annealing at 530 °C for 15 min with the N_2 atmosphere, the currents through the GZO films decrease one order of magnitude at the same voltages, shown in Fig.2b. There are the same curves in specimens with HF or without pretreatments below the voltage of 2 V. However, the current to the specimen without pretreatment increases dramatically than that with HF pretreatment. In the voltage range fluctuating from -3 V to 2 V, the specimen with KOH pretreatment seems to change in the linear curve. The specimen with H_2SO_4 pretreatment has a smooth I-V curve. According to the data tested before and after the annealing process, the KOH solution is more suitable for the p-GaN surface pretreatment to form a better ohmic contact.

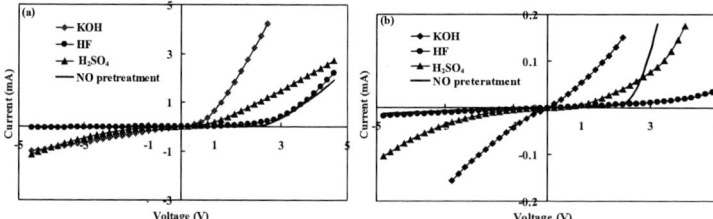

Fig.2 I-V curves of GZO on the p-GaN substrate with different pretreatments:

(a) before annealing, (b) after annealing (530 °C, 15 min, N_2 atmosphere)

Due to the above pretreatments and depositions processes, the LEDs applied GZO as the transparent conductive films and annealed at 530 °C for 15 min in N_2 atmosphere. Fig.4 shows I-V characteristics of the LEDs with different pretreatments. At the same current of 20 mA, the forward voltages (V_f) are 12.5 V, 16.5 V, 17 V and 17.3 V for the LEDs pretreated by KOH, HF, H_2SO_4 and without pretreatment, respectively. If the forward voltage exceeds the threshold, the LEDs have the similar tendency for the current increasing. The LED has non-pretreatment to exhibit the worst electrical properties than those with different pretreatments. This can be attributed to the interface between the GZO film and the p-GaN substrate forming a Schottky contact rather than an ohmic contact. In other words, the GZO/p-GaN interface has a higher resistance to hinder the carrier transportation. Since the LEDs are the devices driven by current, the LED with KOH pretreatment has the smaller forward voltage than other LEDs at the same current. This proves the GZO film on the p-GaN pretreated by KOH presents a better ohmic contact not a Schottky contact.

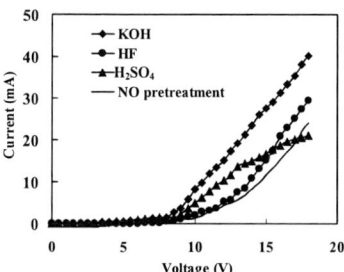

Fig.3 Current-Voltage (I-V) characteristics of LEDs

The luminance for the LEDs with KOH pretreatment and without pretreatment   are showed in Fig.5. At the forward voltage is 19 V, the light luminance are 2830 cd/m^2 and 54.8 cd/m^2 for the KOH pretreatment and without pretreatment, respectively. Due to the area of the pattern is 4 square millimeters, so the EL intensity of the LED with KOH pretreatment is 39.5 mcd At the forward voltage is 19 V. It is found the light luminance of 86.7 cd/m^2 at current of 20 mA is obtained for GZO electrodes of LEDs with KOH pretreatment. Obviously, there is a very significant difference of lighting between the two LEDs. Fig.5b is the real photo to the LED pretreated by KOH (V_f = 15 V). It is true the LED emits the light very bright and uniform. The KOH pretreatment to the p-GaN substrate before GZO deposition can enhance the light luminance efficiently.

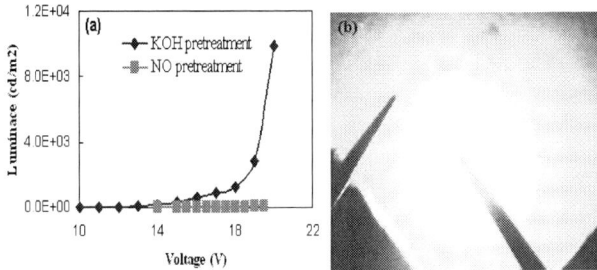

Fig.4. Lighting characteristics of LEDs with KOH pretreatment and without pretreatment: (a) Luminance-Voltage Curve of both LEDs, (b) Real lighting photo of LED with KOH pretreatment (V_f=15 V)

Conclusions

Influence of surface pretreatments of p-GaN on the optical and electrical properties of GZO/p-GaN contacts have been investigated. The results shows that p-GaN surface pretreatment can reduce contact barrier of GZO/p-GaN. Compared with p-GaN surface pretreatment with KOH, H_2SO_4 and HF solution, KOH solution is beneficial to form ohmic contact. Ohmic contact of GZO/p-GaN with KOH pretreatment is achieved after annealing at 530 °C for 15 min. The luminace of the LEDs with KOH pretreatment is far higher than LEDs without pretreatment. The light luminance at current of 20 mA is 86.7 cd/m^2 and LED can emit light uniformly.

Acknowledgments

This research work was supported by the National Natural Science Foundation of China (NSFC) under the grant number of 51072111 and 50675130. The corresponding author, Dr. Jianhua Zhang, would also thank the support from the Program for New Century Excellent Talents in University under the grant number NCET-07-0535, the Science and Technology Committee of Shanghai Municipality under the grant number of 09DZ2292901 and 09DZ1141502.

References

1. S.Q. Guo and Z.J.Liu. etc, *the 7th China international semiconductor lighting BBS, chip and device technology*, ShenZhen, China, 10 (2010).

2. Y.B.Liu, *Vacuum electron technology*, **3**,(2008)

3. J. O. Song, J.S. Ha, and T. Y. Seong, *IEEE T. Electron Dev.*.**57** (1),1 (2010).

4. E. H. Rhoderick and R. H. Williams, *Metal–Semiconductor Contacts*, (1988).

5. J. M. DeLucca, H. S. Venugopalan, S. E. Mohney, and R. F. Karlicek, *J.Appl. Phys. Lett.*, **73**(23), 3402 (1998).

6. T. W. Kang, C. S. Chi, S. H. Park, and T. W. Kim, *Jpn. J. Appl. Phys.*, **39**(3A), 1062 (2000).

7. J. K. Sheu, Y. K. Su, G. C. Chi, P. L. Koh, M. J. Jou, C. M. Chang, C. C. Liu, and W. C. Hung, *Appl. Phys. Lett.*, **74** (16), 2340 (1999).

8. J. K. Kim, J.-L. Lee, J. W. Lee, Y. J. Park, and T. Kim, *Electron. Lett.*, **35**(19), 1676 (1999).

9. P. J. Hartlieb, A. Roskowski, R. F. Davis, and R. J. Nemanich, *J. Appl. Phys.*, **91**(11), 9151 (2002).

10. J. Sun, K. A. Rickert, J. M. Redwing, A. B. Ellis, F. J. Himpsel, and T. F. Kuech, *Appl. Phys. Lett.*, **76**(4), 415 (2000).

11. Jinn-Kong Sheu, Ming-Lun Lee,Y.S.Lu,and K.W.Shu. *IEEE Journal of Quantum Electreonic*, **44**(12),32 (2008)

12. C. J. Tun, J. K. Sheu, M. L. Lee, C. C. Hu,C. K. Hsieh, and G.C. Chi., *J. Electrochem. Soc.*, 153.G296 (2006).

13. V. Gupta and A.Mansigh, *J. Appl. Phys.*,**80**, 1063 (1996).

14. X.A. Cao, S. J. Pearton, A. P. Zhang, G. T. Dang, F. Ren, R.J. Shul, L. Zhang. R. Hickman. And J. M. Hove, *Appl. Phys. Lett.*, **75**, 2569 (1999).

15. C S Chang, S J Chang, Y K Su,Y C Lin,Y P Hsu, S C Shei, S C Chen, C H Liu and U H Liaw, *Semicond. Sci. Technol.*, **18**, L21 (2003).

ECS Transactions, 34 (1) 1053-1057 (2011)
10.1149/1.3567714 ©The Electrochemical Society

A Phase Change Memory Device Fabrication Technology Using $Si_2Sb_2Te_6$ for Low Power Consumption Application

Ying Li[a,b], Xudong Wan[b], Zhitang Song[a], Joseph Xie[b], Bomy Chen[c], Bo Liu[a], Guanping Wu[b], Nanfei Zhu[b], Min Zhong[a], Jia Xu[b], Yifeng Chen[a]

[a] Laboratory of Nano-technology, Shanghai Institute of Microsystem and Information Technology, Chinese Academy of Sciences, Shanghai 200050, China

[b] United Lab Emerging Memory Project I, Semiconductor Manufacturing International Corp., Shanghai 201210, China

[c] Silicon Storage Technology, Incorporated, Sunnyvale, California 94086, USA

The Phase Change Memory (PCM) based on the rapid reversible phase change effect in the chalcogenide film has been regarded as one of the most promising candidates for the next-generation nonvolatile memories. The phase change material of $Si_2Sb_2Te_6$ was newly proposed for PCM application, which has lower power consumption and better endurance than conventional $Ge_2Sb_2Te_5$ materials. In this article, for the first time, the process technology for the phase change memory cell and array fabricated using SST as phase change material replacing conventional GST is presented. By adopting SST/GST PVD process and confined structure, on a small-scale phase change memory array with CMOS access transistors and addressing circuit, we are able to fabricate and demonstrate its full function with low power. The cell electrical testing results are compared between two kinds of films. We also proved a different phase transition mechanism with SST material.

I. Introduction

Phase change nonvolatile memory (PCM) based on the rapid reversible phase change effect in the chalcogenide film has been regarded as one of the promising candidates for the next-generation nonvolatile memories, because of its low cost, low programming voltage, and excellent scalability to nanometer cell sizes[1-2]. Generally, the chalcogenide $Ge_2Sb_2Te_5$ film (GST) is used as phase change media in PCM. However, large writing (RESET current (>1 mA) based on 180 nm lithography is a crucial issue before the commercial application of PCM[3]. In order to reduce the writing current, some approaches are proposed, such as: improving the cell structure to reduce the programming volume[4], or inserting heater layer with high resistivity to improve the heating efficiency, or doping other elements such as nitrogen, silicon and oxygen in the $Ge_2Sb_2Te_5$ film to increase the crystalline resistivity[5,6]. In addition, it is well known that a low melting temperature of the phase-change material means a low RESET current, since the memory cell must be heated to a temperature higher than the melting temperature to realize the phase transition. Si-Sb-Te system phase change material has lower melting temperature and higher crystallization speed than $Ge_2Sb_2Te_5$[7].

Though many efforts have been focused on investigating the characteristics of new phase-change materials, there are very few reports in the literature mention about test chip fabrication integrating CMOS circuitry with post-processing of exotic materials [8-9], and in fact, none of these reports provide a full integration of the foundry work with university BEOL processing to fulfill the engineering of SST materials embedded in device which fabricated in a phase-change memory array.

It is with the goal of developing fabrication and processing abilities for chalcogenide SST materials and technologies potentially useful for industry that we have explored ways to achieve fabricated devices using 0.13 logic technologies. Fabricating front-end-of-line (FEOL) processes at a foundry on the die level and integrating these die with back-end-of-line (BEOL) exotics material processing were successfully achieved.

II Experiments

A. Layout and Mask Design

The design layout included peripheral circuitry, contact and via critical dimension(CD) are 180 and 160 nm with, two $20F^2$ arrays with 160 nm vias and 80 μm x 80 μm bond pads. The design of the test structures contained contacts up to the Metal3 layer and the process is a four layers metal process.

B. Processing for BEOL

The process for FEOL is fabricated by standard CMOS technology. Here the BEOL process which plays the key factor for involve the phase change material into device will show in detail. The BEOL process adopted four layers metal, and the phase change material were inserted between metal3 and top metal, with connect to BEC, top electrode contact to top metal. The schematic of resistance-loop for phase change cell structure of PCRAM device used by SST and GST material are show in Fig1. The fabrication of the phase change units of PCRAM device based on $Si_2Sb_2Te_6$ and $Ge_2Sb_2Te_5$. The transistor were adopted the standard CMOS process flow. The phase change units lay between metal3 and top metal, which bottom connect with tungsten plug and top connect to AlCu alloy. The phase change material was deposited by radio frequency (RF) magnetron sputtering using a single target at room temperature.

Two pieces wafers were prepared adopted by 0.13 micrometer standard logic technology through the fabrication of a phase-change memory test chip which incorporates FEOL CMOS access transistors and array addressing circuitry with the Si/Ge-Sb-Te BEOL materials processing for phase-change memory development. For those two pieces wafers, the FEOL process is the same while for BEOL process, two different sizes of the bottom electrode contact (BEC), 160 and 80 nm for SST and GST respectively.

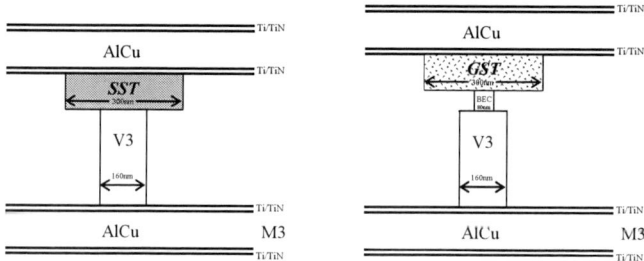

Figure 1 The schematic diagrams for M3 to TM used by SST and GST material

C. Electrical Characterization

Electrical measurements were performed on the die post-BEOL processing, using TEL P8-XL probe station and Agilent 4072A Semiconductor Parameter Analyzer. The operation behaviors of the fabricated devices were obtained through electrical measurement system, in which the voltage pulses for SET and RESET operations were provided by a programmable pulse generator (Agilest 81104A) and characterized by Keithley-2400 digital source-meter. Electrical contact to the die was made using Micromanipulator probes with tungsten tips.

III. Results and Conclusions

Two wafers accompanying each were fabricated, using two different dual chalcogenide stacks. The cross section SEM image post-BEOL process show phase change resistance of BEOL metal and via and 80nm BEC for GST in Figure 2. A completed test key that measure a resistor bit used by two- terminal devices with the 80nm BEC post top metal etch is shown in Figure 3. Alignment of the top electrode mask to the chip on the template wafer was complicated by several factors, including practical resolution limits on the mask aligner and photoresist edge bead issues on the die, which distorted the apparent alignment. Ultimately alignment was achieved by verifying correct placement of the top electrodes over the vias and that no bond pads overlapped. Electrical testing revealed no electrical short or open issues resulting from misalignment. Figure 4 shows the top view of a completed NMOS transistor-controlled Mbit with the 80nm BEC.

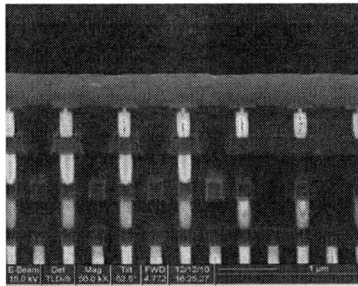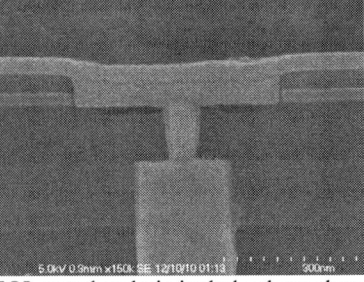

Figure 2. Cross section SEM image a) BEOL metal and via include phase change resistance, b) 80nm BEC for GST post-BEOL process phase change resistance

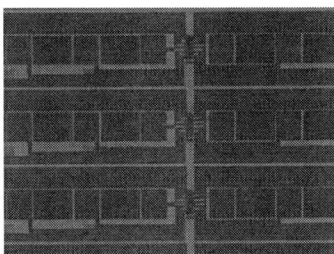

Figure 3. A completed test key with the 80nm BEC post top metal etch

Figure 4. 1M PCRAM device with the test key post top mater etch

Electrical characterization of phase-change devices was performed by forcing the current and measuring the voltage using the Agilent4072A semiconductor parameter analyzer. Two- terminal devices were measured and found to function normally as phase-change devices. Figure 5 shows an IV curve of a two terminal phase change device from a completed test chip with SSS/GST layers. Both the two curves show resistance change obviously, and prove SST with 160nm BEC has the similar phase change ability compare with GST with 80nm BEC, which shows SST has advantage in fabrication as reduce one photo layer, thus lower cost and improve throughput. Furthermore, SST has lower threshold voltage which is 1.2V compare with GST which is 2.6V. SST phase change material device SET window is presented in Figure 6. Before each set operation, a completed reset adopted 0.8mA current which is much lower than GST (>1mA) had done to make sure to fulfill the set. SET window for SST margin, and the best set condition is 0.4mA. The low threshold voltage and low operation current are benefited for power consumption. And low power consumption is preferred as to reduce the size of device and circuits for increasing package density.

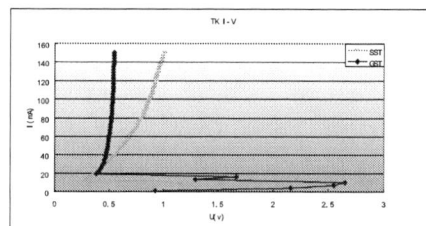

Figure 5. IV curve from $Si_2Sb_2Te_6$/$Ge_2Sb_2Te_5$ two-terminal device

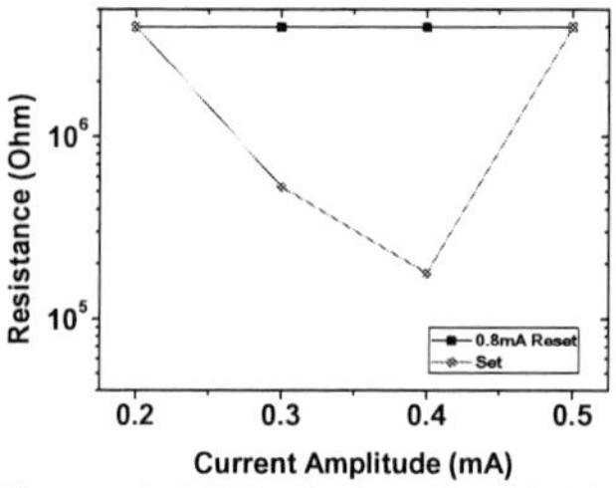

Figure 6. SET window for $Si_2Sb_2Te_6$ phase change material device

From the view of theoretical, the crystallization mechanism for GST and SST is totally different. GST is the nucleation driven, while SST is a diffusion controlled growth with a slowdown of crystallization rate. To decrease the power for GST material, small BEC size is preferred [10,11]. We can consider the SST crystallization mechanism as follows: Some of the doped Si atoms are incorporated into the Sb-Te lattice, while others may precipitate into the grain boundaries and exist in the amorphous phase, which suppresses the grain growth during crystallization and thus the grains are refined. The Si-rich amorphous regions always have far higher resistivity than the crystalline phase [12,13]. It has been reported that because of the high electrical resistance of germanium oxides in the oxygen-doped $Ge_2Sb_2Te_5$ film, the RESET current of PRAM with this material is greatly reduced to 100 mA [14]. Similarly, under RESET pulses, the Si-rich amorphous regions can act as micro-heater and produce much joule heat to melt the neighboring crystalline phase. Therefore, the reduction of RESET current may be realized in PRAM using Si-doped Sb_2Te_3 films. So BEC size not required for small one.

Integration of 0.13um generation universal-based processes with CMOS die of PCRAM by SST material has been successfully fabricated at a foundry and result in functional phase change memory devices was presented for the first time. Post processing of CMOS single die allows for development of new device technologies with great flexibility. SST materials processing on a research and development, which show advantage in fabrication such as lower cost and improve throughput, can be integrated with CMOS circuitry using this technique.

Acknowledgments

This work is supported by National Integrate Circuit Research Program of China (2009ZX02023-003), National Basic Research Program of China (2007CB935400, 2010CB934300, 2011CB309602, 2011CB932800), National Natural Science Foundation of China (60906004, 60906003, 61076121, 61006087), Science and Technology Council of Shanghai (09QH1402600, 1052nm07000).

References

1. S. Lai and T. Lowrey, *Tech. Dig. - Int. Electron Devices Meet.* , **803**, (2001).
2. A. Pirovano, A. L. Lacaita, A. Benvenuti, F. Pellizzer, S. Hudgens, and R. Bez, *Tech. Dig. - Int. Electron Devices Meet.* , **699, (2003)**.
3. S. Lai, *Tech. Dig. - Int. Electron Devices Meet.* , **255, (2003)**.
4. F. Pellizzer et al., *2004 Symposium on VLSI Tech. Dig. of Tech. Papers,* p. 18. (2004).
5. H. Horii et al., *2003 Symposium on VLSI Tech. Dig. of Tech. Papers,* p. 177, (2003).
6. B. W. Qiao, J. Feng, Y. F. Lai, Y. Ling, Y. Y. Lin, T. A. Tang, B. C. Cai, and B. Chen, *Appl. Surf. Sci.* 252,(2006).
7. N. Yamada, E. Ohno, K. Nishiuchi, N. Akahira, M. Takao, *J. Appl. Phys.* 69 2849 (1991).
8. Reeves, N, Liu, Y., Nelson, N. M., Mahlhotra, S., Loganathan, M., Lauenstein, J.-M., Chaiyupatumpa, J., Smela, E., Abshire, P. A., *Proceedings of the 2004 International Symposium on Circuits and Systems* 3 673-6 (2004).
9. Prakash, S. B., Urdaneta, M., Christophersen, M., Smela, E., Abshire, P. *Sensors and Actuators B* 699-704129 (2008).
10. A.L. Lacaita, *Solid-State Electron.* 24, (2006).
11. Y. Yin, H. Sone, S. Hosaka, *Jpn. J. Appl. Phys.* 6177 (2006).
12. N. Matsuzaki, K. Kurotsuchi, Y. Matsui, O. Tonomura, N. Yamamoto, Y. Fujisaki, N. Kitai, R. Takemura, K. Osada, S. Hanzawa, H. Moriya, T. Iwasaki, T. Kawahara, N. Takaura, M. Terao, M. Matsuoka, M. Moniwa, *IEDM Tech. Dig.* 738, (2005).
13. N. Yamada, E. Ohno, K. Nishiuchi, N. Akahira, M. Takao, *J. Appl. Phys.* 69 2849 (1991).
14. W.H. Wang, L.C. Chung, C.T. Kuo, Surf. Coat. *Technol.* 177/178 795, (2004).

Smart Systems

T. Gessner, M. Vogel, T. Otto, S.E. Schulz and R. Baumann

Fraunhofer Institute for Electronic Nano Systems ENAS, Technologie-Campus 3,
09126 Chemnitz, Germany

The Fraunhofer-Gesellschaft is one of the leading organizations for applied research in Germany and Europe. Its core purpose is the pursuit of knowledge of practical utility. The Fraunhofer Institute for Electronic Nano Systems ENAS is focusing on smart systems integration by using micro and nano technologies. Smart systems integration addresses the demand for miniaturized, multi-functional, self-organizing systems with an interface for communication. The technologies for smart systems and their integration significantly influence the competitiveness of different branches like aeronautics, automotive engineering, security applications, logistics, as well as medical and process engineering. Fraunhofer ENAS accesses on a broad variety of technologies and methods for smart systems integration.

Introduction

The actual developments of micro and nanotechnologies are fascinating. Undoubted they are playing a key role in today's product development and technical progress. With a big choice of different devices, different technologies and materials they enable the integration of mechanical, electrical, optical, chemical, biological and other functions into one system using very small space.

Research and development in microsystem technology has lead to a variety of products. Worldwide the MEMS market forecast predicts a further increase [1], Figure 1. However MEMS remains a fragmented market. The development of new MEMS applications is currently taking fewer years to be commercialized than in the past. Simplification of manufacturing remains an objective. The MEMS law "One product, one process, one package" still rules.

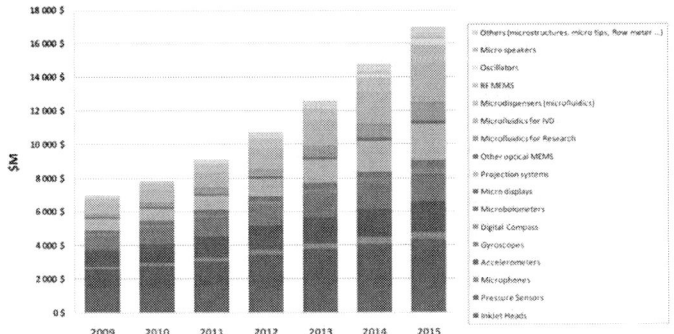

Figure 1: MEMS market forecast by Yole Developpement [1]

In semiconductor technologies/ microelectronics the strategic research agenda of the European Nanoelectronics Initiative Advisory Counsil ENIAC as well as the International Technology Roadmap of Semiconductors ITRS predicts not only a further downscaling of the structural dimensions but also the diversification of technologies [2]. Main topic of this so-called More than Moore strategy is the integration of different components in one system to ensure the multifunctionality of the system itself.

The European Platform on Smart Systems Integration EPoSS takes up this trend to multifunctional devices and smart systems [3,4]. Smart systems go beyond microsystems for single physical, biological or chemical parameter measurements combined with signal processing and actuating functions. Smart systems integration addresses the demand for miniaturized multifunctional devices and specialized connected and interacting solutions. Multidisciplinary approaches featuring devices for complex solutions and making use of shared and, increasingly, self-organising resources are among the most ambitious challenges.

Smart Systems

A lot of smart systems evolved from microsystems. They combine technologies and components from microsystems technology with knowledge, technology and functionality from other disciplines like biology, chemistry, nano sciences or cognitive sciences. Focus is on the design and manufacturing of completely new marketable products and services for specialized applications and for mass market applications. A major challenge in smart systems technology is the integration of a multitude of diverse components, developed and produced in very different technologies and materials. So the smart systems consist of a power source, a communication unit, sensors and actuators and electronic components, Figure 2. All these components need to be integrated and packaged.

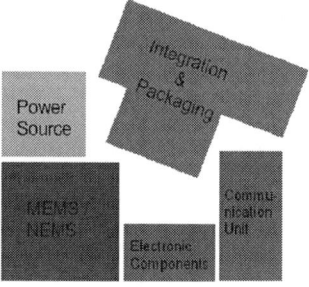

Figure 2: Smart systems

The Fraunhofer Institute for Electronic Nano Systems ENAS in Chemnitz focuses on research and development in the field of smart systems integration by using micro and nanotechnologies with partners in Germany, Europe and worldwide. Based on prospective industrial needs, Fraunhofer ENAS provides services in:

- Development, design and test of MEMS and NEMS (micro and nano electro mechanical systems)
- Wafer level packaging of MEMS and NEMS
- Metallization und interconnection systems for micro and nano electronics as well as 3D integration
- New sensor and system concepts with innovative material systems
- Integration of printed functionalities into systems
- Reliability and security of micro and nano systems

Fraunhofer ENAS accesses on a broad variety of technologies and methods for smart systems integration. The technological basis for the development of components of micro and nano systems are the core competences silicon based technologies for micro and nano systems, polymer based technologies for micro and nano systems as well as printing technologies for functional layers and components. So-called cross-sectional technologies are interconnect technologies and system integration technologies. Design and test of components and systems as well as reliability of components and systems are supporting fields for the other technologies. Based on these basic technologies, the cross-sectional technologies and methods for design, test and reliability Fraunhofer ENAS is able to process complete MEMS/NEMS and to integrate them into challenging smart systems. In the following a system and a component are described more in detail.

Transport label

The example is an active radio frequency identification label for the monitoring of shock, inclination and temperature during transportation processes [5, 6]. This is a joint development with ELMOS Semiconductor AG, KSW Microtec AG, Schenker AG and memsfab GmbH.

The label consists of a polymer substrate, a RF-chip with antenna, a battery for power supply and a sensor system consisting of a micromechanical transducer and the signal processing electronics. All these elements have to be integrated on the thin label substrate fulfilling specific requirements for the sensor system like low power consumption, high signal to noise ratio, high temperature stability and low device / sensor thickness. One packaging solution of such a label is presented in Fig. 3.

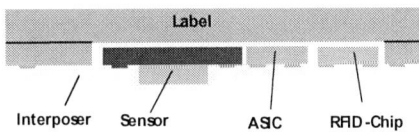

a) Packaging solution with integrated sensor functions for shock, inclination and temperature

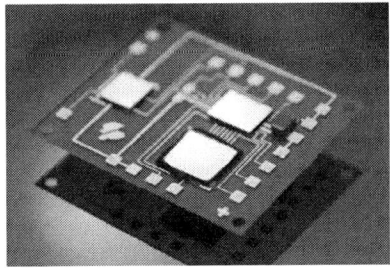

b) Interposer with mounted sensor, ASIC and RFID chip

Figure 3: Smart RFID label

To integrate all the elements into one label a flexible printed board (250 μm thick PET) is used as interposer which carries the sensor, the sensor ASIC and the RFID ASIC. Figure 3b) shows the interposer after its fabrication with conductive path ways and mounted chips. This interposer can be fabricated and tested independently from the label production and assembly and has the special feature of integration ability of further sensor functions.

Printed battery

The integration of nano materials and printed functionalities causes new challenges and requires new approaches in terms of design, testability, and reliability. The main challenge is to integrate different functions and components based on different materials and technologies in one system while maintaining reliability and security at reasonable costs.

Printing technologies are highly productive, cost-efficient manufacturing methods which are usually run under ambient conditions. Today these technologies are employed to manufacture single components of smart systems like antennae and batteries on an increasing scale.

We report on approaches to manufacture batteries employing printing technologies to get ready to meet future demands regarding the autonomy of electronic devices. These might be, e.g. intelligent chip and sensor cards, medical patches and plasters for transdermal medication and vital signs monitoring, as well as lab-on-chip analysis devices. The combination with other thin and flexible modules is intended whereby flexible displays and solar cells may be manufactured in a compatible manner and combined where required.

Methods

For the manufacturing of thin film batteries there are a number of requirements concerning how to process the needed materials. As printing technique screen printing was chosen. Screen printing is a versatile printing technology since it provides the opportunity of printing onto various flexible substrates such as paper, plastic foil or textiles [7-9]. Another advantage of screen printing is its property to produce extra high layer thicknesses (typically 100 μm). Functional materials may be applied in any requested form by formulating functional materials to inks and pastes. Therefore screen printing qualifies perfectly for the fabrication of printed energy storage devices because the amount of chemical material is increased. The principle of screen printing is given in Fig. 4. A mesh is stretched in a metal form to build the screen for the printing process. The mesh is covered by a pattern to determine all areas of the substrate where printing ink shall be deposited (in Fig. 4 open mesh is white, covered mesh is black). The ink is put on the top side of the mesh. Using a squeegee sweeping from one side of the substrate towards the other the ink (in Fig. 4 blue) is deposited on the substrate. Afterwards the ink is dried if needed. The thickness of the applied layer can be controlled by the size of the mesh (gaps, thickness or type of screen).

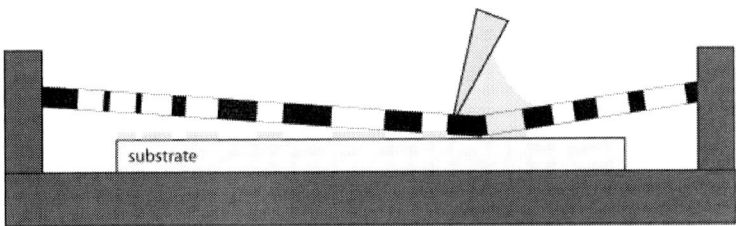

Figure 4: Principle of screen printing

Manufacturing

For the battery system the different battery components were applied layer by layer onto a flexible plastic substrate. The primary battery under discussion is a zinc-manganese dioxide system. As a first step the current collectors were printed onto a 50 µm thick PET (polyethylene-terephthalate) foil. After drying the first layer the positive and negative electrode material was printed onto the regarding current collectors. The electrodes consist of manganese dioxide (positive) and zinc-based (negative) inks which were dried after their application. Subsequently both, positive and negative electrodes were coated with an electrolyte (based on zinc chloride) using a doctor blade. In the finishing step the batteries were encapsulated with high-performance adhesive tape using a proprietary assembling technique. While the current collector consists of a shelf material the other inks were formulated from different materials. The design scheme is shown in Fig. 5.

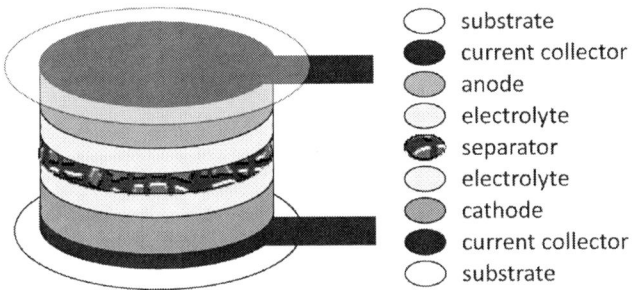

Figure 5: Design scheme of the printed battery

Results

The printed primary batteries are very thin and lightweight, Table 1. A battery with a nominal voltage of 1.5 V has a thickness approx. 0.7 mm and weights less than 1 g. Because the batteries are printed onto plastic foils they have the advantage of being flexible, Figure 6, and the potential to be manufactured in a roll-to-roll printing process. The materials used for their fabrication are easily available, inexpensive and without environmental risk given the fact that. They are free of mercury or other toxic materials.

TABLE I. characteristic values of the battery

parameter	value
Nominal voltage	1.5, 3.0, 4.5 and 6.0 V
capacity	>2 mAh/cm^2
thickness	~0.7 mm

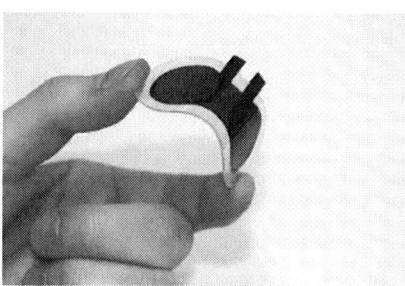

Figure 6: Printed thin film battery, nominal voltage of this battery 3.0 V

Acknowledgments

Many thanks for support to all departments of Fraunhofer ENAS and Center for Microtechnologies of Chemnitz University of Technology.

References

1. J.C. Eloy, MEMS Industry Group, Executive Congress 2010, Scottsdale, Arizona, (2010).
2. International technology Roadmap for Semiconductors, (2009)
3. EPoSS Strategic Research Agenda, (2010)
4. T. Gessner, M. Baum, W. Gessner and G. Lugert, *Chapter Smart integrated systems – from components to products.* in Guo Qi Zhang, Mart Graef, Alfred J. van Roosmalen, *More than Moore: Creating High Value Micro/Nanoelectronics Systems*, (2009)
5. M. Vogel, K. Hiller, A. Bertz, M. Wiemer and T. Gessner, *Microsystems Technology in Germany 2010*, (2010)18-19
6. D. Reuter, M. Nowack, M. Rennau, A. Bertz, M. Wiemer, F. Kriebel and T. Gessner, *Hermetic Thin Film Encapsulation of Mechanical Transducers for Smart Label Applications*, Transducers 09, Denver (USA), (2009)
7. A. Willert, A. Kreutzer, U. Geyer, R. R. Baumann: *"Lab-manufacturing of batteries for smart systems based on printing technologies"*, Smart Systems Integration 2009 - 3rd European Conference & Exhibition on Integration Issues of Miniaturized Systems - MEMS, MOEMS, ICS and Electronic Components, March 10-11, 2009, Brussels, - Heidelberg : AKA Verlag, 2009, pp. 556 – 559.
8. H. Kipphan: *Handbuch der Printmedien - Technologien und Produktions-verfahren*. Springer Verlag Berlin, 2000, p. 401 et seq, ISBN: 3-540-66941-8.
9. J. van Duppen: *Handbuch für den Siebdruck*. Verlag Der Siebdruck Lübeck, 1990, ISBN: 3-925402-20-9.

ECS Transactions, 34 (1) 1065-1070 (2011)
10.1149/1.3567716 ©The Electrochemical Society

Electrical characterization of the MOS (Metal-oxide-semiconductor) system: High mobility substrates

Dennis Lin, Guy Brammertz, Sonja Sioncke, Laura Nyns, Alireza Alian, Wei-E Wang, Marc Heyns, Matty Caymax and Thomas Hoffmann

IMEC vzw, Kapeldreef 75, Leuven, 3001 Belgium

Recent developments on CMOS-driven III-V and Ge MOS (Metal-oxide-semiconductor) technologies provide new opportunities in advancing the performance envelope of MOS device as well as the relevant electrical characterization techniques. Understanding the capacitance-voltage (CV) and conductance voltage (GV) responses of the III-V/Ge MOS devices can lead to better assessments of the oxide-semiconductor material systems and more accurate performance predictions.

Introduction

The high interface defect density nature of the oxide-semiconductor interface states (D_{it}) on III-V and Ge MOS systems has been a major roadblock to the realization of next generation high performance field effect transistors. The interface states, when charged, become fixed charges and greatly reduce the effectiveness of gate control by electrostatically screening the applied electric field. Information regarding the density and energy position of the interface states is of crucial importance in determining the performance potential of the MOS system. In this study, comprehensive $D_{it}(E)$ spectra of germanium, $In_{0.53}Ga_{0.47}As$ and InP capacitors are presented. In addition, we have also conducted a preliminary investigation of oxide traps near the semiconductor interface, i.e., border traps, which are likely to affect the V_{th} stability as well as the reliability of the gate stack

Examples of D_{it} (E) distributions and demonstrations of MOS transistors

In the first part of this study, MOS devices with high mobility substrates such as Ge, $In_{0.53}Ga_{0.47}As$ and InP are investigated. The sample substrates received wet treatment in diluted ammonia sulfide $((NH_4)_2S)$ solution with details described by Sioncke et al. [1] and 8 to 10 nm atomic layer deposited (ALD) Al_2O_3 as the gate dielectric. The Al_2O_3 dielectric layers were grown with trimethylaluminum (TMA) and water precursors at $300°C$. Figure 1 shows the $D_{it}(E)$ distributions of the Ge, $In_{0.53}Ga_{0.47}As$ and InP MOS capacitors. The oxide-semiconductor interface state densities (D_{it}) are evaluated across the semiconductor bandgap using the conductance technique [2] at different temperatures while the trap energy positions are determined according to the charge trapping characteristics [3]. The $D_{it}(E)$ distributions and energy windows suitable for MOSFET operation are identified in figure 1(a) for $In_{0.53}Ga_{0.47}As$ and Ge substrates with a common gate stack. The high D_{it} levels on the Ge conduction band side ($>1x10^{13}/eV-cm^2$) can be ascribed to acceptor-like traps while on the $In_{0.53}Ga_{0.47}As$ valence band side ($>2x10^{13}/eV-cm^2$) the traps can be modeled as donor-like.

1065

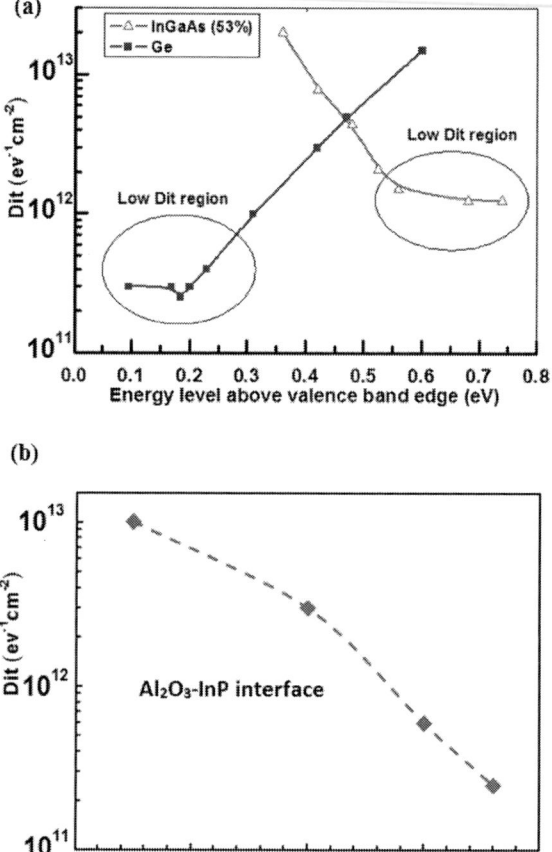

Figure 1 (a) $D_{it}(E)$ distributions of the $In_{0.53}Ga_{0.47}As$ and Ge MOS capacitors with atomic-layer deposited Al_2O_3 as gate dielectric. The marked low D_{it} regions enable the high mobility potentials of these materials. The measured $D_{it}(E)$ distribution at the Al_2O_3-InP interface is shown in (b). Both the Al_2O_3- $In_{0.53}Ga_{0.47}As$ and Al_2O_3-InP interfaces exhibit low D_{it} feature near the conduction band edge.

Across the bandgap, the Ge and $In_{0.53}Ga_{0.47}As$ MOS $D_{it}(E)$ levels in figure 1(a) decrease towards one of the band edges and the $D_{it}(E)$ distributions to a great extent seem to mirror each other with respect to midgap. Relatively low D_{it} levels are found near the Ge valence band ($3x10^{11}$/eV-cm^2) and the $In_{0.53}G_{0.47}As$ conduction band ($1x10^{12}$/eV-cm^2) edges. Low D_{it} levels at the Ge valence band edge suggest less donor-like traps and hence more free holes available to form the a p-channel. On the other hand, low D_{it} levels near the $In_{0.53}Ga_{0.47}As$ conduction band edge mean less acceptor-like traps and hence more free electrons to form the n-channel. Such features not only make transistor operations possible but in fact would unleash the high carrier mobility nature of the Ge p-MOSFET and $In_{0.53}Ga_{0.47}As$ n-MOSFET. The MOS interface assessments are verified by subsequent transistor demonstrations with high drive currents and carrier mobility [4]. Drain currents of 600mA/mm ($In_{0.53}Ga_{0.47}As$) and 200mA/mm (Ge) are obtained at 2.5v gate bias swing while maximum transconductances of 340mS/mm and 95mS/mm are achieved at drain bias of 2V and -2V for 1.5μm $In_{0.53}Ga_{0.47}As$ and Ge MOSFETs respectively. However, the ascending D_{it} levels toward mid-bandgap as well as the corresponding off-state leakage current are responsible for the large sub-threshold swings (>200mv/dec) and relatively low on-off ratio (10^2~10^3). Further efforts on interface passivation are needed to improve the off-state performance.

In addition to Ge and $In_{0.53}Ga_{0.47}As$, our measurements on InP based MOS capacitors with high-k dielectrics concluded a $D_{it}(E)$ distribution similar to that of the $In_{0.53}Ga_{0.47}As$ MOS samples, as shown in figure 1 (b). High interface state densities (>$1x10^{13}$/eV-cm^2) are found below the InP mid-gap energy level. They essentially prevent the InP surface Fermi level from moving closer to the valence band and precludes the accurate extraction of D_{it} figures below mid-gap. On the other hand, low D_{it} levels ($4x10^{11}$/eV-cm^2) are found near the conduction band edge. Such $D_{it}(E)$ distribution favors n-channel operation illustrated by the $In_{0.53}Ga_{0.47}As$ n-MOSFET example. This also supports the concept of inversion mode n-MOSFET fabricated on InP substrate recently demonstrated by Wu et al [5].

Electrical characteristics of border traps

With the knowledge of its $D_{it}(E)$ profile, a MOS transistor with effective gate control can be engineered to meet the basic operation requirements. At the next level, the stability of the threshold voltage (V_{th}) and the reliability of the gate dielectrics become important as well. In the second part of the study, we would like to expand the envelope beyond the oxide-semiconductor interface into the first few mono-layers of the dielectrics. Figure 2 illustrates the multi-frequency CV traces and GV maps of the $In_{0.53}Ga_{0.47}As$ and InP MOS capacitors with emphasis on the accumulation region. With identical surface treatments, gate dielectric and anneals, the $In_{0.53}Ga_{0.47}As$ and InP capacitors differ in two aspects. The first difference is the lack of inversion or weak inversion response on the InP MOS sample, as illustrated by both the CV and GV illustrations of figure 2. This can easily be explained by the higher InP bandgap, relative to $In_{0.53}Ga_{0.47}As$. Secondly, compared to the InP capacitor (figure 2(c)), less frequency dispersion is observed on the $In_{0.53}Ga_{0.47}As$ sample (figure 2(a)) at accumulation. The frequency dispersion of the CV traces at accumulation indicates conductive losses most likely caused by border traps situated just above the oxide-semiconductor interface [6]. The origin, the exact spatial locations and energy levels of these traps cannot be immediately concluded based on limited electrical and physical characterization results. The fact that less frequency dispersion is observed on the $In_{0.53}Ga_{0.47}As$ MOS sample suggests the border traps can be passivated to a greater extend [7] on InGaAs based MOS. In addition, near-flat features at accumulation on the conductance maps of figure 2(b) and 2 (d) illustrate uniform charge trapping at border traps occur within the measurement frequency range. The lack of strong frequency dependence suggests the distribution of border traps could be somewhat uniform in energy, as proposed in figure 3.

And finally, the amount of border traps can be estimated by the spread of CV frequency dispersion. In the case of InP sample in figure 2(c), roughly 10^{12} traps per cm^2 is estimated within the measurement window between 100Hz and 1MHz. Border traps with response time constants outside the regular CV and GV measurement frequency range can be detected through quasi-static CV (slow traps) or pulse IV techniques (fast traps).

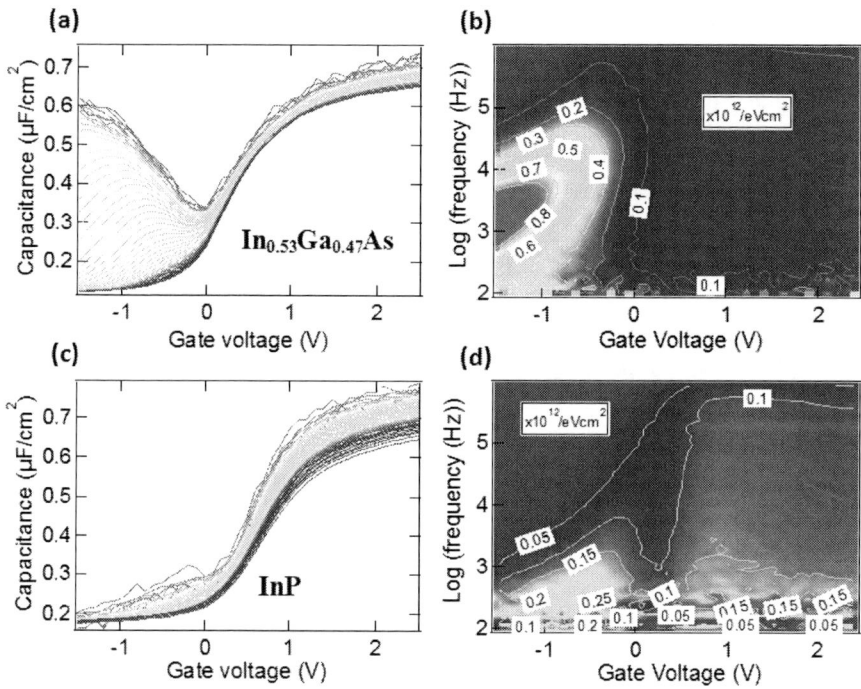

Figure 2 CV and GV plots of $In_{0.53}Ga_{0.47}As$ and InP MOS capacitors measured at room temperature. The conductance maps shown on (b) and (d) are normalized conductance, $G/(\omega Aq)$, where ω is 2π times frequency, A is the device area and q is the elemental charge constant. Larger frequency dispersion (c) as well as more pronounced conductive losses (d) observed at accumulation on the InP MOS sample indicate the presence of border traps.

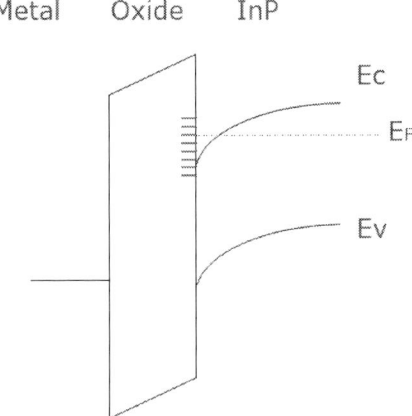

Figure 3 Energy band diagram of an InP MOS device biased into accumulation. The surface Fermi level can advance well beyond the InP conduction band minimum due to low conduction band density of states [8]. The frequency dispersion and conductive loss recorded at accumulation in figure 2 are most likely from the interaction between the conduction band electrons and border traps, as illustrated in the figure. The origin, the exact spatial locations and energy levels of these traps cannot be concluded immediately.

It should be mentioned that the substrate series resistance can also contribute to conductive losses at accumulation and careful distinctions must be made to separate the effects of border traps and series resistance. There are major differences between the border traps and the substrate series resistance in terms of conductive losses at accumulation. The conductive loss with series resistance origin depends strongly on the MOS device capacitance and increases with the measurement frequencies, while the border trap related counterpart does not depend on the device capacitance and tend to be uniform across the frequency range commonly used (100Hz ~1MHz). The oxide border traps can have potential impact on the stability of the threshold voltage V_{th} [9], one of the most critical parameters in terms of transistor operation. Border traps located further in the oxide can cause other reliability issues such as the dielectric breakdown, in addition to the threshold voltage shift, as the oxide thickness is reduced in future device scaling.

Conclusion

In summary we have investigated the electrical characteristics of the oxide-semiconductor interface with high mobility substrates. The implications of these electrical features can be pivotal for the development of the next generation high-performance MOS devices. On one hand, the low interface trap densities near band edges enable the expected large drive current and high transconductance. On the other hand, interaction between conduction band electrons and border traps above the oxide-semiconductor interface will likely introduce threshold voltage instability as well as issues regarding dielectric reliability as the scaling of gate stack continues.

References

1. S. Sioncke et al, UCPSS 2010 proceedings, to be published.
2. E. H. Nicolian and J. R. Brews, MOS Phys. and Tech., *Wiley*, p. 215-216 (2003).
3. G. Brammertz, K. Martens, S. Sioncke, A. Delabie, M. Caymax, M. Meuris and M. Heyns, *Appl. Phys. Lett.* **91**, 133510 (2007)
4. D. Lin, G. Brammertz, S. Sioncke, C. Fleischmann, A. Delabie, K. Martens, H. Bender, T. Conard, W. H. Tseng, J. C. Lin, W. E. Wang, K. Temst, A, Vantomme, J. Mitard, M. Caymax, M. Meuris, M. Heyns and T. Hoffmann, *IEDM Tech. Dig.*, pp. 327-330, (2009).
5. Y.Q. Wu, Y. Xuan, T. Shen, P.D. Ye, Z. Cheng and A. Lochtefeld, *Appl. Phys. Lett.* **91**, 022108 (2007).
6. D. M. Fleetwood, *IEEE Trans. Nucl. Sci.*, **39** (2), 269 (1992).
7. E. J. Kim, L. Wang, P. M. Asbeck, K. C. Saraswat, and P. C. McIntyre, *Appl. Phys. Lett.* **96**, 012906 (2010).
8. E. Lind, Y. M. Niquet, H. Mera, and L. E., *Appl. Phys. Lett.* **96**, 233507 (2010)
9. A. Kerber, E. Cartier, L. Pantisano, R. Degraeve, T. Kauerauf, Y. Kim, A. Hou, G. Groeseneken, H. E. Maes, and U. Schwalke, *IEEE Electron Device Lett.*, **24** (2), 87, (2003)

Characterization and Optical Properties of CdS thin films grown by Chemical Bath Deposition

Weibo Zhang, Shuying Cheng*

Institute of Micro/Nano Devices and Solar Cells, School of Physics & Information Engineering, Fuzhou University, Fuzhou 350108, PR. China
*E-mail: sycheng@fzu.edu.cn

CdS thin films were prepared on soda-lime glass (SLG) by the chemical bath deposition technique. An aqueous solution of cadmium salts, ammonium chloride, ammonium hydroxide and thiocarbamide were used in the deposition. The obtained CdS film samples were characterized by techniques such as X-ray diffraction (XRD), scanning electron microscopy (SEM) and UV−VIS−NIR spectrophotometer. The films are of hexagonal phase with preferred (0 0 2) orientation and the grain size increases with the thickness of the films. The films show good transmittance in the visible region. The band gap of the films is 2.41–2.45 eV and decreases with the increasing the thickness of the films.

1. Introduction

CdS thin films have been widely used as n-type buffer layers in thin film chalcogenide solar cells [1]. It is a II–VI compound semiconductor and has an energy band gap of 2.42 eV at room temperature [2]. Currently, CdS is one of the important materials for application in electro-optic devices such as photoconductive cells, photosensors, transducers, laser materials, optical wave-guides and non-linear integrated optical devices [3]. Nowadays, CdS films can be deposited by low cost techniques such as vacuum evaporation, sputtering, spray paralysis and chemical bath deposition (CBD) techniques [4-5]. Among these techniques, the CBD is successfully used to deposit CdS thin films, and high-efficiency CdTe and CIGS solar cells were developed with CdS buffer layers prepared by CBD [6]. The CBD method is based on the controlled precipitation of the material. Ammonia prevents the undesirable homogenous precipitation by forming complexes with Cd ions, but on the other hand, the effect is to slow down the surface reaction, which implies that an optimal concentration of ammonia can be determined [7]. Due to the simplicity and very economically experimental facilities needed in the film deposition, the CBD is considered as the best method to obtain low-cost CdS thin films which have optimal features for photovoltaic device applications [6]. In this paper, the CBD was also used to prepare CdS films in order to obtain good quality CdS films.

2. Experimental

CdS layers were deposited on the soda-lime glass and conducting glass substrates (ITO/glass) (2 cm×3 cm) by the CBD technique. Before the CBD process, the SLG substrate was ultrasonically washed and cleaned in acetone, alcohol and de-ionized water, and then dried at 90°C in an oven.

Typically, a 100ml reaction solution was prepared in a closed vessel at room temperature. There are aqueous solution of $CdSO_4$ (0.0015M), ammonium hydroxide (28–30%, 2.0 ml), thiocarbamide (0.004M), ammonium sulfate buffer (0.02M $(NH4)_2SO_4$) with pH of 8~10 in the bath. The thiourea was the source of S^{2-} and the cadmium salts was the source of Cd^{2+} in the deposition. The bath was covered and continuously stirred during the deposition to ensure homogeneous distribution of the chemicals. The cleaned SLG substrates were kept immersing in the stirring solution for different deposition times, and the temperature of the substrates was

slowly heated from room temperature to 75°C, 80°C and 85°C respectively, and then, at

predetermined times, the deposited substrates were removed from the bath and sonically washed in de-ionized water, and then they were dried under a stream of nitrogen. The overall reaction is given below [8]:

$$Cd(NH_3)_4^{2+} + (NH_2)CS + 2OH^- \rightleftharpoons CdS + CH_2N_2 + 4NH_3 + 2H_2O \qquad [1]$$

Characteristics were carried out on selected samples, the thicknesses of the CdS thin films were determined by a Dektak 150 profilometer. X-ray diffraction (XRD) patterns were recorded by using a Siemens D500 ($CuK_a \sim 1.5406$ Å). Film structural and morphological aspects were studied by field-emission scanning electron microscopy (FESEM) and atomic force microscopy (AFM). UV−VIS−NIR spectrophotometer measurements were performed to verify optical transmittance of the films.

3. Results and discussions

3.1 The curve of thin films thickness

Fig1(a). shows the curve of film thickness vs deposition time when the aqueous solution is at 85°C.

The curve can be divided into two parts, OA part is approximately linear growth, which is line region; AB part represents growing slowly. When CdS is deposited on the substrate, a large amount of Cd^{2+} and S^{2-} can be consumed, so the precursor ion achieves equilibrium state in a very short period of time. Therefore, the CdS thin film grows slowly. If there are large particles adsorbed on the film, a sharp increase in the thickness will be led, but the film is not strong adhesive to the substrate, and it can be easily removed by using ultrasonic vibration. Fig.1 (a, b, c) give the curves of the film thickness vs deposition time at different solution temperatures. At the same deposition time, CdS film thickness increases with the increase of solution temperature. From the naked eye we can see that the deposited CdS thin films were green, yellow-green and

yellow respectively, when the aqueous solution was at 75°C, 80°C, 85°C respectively. Therefore,

when we prepare CdS thin films, we should make the formation of thin films in the linear region; this will optimize the growth of the CdS thin films. Table 1 lists names of the samples synthesized at different conditions.

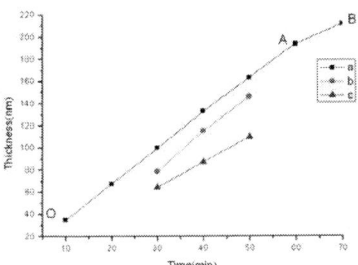

Fig.1.The dependence of CdS films thickness on deposition times. Curves a, b and c correspond to

the solution temperature 85℃, 80℃ and 75℃.

TABLE1. Samples synthesized under different conditions.

Sample	Temperature(℃)	Deposition time(min)	Substrate
A1	85	20	SLG
A2	85	30	SLG
A3	85	40	SLG
A4	85	60	ITO
A5	85	60	SLG
A6*	85	60	SLG

*   stands for annealing after the deposition.

3.2 X-ray diffraction

The crystallographic properties of the films have been investigated by X-ray diffraction (XRD) technique using Cu K_α radiation. As is well known, CdS can exist in two crystalline modifications: the hexagonal (wurtzite) phase and the cubic (zincblende) phase. For solar cell application, hexagonal CdS films are preferable due to its excellent stability [9]. Fig. 2(a) shows the XRD patterns of CdS films deposited on different substrates. The CdS film on the SLG contains several diffraction peaks, indicating that the film is polycrystalline in nature. Cadmium sulfide can exist in a mixed cubic and hexagonal crystalline phase [10]. XRD results show intense and sharp diffraction peaks at 25.0°, 26.6°, 28.4°, 43.9°and 52.1°which are associated with the (100),(002),(101),(110) and (112) reflections of the hexagonal structured of JCPDS-ICDD No. 030653414, respectively. The peaks (002), (110) and (112) of the hexagonal structure are similar to the peaks (111), (220) and (311) of CdS with the cubic structure of JCPDS-ICDD No. 010890440. Thus, the structure of the films is approximate to the predominantly hexagonal, similarly to other reports [11]. The phase of the CdS film on ITO/glass is similar to that on SLG, however there is another peak at 2θ=50.54°associated with the hexagonal (200) and a new peak at 2θ=35.05° corresponding to ITO phase appears in the film.

Fig.2 (b) shows the XRD patterns for CdS thin films before and after thermal annealing (corresponding samples A5 and A6). After thermal annealing, there are no significant changes in the peak position, but the peaks width has changed obviously. The size D_{hkl} of the crystallites is determined from XRD data by the Scherrer formula below.

The Scherrer formula:

$$D_{hkl} = \frac{K\lambda}{\beta\cos\theta}$$ [2]

Where K is a constant, β is FWHM (Full Width at Half Maximum) in radians, λ is the wavelength of X-ray, and θ is the Bragg angle [12]. The K, λ values are taken as 0.9, 1.5406 Å for the calculations, respectively.

The FWHM of H (002) peak at $2\theta = 26.56°$ is 0.2676° for sample A5, however for sample A6, the intensity of H (002) peak is strengthened and the FWHM is decreased to 0.1171°. The average grain size for the as-grown sample is about 30.57 nm, but it reaches to 69.71 nm after thermal annealing at 400°C. The crystallite grains become larger after thermal annealing.

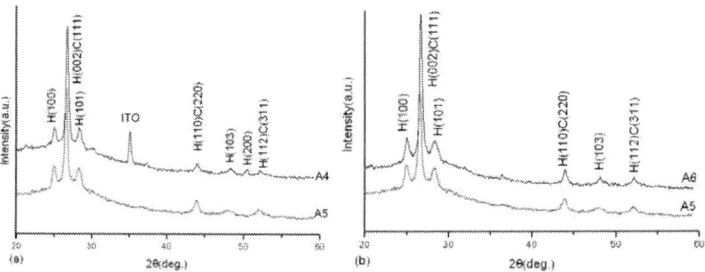

Fig.2 (a).XRD patterns of CdS films deposited at different substrates (ITO and SLG).
(b).XRD patterns of CdS films deposited on SLG (annealed and unannealed).

3.3 Surface morphology

The SEM micro-graphs of samples A2 and A3 are shown in Fig. 3. The average grain sizes of them are about 125nm and 190nm, respectively. We can observe an increasing in the grain size with the increasing of the deposition time from 30 min to 40 min.

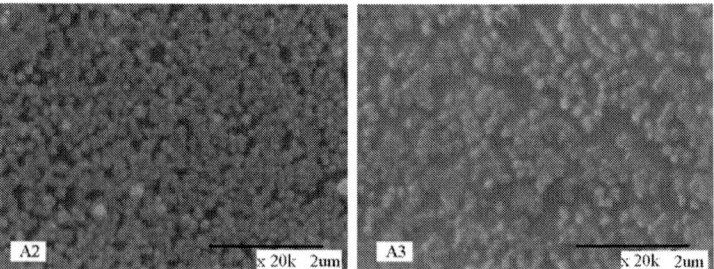

Fig.3. SEM micrographs of samples A2 and A3.

3.4 Optical properties

Fig.4 (a) shows the optical transmittance spectra of samples A1-A3 in the wavelength range from 300 to 1000 nm. It can be seen that the average transmission of all the samples is about 70%-90% in the visible range. The optical band gaps of CdS films are calculated from the optical absorbance measurements in the region of 490–520 nm. At each wavelength, the absorption coefficient α of the films can be calculated by using Eq. [3].

$$\alpha \propto \ln(1/T)/t \qquad [3]$$

Where T is the transmittance and t is the thickness of the film. The relation between the absorption coefficient α and the incident photon energy hv is given by:

$$(\alpha hv)^2 = A(hv - Eg) \qquad [4]$$

Where A is a constant and Eg is the band-gap energy [1]. The optical energy gap Eg could be obtained by extrapolating the straight portion of the curve to zero absorption coefficient, as shown in Fig. 4(a). The band gaps of samples A1, A2 and A3 are 2.41, 2.43 and 2.45eV, respectively. It indicates that the band gaps of the samples decrease with the increasing of deposition time. Since the film thickness increases with the increasing of deposition time, as shown in Fig.1. Therefore, the optical energy gaps of the films decrease with increasing the film thickness. This decrease in the band gap can be due to the influence of various factors, such as grain size, structural parameters, carrier concentration, presence of impurities, deviation from stoichiometry of the film and lattice strain. A detailed analysis is needed to explore the effect of each of these parameters on the value of band-gap energy. However, it has been observed that the lattice parameters, grain size and the strain have a direct dependence on the film thickness. Hence, it is considered that the observed decrease in Eg with increasing thickness is due to the decrease in lattice strain. [13]

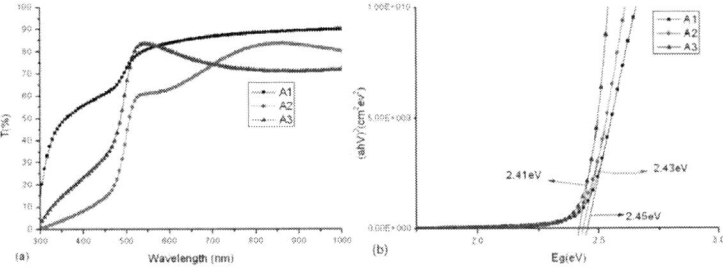

Fig.4. Optical transmittance spectra (a) and $(\alpha hv)^2$ vs. hv (b) of CdS thin films deposited at different deposition time.

4. Conclusion

In this paper, CdS films, which were grown on SLG substrates by CBD at 85°C with stirring, showed good optical properties and had strong adhesion to the substrates. The growth rate of the films was studied in order to obtain optimal thickness as window layers in the solar cells. XRD results of these films indicated that they had mixture phases of both cubic (zincblende) and hexagonal (wurtzite) structures. After thermal annealing, there are no significant changes in the peak position, but the intensity of H (002) peak increases and the FWHM of H (002) peak value decreases. The crystallite grains become larger after thermal annealing. The optical energy gap is about 2.41–2.45 eV and decreases with increasing the film thickness. The optimized CdS films with well adhesion and high transmittance could be used as buffer layers of thin film solar cells.

Acknowledgments

Financial support by the National Nature Sciences Funding of China (61076063) is gratefully acknowledged. The authors also wish to express their gratitude to funding from Fujian Provincial Department of Science & Technology, China (2009J01285).

References

1. J. Hiie, T. Dedova, *Thin Solid Films* 511-512 (2006) 443 – 447.
2. H.Metin, R.Esen, *J. Cryst. Growth* 258 (2003) 141–148.
3. K.Sentil, D.Mangalaraj, Sa.K.Narayandass, *Appl. Surf. Sci.* 169–170 (2001) 476.
4. K.V.Zinoviev, O.Zeleya-Angel, *Mater. Chem. Phys.* 70 (2001) 100.
5. G. Sasikala, P. Thilakan, C. Subramanian, *Sol. Energy Mater. Sol. Cells* 62 (2000) 275–293.
6. J.N.Ximello-Quiebras, G.Contreras-Puente, *Sol. Energy Mater. Sol. Cells* 90 (2006) 727–732.
7. M. Kostoglou, N.Andritsos, A.J. Karebelas, *Ind. Eng. Chem. Res.* 39 (2000) 3272-3283.
8. O. Vigil, A.Arias-Carbajal, F.cruz, *Mater. Res. Bull.* 36 (2001) 521.
9. Li Wenyi, Cai Xun, Chen Qiulong, Zhou Zhibin, *Mater. Lett.* 59 (2005) 1–5.
10. Jae-Hyeong Lee, *Thin Solid Films* 515 (2007) 6089–6093.
11. J.Han,etal,*Sol. Energy Mater. Sol. Cells* (2010), oi:10.1016/j.solmat.2010.10.027.
12. P.K. Nair, O.Gomez Daza, *M.T.S. Nair, Semicond. Sci. Technol.* 16 (2001) 651.
13. Joel Pantoja Enriquez, Xavier Mathew, *Sol. Energy Mater. Sol. Cells* 76 (2003) 313–322

ECS Transactions, 34 (1) 1077-1085 (2011)
10.1149/1.3567718 ©The Electrochemical Society

Electroluminescence of End-Capped
Poly[9,9-di-(2'-ethylhexyl)fluorenyl-2,7-diyl] Blended with F8BT

Qiushu Zhang[a], and Sihui Zhang[b]

[a] School of Mechatronics Engineering, University of Electronic Science and Technology of China, Chengdu, Sichuan 611731, P. R. China
[b] School of Mechanical Engineering & Automation, Xihua University, Chengdu, Sichuan 610039, P. R. China

We report on the fabrication and characterization of light-emitting diodes (LEDs) based on blended end-capped poly[9,9-di-(2'-ethylhexyl)fluorenyl-2,7-diyl] (PF2/6). The light emitting polymer thin film layer was dimethylphenyl (DMP)-end-capped PF2/6 or N,N-bis(4-methylphenyl)-N-phenylamine (TPA)-end-capped PF2/6 doped with the green-emitting alternating copolymer poly(9,9'-dioctylfluorene-alt-benzothiadiazole) (F8BT). Blending end-capped PF2/6 with F8BT enhanced device properties compared to either of the pure components. For the devices based on DMP-end-capped PF2/6 doped with 10% F8BT, the electroluminescence (EL) reached a maximum luminance of 1074 cd/m^2, a maximum external quantum efficiency of 0.16%, and a maximum luminance efficiency of 0.514 cd/A. The improvement is assigned to efficient energy transfer from host end-capped PF2/6 to guest F8BT. However, the energy transfer was incomplete since EL emission did not exclusively come from F8BT, which we think originates from the presence of the regions where the concentration of F8BT is relatively depleted.

Introduction

The blend of poly(9,9'-dioctylfluorene) (PFO) with poly(9,9'-dioctylfluorene-alt-benzothiadiazole) (F8BT) is a well-known and popular polyfluorene (PF) blend system, which has been used as active materials for EL devices in many investigations related to polymer LED that were aimed at a variety of research objectives (1)(2)(3)(4)(5)(6)(7). Blue emitter PFO exhibits a relatively high hole mobility (8) and a high solid state PL efficiency (~60%). F8BT couples a high electron mobility with a high electron affinity of around 3.5 eV below vacuum (4) that brings about easy electron injection. As a result, doping PFO with F8BT gives rise to EL devices with high efficiency, high brightness, and low turn-on voltage (6). Optimal LED performance has been observed for the devices based on the blends containing 95% PFO and 5% F8BT (9)(10). Due to a good spectral overlap between the host PFO blue emission and the guest F8BT absorption (1)(7), efficient energy transfer takes place from PFO to F8BT, which produces an emission spectrum that is essentially that of neat F8BT. However, upon the treatment of the blend film with acetone, a very poor solvent for the PF components, or upon the incorporation of a hole-blocking layer (HBL), 2-(4-biphenylyl)-5-butylphenyl-1,3,4-oxadiazole (PBD), between the cathode and the emissive blend film, the blue emission from the PFO host arose in the PL and EL spectra (7)(5).

As shown in our previous results (11), end-capped PF2/6s are efficient blue emitting polymers. It is likely that in these materials, the hole mobility is much greater than the electron mobility as in many other PF materials (8). Enlightened by polymer LEDs based on PFO blends with F8BT, we conducted research on EL properties of the devices prepared from end-capped PF2/6s blended with F8BT. DMP-end-capped PF2/6 or TPA-end-capped PF2/6 doped with F8BT was used as the light-emitting layer. Blend based devices displayed better performance in terms of luminance and efficiency than those made from either of the pure components. Efficient energy transfer from host end-capped PF2/6 to guest F8BT should be responsible for the improvement. However, the energy transfer was incomplete since EL emission did not exclusively come from F8BT. Although the EL spectra were dominated by the emission of F8BT, the blue emission from the end-capped PF2/6 contributed to the spectra. We tentatively explain the reason for the existence of the blue component in the EL spectra.

Experimental

Molecular structures of end-capped PF2/6s and F8BT we used here are shown in Figure 1. All polymers were synthesized by American Dye Source, Inc., Canada. Both PF2/6 end-capped with DMP and TPA are light yellow powder, while F8BT is yellow powder. They are highly soluble in toluene and tetrahydrofuran. For end-capped PF2/6s, the molecular weight ranges from 40,000 to120,000. The molecular weight of F8BT is around 44,000.

The devices were fabricated at ambient conditions on glass substrates covered by patterned Indium-Tin-Oxide (ITO) electrodes. The ITO substrates were precleaned by two successive ultrasonic baths in acetone and isopropyl alcohol. After drying them with a nitrogen gun, a 45-nm thick film of poly(3,4-oxyethyleneoxythiophene) doped with poly(styrene sulfonate) (PEDOT:PSS) was spin coated over the substrate from a 1.3 wt.% water dispersion. The emissive layers consisted of end-capped PF2/6, F8BT, or end-capped PF2/6 blend with F8BT. They were formed on top of PEDOT:PSS films by spin casting from polymer toluene solutions (10 mg/mL) at a speed of 2000 rpm. The film thicknesses of neat DMP-end-capped PF2/6 and its blend with 10% F8BT are 70 nm. Those for pristine TPA-end-capped PF2/6 and its blend with 5% F8BT are 80 nm. The thickness of pure F8BT film is 55 nm. Devices were dried in a vacuum chamber at room temperature for a minimum of 24 hrs before the deposition of the Al film. The manufacturing was completed with the thermal evaporation of the aluminum cathode (~200 nm) at 2×10^{-6} Torr through a shadow mask. The overlap between the two electrodes gave active device areas of 9 mm^2.

The optical absorption (UV-Vis) spectra were measured with an Agilent-8453 UV-visible spectrophotometer from polymer films spin cast onto a quartz plate. All device testing was implemented in air at room temperature. Current-voltage characteristics were measured on a Keithley 236 source-measure unit. The power of EL emission was measured through using an ILX Lightwave OMM-6810B optical multimeter equipped with a silicon power/wavehead (OMH-6722B). By assuming Lambertian distribution of the EL emission, luminance (cd/m^2) was calculated via utilizing the forward output light power and the EL spectra of the devices.

(a)

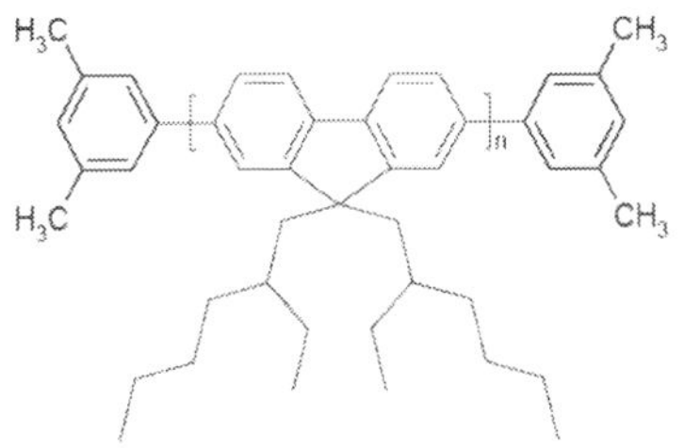

(b)

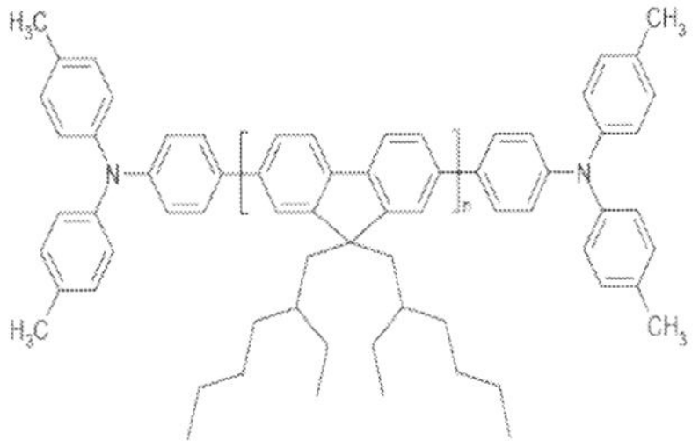

(c)

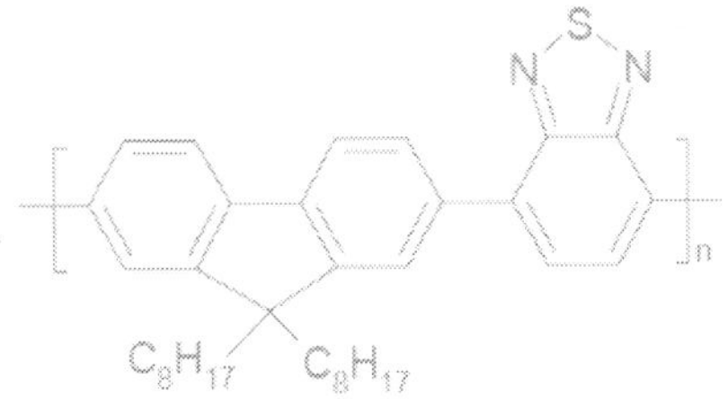

Figure 1. Chemical structures of the polymers: (a) PF2/6 end capped with DMP, (b) PF2/6 end capped with TPA, (c) F8BT.

Results and Discussion

The optical absorption (UV-Vis) spectra of thin films of end-capped PF2/6s and F8BT are shown in Figure 2. The UV-Vis spectra of DMP- and TPA-terminated PF2/6 are very similar to each other. The absorption of DMP-terminated PF2/6 had an onset at 425 nm, the same as that of TPA-terminated PF2/6. And both spectra exhibited two peaks at nearly the same wavelengthes.The optical absorption of F8BT had three peaks at 209, 322, and 464 nm with two shoulders at 254, and 340 nm. Its onset was at 520 nm. From the UV-Vis spectra, the band-gap energies were determined to be 2.92 eV for end-capped PF2/6s and 2.38 eV for F8BT, respectively.

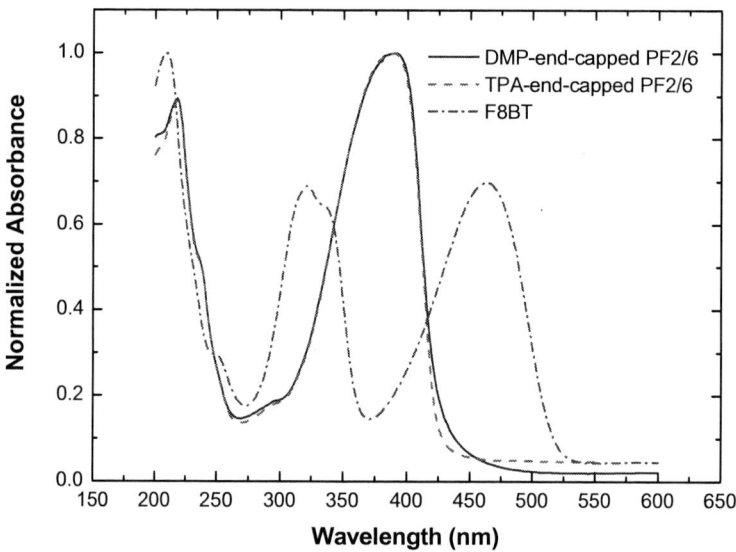

Figure 2. UV-Vis spectra of DMP-end-capped PF2/6, TPA-end-capped PF2/6 and F8BT.

Figure 3 displays current density as a function of the electric field for devices with DMP-end-capped PF2/6, TPA-end-capped PF2/6, F8BT and their blends as active layers. The F8BT device operates under higher electric fields than any other devices and it has the highest turn-on voltage of ~13.5 V. The devices based on DMP-end-capped PF2/6 and its blend with 10% F8BT have approximately the same turn-on voltages at ~5 V. Although, under low electric field, the current passing through the DMP-end-capped PF2/6 device is higher than that through the 10% F8BT device, the latter increase faster with electric field. The operating electric fields for the device based on TPA-end-capped PF2/6 blend with 5% F8BT are lower than those for the devices based on either of the pristine components. The turn-on voltages for TPA-end-capped PF2/6 and its blend devices are ~5 V.

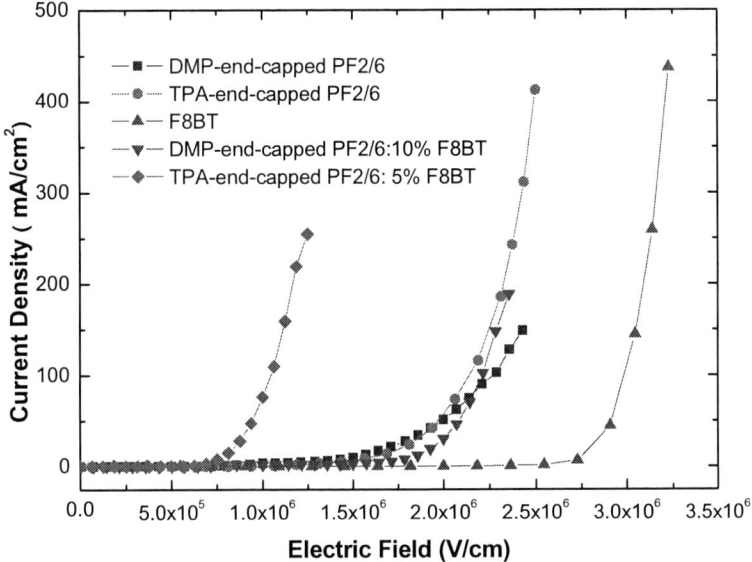

Figure 3. Current density-electric field characteristics of polymer LEDs based on end-capped PF2/6s, F8BT and their blends.

The variation of the light power with current for the LEDs under study is presented in Figure 4. By using the data shown in Figure 4, we calculated performance including threshold current density for light emission, luminance, external quantum efficiency, and luminance efficiency, which are summarized in Table I. Compared to the devices based on either of the components, blend based devices exhibited better properties in light of brightness and efficiencies. The best performance was observed from DMP-end-capped PF2/6 blend containing 10% F8BT with a maximum luminance of 1074 cd/m^2 at 213.6 mA/cm^2, a maximum external quantum efficiency of 0.16% and a maximum luminance efficiency of 0.514 cd/A at 205.2 mA/cm^2.

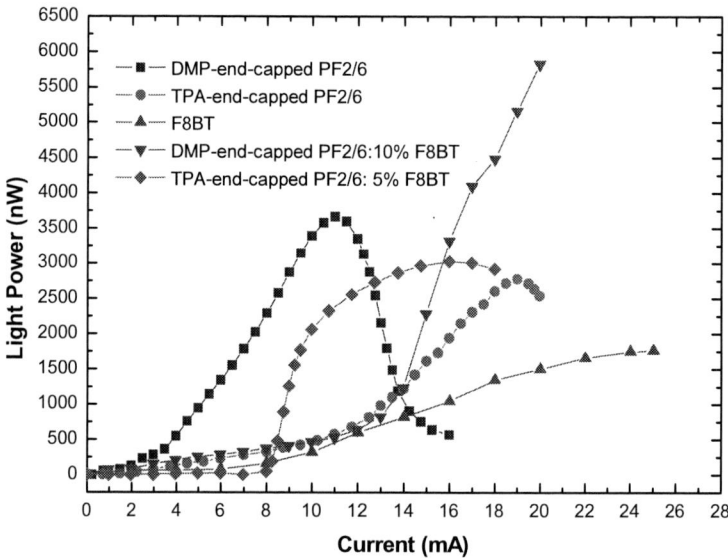

Figure 4. Light power-current characteristics for the LEDs made from end-capped PF2/6s, F8BT and their blends.

TABLE I. EL performance for the LEDs based on end-capped PF2/6s, F8BT and their blends.

Emission layers	DMP-end-capped PF2/6	TPA-end-capped PF2/6	F8BT	DMP-end-capped PF2/6:10% F8BT	TPA-end-capped PF2/6:5% F8BT
d_{EL} (nm)	70	80	55	70	80
$J_{threshold}$ (mA/cm^2)	2.4	1.2	16.6	8	30.2
V_{on} (V)	5	5	13.5	5	5
B_{max} (cd/m^2)	381	327	360	1074	474
η_{max} (%)	0.16	0.069	0.037	0.16	0.103
LE_{max} (cd/A)	0.319	0.156	0.138	0.514	0.308
λ_{peak} (nm)	420, 445, 485 (maximum)	422, 443, 512 (maximum)	541	420, 442, 528 (maximum)	420, 441, 520 (maximum)

d_{EL} : Thickness of emissive layer.

$J_{threshold}$: Threshold current density for light emission. It is defined as the minimum current density at which the resulting light power can be detected.

V_{on} : Turn-on voltage for the current at which the current reaches ~8.5×10^{-5} A.

B_{max} : Maximum luminance.

η_{max} : Maximum external quantum efficiency.

LE_{max} : Maximum luminance efficiency.

λ_{peak} : EL peak wavelength.

Figure 5 shows the EL spectra for the devices made from end-capped PF2/6s, F8BT and their blends. Although the emission of F8BT dominates the EL spectrum of DMP-end-capped PF2/6 blend, that of DMP-end-capped PF2/6 contributes to it as well. Two emission peaks representing the blue component obviously correspond to those at 420 nm and 445 nm in the EL spectrum of DMP-end-capped PF2/6. The maximum emission peak is found to be at 528 nm, which are blue-shifted by 13 nm relative to that of pure F8BT. A similar phenomenon was observed from other blend systems (12). This could be due to the reduced interchain interactions between the F8BT chains arising from the dilution effect of the host DMP-end-capped PF2/6. The EL spectrum of TPA-end-capped PF2/6 blend with F8BT is similar to that of DMP-end-capped PF2/6:F8BT blend. Although the F8BT emission is predominant, that from TPA-end-capped PF2/6 makes contributions to the EL spectrum of the blend. Two blue emission peaks evidently originate from the peaks at 422 nm and 443 nm in the EL spectrum of TPA-end-capped PF2/6. Like DMP-end-capped PF2/6:F8BT blend, the maximum emission peak is blue-shifted compared to that of pure F8BT.

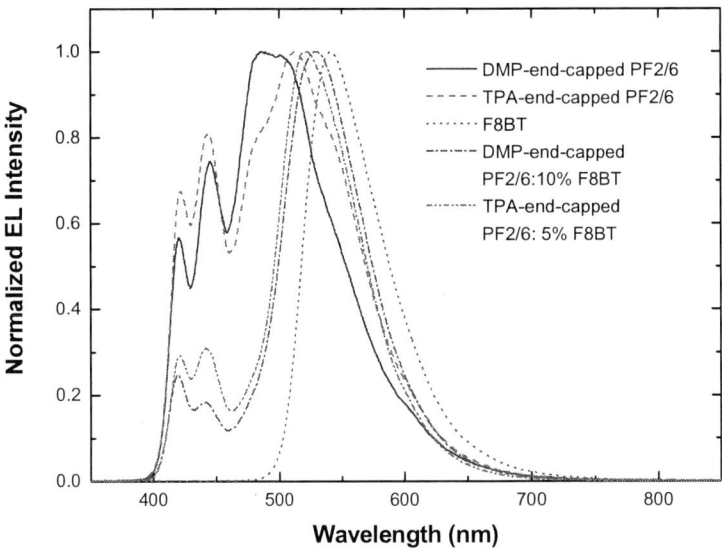

Figure 5. EL spectra of the LEDs made from end-capped PF2/6s, F8BT and their blends.

We believe that efficient energy transfer of excitations is the primary reason for the improvement in device luminance and efficiency which was observed from end-capped PF2/6 blends with F8BT. End-capped PF2/6s are efficient blue emitters with a wide energy gap of ~2.92 eV. They can therefore act as an adequate host material to achieve efficient energy transfer to a guest. F8BT possesses a band-gap energy of ~2.38 eV. The relatively large band-gap difference between end-capped PF2/6 and F8BT favors energy transfer (4). On the other hand, it is probable that end-capped PF2/6s have a much larger hole mobility than electron mobility. Doping end-capped PF2/6s with electron transporting F8BT could result in a better balance between the transports of both types of carriers.

Interestingly, the energy transfer from end-capped PF2/6s to F8BT is incomplete because the residual blue component appeared in the blend EL spectra. Actually, blue emission was also observed from PFO:F8BT blends when HBL, PBD, was incorporated between the emissive layer and the cathode, or when the blend film was treated with acetone (5)(7). In the first case, the blue emission was explained by confining the recombination to a region close to the emissive polymer/PBD interface, where the concentration of F8BT was relatively depleted (9). In the second case, it was suggested that acetone drove the blend system towards further de-mixing. We propose that in the present blend systems the residual blue emission resulted from the presence of depleted regions where the concentration of F8BT is very low and not sufficient for complete quenching of end-capped PF2/6 luminescence. In these regions, relatively complete phase separation of the two components occurs on a larger scale than the exciton diffusion length and the Forster transfer radius.

Conclusion

We find that an enhancement in brightness and efficiencies can be realized via doping end-capped PF2/6 with F8BT. With an active layer of DMP-end-capped PF2/6 blended with 10% F8BT, the EL reached a maximum luminance of 1074 cd/m^2, a maximum external quantum efficiency of 0.16%, and a maximum luminance efficiency of 0.514 cd/A. The improvement is assigned to efficient energy transfer from host end-capped PF2/6 to guest F8BT. However, the energy transfer is incomplete because the blue component is observed in blend EL spectra. We think that the presence of the regions where the concentration of F8BT is relatively depleted accounts for the residual blue emission.

References

1. J. Morgado, R. H. Friend and F. Cacialli, *Appl. Phys. Lett.*, **80**, 2436 (2002).
2. R. B. Fletcher, D. G. Lidzey, D. D. C. Bradley, M. Bernius and S. Walker, *Appl. Phys. Lett.*, **77**, 1262 (2000).
3. J. Morgado, N. Barbagallo, A. Charas, M. Matos, L. Alcacer and F. Cacialli, *J. Phys. D: Appl. Phys.*, **36**, 434 (2003).
4. E. Moons, *J. Phys. Condens. Matter*, **14**, 12235 (2002).

5. J. Morgado, E. Moons, R.H. Friend and F. Cacialli, *Synth. Met.*, **124**, 63 (2001).
6. A. J. Campbell, H. Antoniadis, T. Virgili, D. G. Lidzey, X. Wang and D. D. Bradley, *Proc. SPIE Int. Soc. Opt. Eng.*, **4464**, 211 (2002).
7. J. Morgado, E. Moons, R. H. Friend and F. Cacialli, *Adv. Mater.*, **13**, 810 (2001).
8. M. Redecker, D. D. C. Bradley, M. Inbasekaran and E. P. Woo, *Appl. Phys. Lett.*, **73**, 1565 (1998).
9. J. Chappell, D. Lidzey, P. Jukes, A. Higgins, R. Thompson, S. O'Connor, I. Grizzi, R. Fletcher, J. O'Brien, M. Geoghegan and R. Jones, *Nat. Mater.*, **2**, 616 (2003).
10. A. M. Higgins, S. J. Martin, R. L. Thompson, J. Chappell, M. Voigt, D. G. Lidzey, R. A. L. Jones and M. Geoghegan, *J. Phys. Condens. Matter*, **17**, 1319 (2005).
11. Q. Zhang, *Chin. J. Chem.*, **28**, 1482 (2010).
12. H.-J. Su, F.-I. Wu and C.-F. Shu, Macromol., **37**, 7197 (2004).

ECS Transactions, 34 (1) 1087-1094 (2011)
10.1149/1.3567719 ©The Electrochemical Society

Enhancement of Luminance via Blending F8BT with Tetraphenyldiaminobiphenyl-Containing Hole Transport Polymer

Qiushu Zhang[a]

[a] School of Mechatronics Engineering, University of Electronic Science and Technology of China, Chengdu, Sichuan 611731, P. R. China

We demonstrated polymer light-emitting diodes (PLEDs) based on poly(9,9'-dioctylfluorene-alt-benzothiadiazole) (F8BT) doped with a nonpolyfluorene tetraphenyldiaminobiphenyl-containing hole transport material, poly(N,N'-bis(4-butylphenyl)-N,N'-bis(phenyl) benzidine) (PPB). The devices have a configuration of ITO/PEDOT:PSS/emissive layer/Al. The emissive layers consisted of F8BT blended with PPB at varied weight ratio. The brightness of the blend devices was significantly increased in comparison with pure F8BT. For the 7% PPB blend, the luminance reached 1078 cd/m^2 at 47.5 mA (528 mA/cm^2), improved by 3 fold. We propose that increased charge carrier injection and transport is responsible for the enhanced electroluminescence (EL) performance. EL spectrum was hardly influenced by blending.

Introduction

There are much fewer n-type polyfluorenes (PFs) than p-type ones because of the susceptibility of polymers to oxidation and the typically smaller injection barriers at the anode/polymer interface (1). Chemically some modifications have been carried out to improve electron transport in PFs, which include the incorporation of electron-transporting moieties into the PF polymer chain (2), as well as end-capping the polymer with electron-transporting moieties (3). The green-emitting alternating copolymer poly(9,9'-dioctylfluorene-alt-benzothiadiazole) (F8BT) is a known electron transporting polyfluorene copolymer. It combines high electron mobility with high electron affinity, which leads to easy electron injection (4). In recent years, F8BT has attracted considerable academic interests in PLED applications. Although PLEDs featuring neat F8BT displayed good EL performance (5)(6), more commonly F8BT was blended with other polymers such as poly(9,9-dioctylfluorene) (PFO) and poly((9,9-dioctylfluorene)-alt-N-(4-butylphenyl)diphenylamine) (TFB) to act as the active layer in EL devices because the device properties were greatly enhanced upon doping (7)(8). Some researchers conducted investigations on film morphology of F8BT blends, trying to find out the relationship between PLED device performance and the phase-separated morphology of the blends (8)(9)(10)(11)(12)(13).

For PLED applications, F8BT is mostly blended with polyfluorenes. However, in this paper, we report on the fabrication and characterization of EL devices based on F8BT doped with a nonpolyfluorene hole transport material, poly(N,N'-bis(4- butylphenyl)-N,N'-bis(phenyl)benzidine) (PPB). Blending almost did not affect the EL spectrum, but

1087

an enhancement in luminance was observed for blend based devices. The origin of the EL improvement is discussed.

Experimental

Molecular structures of F8BT and PPB are shown in Figure 1. Both polymers were obtained from American Dye Source, Inc., Canada. F8BT is yellow powder, while PPB is light yellow powder. They are highly soluble in toluene and tetrahydrofuran. Their molecular weights are around 44,000 and 12,000, respectively, as determined by gel permeation chromatography using polystyrene standards.

The EL devices have a configuration of ITO/PEDOT:PSS/emissive layer/aluminum, which were fabricated at ambient conditions on glass substrates covered by patterned indium-tin-oxide (ITO) electrodes. The ITO substrates were precleaned by two successive ultrasonic baths in acetone and isopropyl alcohol. After drying them with a nitrogen gun, a 45-nm thick film of poly(3,4-oxyethyleneoxythiophene) doped with poly(styrene sulfonate) (PEDOT:PSS) was spin coated over the substrate from a 1.3 wt.% water dispersion. The emissive layer (~ 55 nm) consisted of pristine F8BT or F8BT blends with PPB at varied weight ratio. It was spin coated from polymer toluene solutions (10 mg/mL) at 2000 rpm. Devices were dried in a vacuum chamber at room temperature for over 24 h before the deposition of aluminum film. The manufacturing was completed with thermal evaporation of the aluminum cathode (~ 200 nm) at 2×10^{-6} Torr through a shadow mask. The overlap between the two electrodes gave device active areas of 9 mm^2.

The optical absorption (UV-Vis) spectra were measured with an Agilent-8453 UV-visible spectrophotometer from polymer films spin cast onto a quartz plate. All device testing was implemented in air at room temperature. Current-voltage characteristics were measured on a Keithley 236 source-measure unit. The power of EL emission was measured using an ILX Lightwave OMM-6810B optical multimeter equipped with a silicon power/wavehead (OMH-6722B). By assuming Lambertian distribution of the EL emission, luminance (cd/m^2) was calculated utilizing the forward output light power and the EL spectra of the devices.

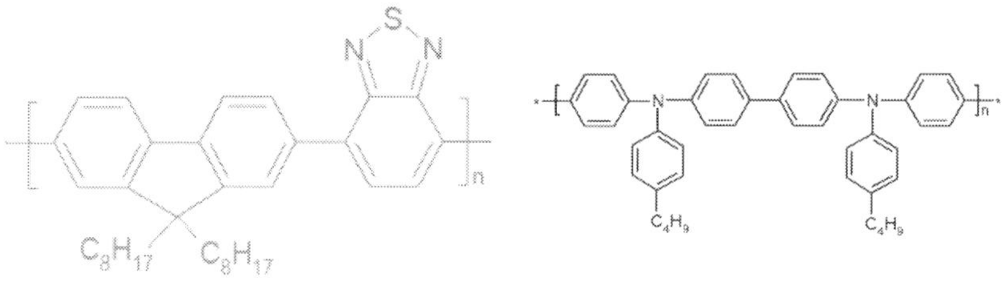

F8BT   PPB

Figure 1. Molecular structures of F8BT and PPB.

Results and Discussion

Figure 2 shows the UV-Vis spectra for thin films of F8BT and PPB. The optical absorption of F8BT had three peaks at 209, 322, and 464 nm with two shoulders at 254, and 340 nm. From the extrapolation of the optical absorption spectrum, the band-gap energy of F8BT was calculated to be 2.38 eV ($\lambda_{onset}=520$ nm), which is close to the value reported in Ref.(8). The UV-Vis spectrum of PPB exhibited peaks at 214 and 372 nm. Its band-gap energy was determined to be 2.88 eV ($\lambda_{onset}=430$ nm).

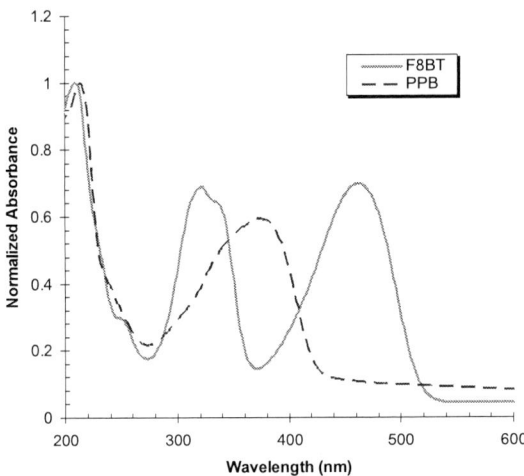

Figure 2. Optical absorption spectra of thin films of F8BT and PPB.

Figure 3 displays the current-voltage characteristics of PLEDs based on F8BT and its blends with PPB with varied blend ratio. In comparison with the pure F8BT, the turn-on voltage is decreased for all blend cases and it is shown that although below 17.5 V the current through pure F8BT device is lower than those though the blend emission layers, it increases faster with voltage.

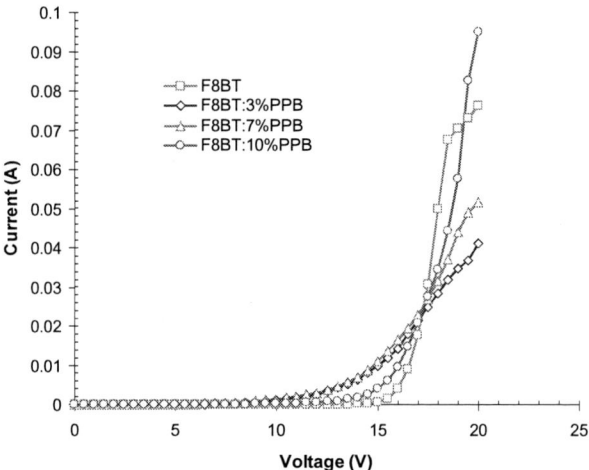

Figure 3. Current as a function of the applied voltage for PLEDs based on F8BT and its blends with PPB with varied blend ratio.

Figure 4 compares the light power-current characteristics of the PLEDs fabricated with F8BT and its blends with PPB. By utilizing the data shown in Figure 4, we calculated brightness, luminance efficiency, and external quantum efficiency which are summarized in Table I. The introduction of PPB into the emissive layer reduces the threshold current for light emission, which is indicative of an improved hole/electron balance within the active layer. Enhanced performance was found in the blend devices with a maximum brightness of 1078 cd/m^2, a maximum luminance efficiency of 0.207 cd/A, and a maximum external quantum efficiency of 0.055% for the 7% PPB blend.

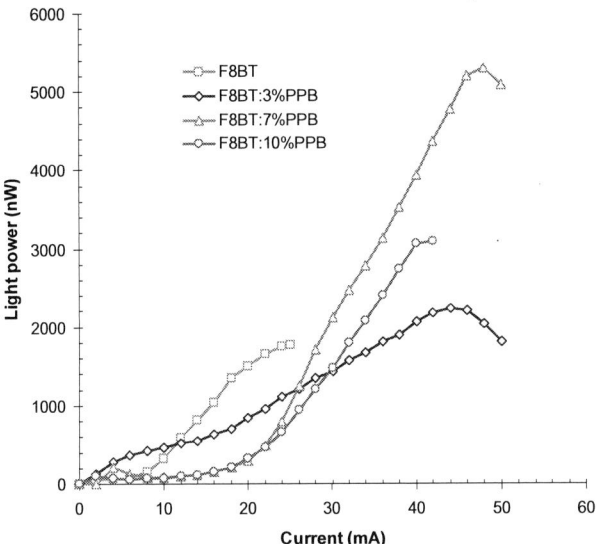

Figure 4. Variation of the light power with current for PLEDs based on F8BT and its blends with PPB with varied blend ratio.

TABLE I. EL performance for the PLEDs based on F8BT and its blends with PPB.

Emission layers	F8BT	F8BT:3% PPB	F8BT:7% PPB	F8BT:10% PPB
$J_{threshold}$ (mA/cm^2)	16.6	10.9	10.9	11
V_{on} (V)	13.5	6	5.5	8
B_{max} (cd/m^2)	360	455	1078	627
η_{max} (%)	0.037	0.038	0.055	0.037
LE_{max} (cd/A)	0.138	0.143	0.207	0.14

$J_{threshold}$: Threshold current density for light emission. It is defined as the minimum current density at which the resulting light power can be detected.

V_{on} : Turn-on voltage for the current at which the current reaches $\sim 8.5 \times 10^{-5}$ A.

B_{max} : Maximum luminance.

η_{max} : Maximum external quantum efficiency.

LE_{max} : Maximum luminance efficiency.

There are a few possible reasons for the improved EL properties seen in the blend cases. Since the band-gap difference in F8BT:PPB is 0.5 eV, it is likely that some energy was transferred from the higher band gap PPB to F8BT (8). However, the efficiency of energy transfer might be low because of weak EL observed from the neat PPB devices we fabricated (PPB is a hole transport polymer, as well as a light emitting polymer) (14). More possibly, the improvement of device performance may be due to enhanced charge carrier injection and transport. As a hole transporting component, PPB that is added to the emissive layer can improve the hole injection from the anode and the hole transport through the layer. On the other hand, F8BT has a high electron affinity, and exhibits strong electron transport. The combination of both is advantageous to EL enhancement. Another possibility is that spatial confinement of excitons made some contributions to the improvement of device properties. Phase-separated domains in the thin film morphology of the 7% PPB blend were observed in the SEM micrograph (Figure 5). The isolated domains may help confine the excitons in the blends (14). The possible reason why the best performance was found in the 7% PPB blend may be that this blend ratio corresponds to a more balanced ratio between the electron and hole currents, and/or that this blend ratio produces a phase separation morphology that can give rise to a better spatial confinement effect.

Figure 5. SEM image of a thin film of F8BT:7% PPB blend spin-cast from 10mg/mL solution in toluene onto a PEDOT:PSS-covered ITO substrate.

Doping F8BT with PPB had no effect on the EL spectrum. For all devices under study, EL emission exclusively came from the contribution of F8BT. EL peaks at 541 nm (Figure 6), which is close to the value reported in Ref.(5). The light intensity increased with an increasing of the applied voltage, and the EL spectrum was reproducible and remained stable during device characterization.

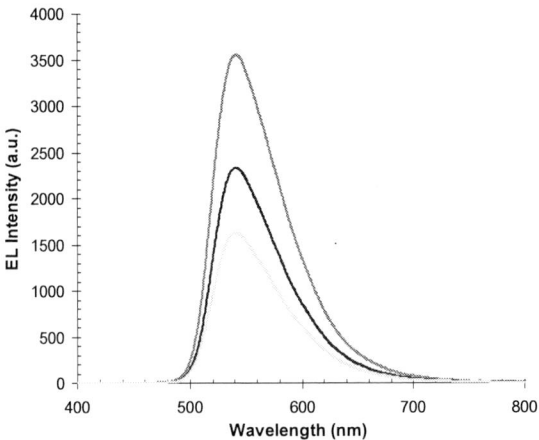

Figure 6. EL spectra of the pristine F8BT devices under different forward bias.

Conclusion

Addition of hole transporting polymer PPB to F8BT is shown to bring about an increase in PLED device properties. The devices with the 7% PPB blend as the emissive layer exhibited the best performance in light luminance and efficiency. The maximum brightness reached 1078 cd/m^2 at a current density of 528 mA/cm^2, which is three times higher than that of the pristine F8BT PLEDs. EL spectrum was not influenced by blending. A few possible reasons are given to explain the enhanced performance observed from the blend devices. While F8BT are generally blended with polyfluorene or polyfluorene copolymer for PLED applications, our results indicate that the addition of nonpolyfluorene hole transport polymer can improve the device properties of PLEDs based on F8BT.

References

1. W. Wu, M. Inbasekaran, M. Hudack, D. Welsh, W. Yu, Y. Cheng, C. Wang, S. Kram, M. Tacey, M. Bernius, R. Fletcher, K. Kiszka, S. Munger and J. O'Brien, *Microelectron. J.*, **35**, 343 (2004).
2. J. F. Ding, M. Day, G. Robertson and J. Roovers, *Macromol.*, **35**, 3474 (2002).
3. X. Gong, W. Ma, J. C. Ostrowski, K. Bechgaard, G. C. Bazan, A. J. Heeger, S. Xiao and D. Moses, *Adv. Funct. Mater.*, **14**, 393 (2004).
4. J. Morgado, R. H. Friend and F. Cacialli, *Appl. Phys. Lett.*, **80**, 2436 (2002).
5. Y. He, S. Gong, R. Hattori and J. Kanicki, *Appl. Phys. Lett.*, **74**, 2265 (1999).
6. T. Kawase, P. K. H. Ho, R. H. Friend and T. Shimoda, *Mater. Res. Soc. Symp. Proc.*, **598**, BB11.49.1 (2000).
7. A. J. Campbell, H. Antoniadis, T. Virgili, D. G. Lidzey, X. Wang and D. D. Bradley, *Proc. SPIE Int. Soc. Opt. Eng.*, **4464**, 211 (2002).

8. E. Moons, *J. Phys. Condens. Matter*, **14**, 12235 (2002).
9. J. Morgado, E. Moons, R.H. Friend and F. Cacialli, *Synth. Met.*, **124**, 63 (2001).
10. J. Chappell, D. Lidzey, P. Jukes, A. Higgins, R. Thompson, S. O'Connor, I. Grizzi, R. Fletcher, J. O'Brien, M. Geoghegan and R. Jones, *Nat. Mater.*, **2**, 616 (2003).
11. A. M. Higgins, S. J. Martin, R. L. Thompson, J. Chappell, M. Voigt, D. G. Lidzey, R. A. L. Jones and M. Geoghegan, *J. Phys. Condens. Matter*, **17**, 1319 (2005).
12. J. Morgado, E. Moons, R. H. Friend and F. Cacialli, *Adv. Mater.*, **13**, 810 (2001).
13. R. Stevenson, R. Riehn, R. G. Milner, D. Richards, E. Moons, D.-J. Kang, M. Blamire, J. Morgado and F. Cacialli, *Appl. Phys. Lett.*, **79**, 833 (2001).
14. A. P. Kulkarni and S. A. Jenekhe, *Macromol.*, **36**, 5285 (2003).

CHAPTER 10

SILICON MATERIALS FOR ELECTRONIC
AND PHOTOVOLTAIC APPLICATIONS

ECS Transactions, 34 (1) 1097-1101 (2011)
10.1149/1.3567720 ©The Electrochemical Society

Improvements on The Uniformity of a-Si Solar Thin Films by Using Auxiliary Magnetic Field

L.C. Hu[a], Y.P. Chen[a], J.Y. Chang[b], J.J. Lee[b], I.C. Chen[c], and Tomi T. Li[a*]

[a] Department of Mechanical Engineering, National Central University, Taoyuan 320, Taiwan (R.O.C.)
[b] Optical Science Center, National Central University, Taoyuan 320, Taiwan, (R.O.C.)
[c] Institute of Materials Science and Engineering, National Central University, Taoyuan 320, Taiwan, (R.O.C.)

The uniformity of large area thin films deposition is a crucial process as we widely apply electron cyclotron resonance chemical vapor deposition (ECR-CVD) in solar industry. In this work, we installed sub-magnetic (auxiliary) fields for inner and outer coils under ECR-CVD process chamber to improve the deposition uniformity of a-Si solar thin films. Next, we measured the distribution of magnetic field along the central axis of chamber and the diameter of substrate surface. By this approach, we investigated the effect of sub-magnetic field to the uniformity of electron cyclotron resonance a-Si solar thin films deposition. We succeeded in obtaining an excellent deposition uniformity of a-Si solar thin films over 150mm diameter on glass substrates by adjusting the magnetic field distribution from inner and outer magnetic coils. The uniformity is within 10%. Moreover, We obtained optimal conditions for solar cell fabrication between the rapid process deposition rates (>10 Å/sec) and magnetic parameters.

Introduction

Electron cyclotron resonance (ECR) plasma possesses a high plasma density under low pressure operation. It has several advantages such as high deposition rate, low contamination and low ion bombardment. Most importantly, it can provide large area plasma, so it has been used for thin film coating and surface etching in the semiconductor process[1,2]. In order to develop a wide use of application for electron cyclotron resonance chemical vapor deposition (ECR-CVD) in solar industry, the uniformity of large area thin films deposition will be a crucial process.

Previously, many studies have already successfully improved the uniformity of ECR thin films by controlling process parameters like microwave power, working pressure, substrate tempature and magnetic field distribution of deposition chamber[3-8]. In the magnetic field distribution study on uniformity, it usually needs to adjust the permanent magnet rings or many magnetic coils which are huge, heavy and high manufacturing cost.

In the experiment, we studied the relation between the magnetic field distribution on substrate surface and deposition uniformity. Furthermore, we tried to find a more effecive way but less work to keep good uniformity of large area for thin films deposition. By installing a sub-magnetic (auxiliary) field for inner and outer coils under process

1097

chamber, we improved the deposition uniformity with high deopsition rate.

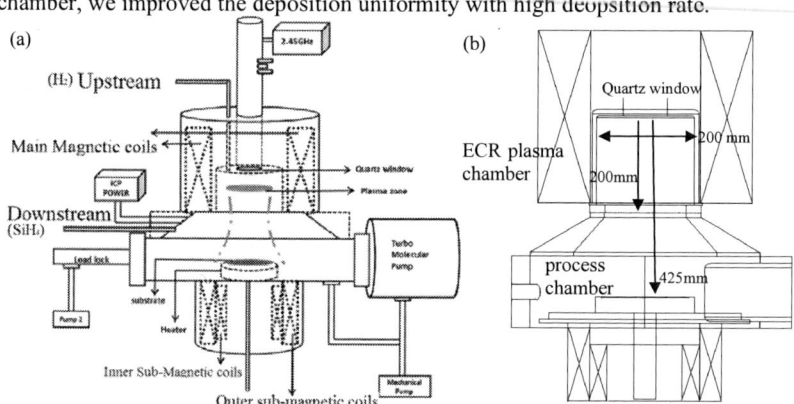

Fig. 1. Schematic diagram of the (a) experimental equipment and (b) plasma and process chamber

Experimental

The experimental setup is schematically shown in Fig. 1. The microwaves power supply has the frequency of 2.45 GHz and output power ranges from 350W to 2kW. Microwaves are radiated into a cylindrical plasma chamber (D200mm*H200mm) across a quartz glass plate 90mm diameter and 15mm in thickness. From quartz window to substrate surface is 425mm. Argon, silane and hydrogen are used as working gases. The working gases are injected into the plasma chamber from upstream(H_2) and downstream (SiH_4). In order to optimize a-Si process requirements, the microwave power 800 watts, process time 600 seconds and substrate temperature 225°C were adopted. The hydrogen dilution ratio is defined as R= H_2/SiH_4 = 1 and chamber working pressures kept on 5 mTorr. The a-Si film thickness uniformity was measured by Dektak Surface Profiler (Veeco/Dektak 6M) for a total of 17 points.

The deposition of a-Si thin films is under different magnetic field combination. Aside from the resonant main coils, we installed a sub-magnetic (auxiliary) field for inner and outer coils under ECR-CVD process chamber. Note that our experiment condition is that the outer magnetic field is the same direction as the main coil whereas the inner magnetic field is opposite direction to the main coil. To measure the distribution in magnetic profiles on substrate surface and plasma chamber, we employed Gaussmeter (F.W. Bell 9900) with a sensing probe extending in the space of ECR chamber along the center axis and substrate surface. For obtaining the various combinations in magnecic field configurations, we adjusted the current of inner and outer magnetic coils in conjunction with main magnetic coils to achive the best control of magnetic distribution on substrate (Engle-XG glass) surface.

Results and Discussion

a.Magnetic Coils Characteristics

Fig. 2 shows the magnetic field distribution of different currents in main coil. As anticipated, the microwave resonance zone (875 Gauss) obviously moves downward to the substrate as the current of main coil increase. The currents of main coil affect the thickness of the plasma zone and resonance position. In order to let the resonance plasma close to the exit of the downstream for increasing dissociation of SiH_4, 55A was chosen as a main magnetic field current parameter. At 55A main coil current, plasma zone thickness is about 5mm. From Fig. 3, Off and on for sub-magnetic coils was almost no impact on plasma resonance zone.

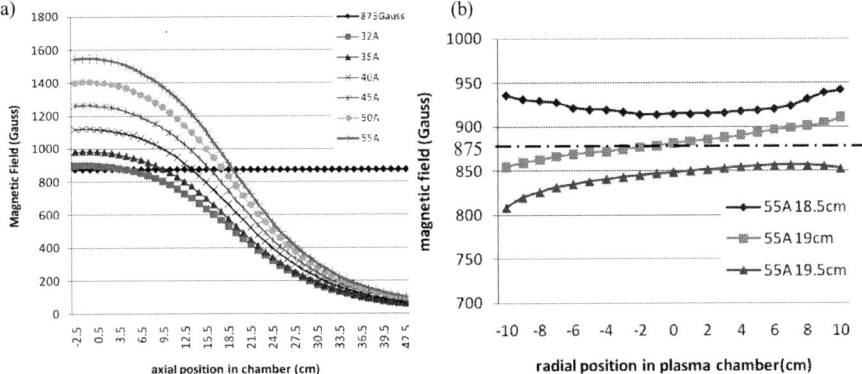

Fig. 2. (a) The magnetic field distribution of different main coil currents (b) Radial magnet profile of 55A main coil with different distances from substrate

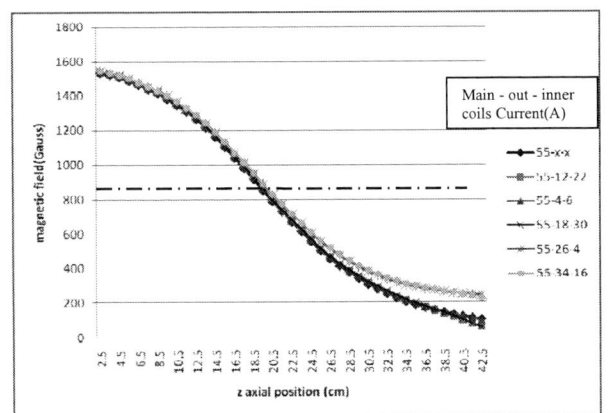

Fig. 3. Magnetic field configuration along central axis of chamber

b. Magnetic Field Effect on Substrate Surface

On the other hand, the magnetic field on the substrate surface is always less than 110 Gauss when only the main coil is used. Toward to the substrate, the impact from the main coils magnetic field will be minimal. Thus, we can easily control the contribution of the magnetic field on substrate surface by adjusting sub-magnetic field from the inner and outer coils. The results about different magnetic fields on substrate surface with the same resonance zone is listed in Table 1 and shown in Fig 4. In light of experimental results, we found that deposition rates increase when we adjusted the magnetic field becoming stronger. More electrons and ions are expected to be attracted onto the substrate surface as z component increases. From Table 1, the uniformity of thin films has dramatically influenced by z component of magnetic field. The x component of magnetic field shows it is a dependent adjustment related to the z component. Specifically, the x component magnetic field heavily relies on the difference of z component magnetic field between the center and edge on the substrate.

By adjusting the z magnetic strength of the edge of the substrate higher than center, a good uniformity of deposition is obtained. Furthermore, we also observed the magnetic strength of the substrate edge had approximately 20% greater than that of the substrate center, the optimal and best uniformity could be obtained.

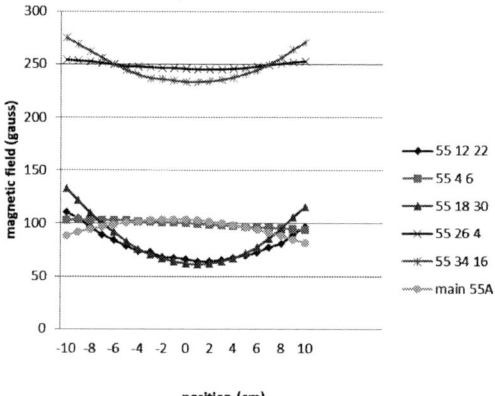

Fig. 4. Z component of magnetic field distribution on substrate surface

TABLE I. Types of Thin Films Sample

Sample No.	Strength of Z* Component Magnetic Field** (Gauss)	Max. Strength of X*** Component Magnetic Field (Gauss)	Average Deposition Rate(Å/sec)	Uniformity of Thin Films (%)
55-x-x	104~80	40	7.2	18.4%
55-12-22	64.5~111	70	7.5	17.4%
55-4-6	94~103	48	7.8	13.7%
55-18-30	62~132.7	77	8.2	15.6%
55-26-4	245~253	13	9.7	12.7%
55-34-16	233.5~274	30	10.1	11.1%

*Z component indicate the vertical direction of magnetic field to the substrate
**indicates the strength of magnetic field on the substrate from center to edge
***X component indicate the vertical direction of magnetic field to the substrate

Conclusion

Based on the results, we successfully improve the uniformity of thin films deposition with high deposition rate by sub-magnetic (auxiliary) coils. We found the magnetic strength of the substrate edge had approximately 20% greater than that of the substrate center, the optimal and best 10% uniformity could be obtained in 150mm diameter substrate. This study will provide a baseline of a-Si thin film for solar cell fabrication.

Acknowledgments

This work was supported by a grant from the fund for "National Science and Technology Nano Program" (NSC-99-2120-M-008-003) of the National Science Council of Taiwan (R.O.C.).

References

1. T. Ono, M. Oda, C. Takahashi and S. Matsuo, *J. Vac. Sci. Technol. B*, **4**(3), 696 (1986).
2. S. Matsuo and M. Kiuchi, *Jpn. J. Appl. Phy.*, **22**, L210 (1983).
3. R. Kar, S.A. Barve, S.B. Singh, D.N. Barve, N. Chard and D.S. Patil, *J. Vacuum*, **85**, 151 (2010).
4. A. Murai, I. Ohya, T. Yasui, H. Tahara and T. Yoshikawa, *Thin Solid Films*, **281-282**, 146 (1996).
5. K. Nakase, T. Shibata, T. Yasui, H. Tahara and T. Yoshikawa, *Thin Solid Films*, **281-282**, 152 (1996).
6. C. S. Ren, D. Z. Wang, J. Zhang, X. L. Qi and Y. N. Wang, *J. Vacuum*, **83**, 423 (2009).
7. H. Nishimura, M. Kiuchi and S. Matsuo, *Jpn. J. Appl. Phys.*, **32**, 322 (1993).
8. Y. Kawai, K. Uchino, H. Muta, S. Kawai and T. Röwf, *J. Vacuum*, **84**, 1381 (2010).

Hydrogenated Silicon Thin Film and Solar Cell Prepared by Electron Cyclotron Resonance Chemical Vapor Deposition Method

Chien-Chieh Lee[a], Jenq-Yang Chang[b], Yen-Ho Chu[b], Chung-Min Lien[b], I-Chen Chen[c], and Tomi Li[d]

[a]Optical Science Center, National Central University, Taiwan, R.O.C.
[b]Department of Optics and Photonics, National Central University, Taiwan, R.O.C.
[c]Institute of Materials Science and Engineering, National Central University, Taiwan, R.O.C.
[d]Department of Mechanical Engineering, National Central University, Taiwan, R.O.C

Hydrogenated silicon thin film was deposited on glass substrate using the electron cyclotron resonance chemical vapour deposition (ECR-CVD) system. Fourier transform infrared spectroscopy (FTIR) and scanning electron microscope (SEM) were used to measure the film properties. It showed that higher deposition rate (>2nm/sec) and lower microstructure parameter (20%<R*<30%) could be achieved with increasing input power and decreasing hydrogen dilution ratio (H2/SiH4). In addition, p-type and n-type hydrogenated amorphous silicon thin films were grown using SiH4/Ar/H2/B2H6 or PH3 gases. Hall mobility decreased with increasing the doping gas flow rate. The carrier concentration exhibited an entirely converse behavior. With increasing the doping gas flow rate, the concentrations increased and the highest values of 5.2×10^{19} cm-3 (3.9×10^{19} cm-3) at H_2/B_2H_6 (H_2/PH_3) of ~20 for p-type (n-type) hydrogenated amorphous Si thin films were achieved. These films are suitable for the amorphous Si thin film solar cell.

Introduction

Hydrogenated amorphous silicon (a-Si:H) has shown excellent characteristics for use not only in solar cells [1], but also in photosensitive devices and thin-film transistors [2]. One of the key issues for a large scale production of thin-film silicon solar cells is to find deposition techniques that allow high deposition rates and at the same time yield material of high quality. In addition, low-cost substrates such as glass limit the deposition temperature to range up to a few hundred °C at maximum. One technique that seems to fulfill the requirements is the electron-cyclotron resonance (ECR) chemical vapor deposition [3-5]. The plasma generated in the ECR process is via resonant absorption of a microwave by electrons in a magnetic field and gas ionization via subsequent electron-atom collisions. By the application of this plasma-generating process, the total working pressure may typically lie in the mTorr range, which is 1~2 orders of magnitude lower than that in the usual parallel-plate plasma-enhanced chemical vapor deposition (PECVD). Because the average mean free path of particles in the gas-plasma phase exceeds the thickness of the plasma sheath, the generated ions are not affected by collisions on their way to the substrate, and they impinge with a rather sharp and well defined energy distribution. Thus, ECR CVD enables an improved control of the deposition process. Moreover, the high plasma density achieved by the ECR techniques also results in high deposition rate. The advantages of ECR have mainly been demonstrated in the field of anisotropic etching so far [6]. Thin films of hydrogenated amorphous and microcrystalline Si (a-Si:H, μc-Si:H) are attracting considerable interest for solar cells. Hence the growth of these films is an important area of research.

Here, we demonstrate that with ECR CVD it is possible to deposit intrinsic, phosphine, and diborane doping hydrogenated amorphous thin films which are suitable as absorber, p-type and n-type layers in a-Si solar cells. The aim of this work was to find optimized deposition parameters through a factorial analysis resulting in a high deposition rates of the a-Si:H thin films.

Result and Discussion

An ECR system was employed for the depositions, and its plasma is excited by using a 2.45 GHz microwave poser . The deposition chamber is situated underneath the cylindrical ECR source. The chamber walls were water cooled during depositions. Evacuation of the chamber down to high-vacuum regime is performed by an turbomolecular pump providing a residual gas pressure nearly 5×10^{-6} Torr. Gases of $SiH_4/H_2/Ar/B_2H_6/PH_3$ are supplied via an inlet valve upon the ECR region. The total pressure was regulated via a throttle valve to 5 mTorr, and the substrate temperature was set to 325°C during most depositions presented in this work.

Figure 1 shows the relation between the deposition rate and microwave power for three H_2/SiH_4 ratio. Deposition rates increased as the plasma power increased from 500W to 1000W, but deposition rate decreased when the power continue increased above 1200W. This may be due to hydrogen dilution effects of the radicals (reaction precursor) to substrate in the plasma reaction system. Plasma with high energy also makes ion bombardment on substrate, resulting in the etching effect. When the plasma power increased from 500W to 1000 W, the amount of radical ionized from reaction gases SiH_4 increased in the plasma region. The deposition effect was more effective than bombardment effect, and then the deposition rate of silicon thin film was increased. When the plasma power increased above 1200W, the ion bombardment of H^+ was enhanced, resulting in that total deposition rate was declined. Therefore, by way of deposition and etching effects, we can optimize the deposition rate of hydrogenated silicon thin film by controlling plasma power and dilute ratio. We also found that most of the structures of deposited silicon thin film are amorphous, but there is a trend of slight crystallization when plasma power was increased from 1200W to 1500 W.

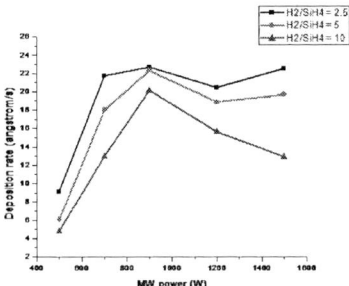

Figure 1. The deposition rate of a-Si:H thin films deposited with various microwave power for different hydrogen dilution ratio by using with ECRCVD.

The Fourier-transform infrared spectroscopy (FTIR) determines the content of bonded H and the silicon-hydrogen bonding configurations. The FTIR spectrum was deconvolued with Gaussian absorption profiles for the desired vibrational modes. The hydrogen-bonding configuration is characterized by the microstructure parameter R^*, defined as $R^*=I_{2000}/(I_{2000}+I_{2090})$. High microstructure value is usually used a sign of high void density in the material. Figure 2 shows the relation of microstructure parameters of silicon thin films (R^*) on the various microwave power.

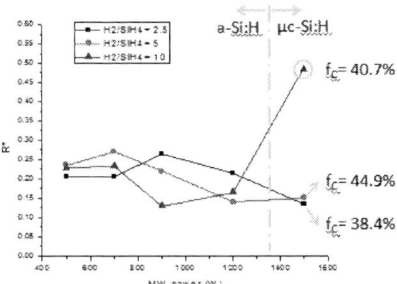

Figure 2. The microstructure factor (R^*) of a-Si:H thin films deposited with various microwave power for different hydrogen dilution ratio.

Fig. 3 shows the schematic illustration and SEM image of hydrogenated amorphous silicon thin films for different input power. It can be clearly seen that as power increases, the grain aggregation of silicon film gradually reduced, and surface morphology of silicon thin film become somooth as input power upto 1200W.

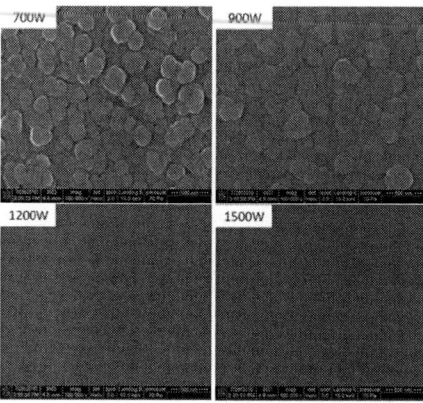

Figure 3. The SEM pictures of a-Si:H thin films deposited with various microwave power by using with ECRCVD.

Gases used for the in situ doping were H_2, SiH_4 and PH_3/H_2 or B_2H_6/H_2 mixed gases. The deposition parameters for this study were 100, 150 sccm for the H_2 gas flow rate, 2 sccm for the SiH_4 flow rate, 5~15 sccm for the PH_3/H_2 flow rate, 5~10 sccm for the B_2H_6/H_2 flow rate , 350°C for the substrate temperature, 1500 W for the microwave power and 5 mTorr for the process pressure.

Figs. 4 and 5 show the change of carrier Hall mobility and carrier concentration as a function of the doping gas flow rate ratio. It can be seen that the carrier mobility decreased with increasing the gas phase doping ratio for p-type and n-type doping. The mobility of a-Si:H thin films decreased from 0.66 to 0.1 cm^2-Vs for p-type samples and 0.66 to 0.27 cm^2-Vs for n-type samples as the hydrogen dilution ratio increased from 93% to 98%, but it increased with further hydrogen dilution. On the other hand, the carrier concentration exhibited an entirely different behavior. It increased from 7.5×10^{18} cm^{-3} to 5.2×10^{19} cm^{-3} for p-type samples and 1.3×10^{19} cm^{-3} to 3.9×10^{19} cm^{-3} for n-type samples as the hydrogen dilution ratio increased from 93% to 98%.

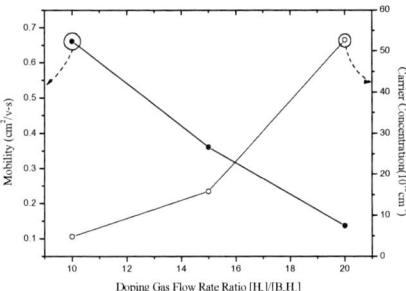

Figure 4. Hall mobility and carrier concentration of heavily doped Si thin films as a function of $[H_2]/[B_2H_6]$ gas flow rate ratio (p-type).

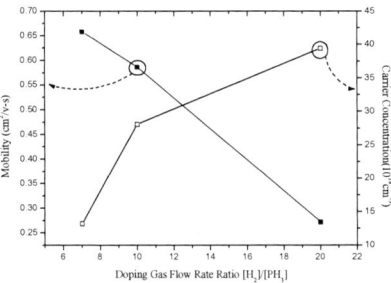

Figure 5. Hall mobility and carrier concentration of phosphorus-doped as Si thin films as a function of $[H_2]/[PH_3]$ gas flow rate ratio (n-type).

Conclusions

Hydrogenated silicon thin film was deposited on glass substrate using the technique of electron cyclotron resonance chemical vapour deposition (ECR-CVD). It showed that higher deposition rate (>2nm/sec) and lower microstructure parameter (20%<R*<30%) could be achieved with increasing applied microwave power and decreasing hydrogen dilution ratio (H_2/SiH_4). Using $SiH_4/Ar/H_2/B_2H_6$ or PH_3 gases, the concentrations increased and the highest values of 5.2×10^{19} cm-3 (3.9×10^{19} cm^{-3}) at H_2/B_2H_6 (H_2/PH_3) of ~20 for p-type (n-type) hydrogenated amorphous Si thin films were achieved. These films are suitable for the amorphous Si thin film solar cell.

Acknowledgments

The authors are grateful for the financial support of this research received from the National Science Council of Taiwan, R.O.C. under the grant number NSC-98-2120-M-008-005.

References

1. D.E. Carlson and C.R. Wronski, Appl. Phys. Lett. **28** 671 (1976).
2. P.G. Le Comber and W.E. Spear, in semiconductor and semimetals, edited by J. PanKore Vol. 21, Part D, Chapter 6, Academic, New York (1984),.
3. T. Lagarde, Y. Arnal, A. Lacoste, J. Pelletier, Plasma Sources Sci. Technol. **10**, 181 (2001).
4. P. Bulkin, N. Bertrand, B. Drevillon, Thin Solid Films, **296,** 66 (1997).
5. A. Lacoste, T. Lagarde, S. Béchu, Y. Arnal, J. Pelletier, Plasma Sources Sci. Technol. **11**, 407 (2002).
6. S. Watanabe, in *Plasma Etching*, edited by M. Sugawara (Oxford University Press, Oxford, p. 252 (1998).

Properties of multicrystalline silicon wafers based on UMG material

Tingting Jiang, Xuegong Yu,[*] Xiaoqiang Li, Xin Gu, Peng Wang, Deren Yang

State Key Laboratory of Silicon Materials and Department of Materials Science and Engineering, Zhejiang University, Hangzhou, 310027, People's Republic of China

In this paper, we have studied various properties of UMG silicon wafers by combining four-point probe, microwave photo-conductance decay (μ-PCD), Fourier transform infrared (FTIR) spectroscopy , Inductively coupled plasma-mass spectrometry (ICP-MS) and optical microscopy techniques. The results show that the resistivity and the minority-carrier lifetime of UMG silicon are lower than that of standard multicrystalline material, while the detrimental interstitial iron concentration is larger. The concentration of substitutional carbon, interstitial oxygen and dislocation density are close to that in the standard multi-crystalline silicon. Furthermore, a phosphorous gettering has been used to improve the quality of UMG samples. It is found that phosphorus gettering in UMG silicon can increase the minority carrier lifetime and reduce the interstitial iron concentration. But, the minority-carrier lifetime is still not as high as the conventional/standard silicon counterpart after gettering. These results will help us to better understand the properties of UMG silicon wafer.

Introduction

The serious energy and environment crisis has attracted much interest in researches related to sustainable energy. Of all the renewable energy sources, crystalline silicon solar cell has so many advantages that it has been intensively investigated in recent years. Considering the fact that the high cost restricted the commercial development of solar cell, there are two ways to meet the demand for the reduction of power generation cost. One is to develop high efficiency solar cells[1], and the other is the application of low cost raw silicon material[2].

The upgraded metallurgical grade (UMG) silicon material has a lower production cost than that of conventional silicon material. The application of this raw material can realize the demand of low cost solar cell production. It is estimated that the raw silicon material cost share more than 50 percent in the whole cost of silicon solar cell.[3] Generally, the UMG raw material can be refined by several low-cost steps such as pyrometallurgical refinement, acid or alkali leaching, plasma torch treatment as well as directional solidification[4]. These years, the application of UMG silicon has received specific attention. However, several problems still remain unsolved. One is that the solar cell fabricated with UMG silicon has a lower photo-electric conversion efficiency due to the high concentration of impurities such as boron, phosphorous and metallic impurities. Especially, transition metallic impurities located in substitutional or interstitial sites in the

---

[*] Electronic mail: yuxuegong@zju.edu.cn

silicon lattice or existing even as precipitates, are detrimental to the properties of UMG silicon[5]. Of all the metallic impurities, iron, possibly introduced during the ingot casting and cell producing process, is one of the most common metallic impurities in solar grade silicon. Iron in silicon material often exists as precipitates or in interstitial state, and both states do harm to the material quality and hence the cell performance[6]. Most of the time, the electrical property of the silicon wafer is determined by the content of the Fe impurity. However, to date, little systematical work has been reported on behavior of UMG silicon, and it is necessary to find out some basic properties of the material.

An effective way to remove these metallic impurities and improve the UMG silicon quality is via gettering[7-10]. Gettering treatments greatly improve electronic grade silicon, while solar grade silicon does not respond as well to the gettering treatments. As for solar cell, the effective region is the whole thickness of the material. As a result, only extrinsic gettering is available for UMG silicon. The phosphorous in-diffusion process, which is also the step of p-n junction formation for solar cell fabrication[11], has been widely used to getter metallic impurities for solar grade silicon. However, few papers has been published concerning about the effect of phosphorus gettering in the UMG silicon. In this paper, we have investigated the electrical properties of the UMG silicon material and then the phosphorous gettering effect on the UMG silicon with the comparison of conventional mc-Si. These preliminary results are expected to be a reference for future researches on UMG silicon and its application.

Experimental details

Firstly, the raw p-type UMG silicon material was cast into an ingot, which was then cut into 200 μm thick wafers. Multicrystalline silicon (mc-Si) wafers from the middle part of a standard ingot (produced from high purity mc-silicon) were taken as reference sample. The impurity content was detected by inductively coupled plasma-mass spectrometry (ICP-MS). The resistance of these samples was measured by the four-point probe technique. After removing the surface damage layer by chemical polishing in a mixture of HNO_3 and HF, the minority carrier lifetime was measured by the microwave photo conductance decay method (μ-PCD, WT-2000/2M). The interstitial iron concentration was determined comparing the minority carrier lifetime before and after Fe-B pair dissociation. The concentrations of substitutional carbon and interstitial oxygen were determined by a Bruker IFS 66 V/S Fourier Transform Infrared Spectroscope (FTIR) using calibration coefficients of $1 \times 10^{17} cm^{-2}$ and $3.14 \times 10^{17} cm^{-2}$, respectively. In order to study the dislocation density, an optical microscopy (Olympus MX50) was used to observe the mechanically polished and Secco (2 HF: 1 $K_2Cr_2O_7$ (0.15 mol·L^{-1})) etched wafers.

Moreover, phosphorus gettering was performed on the as-grown wafers. After cleaning and chemical polishing the wafers, a liquid P diffusion source (p-854, Honeywell) was deposited on both sides of the samples by spin-on coating (3000 rpm/s), followed by pre-baking at 200 °C for 10 min. Phosphorus in-diffusion was then performed in a rapid thermal process furnace (RTP 300) at 900 □ for 120 s[12]. After removing the phosphorus-silicon glass layer in a dilute hydrofluoric solution, the carrier lifetime and the concentration of interstitial iron in these samples were measured using μ-PCD.

Results and discussions

Basic properties of the UMG silicon material

The resistance of the reference mc-Si samples is in the range of 0.8-1.4 Ω•cm which is suitable for solar cell fabrication. The resistivity of UMG mc-Si samples is about 0.3 Ω•cm, which is much lower than that of the reference mc-silicon wafers. This is due to the fact that in UMG silicon raw material, the impurity content, especially boron and phosphorous is higher[13]. If not compensated, the boron concentration can be calculated as majority carrier concentration, while in UMG silicon, boron dopant is partially compensated by phosphorus. Table 1 presents the concentration of boron, phosphorous and metallic impurities in the UMG silicon and a reference mc-Si. The concentration of boron, aluminum and phosphorous are higher than that in the conventional solar cell grade wafers. Considering the ionization of boron, phosphorous and aluminum, the concentration of the above elements is in accordance with the resistivity obtained by the four-point probe. The content of metallic impurities in UMG silicon is higher than that in the conventional mc-Si such as iron, but still in the level that is insufficiently detrimental to the electrical performance of silicon wafers. An exception is titanium. The concentration of Ti in UMG sample is quite high, which is probably due to the richness of Ti in the raw material or low purification efficiency of Ti because of the substitutional location property of Ti in the Si matrix.

TABLE I. Impurities concentration in UMG silicon wafers.

Impurity	Content in UMG sample ($10^{-9} cm^{-3}$)	Content in reference sample ($10^{-9} cm^{-3}$)
B	2016	820
P	1926	23
Al	842	75
Ti	351	20
Cr	12	13
Fe	88	12
Cu	Under detection limit	Under detection limit
Ni	16	11
Co	20	16
Zn	162	108

Figure 1 shows the result of minority carrier lifetime mapping in a UMG and a reference sample. Note that all of the samples were measured after a 10 min passivation in dilute HF solution (10%). It is clear that the distribution of the minority carrier lifetime in these two mc-Si sample is not uniform. The carrier lifetime in these two samples is influenced by the non-uniformly distributed defects and impurities. Moreover, we can find that the average lifetime of the reference sample is 5.4μs, about three times that of the UMG wafer. The lower lifetime in UMG silicon is due to the higher impurity content. The minority carrier lifetime thus reflects the quality of silicon material in a direct way[14]. The poor electrical properties of UMG Si may lead to low efficiency of the solar cells made from this type material.

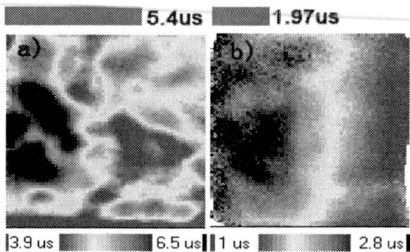

Figure 1. As-grown minority carrier lifetime mapping of (a) a UMG and (b) a reference. mc-Si wafers.

Figure 2(a) shows the minority carrier lifetime mapping of an as-grown UMG sample, and figure 2(b) presents the distribution of interstitial iron concentration for the same sample. It is known iron and boron in silicon will form Fe-B pairs, which will dissociate under illumination[15]. The two different sates of the Fe in silicon leads to different carrier lifetimes, and thus the interstitial iron concentration can be determined from lifetime measurements before (τ_{before}) and after Fe-B pair dissociation (τ_{after}) using:

$$[Fe_i]=K \cdot (1/\tau_{before} - 1/\tau_{after}) \qquad [1]$$

where $K_{\mu\text{-PCD}} = 3.4 \times 10^{13}$ s/cm^3.[16] The measured concentration of interstitial iron in the UMG wafer is around 3×10^{12} cm^{-3}. The total concentration of iron measured by ICP-MS is around 10^{15} cm^{-3}, while the local distribution of minority carrier lifetime is not fully correlated with the interstitial iron concentration. This indicates that the iron precipitates and the structural defects, such as GBs and dislocation, may also have a significant influence on the lifetime. We also measured the interstitial iron concentration in reference mc-Si samples, which is about 1.4×10^{12} cm^{-3} , less than half of that in the UMG sample. It reveals that there is more iron impurities in UMG silicon, serving as deep energy level recombination center and degrading the performance of UMG silicon wafers.

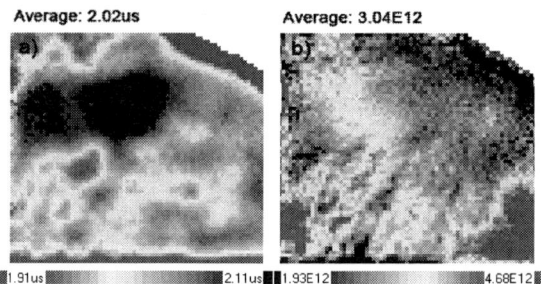

Figure 2. (a) Minority carrier lifetime and (b) interstitial iron concentration mapping in an as-grown UMG sample.

Figure 3(a) and (b) show the concentrations of substitutional carbon $[C_s]$ and interstitial oxygen $[O_i]$ in both types of the samples, respectively. The substitutional carbon and the interstitial oxygen concentrations in the UMG samples are higher than that in the reference samples. Considering that the UMG ingot and the reference ingot have been fabricated using similar process parameters, the concentration of the process-introduced carbon and oxygen should be almost the same for the two types of the samples. Thus we can deduce that the higher concentration of carbon and oxygen in UMG samples can be attributed to the UMG feedstock material.

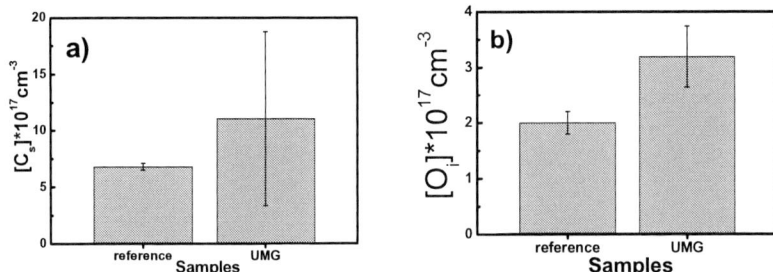

Figure 3. The concentration of (a) substitutional carbon and (b) interstitial oxygen in UMG and reference mc-Si wafers.

Figure 4 shows typical optical microscopy (OM) micrographs of a preferentially etched reference and UMG sample. Based on these OM micrographs, one can calculate the dislocation densities and obtain in both cases a density that is close to 4×10^5 cm^{-2}. This shows that the dislocation formation mainly depends on the solidification process, rather than on the impurity content. While it should be noted that in UMG silicon, defects may be decorated by a higher concentration impurities than in the case of reference mc-Si, probably leading to a stronger recombination activity of the defects in UMG wafers, and thus resulting in low carrier lifetime.

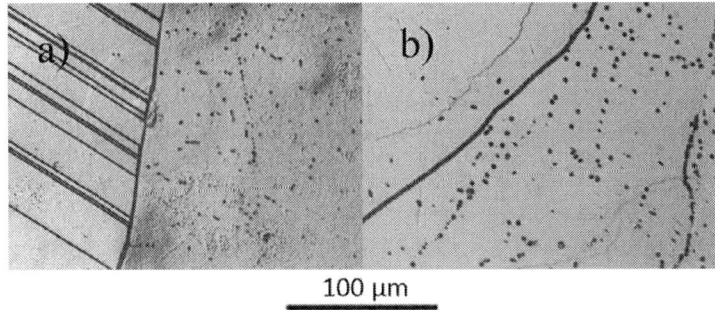

Figure 4. OM images of etched UMG and reference mc-Si samples. (a) UMG samples (b) reference mc-Si.

Phosphorous in-diffusion gettering on UMG silicon

Figure 5 shows minority carrier lifetime maps of both an UMG and a reference samples before and after gettering. A redistribution of carrier lifetime in both standard mc-Si and UMG wafers is obtained after gettering. For the standard mc-Si sample of which the as-grown carrier lifetime is shown in figure 5(a), after gettering, most part of the sample has an improved lifetime as shown in figure 5(b). For the UMG sample, the as-grown lifetime map can be seen in figure 5(c), and a lifetime improvement can be observed in some region of the whole wafer, while in some other region, the lifetime has degraded after gettering. We have mentioned that metallic impurities in silicon can be present as precipitates, and during the gettering process, some of these precipitates can be dissolved, while if these dissolved impurities cannot be effectively gettered, they may introduce higher concentration of recombination centers after gettering. This may be the reason that some region of the UMG sample shows lifetime degradation after gettering. Figure 6 shows the average minority carrier lifetime before and after gettering in both UMG and conventional mc-Si samples. We can find that the improvement of the lifetime in reference mc-Si is much more pronounced compared to UMG samples. Minority carrier lifetime in reference mc-Si increases sharply from 4 µs to 7 µs, while for the UMG samples, the minority carrier lifetime after gettering is about 2 µs, still much lower than that of the reference mc-Si.

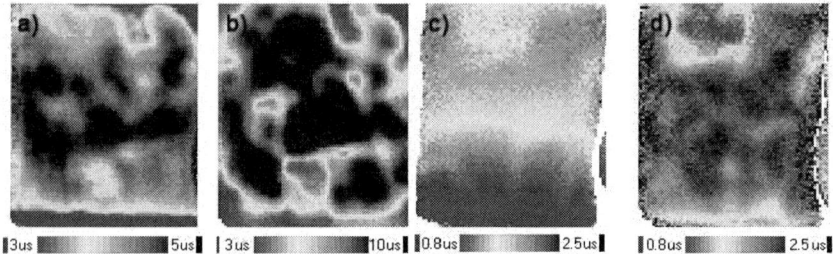

Figure 5.   Minority carrier lifetime maps before and after phosphorous in-diffusion gettering of UMG and reference mc-Si wafers. (a) reference mc-Si, as-grown (b) reference mc-Si, after gettering (c) UMG samples, as-grown (d) UMG samples, after gettering.

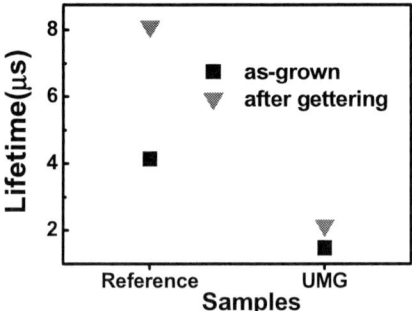

Figure 6. Minority carrier lifetime before and after phosphorous in-diffusion gettering of UMG and the reference mc-Si wafers.

Figure 7 presents the interstitial iron concentration in the UMG and reference samples before and after the phosphorous gettering. The concentration of interstitial iron was obviously reduced by gettering. For the UMG silicon, phosphorous gettering can remove metallic impurities to some extent, but is not effective to eliminate sufficiently the influence of the metallic impurities.

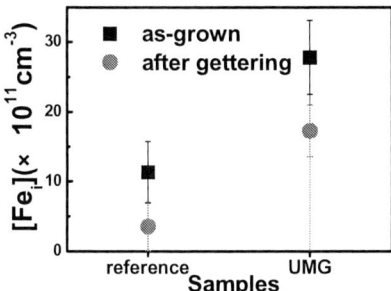

Figure 7. Interstitial iron concentration before and after phosphorous in-diffusion gettering of UMG and reference mc-Si wafers.

Gettering of metallic impurities involves three steps, which are dissolution of the metallic impurities, diffusion of the impurities and capturing in the gettering layer. [9] Firstly, there is a much higher concentration of metallic impurities in the raw material of the UMG wafers. Thus, during the casting process, there is a much higher density of metallic precipitates formed in the ingot. During the gettering process, it is very hard to totally dissolve these grown-in precipitates. Moreover, there a much higher concentration of dissolved metallic impurities in the UMG wafer, in particular interstitial iron. Furthermore, the RTP gettering process is rather short; there may not be enough time for some impurities to out-diffuse to the phosphorus in-diffused layer. Based on these facts, RTP gettering shows little effect on improving the electrical properties of UMG wafers. Further work is needed therefore to optimize use of UMG silicon.

Conclusion

The properties of UMG silicon wafers were observed by comparing them to those of standard mc-silicon wafers. It is found that the concentration of boron and phosphorous in UMG silicon is higher than that of the standard sample. The resistivity and minority carrier lifetime are lower for UMG silicon wafers while they contain higher interstitial iron content. The substitutional carbon and interstitial oxygen concentrations are approximately the same for the two types of silicon wafers. As for the dislocation concentration, also little differences exist between the UMG silicon and conventional mc-silicon samples. Based on these crystal properties, a phosphorous in-diffusion gettering has been applied to improve the quality of UMG samples. It is obviously that phosphorus gettering in UMG silicon can increase the minority carrier lifetime and reduce the interstitial iron concentration. However, the minority carrier lifetime is still not as high as the conventional silicon counterpart after gettering. The results will contribute to a better understanding of the properties and the future applications of UMG silicon in silicon solar cells.

Acknowledgments

This project is supported by the National Natural Science Foundation of China (No. 60906002 and 50832006). The authors also appreciate the valuable suggestions of Prof. Jan Vanhellemont.

References

[1] M.A. Green, J. Zhao, A. Wang, S.R. Wenham, Solar Energy Materials and Solar Cells, 65 (2001) 9-16.
[2] A.F.B. Braga, S.P. Moreira, P.R. Zampieri, J.M.G. Bacchin, P.R. Mei, Solar Energy Materials and Solar Cells, 92 (2008) 418-424.
[3] W.S.C. Li J F, China Solar PV report, China Environmental Science Press, Beijing, 2008.
[4] C. Alemany, C. Trassy, B. Pateyron, K.I. Li, Y. Delannoy, Solar Energy Materials and Solar Cells, 72 (2002) 41-48.
[5] J. Degoulange, I. P 開 ichaud, C. Trassy, S. Martinuzzi, Solar Energy Materials and Solar Cells, 92 (2008) 1269-1273.
[6] A.A. Istratov, T. Buonassisi, R.J. McDonald, A.R. Smith, R. Schindler, J.A. Rand, J.P. Kalejs, E.R. Weber, Journal of Applied Physics, 94 (2003) 6552-6559.
[7] S.A. McHugo, H. Hieslmair, E.R. Weber, Appl. Phys. A-Mater. Sci. Process., 64 (1997) 127-137.
[8] A. Bentzen, H. Tathgar, R. Kopecek, R. Sinton, A. Holt, in: Photovoltaic Specialists Conference, 2005. Conference Record of the Thirty-first IEEE, 2005, pp. 1074-1077.
[9] J.S. Kang, D.K. Schroder, Journal of Applied Physics, 65 (1989) 2974-2985.
[10] A. Cuevas, M. Stocks, S. Armand, M. Stuckings, A. Blakers, F. Ferrazza, Applied Physics Letters, 70 (1997) 1017-1019.
[11] M. Seibt, A. Sattler, C. Rudolf, O. Voß, V. Kveder, W. Schröter, physica status solidi (a), 203 (2006) 696-713.
[12] D. Mathiot, A. Lachiq, A. Slaoui, S. Noël, J.C. Muller, C. Dubois, Materials Science in Semiconductor Processing, 1 (1998) 231-236.
[13] D. Sarti, R. Einhaus, Solar Energy Materials and Solar Cells, 72 (2002) 27-40.
[14] R.N. Hall, Physical Review, 87 (1952) 387.
[15] G. Zoth, W. Bergholz, Journal of Applied Physics, 67 (1990) 6764-6771.
[16] O. Palais, S. Martinuzzi, J.J. Simon, Materials Science in Semiconductor Processing, 4 (2001) 27-29.

Defect Evaluation by Photoluminescence for Uniaxially Strained Si-On-Insulator

Dong Wang[a], Keisuke Yamamoto[b], Hongye Gao[c], Haigui Yang[a], and Hiroshi Nakashima[a]

[a] Art, Science and Technology Center for Cooperative Research, Kyushu University, 6-1 Kasuga-koen, Kasuga, Fukuoka 816-8580, Japan
[b] Interdisciplinary Graduate School of Engineering Sciences, Kyushu University, 6-1 Kasuga-koen, Kasuga, Fukuoka 816-8580, Japan
[c] Faculty of Engineering Sciences, Kyushu University, 6-1 Kasuga-koen, Kasuga, Fukuoka 816-8580, Japan

> Uniaxial strain was introduced to Si-on-insulator (SOI) substrate by SiN deposition using electron cyclotron resonance sputtering followed by gate-opening using lift-off technique. Then thermal treatments were performed at different temperatures. Strain-relaxation was observed by Raman spectroscopy. Photoluminescence (PL) was used to evaluate defects generated during strain-relaxation. Defect-related PL signal was observed for the thermally-treated strained channel. The intensity of defect-related PL signal increased with increasing annealing temperature. The energy position and profile of defect-related PL signal also varied with annealing temperature and SiN thickness.

1. Introduction

It is well known that carrier mobility can be effectively enhanced by lattice-strain for semiconductor materials (1). The strained-Si technology has been demonstrated for improving performance of metal-oxide-semiconductor field effect transistors (MOSFETs) (2). Compare with biaxial strain, uniaxial strain is more effective on carrier-mobility enhancement. Thus far, uniaxial strain technology has been practical for mass production of complementary metal-oxide-semiconductor devices (3). Since a MOSFET fabricated by Si-on-insulator (SOI) substrate has advantages of suppressed parasitic capacitance and leakage current, it is promising to apply uniaxial strain to a SOI substrate. To optimize strain technology, it is important to evaluate strain ratio and strain distribution, as well as strain-induced defects for a strained MOSFET-channel. To evaluate strain, Raman spectroscopy is a sensitive and nondestructive method (4). To fabricate a MOSFET with strained channel, after stressor fabrication, generally subsequent thermal process should be carried out, i.e., a gate processing. During the thermal process, strain relaxation and correspondingly defect-generation may happen. To evaluate defects generated by strain-relaxation, a nondestructive method is highly desired because a destructive method can alter strain and cause additional defect generation or transformation. In this work, photoluminescence (PL) method was used to evaluate strain-related defects for the strained channel fabricated by SOI, which is nondestructive and more sensitive compare with structural analysis methods (5).

2. Experimental

In this work, uniaxial strain was introduced to SOI substrate by SiN deposition using electron cyclotron resonance (ECR) sputtering followed by gate-opening using lift-off technique. The (100) surface-oriented SOI substrate was low-dose separation by implanted oxygen substrate with SOI thickness of 150 nm and buried oxide (BOX) thickness of 100 nm. The channel direction was taken along <110> direction. The ECR SiN-deposition was performed at room temperature with a Si target, for which the microwave power, the radio frequency power, the Ar/N_2 flow rates, and the chamber presser were 500 W, 500 W, 16/3 sccm, and 0.11 Pa, respectively, resulted in a deposition rate of approximately 7.5 nm/min. The residual compressive stress in SiN film was approximately 0.7 GPa. We deposited SiN films with varied thicknesses of 50, 110, 150, 190, and 230 nm. Cross-sectional and plan views of the fabricated sample structure are shown in Figs. 1(a) and 1(b), respectively. After SiN patterning, SiO_2 was fabricated on the strained channel by ECR deposition or dry oxidation, as shown in Figs. 1(c) and 1(e), respectively. The dry oxidation was performed at 900 °C for 30 min. The ECR SiO_2-deposition was performed at 130 °C with the microwave power, the radio frequency power, the Ar/O_2 flow rates, and the chamber presser of 500 W, 500 W, 16/8 sccm, and 0.14 Pa, respectively, resulted in a deposition rate of approximately 3.5 nm/min. Forming gas annealing (FGA) was performed at 400 °C for 30 min after ECR SiO_2-deposition to improve SiO_2 quality. Then thermal annealing was carried out in N_2 for the FGA-sample at temperatures of 500, 700 and 900 °C for 30min, as shown in Fig. 1(d). In this experiment, the SiO_2 formation plays roles in surface passivation for PL measurement and simulating the thermal process after stressor patterning for a real MOSFET fabrication.

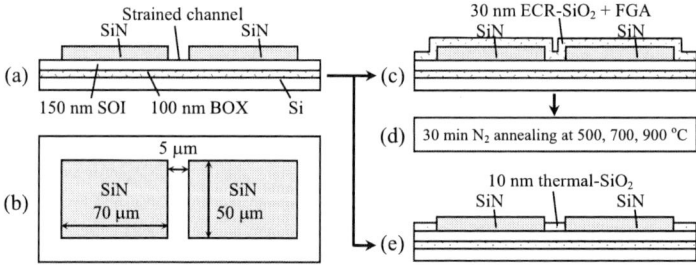

Figure 1. (a) cross-sectional view of the sample structure after SiN patterning; (b) plan view of the sample structure after SiN patterning; (c) cross-sectional view of the SiN-patterned sample followed by ECR SiO_2-deposition and FGA; (d) thermal annealing condition for the sample in Fig. 1(c); (e) cross-sectional view of the SiN-patterned sample followed by dry oxidation.

The micro-PL measurements were carried out using a continuous waved 378 nm UV line of a semiconductor laser focused on the sample surface with a power of 4 mW and a spot diameter of approximately 2 μm. The sample temperature was cooled to 8.5 K using a cryostat. In this work, only electron-hole-plasma (EHP) related PL signal could be observed for the SOI substrate (the part without SiN pattern), but not free-exciton (FE) related PL signal for the Si layer under BOX, indicated a penetration depth of 378 nm line less than SOI thickness of 150 nm (6). Although the diffusion length of FE is several micrometers in Si at 8.5 K (7), the BOX layer terminates FE diffusion. Therefore, PL should only receive contributions from the SOI layer. PL from the samples was dispersed by LabRAMHR-PL (Horiba Ltd.) system and detected by an InGaAs array sensor cooled at 180 K. We also evaluated strain for the samples by backscattering micro-Raman spectroscopy at room temperature, which was excited by continuous waved 532 nm-line of a YAG laser with a power of 5 mW and a spot diameter of approximately 1 μm. Raman scattering signal was also dispersed by the LabRAMHR-PL (Horiba Ltd.) system and detected by a CCD array sensor cooled at 200 K. We used the Si-TO (TO: transverse optical phonon) Raman line of the unstrained Si substrate as a reference and defined the line shift relative to it. It is possible to improve the accuracy of the Raman-shift measurement by using a shorter laser wavelength. However, we used the 532 nm-line of the YAG laser because it provides best stability and signal-noise ratio among all the available lasers in our laboratory. The actual strain ratio for the samples may slightly greater than the measured result. In this work, since we focused on strain relaxation, the variation of strain ratio is important but not the absolute values of strain ratio.

3. Results and discussion

Since the strained layer was very close to sample surface, PL measurement could be strongly influenced by the surface nonradiative recombination. The strained channel surface also can be easily influenced by surrounding environment, resulting in unstable surface states. Therefore, the PL measurement is not meaningful for the sample without surface passivation, i.e., the structure shown in Fig. 1(a). By forming surface SiO_2, as shown in Figs. 1(c) and 1(e), a very good surface passivation can be realized. PL measurements showed very strong and similar band-band transition signal intensities for the position far away (> 50 μm) from the SiN pattern for these samples, implied effective and uniform surface passivation.

First, we focused on the sample with 230 nm-thick SiN deposition, which has the maximum strain ratio among all the samples. Two thermal treatments were performed: ECR-SiO_2 deposition followed by FGA at 400 ^{0}C and dry oxidation at 900 ^{o}C. These two temperatures approximately represent typical minimum and maximum temperatures during gate process for a modern MOSFET fabrication. Figure 2(a) shows cross-sectional structure of the sample, for which Raman-shift was measured in details, as shown in Fig. 2(b). Compressive strain was observed for the strained channel. The sample with dry oxidation showed lower strain ratio, implied stronger strain relaxation, for which a clear and broad signal was observed at around 1.0 eV, as shown in Fig. 2(c). This signal was confirmed to be defect-related by observing its blue-shift with increasing laser power. In this experiment, for the PL observation, the laser beam always focused on the center of strained channel.

Under the strained channel, SOI may not be completely strained. Therefore, both strained and unstrained parts could coexist, as illustrated by the inset of Fig. 2(c). The excitons generated in the strained part would drift toward sample surface by the strain-induced band-gap gradient (8). In general, these excitons would mainly recombine in three ways: 1) to be bound to defects and recombine by emitting defect-related PL signal; 2) recombine by nonradiative surface recombination; and 3) recombine by band-band transition. Therefore, the signal intensity of band-band transition for the strained-part depends on densities of lattice defects and surface states. On the other hand, the excitons generated in unstrained part would freely diffuse to any direction, which would have great chance to recombine as band-band transition in the high-crystallinity unstrained-part.

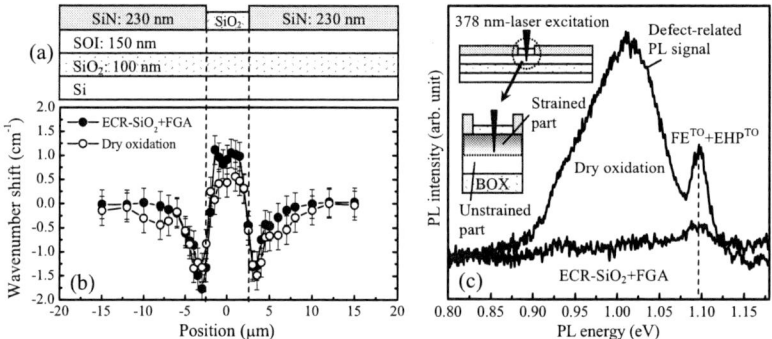

Figure 2. (a) Cross-sectional structure for the sample with 230 nm-thick SiN deposition; (b) strain distribution measured by Raman spectroscopy for the samples with different thermal treatment; (c) PL signals for the strained-channel center of samples in Fig. 1(b). Defect-related PL signal was clearly observed for the sample with 900 °C dry oxidation, but not observed for the sample with ECR-SiO$_2$ deposition followed by FGA at 400 °C. FE^{TO} and EHP^{TO}: TO assisted band-band recombination for FE and EHP, respectively.

From the PL signal for the sample with FGA shown in Fig. 2(c), it can be found that: 1) since the band-band transition signal (FE^{TO}+EHP^{TO}, see the caption of Fig. 2) has very weak intensity, there was almost no exciton generated in the unstrained part, implied a penetration depth of 378 nm-line smaller than the strained-part thickness; 2) the defect-related luminescence centers have small density in our observation range from 0.8 eV to band-gap, implied low lattice-defect density; 3) most of the excitons recombined by surface nonradiative recombination because they would drift toward sample surface by the strain-induced band-gap gradient. As to the sample with dry oxidation, strain was further relaxed, resulted in thinner strained-part thickness. Therefore, some of the 378 nm-line penetrated through the strained part and generated excitons in the unstrained part, resulted in a relatively strong intensity of the band-band transition signal. The strong strain-relaxation should induce high-density defects, which caused a very clear and broad defect-related signal at round 1.0 eV.

Second, we studied defect generation dependence on thermal history for 230 nm-SiN deposited sample. Four ECR-SiO$_2$-deposited samples were measured by PL, which were the sample just after FGA, and the samples with FGA followed by N$_2$ annaling at 500, 700, and 900 °C, respectively. The Raman-shifts for the strained-channel center of these samples were measured, as shown in Fig. 3(a), where the Raman-shift for the as SiN-patterned sample (see Fig. 1(a)) is also shown. It is very clear that the strain gradually relaxed by increasing annealing temperature (T_A). Along with the strain relaxation, defects also gradually generated, which was confirmed by PL observation, as shown in Fig. 3(b). It was found that the defect density remained relatively low value for the sample with T_A of 500 °C and drastically increased for the samples with T_A greater than 700 °C. With increasing defect density, energy position of the defect-signal goes to blue-direction, implying defect transformation. For the sample with T_A of 900 °C, defect-signal showed very similar profile and intensity to those of the sample with dry oxidation, as shown in Fig. 2(c), due to the similar strength of thermal treatment for them. The increase in signal intensity of band-band transition should originate from decrease in strained-part thickness, which has been discussed before.

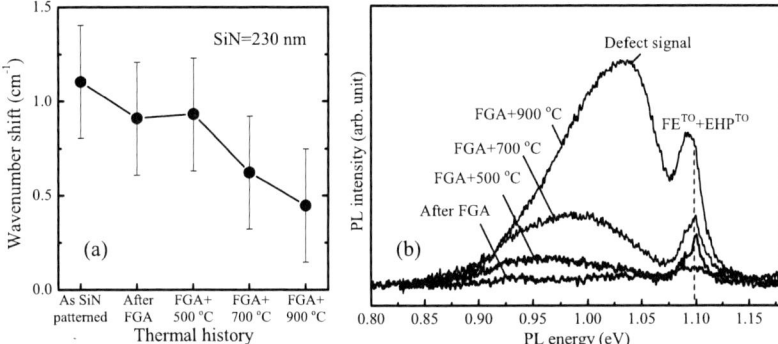

Figure 3. (a) Raman shifts for the strained-channel center of samples with 230 nm-SiN deposition and different thermal history. FGA+500 °C, FGA+700 °C, and FGA+900 °C represent the sample with FGA followed by 500, 700 and 900 °C annealing in N_2 for 30 min, respectively; (b) PL signals for the strained-channel center of samples in Fig. 3(a).

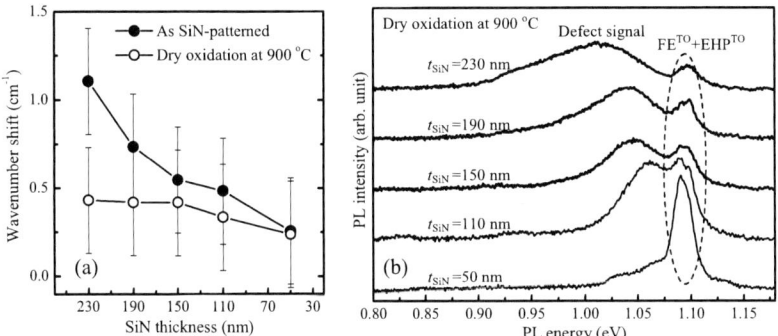

Figure 4. (a) Raman shifts for the strained-channel center of samples with different SiN-thickness, for which both as-SiN-patterned samples and dry-oxidized samples were measured; (b) PL signals for the strained-channel center of the samples with dry oxidation in Fig. 4(a).

Finally, we studied SiN-thickness (t_{SiN}) effect on defect generation. Figure 4(a) shows Raman shifts for the strained-channel center of samples with different t_{SiN}, for which both as-SiN-patterned samples and dry-oxidized samples were measured. It was found that strain relaxation ratio decreased with decreasing t_{SiN}, which is reasonable because the smaller t_{SiN}, the lower stress in SOI. For the sample with minimum t_{SiN} of 50 nm, strain relaxation was difficultly observed. The PL results are shown in Fig. 4(b) for the samples with dry oxidation. Defect signal intensity was very weak for the sample with t_{SiN} of 50 nm, which consisted with the Raman-shift result. The integration intensity of defect signal increased with increasing t_{SiN}, which positively depends on strain relaxation ratio. However, the energy position of defect signal goes to red-direction with increasing t_{SiN}, simultaneously the peak width becomes broad, implied varied defect composition.

4. Summary

Uniaxial strain was introduced to SOI substrate by SiN deposition using ECR sputtering followed by gate-opening using lift-off technique. Surface SiO_2-passivation was fabricated on the strained channel by ECR deposition or dry oxidation. FGA and N_2 annealing was also performed for ECR-SiO_2 deposited sample. Strain-relaxation was observed by Raman spectroscopy for the thermally-treated samples, for which defect-related PL signals were observed. The intensity of defect signal positively depends on thermal treatment temperature and SiN thickness. For the sample with 230 nm-SiN deposition, the defect density remained relatively low value after N_2 annealing at 500 °C. There was almost no defect signal observed for the sample with 50 nm-SiN deposition after dry oxidation at 900 °C. Defect transformation was also confirmed during strain relaxation.

Acknowledgments

This study was partially supported by JSPS Grant-in-Aid for Young Scientists (B) (21760011) and a grant of the Knowledge Cluster Initiative implemented by MEXT of Japan.

References

1. Y. Sun, S. E. Thompson, and T. Nishida, *J. Appl. Phys.*, **101**, 104503 (2007).
2. S. Takagi, J. L. Hoyt, J. J. Welser, and J. F. Gibbons, *J. Appl. Phys.*, **80**, 1567 (1996).
3. S. E. Thompson et al., *IEEE Electron Device Lett.*, **25**, 191 (2004).
4. M. Takei, D. Kosemura, K. Nagata, H. Akamatsu, S. Mayuzumi, S. Yamakawa, H. Wakabayashi, and A. Ogura, *J. Appl. Phys.*, **107**, 124507 (2010).
5. D. Wang, S. Ii, H. Nakashima, K. Ikeda, H. Nakashima, K. Matsumoto, and M. Nakamae, *Appl. Phys. Lett.*, **89**, 041916 (2006).
6. M. Tajima and S. Ibuka, *J. Appl. Phys.*, **84**, 2224 (1998).
7. N. Usami, K. Leo, and Y. Shiraki, *J. Appl. Phys.*, **85**, 2363 (1999).
8. D. Wang, H. Yang, T. Kitamura, and H. Nakashima, *J. Appl. Phys.*, **107**, 033511 (2010).

Effects of Transverse Magnetic Field on Thermal Fluctuations in the Melt of a Cz-Si Crystal Growth

Xin Liu, Lijun Liu [1], Yuan Wang

School of Energy and Power Engineering, Xi'an Jiaotong University, Xi'an, Shaanxi 710049, China

Abstract: Three dimensional (3D) unsteady computations were carried out to understand the effects of transverse magnetic field (TMF) on thermal fluctuations in the melt of a 300 mm Cz-Si crystal growth. A developed LES program with the dynamic SGS model was employed to predict the melt turbulence with or without TMF. The effects of TMF on statistical behaviors of the melt convection were studied. It was found that the thermal fluctuations in the crystallization zone decreased significantly when the TMF was applied. This indicates that the TMF can suppress the flow instability in the melt of the Cz–Si crystal growth.

1. Introduction

Nowadays, the Czochralski (Cz) process is the dominant technique for the production of bulk single crystals of a wide range of electronic and photovoltaic silicon. The melt flow in the crucible has significant effects on the formation of micro-defects and impurities concentration through convective heat and mass transfer. With the scaling-up of crystal, the melt flow in crucible of the industrial Cz-Si growth is characterized by turbulent large-scale velocity and thermal fluctuations [1]. With a fluctuating local heat transfer, a point on the melt-crystal (m-c) interface may alternate between periods of growth and remelting, leading to large densities of dislocations and other micro-defects in the crystal. Therefore, the simulation of melt turbulence in industrial Cz-Si growth processes has become a prerequisite for an economical and timely development of new or improved processes.

Application of static magnetic fields in Cz-Si growth is an effective method for controlling the melt turbulence in an industrial-scale crucible and therefore for improving crystal quality. The main effects of static magnetic fields are the damping of velocity and thermal fluctuations and the suppression of the impurities transport from the crucible to the crystal. They also control impurities and inhomogeneities at microscopic levels by producing better conditions in the vicinity of the m-c interface. Currently, the most frequently applied field is probably the transverse magnetic field (TMF). TMF particularly suppresses the melt flow in the vertical direction. In this aspect, many research works [2-5] have been published. However, a thorough understanding of the thermal fluctuations is still missing in industrial Cz-Si growth with TMF due to that the magnetohydrodynamic (MHD) effects give an additional complexity on this turbulent melt convection problem. Owing to the asymmetry of the field and the turbulence nature of the melt flow in industrial Cz-Si growth, the melt turbulence in a real Cz-Si growth can only be described by 3D calculations with some feasible turbulence models. It appears that the Large Eddy Simulation (LES) technique with a moderate number of

---

[1] Corresponding author: tel./ fax: 86-29-82663443, email: ljliu@mail.xjtu.edu.cn

computational grids can be a compromise choice between the accuracy of Direct Numerical Simulation (DNS) and the efficiency of Reynolds-Averaged Navier–Stokes (RANS) methods [6]. Some pioneer works of LES for the Cz-Si melt convection in static magnetic fields have been conducted by using the classical Smagorinsky subgrid-scale (SGS) stress model, in which the model coefficient was empirically assigned [7-8]. Such a SGS model with empirical coefficients can't guarantee its universality for various flow problems. On the other hand, MHD effects on the statistical characteristics of melt turbulence in a Cz-Si crystal growth haven't been assessed in detail.

In this work, 3D unsteady computations were carried out to understand the effects of TMF on thermal fluctuations in the melt of a 300 mm Cz-Si crystal growth. A developed LES program with the dynamic SGS model for body-fitted grids was employed to predict the turbulent melt convection. The effects of TMF on dynamic behaviors and statistic features of the melt convection were studied.

2. Model formulation

The melt flow is calculated by solving the 3D time-dependent equations of mass, momentum and heat conservation applying the Boussinesq approximation for an incompressible Newtonian fluid. These equations are filtered implicitly in space by a second-order, finite-volume solution methodology which is equivalent to the box filtering [9]. The filtered equations for the resolvable scales of melt turbulence in the TMF could be written as follows in a rotating reference frame:

$$\nabla \cdot \overline{\mathbf{V}} = 0,\qquad(1)$$

$$\rho \frac{D\overline{\mathbf{V}}}{Dt} = -\nabla \overline{p} + \nabla \cdot \left[\mu_{eff} \left(\nabla \overline{\mathbf{V}} + \nabla \overline{\mathbf{V}}^{T} \right) \right] - \rho \mathbf{g} \beta_{T} \left(\overline{T} - T_{0} \right) - 2\rho \Omega \times \overline{\mathbf{V}} + \overline{\mathbf{J}} \times \overline{\mathbf{B}},\qquad(2)$$

$$\frac{D\overline{T}}{Dt} = \nabla \cdot \left(k_{eff} \nabla \overline{T} \right),\qquad(3)$$

where $\overline{\mathbf{V}}, \overline{p}, \overline{T}, \overline{\mathbf{J}}$ and $\overline{\mathbf{B}}$ are the melt velocity, melt pressure, melt temperature, electrical current density and magnetic flux density, respectively. The overbar denotes the implicit grid filter operation. Ω is the angular velocity of the reference system. The last term in Eq. (2) is the Lorentz force due to the applied magnetic field. The effective eddy viscosity and the effective thermal diffusivity are defined as $\mu_{eff} = \mu + \mu_{SGS}$ and $k_{eff} = k + \mu_{SGS} / \sigma_{SGS}$. The SGS eddy viscosity μ_{SGS} is calculated with the dynamic SGS model and σ_{SGS} is the SGS turbulent Prandtl number, which was taken as 0.9. At the solid walls, μ_{SGS} were estimated with a generalized three layer wall function [10].

The dynamic eddy-viscosity relationship is given as:

$$\mu_{SGS} = 2\rho C_{D} \left(\mathbf{x}, t \right) \overline{\Delta}^{2} \left| \overline{\mathbf{S}} \right|,\qquad(4)$$

where the grid-filter scale is taken as $\overline{\Delta} = \left(Volume \right)^{1/3}$ in the curvilinear grids. The magnitude of the resolvable strain rate tensor is $\left| \overline{\mathbf{S}} \right| = \sqrt{2\overline{S}_{ij}\overline{S}_{ij}}$ and $\overline{S}_{ij} = \left(\partial \overline{u}_{i} / \partial x_{j} + \partial \overline{u}_{j} / \partial x_{i} \right)/2$. The dynamic model coefficient $C_{D} \left(\mathbf{x}, t \right)$ was evaluated basing on the procedure proposed by Germano [11] and Lily [12]. The construction and validation of this dynamic SGS model for body-fitted grids were described in our previous paper [13].

The m-c interface geometry and the thermal boundary conditions at the crucible wall were obtained from a 2D axisymmetric model of global heat transfer [14]. The non-slip conditions were applied for velocities at solid boundaries. At the melt free surface, the radiation loss is taken into account by the Stefan-Boltzmann equation. The shear stress is balanced with the surface tension induced by the radial temperature gradient.

In order to obtain the Lorentz force exerted on the melt, the electromagnetic field in the melt domain was solved. On all of the external boundaries of the melt domain, the non-penetration conditions for electrical current density were imposed.

3. Results and discussions

The computational domain of LES is shown in Fig. 1. The magnetic field is homogeneous with an intensity of 0.4 T oriented in the x-direction. The rotation rates of crucible and crystal are $\omega_C = -4\,\text{rpm}$ and $\omega_S = 8\,\text{rpm}$, respectively. A grid number of 175,000 and a time step of 0.02 s were used.

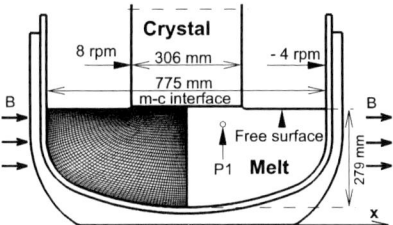

Fig. 1 Hot zone of an industrial-scale Cz-Si growth and the computational domain of LES.

3.1 TMF effects on the time-averaged fields
To analyze the statistical behaviors of the melt flow, the velocity and temperature fields were averaged for a long time period. The comparison of the time-averaged flow fields between the case with a TMF and the case without a TMF is shown in Fig. 2. An additional averaging in the circumferential direction was conducted for presenting different quantities in the meridional plane. In this figure, a cellular motion with '+' corresponds to a clock-wise motion and '-' corresponds to a counter clock-wise motion. Contours of stream function for B=0 revealed three major vortexes in the radial direction in Fig. 2 (a). The strongest vortex induced by thermal buoyancy was pushed to the sidewall by the crucible rotation. The dissolution rate of oxygen from the wetted crucible surface may be enhanced by this convective transport. This flow structure resembles closely with the result of Raufeisen [15] who conducted DNS for the melt flow in a cylindrical crucible with a diameter of 340 mm. When the TMF of 0.4 T was applied, two vortexes emerge in the vertical direction, as shown in Fig. 2 (b). This is due to the fact that the melt flow in the vertical direction was suppressed significantly by the Lorentz force induced by the TMF. This suppression effect on the buoyant convection near the crucible sidewall can reduce the oxygen dissolution from the crucible. It is favorable for the reduction of oxygen concentration in the Cz-Si crystals.

Fig. 3 presents the comparison of the time-averaged temperature distributions at the melt free surface. For the case absent of TMF, as shown in Fig. 3 (a), a non-axisymmetric temperature distribution could be found even with axisymmetric thermal boundary conditions at the crucible wall. It demonstrates that the geostrophic turbulence is generated in the melt due to the rotation of the large diameter crucible [16]. With introduction of the TMF, the temperature distribution at the melt surface changes to a quite regular and stable pattern with a rotating axisymmetric feature, as shown in Fig. 3(b).

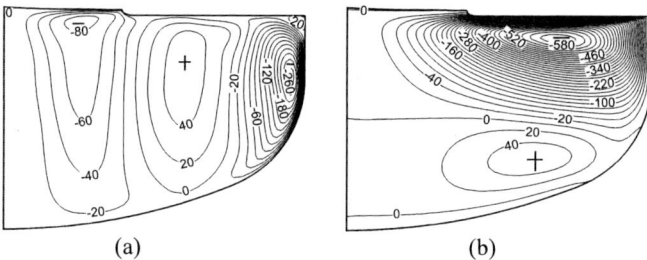

(a)   (b)

Fig. 2. The time-averaged stream functions in the meridional plane. (a) B=0. (b) B= 0.4 T. Contour lines are plotted every 20 g/s.

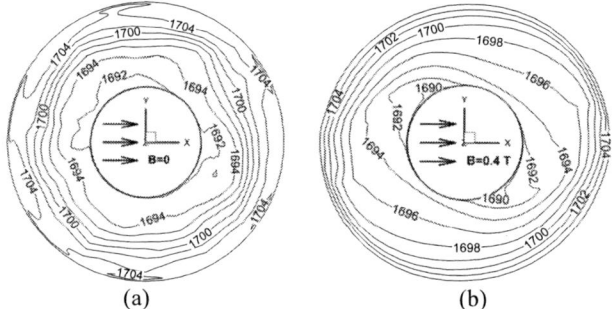

(a)   (b)

Fig. 3. The time-averaged temperature distributions at the melt surface. B=0. (b) B= 0.4 T. Isothermals are plotted every 2 K.

3.2 TMF effects on the thermal fluctuation field

Fig. 4 presents the comparison of the thermal fluctuations in the meridional plane, which were averaged in the circumferential direction. The maximal thermal fluctuation in the crystallization zone under the m-c interface is 3.9 K without TMF, as shown in Fig. 4(a). This may be one of the important origins of growth striations and dislocations in the Cz-Si crystal. When the TMF was applied, the maximal thermal fluctuation in the crystallization zone decreased to 1.2 K as shown in Fig. 4(b). It is favorable to reduce undesirable striations and large dislocation densities in grown crystals. Thermal fluctuations in the meridional plane are wakened significantly by TMF effects. Strong thermal fluctuations were only found at the meeting interface of the two time-averaged vortexes with a maximum of 3.6 K.

The comparison of thermal fluctuation distributions at the melt free surface are presented in Fig. 5. Strong thermal fluctuations were found in Fig. 5 (a) for the case absent of TMF. The maximal value is 6.5 K. In the case with the TMF, thermal fluctuations at the melt free surface were suppressed significantly, as shown in Fig. 5(b). The flow instability caused by the geostrophic turbulence was damped significantly by the TMF. Its distribution is rotating axisymmetric, similar to the time–averaged temperature distribution shown in Fig. 3(b).

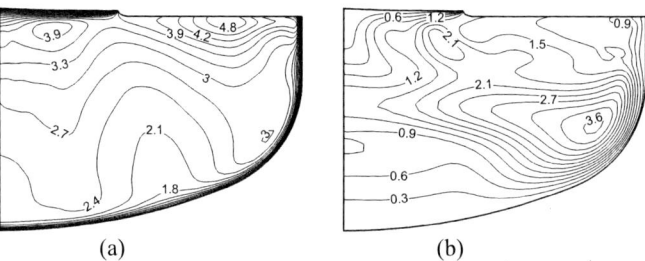

(a)   (b)

Fig. 4. The thermal fluctuations in the meridional plane. (a) B=0. (b) B= 0.4 T. Contour lines are plotted every 0.3 K.

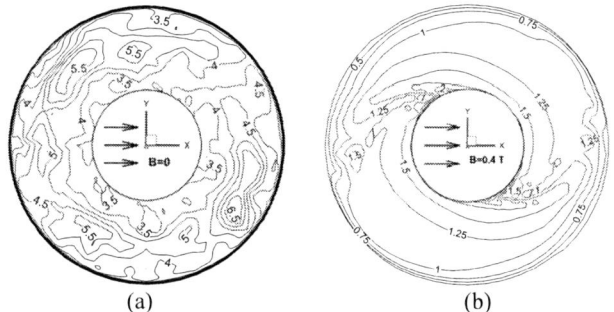

(a)   (b)

Fig. 5. The thermal fluctuations at the free surface. (a) B=0. (b) B= 0.4 T. Contour lines are plotted every 0.5 K for (a) and 0.25 K for (b).

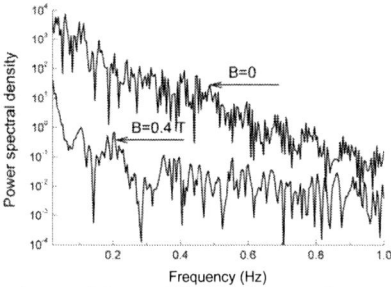

Fig. 6. Spectral analyses of thermal fluctuations at the reading point P1.

3.3 TMF effects on the spectral characteristics of thermal fluctuations

Temperatures at a reading point in the crystallization zone were sampled for spectral analysis of thermal fluctuations. The reading point is located 40 mm below the m-c interface at radius R=100 mm, as indicted in Fig. 1. Spectral analyses for thermal fluctuations at P1 for B=0 and B=0.4 T are presented in Fig. 6. We can see that the thermal fluctuations were damped dramatically by the TMF effects. The power spectral density in the low-frequency band decreased over two orders.

4. Conclusion

The effects of TMF on thermal fluctuations in the melt of a 300 mm Cz-Si crystal growth were investigated numerically by LES method. The comparisons of the time-averaged fields indicate that the Lorentz force induced by TMF could suppress the thermal buoyant convection in the melt. The geostrophic turbulence in the azimuthal plane could be suppressed significantly by the introduction of TMF. The thermal fluctuations in the crystallization zone were also damped effectively by TMF. It reveals the stabilizing effects of statistic magnetic fields on the melt turbulence in the industrial Cz-Si growth. Optimization with respect to the combination of the magnetic field intensity and the rotation rate of crucible may create more favorable growth conditions for an industrial Cz-Si process.

Acknowledgments

This research was supported by NSFC (No. 50876084), NCET-08-0442 , RFDP (No. 20100201110016) and Fundamental Research Funds for the Central Universities of China.

References

1. J. Virbulis, T. Wetzel and E. Tomzig et al., *Mat. Sci. Semicon. Proc.*, **5**(4-5), 353, (2002).
2. K. Kakimoto, *J. Cryst. Growth*, **230**(1-2), 100, (2001).
3. D. Vizman, J. Friedrich and G. Müller, *J. Cryst. Growth*, **230**(1-2), 73, (2001).
4. A. Krauze, A. Muižnieks and A. Mühlbauer et al., *J. Cryst. Growth*, **262**(1-4), 157, (2004).
5. A. Muiznieks, A. Krauze and B. Nacke, *J. Cryst. Growth*, **303**(1), 211, (2007).
6. G. Müller and J. Friedrich, *J. Cryst. Growth*, **266**(1-3), 1, (2004).
7. N. Ivanov, A. Korsakov and E. Smirnov et al., *J. Cryst. Growth*, **250**(1-2), 183, (2003).
8. I. Evstratov, V. Kalaev and A. Zhmakin et al., *J. Cryst. Growth*, **237-239**(3), 1757, (2002).
9. S. Jordan, *Int. J. Heat Fluid Fl.*, **23**(1), 1, (2002).
10. I. Evstratov, V. Kalaev and A. Zhmakin et al., *J. Cryst. Growth*, **230**(1-2), 22, (2001).
11. M. Germano, U. Piomelli and P. Moin et al., *Physics of Fluids A: Fluid Dynamics*, **3**(7), 1760, (1991).
12. D. Lilly, *Physics of Fluids A: Fluid Dynamics*, **4**(3), 633, (1992).
13. X. Liu, L. Liu and K. Kakimoto, *The 16th International Conference on Crystal Growth (ICCG-16)*, August 8-13, 2010, Beijing, China.
14. L. Liu and K. Kakimoto, *Int. J. Heat Mass Tran.*, **48**(21-22), 4481, (2005).
15. A. Raufeisen, M. Breuer and T. Botsch et al., *Int. J. Heat Mass Tran.*, **51**(25-26), 6219, (2008).
16. Y. Kishida and K. Okazawa, *J. Cryst. Growth*, **198-199**(1), 135, (1999).

ECS Transactions, 34 (1) 1129-1134 (2011)
10.1149/1.3567725 ©The Electrochemical Society

Light Trapping for High Efficiency Heterojunction Crystalline Si Solar Cells

Qi Wang, Yueqin Xu, Eugene Iwaniczko, and Matthew Page

National Renewable Energy Laboratory. Golden, CO 80401, USA

Light trapping plays an important role to achieve high short circuit current density (J_{sc}) and high efficiency for amorphous/crystalline Si heterojunction solar cells. Si heterojunction uses hydrogenated amorphous Si for emitter and back contact. This structure of solar cell posses highest open circuit voltage of 0.747 V at one sun for c-Si based solar cells. It also suggests that over 25% record-high efficiency is possible with further improvement of J_{sc}. Light trapping has two important tasks. The first one is to reduce the surface reflectance of light to zero for the solar spectrum that Si has a response. The second one is to increase the effective absorption length to capture all the photon. For Si heterojunction solar cell, surface texturing, anti-reflectance indium tin oxides (ITO) layer at the front and back are the key area to improve the light trapping.

Introduction

Crystalline Si solar cells with hydrogenated amorphous silicon (a-Si:H) emitters and back contacts (1-9) possess high open-circuit voltage and high efficiency in addition to low temperature (<250°C) and manufacturable processes. The best cell efficiency has reached 23% from Sanyo R&D lab (1). High open-circuit voltage of over 0.700 V is a key to obtain low temperature coefficient of the power (2). Processing temperature below 250°C in crystalline Si (c-Si) solar cell production enables a low thermal budget and avoids bowing of thin wafers: this is a promising approach for the future thin c-Si wafer manufacturing. To further improve the performance of Si heterojunction solar cell, light trapping will be essential. Si heterojunction solar cell suffers low short circuit current density in compared to the other high performance c-Si solar cell. The difference may come from the absorption of amorphous Si and ITO layers, and light trapping scheme. In this paper, we will present our recent process and understanding in surface texturing and anti-reflectance layer for high efficiency Si heterojunction solar cells. Random pyramidal surface texturing of c-Si is one of the effective means. With the help of ITO layer, the reflectance at optimized surface can be reduce to less the 4% for a wide range spectrum that Si has a response.

Experimental

High quality float-zone (FZ) both n-type and p-type crystal Si wafers (either polished or anisotropically textured) were used for high-efficiency cell development and light trapping study. The Si wafers are (100) orientated, about 250 μm in thickness, and 1-4 $\Omega \cdot$cm in resistivity. The minority carrier lifetime is in the order of 1 ms measured by the

1129

Sinton PCD lifetime tester. The c-Si wafer is either polished or textured. KOH with IPA were used for random pyramid texturing. ITO was thermally evaporated from 90% In and 10% Sn source in the presence of O_2 atmosphere. The reflectance measurements were made using an n&k analyzer model 1280 from the n & k Technology, Inc. and Cary G5 for integrated reflectance measurement. Scanning electron microscopy (SEM) was used to study the structure of the textured surfaces. All Si film layers were deposited using the hot-wire CVD process. We use the multi-chamber T-system (11) at NREL to fabricate the intrinsic passivation layer, the emitter and back a-Si:H contacts and a final 2-5% HF cleaning before being loaded into a HWCVD a-Si:H deposition chamber.

Results and Discussions

Both crystalline and amorphous Si material have a highly reflective surface, more than 35% of visible light reflects off the surface because of its high index of refraction. Reflected light translates to a direct loss in the solar cell performance, especially in the short circuit current. To recover this loss, many light trapping methods are applied. Anti-reflectance coatings and surface texturing are a few very effective means.

Figure 1 shows a light-trapping scheme for c-Si solar cell. The Si material is sandwiched between two optical dielectric layers. When the light is shining on the front surface, light will be partially reflected. Choosing the proper index of optical layer and adjusting its thickness can minimize the reflected light. For the light through the Si, the light can have a total internal reflectance on the other side surface of Si so that the light will bounce more than twice and increase the effective absorption length. Figure 2 shows the photo-generated current density as a function of Si thickness. The maximum J_{sc} for an infinite thickness of Si is 44 mA/cm^2. One can use this chart to figure out the effective absorption length.

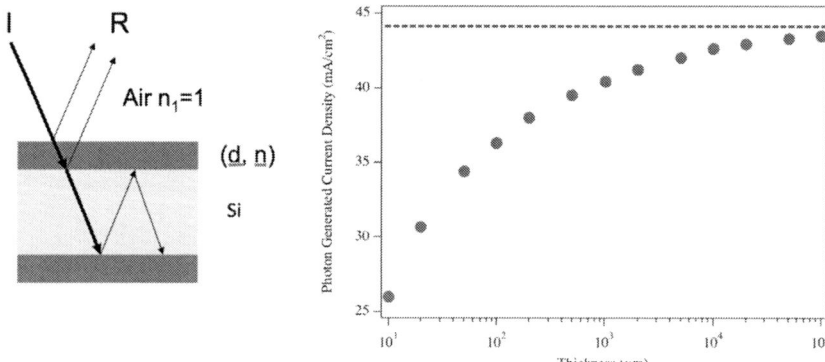

Figure 1, A scheme of the light trapping for c-Si solar cell.

Figure 2, Photon generated current density as a function of c-Si thickness. The current density is calculated from the integration of standard solar flux at one sun with assumption of 100% QE.

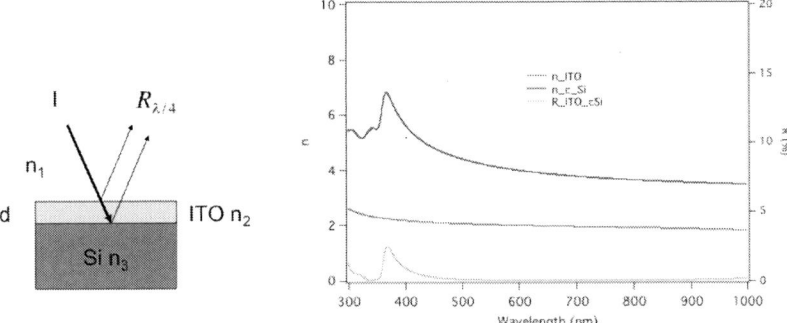

Figure 3. Schematic of single layer anti-reflectance on c-Si.

Figure 4. The index of refraction of c-Si and ITO, and calculated R a quarter wavelength as a function of wavelength.

For example, for a Si solar cell with a thickness of 200 μm, if the J_{sc} with the light trapping is at 40 mA/cm², the effective length will be equal to 700 μm, a gain of 3.5. The best reported J_{sc} is at 43 mA/cm², a 98% capture of all the photons for c-Si materials. However, the best short circuit current density for a c-Si heterojucntion solar cell is at 39 mA/cm², which is only 89% capture of all the photons. There is still a room for further improvement of J_{sc} for a heterojunction c-Si solar cell.

To recover this loss of reflectance at the front surface, anti-reflectance coating is applied. We will present our study of ITO as the anti-reflectance coating and its optimization to the Si heterojunction solar cells. Figure 3 shows the structure of a single ITO layer to c-Si to reduce the reflectance. Optically, light coming from the air will bounce at the ITO surface and the c-Si surface. By adjusting the thickness of ITO layer, the reflectance can be reduced. At the optimized condition, the reflectance can be near zero. From Equation 1, it is clear that the reflectance (R) will be at zero if n_2 = square root of n_1 x n_3, where, n_1 is the index of air, n_2 is the index of ITO, and n_3 is the index of c-Si.

$$R_{\frac{\lambda}{4}} = \frac{(n_1 n_3 - n_2^2)^2}{(n_1 n_3 + n_2^2)^2} \qquad [1]$$

Figure 4 plots the index of ITO and c-Si as a function of wavelength. Fortunately, the reflectance of quarter wavelength at the structure of ITO layer on c-Si, according to equation 1, can be very smaller for wide range of wavelength except the short wavelength, around 380 nm, the highest reflectance is less than 3%, which is still much smaller than 30% reflectance of c-Si surface.

Experimentally, we deposit various ITO thicknesses from 515 to 1039 nm to c-Si and measure the reflectance. For every ITO layer, there is a minimum reflectance at certain wavelength. Figure 5 shows the reflectance of two ITO thicknesses as a function of wavelength. We also checked the relationship between the wavelength at minimum reflectance and ITO thickness and it does follow equation 2.

Another interesting result is that the ITO thickness on c-Si is color sensitive. The calibration of color to the ITO thickness is inserted in the Figure 5. In general, the thinner ITO shows a brown color, and thicker ITO layer appears light blue. Most c-Si solar cells show a dark blue or purple appearance, which corresponding to the ITO thickness around 720 nm. In the later section, we will explain why most cells appear dark blue color.

$$\lambda_{min} = 4n(\lambda)d$$

[2]

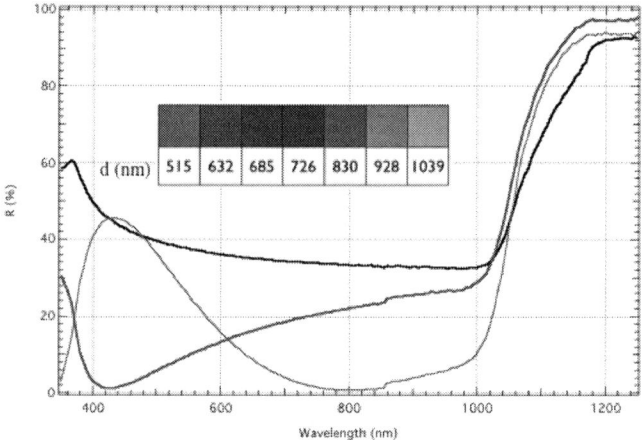

Figure 5. Reflectance of c-Si and two ITO thicknesses (460 nm and 930 nm) on c-Si as a function of wavelength. Insert is sample color that appears from the various thicknesses of ITO on planar c-Si wafer.

One caveat of single layer ITO on c-Si is that there is only one minimum in the solar spectrum. An ideal solar cell would have a zero reflectance in the wide range of wavelength, for example, from 300 to 1200 nm for c-Si based solar cells. Applying a single layer anti-reflection coating cannot meet the goal. Therefore, double layers anti-reflectance coating (ARC) will have a better result than a single layer ARC. There are many references about the double-layer coating, interested readers can refer to the reference (12).

Surface texturing is other technique used to reduce the surface reflectance. For c-Si, KOH etching with IPA can very effectively achieve random pyramid surfaces. A SEM picture of a textured c-Si surface is insert in Figure 6. The surface is populated with various sizes of small pyramids. The largest feature can be at 10 μm. The facet of each pyramid is (111) oriented. In addition, amorphous Si layer can be conformal deposited on the textured surfaces (10). Cleaning and conformal deposition of a-Si:H are the key components that enable the use of random pyramidal texturing for Si heterojunction solar cells.

Figure 6 shows the results of reflectance from various surfaces. With this random pyramidal surface, the reflectance is reduced to only 15% from 35% on a flat surface. With an ITO coated layer, the reflectance can be further reduced to only less than 4%. This is much better result than single layer ITO on planar c-Si in Figure 5. Therefore, by texturing the front surface with anti-reflection coating or applying double layer anti-reflectance coating, the reflectance can be reduced. Another approach to reducing the reflectance is to use a graded-index of Si. The reflectance is reduced to less than 1% without anti-reflection layer. The graded index layer with nano-size cones are made either by chemical or dry etching or photolithographic process.

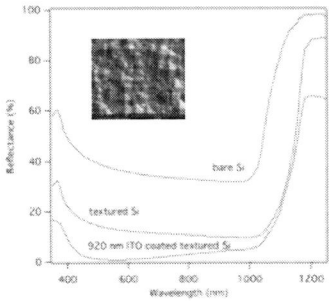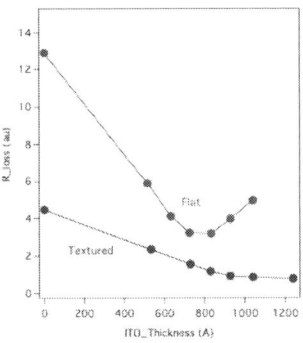

Figure 6, Reflectance of ITO coated textured c-Si, textured c-Si and bare c-Si.

Figure 7. Reflectance loss as a function of ITO thickness for ITO on planar and textured c-Si.

We use the integration of the reflectance as a loss over the solar spectrum at AM 1.5 to optimize the ITO thickness. In Figure 5, we show that, for a thin ITO, one can reduce the blue loss but not in the red; one can reduce the red loss but not in the blue. By integrating the reflectance over the visible spectrum, we can find the optimal thickness of ITO to achieve the best spectral response from the solar cell. Figure 7 shows the summary plot of relative loss from reflectance for ITO on planar and textured c-Si as a function of ITO thickness. Clearly, the optimized ITO thickness is at 730 nm for ITO on planar c-Si. This is the thickness that corresponds to a dark blue color in Figure 3. Most c-Si solar cells should appear in the dark blue following an optimize thickness for single layer anti-reflectance coating. For ITO on textured c-Si, we found that there is a broad minimum, we use an ITO thickness of 900 nm. But for thicker ITO than 900 nm, it does not seem to harm the cell's photon collection.

For photons reach the back of c-Si, they can be totally reflected if the incident angle greater than internal reflectance angle defined by the c-Si and optical layer such as ITO. For a normal incident light on textured surface, it changes its path at the front pyramidal surface. Internal reflectance increases the effective absorption length so that thinner Si wafer can be used. For c-Si, a gain about 7 has been reported (13).

Acknowledgements

We would like to express sincere thanks to Anna Duda and Scott Ward at NREL for their help in device processing, and David Young for assistance on integrated QE software. This research was supported by the U.S. Department of Energy under Contract No. DE-AC39-98-GO10337.

Reference

1. http://sanyo.com/news/2009/05/22-1.html (5/22/09). 2009.
2. S. Taira, Y. Yoshimine, T. Baba, M. Taguchi, H. Kanno, T. Kinoshita, H. Sakata, E. Maruyama, and M. Tanaka, Proceedings of the 22nd EU PVSEC: p. 932, 2008.
3. H. Fujiwara and M. Kondo, WCPEC-4, Hawaii. 2006.
4. E. L.C. Korte, K. v. Maydell, H. Angermann, C.Schubert, R. Stangl, R and M. Schmidt; 21st European Photovoltaic Solar Energy Conference: proceedings of the international conference held in Dresden, Germany, 4 - 8 Sept. 2006 p. 2DO.3.5. 2006.
5. Stefaan De Wolf and M. Kondo, Appl. Phys. Lett., 90: p. 042111. 2007.
6. T.H. Wang, M.R. Page, E. Iwaniczko, Y.Q. Xu, Y.F. Yan, L. Roybal, D. Levi, R. Bauer, H.M. Branz, and Q. Wang. WIP-Renewable Energies. 21st European Photovoltaic Solar Energy Conference, p. 781. 2006.
7. T.H. Wang, E. Iwaniczko, M.R. Page, D.H. Levi, Y. Yan, V. Yelundur, H.M. Branz, A. Rohatgi, and Q. Wang. IEEE.Proceedings of the 31st IEEE Photovoltaic Specialists Conference, p. 955. 2005.
8. U. K. Das, M. Z. Burrows, M. Lu, S. Bowden, and R.W. Birkmire, Appl. Phys. Lett., **92**, 063504 (2008).
9. M.R. Page, E. Iwaniczko, Y. Xu, Q. Wang, Y. Yan, L. Roybal, H.M. Branz, and T. H. Wang. the 2006 IEEE 4th World Conference on Photovoltaic Energy Conversion (WCPEC-4) p. 6. 2006.
10. Q. Wang, M.R. Page, E. Iwaniczko, Y.Q. Xu, L. Roybal, R. Bauer, B. To, H.C. Yuan, A. Duda, and Y.F. Yan, in *Proceedings of the 33rd Photovoltaic Specialists Conference (IEEE), San Diego, CA, USA* 118 (2008).
11. Q. Wang and E. Iwaniczko, Mat. Res. Soc. Proc., Vol715: p. 547. 2002.
12. http://pvcdrom.pveducation.org/DESIGN/NORFLCTN.HTM
13. K. R. McIntosh, N. C. Shaw and J.E. Cotter, " Proceedings of the 19th European. Photovoltaic Solar Energy Conference, Paris, France 2004.

ECS Transactions, 34 (1) 1135-1143 (2011)
10.1149/1.3567726 ©The Electrochemical Society

Fabrication and Quantum Confinement Investigation of Ge Multiple Quantum Wells with Si_3N_4 Barriers

Jian Chen, Sammy Lee, Shujuan Huang

ARC Photovoltaics Centre of Excellence
University of New South Wales, Sydney, NSW2052, Australia

Fabrication process and quantum confinement effect of Ge multiple quantum wells (MQWs) structure with Si_3N_4 barriers were investigated for its potential application in 3rd-Generation tandem solar cells. The structure was fabricated by alternating deposition of Ge and Si_3N_4 layers using magnetron co-sputtering and crystallized by rapid thermal annealing (RTA). The Ge MQWs structures were characterized by X-ray reflection, Raman microscopy and glancing incidence X-ray diffraction. Optical properties were analyzed by transmission and reflection measurements. Above characterizations demonstrated that Ge has formed polycrystalline layers restrained between amorphous Si_3N_4 barriers after annealing. The crystallization temperature increased with decreasing Ge layer thickness; in other words, thicker Ge layer is prone to crystallize. And this Ge layer crystallinity is found to be important in deciding quantum confinement effect achieved in the multilayer structure.

1. Introduction

"Third generation" photovoltaic approaches are designed to minimize the energy loss in solar cells and to achieve ultra-high energy conversion efficiency. One of the potential methods for "Third-generation" approachs is to apply different band gap solar cells combined in a multi-junction tandem structure (1). To achieve this, nanoscale structures such as quantum wells, quantum wires and quantum dots have been investigated to modify the effective band gaps of photovoltaic materials. A successful example of low dimension photovoltaic material engineering is "all-Si" tandem cell based on Si quantum dots fabrication (2).

Compared with quantum wire and quantum dot, quantum well structure has only one-dimension confinement in the structure. Therefore, it has the advantage of easy control of quantum confinement effect by controlling deposition rate and time, but also has the disadvantage of less obvious quantum confinement effect. To achieve effective quantum confinement from this one-dimension quantum well structure, Ge is chosen as the quantum well material. It has smaller electron and hole effective masses and a larger dielectric constant than Si and GaAs. The Bohr radius of bulk Ge (~24nm) is larger than Si (~5nm), which allows an easier achievement of quantum confinement effect. And the small band gap of Ge (0.66eV) promises the fabrication of bottom cell for tandem structure (3).

1135

In this work, a structure of Ge multiple quantum wells with Si_3N_4 barriers was fabricated to achieve quantum confinement effect, and its manufacturing conditions were investigated for their impacts on quantum confinement.

2. Experimental Details

2.1 Sample preparation

The preparation of Ge multiple quantum wells with Si_3N_4 barrier layers was conducted by RF magnetron sputtering deposition of Ge/Si_3N_4 multilayer structure followed with Ge layer crystallization through rapid thermal annealing (RTA).

Firstly, RF magnetron sputtering system (AJA International) with a radio-frequency supply of 13.56 MHz was used to deposit Ge/Si_3N_4 multilayers on standard (100) silicon substrates and quartz substrates by alternate deposition of Ge and silicon nitride thin layers at room temperature. The deposited layer thicknesses were controlled through deposition rate and time. Secondly, the deposited samples were rapidly annealed in the RTA machine at selected temperatures to crystallize the Ge quantum well layers. Table I. below shows detailed processing parameters of the samples.

TABLE I. Sample Processing Parameters

Sample ID	Ge/Si₃N₄ layer thickness (nm)	Ge/Si₃N₄ bi-layer number	Si₃N₄ capping layer thickness (nm)	Annealing temperature (°C)	Annealing time (min)
G2S2-RT	2.0/2.0	15	10.0	Non annealed	0.0
G2S2-RTA700	2.0/2.0	15	10.0	700	2.0
G2S2-RTA800	2.0/2.0	15	10.0	800	2.0
G3S2-RT	3.0/2.0	10	10.0	Non annealed	0.0
G3S2-RTA700	3.0/2.0	10	10.0	700	2.0
G3S2-RTA800	3.0/2.0	10	10.0	800	2.0

2.2 Characterisation

X-ray reflection and Glancing incidence X-ray diffraction (GIXRD, Philips X'Pert Pro) using Cu Kα radiation (λ=0.154nm) and Raman microscopy (Renishaw inVia Raman Microscope) operating at 2.5mW and 12A were used to study the structural properties of Ge thin layers sandwiched between Si_3N_4 barriers. The internal structural properties of multilayers such as thickness and interfacial characteristics were studied by X-ray reflection (XRR). The XRR measurements were carried out using the same apparatus that used for the GIXRD measurements. In addition optical transmission and reflection were measured to estimate the optical absorption properties of the samples, including confined bandgap energy values

3. Results

3.1 X-ray Reflection (XRR) study

XRR is a tool that commonly used to investigate the structural properties of layered structures (4). This technique enables the study of interface characteristics, such as layer thickness, interface roughness and change in structural parameters for multilayered thin film structures. Also, XRR is a non-destructive technique which can provide statistical information averaged in a large sample area. In this work, XRR was applied to investigate the multilayer thickness and periodicity changes after RTA processing. The reflected X-ray intensity was measured as a function of small X-ray incidence angle θ from 0.1° to 2.5°. When θ is smaller than a critical angle θc (~0.25°), total external reflection at the air/ film interface occurs as shown by the constant reflection intensity part in Figure 1. Above θc , the oscillating reflection intensity decays asymptotically due to the interference effect of the X-ray waves that reflected at different interfaces within the multilayer (4). The high frequency osciallation that is called Kiessig fringe corresponds to the total thickness of the samples, whereas the strong peaks, the so called Bragg peaks are caused by the constructive interferences and hence give the thickness of the multilayer period.

In Figure 1 below, all the spectra clearly show Bragg peaks indicating that multilayer structures formed and remained in the as-deposited and annealed samples. The XRR spectra of both G2S2 and G3S2 show little change after 700 ℃ and 800 ℃ RTA processing, demonstrating that no obvious layer thickness change caused by RTA process, and 2nm and 3nm Ge quantum well layers' thickness remain unchanged after annealing. The bilayer thickness of Ge/Si3N4 can be calculated using the position of two Bragg peaks. They are 4.2 and 5.2 nm for the G2S2 and G3S2 samples respectively, which is in good agreement with the designed thicknesses.

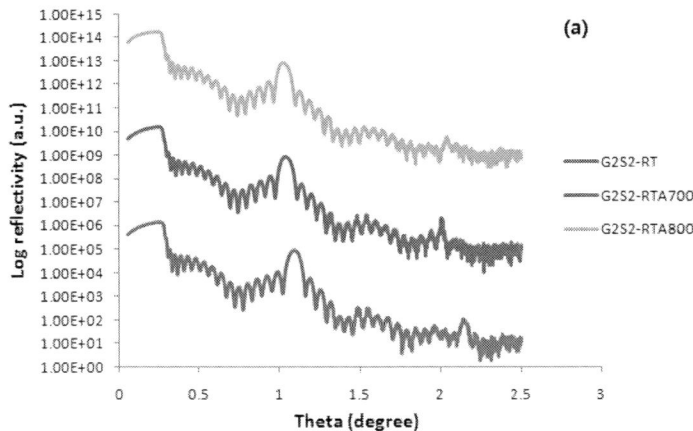

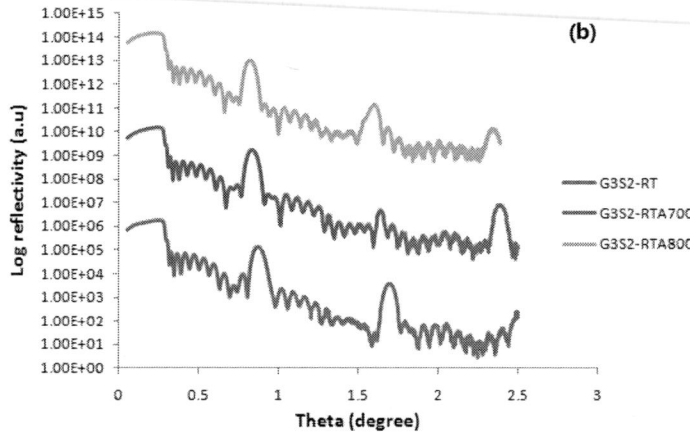

Figure 1: XRR spectra of 2nm Ge MQWs (a) and 3nm Ge MQWs (b) annealed at different temperatures.

3.2 Raman Study

Raman microscopy was employed to study the crystallization of Ge thin layers under different processing conditions. Figure 2 shows Raman spectra of 2nm and 3nm Ge MQWs prepared at room temperature and annealed at 700 and 800 °C. The Raman spectra of both 2nm and 3nm as-deposited Ge quantum well samples show no peaks but almost flat intensity, which represented amorphous Ge. After annealing, one appeared for each annealed sample at around 300cm^{-1}, closely matching with the Raman peak position of Ge crystalline (300.7±0.5cm^{-1}) (5). Besides, this peak becomes shaper and narrower at higher annealing temperature, representing the growth of Ge nanocrystalline size (6). This implies that higher annealing temperature should be used to improve crystallinity of the Ge MQWs.

Adding to the statement above, thickness of Ge also affected the crystallinity; 3nm Ge quantum wells had higher peak intensity than 2nm quantum wells at same annealing temperature. Therefore, thicker Ge quantum well layer was found to be required for well-crystallization.

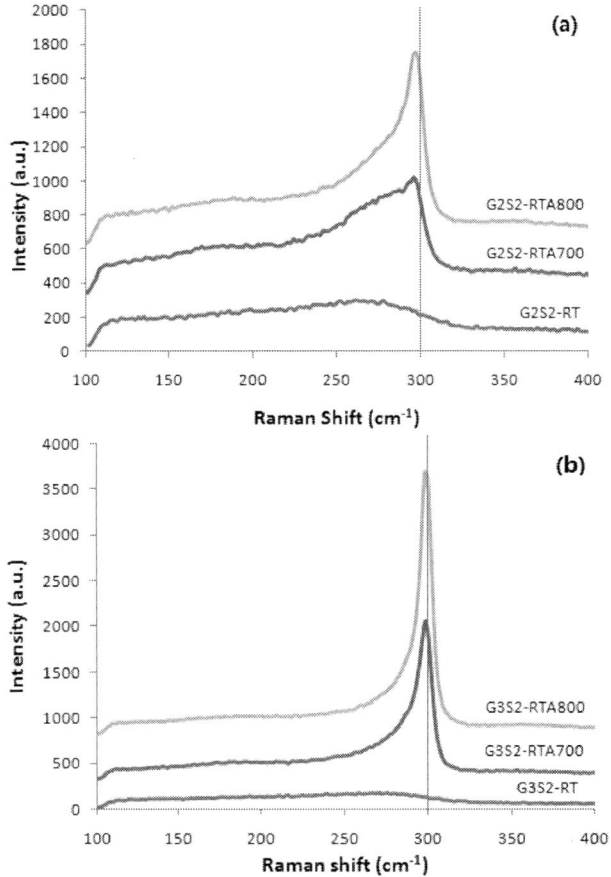

Figure 2: Raman spectra of 2nm Ge MQWs (a) and 3nm Ge MQWs (b) annealed at different temperatures.

3.3 GIXRD Study

GIXRD was used to confirm the formation of Ge nanocrystalline in the samples. 2 theta scan measurements were performed with the samples fixed at a small incidence angle which is close to the critical angle. In Figure 2, bragg peaks observed at 2Theta=27.3°, 45.3° and 53.8° represented the formation of Ge nanocrystalline. For as-deposited samples, only two broad peaks were observed meaning no crystallization occurred but remaining as amorphous state. At higher annealing temperature, the Bragg peaks became sharper and narrower, which reconfirmed the conclusion drawn from the Raman spectra that high annealing temperature is a crucial factor for well-crystalization

By comparing Figure 2(a) and 2(b), it was also concluded a thicker Ge layer is required for easier crystallization.

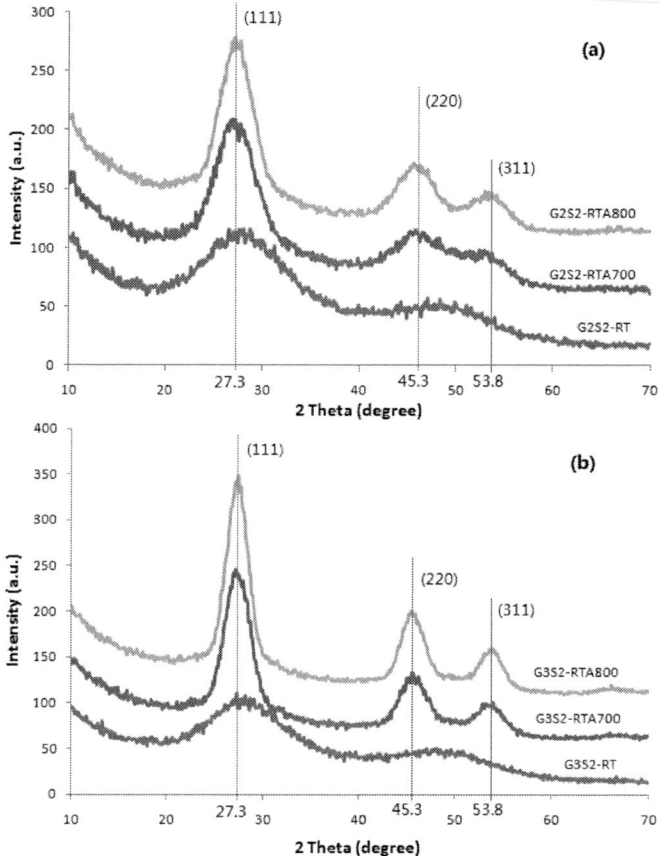

Figure 2: GIXRD spectra of 2nm Ge quantum wells (a) and 3nm Ge quantum wells (b) annealed at different temperatures.

3.4 Optical absorption study

To investigate the quantum confinement effect of the structure of Ge multiple quantum wells with Si_3N_4 barriers, optical transmission (T) and reflection (R) measurements were conducted to obtain the optical absorption properties of these samples. Figure 3 shown below was calculated from the measured transmission and reflection with the formula:

$$Absorption = 1 - T - R \qquad [1]$$

In Figure 3(a) and 3(b), the absorption edge of the structures showed blue shift when annealed meaing higher energy photons can be absorbed by annealed samples than as-deposited sample. This implies that quantum confinement effect is improved by annealing process and a bigger effective band gap is resulted from rapid thermal annealing. To get deeper insight, the improvement of quantum confinement effect after annealing is resulted from the improvement of Ge layer crystallinity. As stated by A. F Khan *et al*, the crystallization of Ge quantum well layers is a necessary precursor to achieve quantum confinement effect, and with better quantum well layer crystallinity the quantum confinement effect is improved (7).

However, it is observed that there is no apparent absorption difference between 700°C and 800°C annealing. For this phenomenon, a possible reason is that the crystallinity in 800°C annealed samples has little difference with 700 °C, so the effective band gap is unchanged.

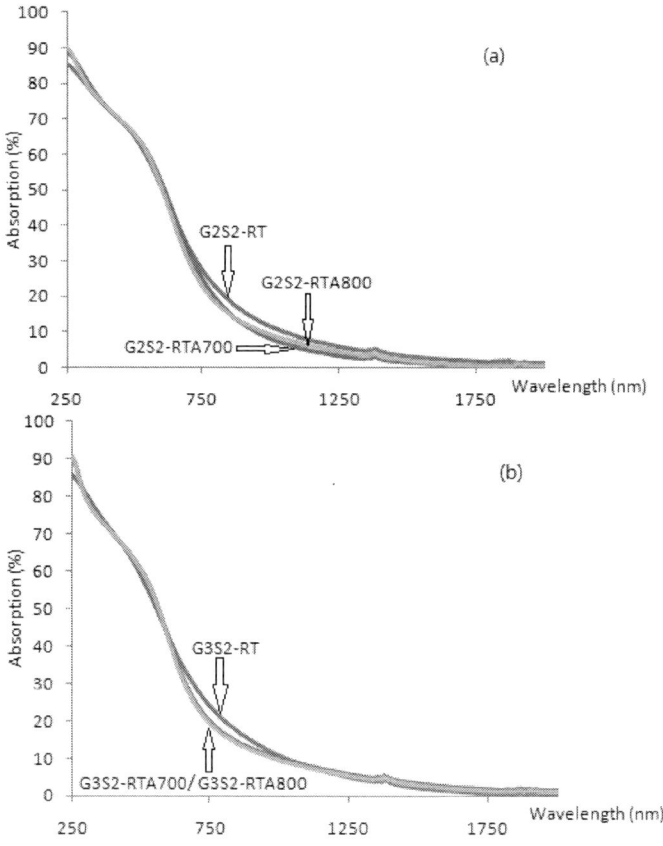

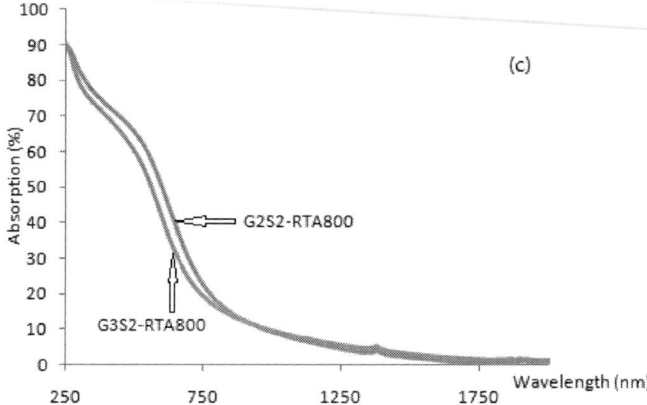

Figure 3: Optical absorption of 2nm Ge quantum wells (a) and 3nm Ge quantum wells (b) annealed at different temperatures; (c) absorption of G2S2-RTA800 and G3S2-RTA800

In Figure 3(c), another interesting phenomenon was observed. Based on Schrödinger equation:

$$\Delta E_n = \frac{\pi^2 \hbar^2}{2m^* a^2} \bullet n^2 \qquad [2]$$

The thinner quantum well layers, the higher effective band gaps, so 2nm Ge multiple quantum well structure should have larger effective band gap than 3nm structure. However, in Figure 3(c) 3nm structure actually shows bigger effective band gap than 2nm structure. To explain this, the influence of Ge layer crystallinity on quantum confinement should again be investigated. According to the structural study from Raman spectroscopy and GIXRD, the crystallinity of 3nm Ge quantum well structure is much better that 2nm, so it is highly possible that the impact of Ge layer crystallinity becomes the primary factor, and better crystallinity leads to better quantum confinement effect, regardless of Ge layer thickness.

4. Conclusion

Based on the experimental results discussed above, both RTA and layer thickness are proved to have impact on Ge thin layer crystalisation, and therefore influence the quantum confinement effect of the structure:

● It is demonstrated that annealing at higher temperature is effective in improving the crystallinity of Ge quantum well layers, and the quantum confinement effect is improved by rapid thermal annealing. The Raman and GIXRD results show clear evidence for the crystallization of Ge multiple quantum well layers and the increase of layer crystallinity with higher annealing temperature. The blue shift in optical absorption results indicates the increase of effective band gap after annealing. Therefore better quantum confinement effect is resulted from RTA.

- It is demonstrated that thicker Ge quantum well layer is prone to crystallize. This is evident by comparing the Raman and GIXRD spectra of 2nm and 3nm Ge multiple quantum wells structures with same processing conditions. And this impact is thought to be the main reason for better quantum confinement effect in 3nm Ge quantum wells rather than 2nm.

Acknowledgments

Thanks to the ARC Photovoltaics Centre of Excellence in the University of New South Wales for the state of art photovoltaic facilities used in this research; Thanks to the Solid State & Elemental Analysis Unit in the University of New South Wales for the XRR, GIXRD and Raman characterization.

References

1. Martin Green, *Third Generation Photovoltaics*, p. 59, Springer (2003)
2. Conibeer, G., M. Green, et al. "Silicon quantum dot nanostructures for tandem photovoltaic cells", *Thin Solid Films* **516**(20): 6748-6756 (2008).
3. D. Bellet, E. Bellet-Amalric, T. Hanley, A.Nelson, D. Song, S. Huang, T. Fangsuwannarak, S. W. Park, E. Pink, G. Scardera, E.-C.Cho, G. Conibeer, D. König, and M. A. Green, Proceedings of the 22nd European PV Solar Energy Conference, p.472, Milan, Italy(2007)
4. J. Daillant and A. Gibaud, *X-ray and Neutron Reflectivity: Principles and Applications*, Springer, New York (1999).
5. J. H. Parker, Jr, D. W. Feldman and M. Ashkin, "Raman Scattering by Silicon and Germanium", *Phys. Rev.*, Vol 155, No. 3 (1967).
6. Akhilesh K. Arora, M. Rajalakshmi, T. R. Ravindran and V. Sivasubramanian, "Raman spectroscopy of optical phonon confinement in nanostructured materials", *J. Raman Spectrosc.* **38**: 604-617 (2007).
7. Khan, A. F., M. Mehmood, et al. "Nanostructured multilayer TiO2-Ge films with quantum confinement effects for photovoltaic applications." J *of Colloid and Interface Science*, **343**(1): 271-280 (2010)

ECS Transactions, 34 (1) 1145-1149 (2011)
10.1149/1.3567727 ©The Electrochemical Society

Structural and Optical properties of porous SiGe/Si Multilayer Films

Bi Zhou[a], Xuemei Li[a], Shuwan Pan[b], Songyan Chen[b], Cheng Li[b]

[a] Department of Physics and Electronic Information Engineering, Minjiang University,
Fuzhou 350108, People's Republic of China
[b] Semiconductor Photonics Research Center, Department of Physics, Xiamen University,
422 South Siming Road, Xiamen, Fujian 361005, People's Republic of China

The structural and optical properties of heterogeneous SiGe/Si
multilayered films, which was prepared through a combination of
ultrahigh vacuum chemical vapor deposition and electrochemical
anodization, have been investigated. The structural parameters of
as-grown multilayer films were determined by double crystal X-
ray diffraction and the simulation. The visible luminescence
spectra with multiple emission peaks have been detected in the
temperature range from 10 K to room temperature. The origins of
PL with multiple peaks have been discussed in detail.

1. Introduction

The observation of visible photoluminescence (PL) from porous Si or SiGe has drawn an
extensive attention due to the feasibility of Si-based light emission (1, 2). However, some
problems related to the porous Si or SiGe physics have emerged: low external quantum
efficiency, wide emission band, and long recombination lifetimes. For the understanding
of fundamental physics and potential device applications, the size control and high
density of Si-based nanocrystals are essential. Recently, a new strategy has been
successfully demonstrated good controllability of the size of nanocrystals, which utilized
the phase separation between Si-rich SiO_x/SiO_2 (3), or a-Si/a-SiN$_x$ (4, 5) multilayer
structure. The size of the nanocrystals is effectively controlled by the thickness of Si-rich
SiO_x or a-Si sublayers.

Alternatively, in this paper, porous SiGe/Si films by electrochemical anodization
(ECA) of SiGe/Si multiple layers grown by Ultra-high vacuum chemical vapor
deposition (UHVCVD) are fabricated. The visible PL with multiple peaks from porous
SiGe/Si multilayer films has been observed. The luminescence mechanism has been
studied through temperature-dependence PL.

2. Experimental details

2.1 Growth of GeSi/Si multiple layers

The sample was grown in a UHVCVD system with a base pressure of 5×10^{-8} Pa.
Pure Si_2H_6 and GeH_4 were used as precursors. A 4-inch N-type Si (100) wafers was used
as a substrate. The wafer was cleaned by the Radio Corporation of American (RCA)
method and dried by N_2 before loading into the growth chamber. The wafer was baked at
850 °C for 30 min to deoxide it, followed by growing a Si buffer at 750 °C. Next, strained
SiGe/Si multiple layers and the Si-cap layer were grown at 650 °C. The sample consists

1145

of 6 periods $Si_{0.87}Ge_{0.13}$/Si multiple layers with a ~7-nm thick SiGe layer and a ~28-nm-thick Si layer alternating and a final 160-nm-thick Si cap layer. The parameters were determined by X-ray diffraction (XRD). The Ge content in the sample was measured by sputtering Auger electron spectroscopy (not shown here) (6). The total pressure in the chamber during growth was around 2×10^{-2} Pa. For the growth of SiGe layers, a flow rate ratio of Si_2H_6 and GeH_4 was controlled to be 3:1.

2.2 ECA process

In order to obtain good ohmic contact and ensure a uniform anodic current distribution, thin Al films with ~300 nm thickness were deposited on the back of the wafers. A hydrofluoric (HF)–ethanol solution was formed by a mixture of 40% HF acid solution with ethanol at a volume ratio of 1:2. The sample was etched with a current density of 20 mA/cm^2 for 3 min. The experiment was performed under indoor-fluorescent lights. After the etching process, the sample was rinsed with deionized water and blown dry in air before the scanning electron microscopy (SEM) and PL characterizations.

2.3 Material characterization

XRD patterns were recorded by a double crystal XRD measurement (Bede, D1 system), using Cu Kα1 ($\lambda = 0.15406$ nm) as the x-ray source. The top morphology was analyzed by SEM (model FE-SEM LEO 1530). The PL measurements were performed using a cw He–Cd laser emitting at 325 nm as the excitation source, the signal was dispersed by a 750 mm monochromator combined with suitable filters and detected by a photomultiplier (PMT) using the standard lock-in amplifier technique.

3. Results and discussion

3.1 XRD.

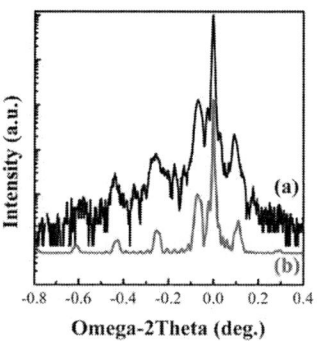

Fig. 1 (a) XRD rocking curve (b) simulated rocking curve for $Si_{0.87}Ge_{0.13}$/Si multiple layers. consisting of 6 periods of a 6.8-nm thick $Si_{0.87}Ge_{0.13}$ and a 28. 0-nm-thick Si layer alternatively first, and a final 160-nm-thick silicon cap layer.

Fig. 1 (a) shows the XRD rocking curve for the as-grown $Si_{0.87}Ge_{0.13}$/Si multilayer sample grown on Si (001) substrate. The simulated rocking curve is also shown in Fig. 1 (b), which matches well with the experimental data in terms of peak position and peak intensity of each main peak. Apart from the Si (004) diffraction peak from Si substrate, the satellite peaks from the SiGe/Si multiple layers are clearly observed up to the third order. And some weak secondary satellite peaks are also visible. It is indicated that SiGe/Si multiple layers with clear SiGe/Si interfaces and high crystalline quality had been achieved.

3.2 SEM.

SEM was used to determine the surface features of the as-etched sample. Fig. 2 shows typical SEM top-view image of ECA SiGe/Si multiple layers film. The surface is covered mostly by square isolated islands. The island size was estimated, and its average diameter was ~ 20 nm.

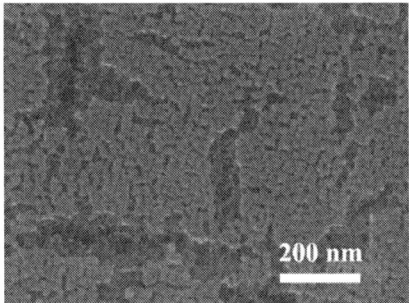

Fig. 2 Typical SEM top-view image of as-etched SiGe/Si multilayer films, etching time 3 min, current density ~20 mA/cm².

3.3 Visible PL with multiple peaks.

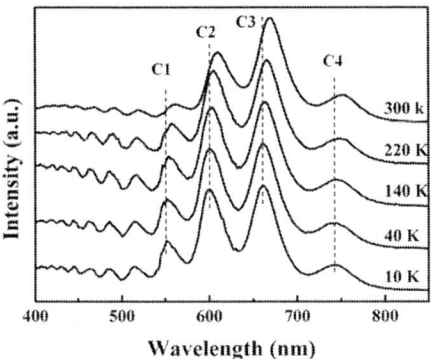

Fig. 3 PL spectra at various temperatures from 10 K to 300 K under an excitation wavelength of 325 nm.

Fig. 3 shows temperature evolution of PL of the sample in the temperature range from 10 K to 300 K. A broad spectral feature is observed in the wavelength range of 400~800 nm with multiple peaks, which is significantly different from the characteristic spectra of porous Si (PS) (7) or porous SiGe alloy(8), in which there exists a single PL peak mostly. No distinct change of PL line-shape induced by the change of temperature is observed. The intervals between two adjacent peaks arise with the rising to longer wavelength of the center of the peaks. As temperature rises, the positions of PL peaks shifts gradually toward longer wavelengths, .i.e., the red-shift.

In addition, Fig. 3 shows a complicated temperature-dependence of the intensity of these PL peaks. As the temperature rises, the relative intensity of the peaks in shorter wavelength range (i.e., the left side of main peak C3) decreases, while that in longer wavelength range (i.e., the right side of main peak C3) increases. For example, compared with the intensity of the main peak C3, the relative intensity of peak C1 and C2 weaken respectively from 0.482 to 0.177 and 0.967 to 0.674, while that of peak C4 strengthens from 0.248 to 0.268.

Since porous Si or SiGe only shows a broad PL spectrum with a single peak, the sharp and multi-peak PL characteristic must be related to the cavity effect, as shown in Fig. 3. There is a Fabry–Perot (F–P) cavity between the mirrors formed by the porous layer/Air and the porous layer/substrate interface that has a strong effect on the light emission spectra from the porous structure, which results in the multi-peak structure of the spectra.

As is well known, only when the wavelength of the luminescence meets the Eq. [1], it can transmit the F-P cavity and be collected(9).

$$2\frac{2n\pi}{\lambda}L + \varphi_1 + \varphi_2 = 2m\pi \quad (m = 1, 2, 3, \ldots) \tag{1}$$

Here n is the refractive index of porous layer, λ is the vacuum wavelength of the luminescence, L is the cavity length, and φ_1 and φ_2 denote phase shifts due to the penetration into the mirrors.

The intervals of adjacent peaks increase with the increasing of the wavelength with coincide with the results obtained by the free spectral range (FSR) Eq. [2] induced by the wavelength characteristic of the F-P cavity (9).

$$FSR = \frac{\lambda^2}{2n_{eff}L_{eff}} \tag{2}$$

Here λ is the wavelength of the luminescence, n_{eff} and L_{eff} denote the effective refractive index and the effective cavity length, respectively.

The two temperature-related behaviours of the PL spectra under the cavity effect, as shown in Fig. 3, result from the coefficient of the thermo-optic and thermal expansion and contraction effects of the cavity and the shrinkage of band gap (10, 11).

As the temperature is increased from 10 K up to 300 K, the thermo-optic effect and the thermal expansion effect lead to the increase of the refractive index and the cavity length L, respectively. According to the Eq. [1], these lead to the increase in the cavity wavelength λ, i.e. the red-shift of the resonance peaks. On the other hand, the band-gap shrinkage with temperature increasing result in the red-shift of the broad PL spectra from the porous structure under no cavity effect, which should not change the multi-peak positions. The two red-shift effects above result in the above-mentioned temperature dependence in relative intensity of PL peaks.

4. Conclusion

To summarize, multilayer SiGe/Si porous films have been fabricated through a combination of UHVCVD and ECA and characterized by double crystal X-ray diffraction and SEM. Visible PL spectra with multiple emission peaks at various temperatures from 10 K to 300 K have been observed. The multi-peak PL origins related to the F-P cavity effect have been discussed in detail.

Acknowledgments

This work was supported by the Nation Natural Science Foundation of China under grant Nos: 50672079, 60837001 (Key Program) and National Basic Research Program of China (973 Program) under grant Nos: 2007CB613404 and Natural Science Foundation of Fujian Province under grant Nos: 2008J0221 and Foundation of Education Committee of Fujian Province (JB10124) and Technology Startup Project of Minjiang University (KQ1004).

References

1. L. T. Canham, *Appl Phys Lett*, **57**, 1046 (1990).
2. S. Gardelis, J. S. Rimmer, P. Dawson, B. Hamilton, R. A. Kubiak, T. E. Whall and E. H. C. Parker, *Appl Phys Lett*, **59**, 2118 (1991).
3. L. Tsybeskov, K. D. Hirschman, S. P. Duttagupta, M. Zacharias, P. M. Fauchet, J. P. McCaffrey and D. J. Lockwood, *Appl Phys Lett*, **72**, 43 (1998).
4. M. Wang, X. Huang, J. Xu, W. Li, Z. Liu and K. Chen, *Appl Phys Lett*, **72**, 722 (1998).
5. W. K. Tan, M. B. Yu, Q. Chen, J. D. Ye, G. Q. Lo and D. L. Kwong, *Appl Phys Lett*, **90** (2007).
6. Y. Zhang, K. Cai, C. Li, S. Chen, H. Lai and J. Kang, *J Electrochem Soc*, **156**, H115 (2009).
7. S. Y. Chen, Y. H. Huang and B. N. Cai, *Solid State Electron*, **49**, 940 (2005).
8. M. Schoisswohl, J. L. Cantin, M. Chamarro, H. J. von Bardeleben, T. Morgenstern, E. Bugiel, W. Kissinger and R. C. Andreu, *Thin Solid Films*, **276**, 92 (1996).
9. M. S. Unlu and S. Strite, *J Appl Phys*, **78**, 607 (1995).
10. Y. P. Varshni, *Physica*, **34**, 149 (1967).
11. H. Rinnert, O. Jambois and M. Vergnat, *J Appl Phys*, **106**, 023501 (2009).

ECS Transactions, 34 (1) 1151-1157 (2011)
10.1149/1.3567728 ©The Electrochemical Society

On the Impact of Heavy Doping on Grown-in Defects in Czochralski-grown Silicon

X. Zhang[1], W. Xu[1], J. Chen[1], X. Ma[1], D. Yang[1], L. Gong[2], D. Tian[2], and J. Vanhellemont[3]

[1]State Key Laboratory of Silicon Materials and Department of Materials Science and Engineering, Zhejiang University, Hangzhou 310027, China
[2]QL Electronics, Ningbo 315800, China.
[3]Department of Solid State Sciences, Ghent University, 9000 Ghent, Belgium

The effects of heavy doping of Si on grown-in void size-density distributions and on Flow Pattern Defect (FPD) and Secco Etch Pit Defect (SEPD) density are discussed. Grown-in defects are studied using Scanning Infra Red Microscopy (SIRM) and Secco etching. Doping with 10^{20} Ge atoms cm^{-3} has a limited effect on the grown-in void size-density distribution but has a clear effect on the FPD density. The observed lower FPD density is most probably related to a decrease of the multiple void density.
Co-doping with 10^{20} B atoms cm^{-3} leads to strong a suppression of the void density by nearly two orders of magnitude in agreement with the reported strong reduction of Crystal Originated Particle (COP) density in low resistivity p-type Si.

Introduction

Heavily doped Si wafers are used routinely as substrate for the production of epitaxial wafers. Epitaxial wafers have the advantages of an improved gettering performance both by segregation gettering due to the enhanced solubility of metals in the highly doped Si bulk and also due to the enhanced interstitial oxygen precipitation due to the high level of doping leading to improved internal gettering. Especially so called p++ substrates that are heavily B doped down to resistivities in the mΩcm range, are very good getterers for Fe contamination leading to very high quality epitaxial layers that are moderately doped with B and that are very resistant against metal contamination during device processing. A problem that arises in case of high concentrations of B is that the misfit between the substrate and the epitaxial layer becomes very high which can lead to dislocation generation and wafer bow. This high strain level can be decreased by doping the heavily B doped substrate also with Ge. The Ge atom is larger than the Si atom and thus the compressive strain due to B doping can effectively be compensated. Further beneficial effects of co-doping with Ge, are that the mechanical strength of the wafer and crystal is increased and at the same time oxygen precipitation can be enhanced. For those reasons doping of Czochralski-grown Si crystals with Ge concentrations between 10^{16} and 10^{20} cm^{-3} have been studied intensively during the last years. Little work has been published however on the effect of co-doping with Ge in very low resistivity Si.

Recently, Londos et al. (1) performed an extensive study of the influence of Ge doping on the behavior of oxygen and carbon impurity related complexes in electron irradiated Si. Their observations were explained by assuming that for Ge concentrations below 10^{20} cm^{-3}, Ge atoms act as temporary traps for vacancies and as such reduce the

1151

recombination rate of intrinsic point defects and Frenkel pairs. This leads to an increase of vacancy, self-interstitial, interstitial-oxygen and interstitial-carbon related defects. For Ge doping above 10^{20} cm^{-3}, an opposite behavior was however observed which was assumed to be due to the formation of Ge clusters acting as recombination centers for the intrinsic point defects. The described effects of Ge doping can have important consequences not only for radiation-induced defect formation and population dynamics but also for defect formation during crystal pulling and device fabrication (2,3).

In the present paper, the effects of heavy doping of Si during Czochralski (Cz) crystal pulling on grown-in void size-density distributions and on the FPD and SEPD density are discussed and illustrated. The grown-in defects are studied using SIRM and Secco etching of polished wafers prepared from different crystal positions.

Doping with 10^{20} Ge atoms cm^{-3} has a limited effect on the grown-in void size-density distribution but has a clear effect on the FPD density that is lowered by increasing the Ge doping (4-6). The observed lower FPD density is most probably related to the decrease of the multiple void density with increasing Ge doping (7).

Co-doping with 10^{20} B atoms cm^{-3} leads to strong a suppression of the void density by nearly two orders of magnitude as illustrated in Table I. The observed reduction in void concentration is in excellent agreement with the reported strong reduction of COP density observed in p-type Si when the resistivity drops below 10 mΩcm (8).

Experimental

Wafers with known positions in 4 in. Cz-grown crystals produced by QL Electronics, are used, with specifications as listed in Table I. The average pulling rate was about 1.25 mm/min. The Ge concentration in the GGCZ crystal was determined using Secondary Ion Mass Spectroscopy while the other values are nominal values. The interstitial oxygen concentration was determined by Fourier Transform Infrared Spectroscopy using the 3.14 calibration factor.

The defect distribution at a depth below the surface of 80 μm for standard doped and 40 μm for the heavily boron doped samples below the wafer surface is investigated with a SIRM-300 instrument of Semilab using the 980 nm wavelength laser. Each measurement records the scattered light image of a 180×180 μm^2 area, revealing typically between zero and ten defects. Due to this low defect density, a large number of sampling areas have to be measured in order to have enough statistical relevance as illustrated in Fig. 1 comparing a 1.8×1.8 mm^2 area in the center of two 4 in. wafers (left: GGCZ, right: BGCZ) . A nearly two orders of magnitude lower defect density is observed in the heavily B doped sample.

TABLE I. Material characteristics of the 4 in. wafers investigated in the present study.

Label	C_{Ge} $(\times 10^{20} \text{cm}^{-3})$	Doping	Resistivity (Ωcm)	C_{OI} $(\times 10^{17} \text{ cm}^{-3})$
GGCZ	1.22 ± 0.39	P	8 to 15	11.3
GCZ	0.1 (nominal)	P	15 to 30	11
CZ	0	P	15 to 30	10
BGCZ	1 (nominal)	B	6×10^{-3}	25
HB	0	B	$3\text{-}5 \times 10^{-3}$	-

FPD etching was performed for 30 min with the etching bath kept at 30 °C. Two samples were cut per wafer with a sample size was about 1x2 cm^2. After etching, FPD and SEPD defects were counted using an optical microscope. 15 areas of 0.039 cm^2 each, or in total 0.585 cm^2, were investigated on each sample. To check the reproducibility, the same procedure was repeated on a second set of samples. The Si layer thickness removed by the Secco etching was determined by weighing the samples before and after the etching treatment using an electronic balance with an accuracy of 0.1 mg and was typically between 10 and 16 μm after 30 min etching. For the heavily boron doped samples the etching rate of the standard Secco etch with composition K$_2$CrO$_7$(0.15mol/L):HF=1:2, is too high and a diluted version with composition K$_2$CrO$_7$ (0.0375mol/L):HF=1:2 was used in order to remove a similar layer thickness by 1h etching.

Observations and discussion

Figure 1 shows a typical composite SIRM image of samples taken from crystals GGCZ and BGCZ and illustrates the nearly two orders of magnitude difference in void density due to the high doping concentration.

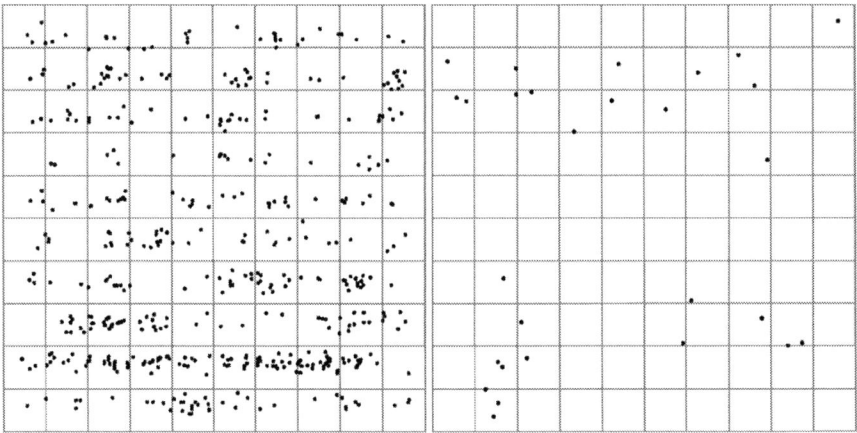

Figure 1. Typical void maps obtained in a 1.8×1.8 mm^2 area in the center of two 4 in. wafers (left: GGCZ, right: BGCZ) after measuring a matrix of 10×10 adjacent areas of 180×180 μm^2.

TABLE II. Measured void diameter (D_{void}) and density (n_{void}) as well as the total number of voids (N_{tot}) characterized in the central area of wafers from two crystals doped with 10^{20} Ge atoms cm^{-3}.

Label	n_{void} (10^7 cm^{-3})	D_{void} (nm)	N_{tot}
GGCZ	1.7 ± 1.3	53.1 ± 3.0	1646
BGCZ	0.033 ± 0.079	36.4 ± 1.9	25

Table II lists average values of the defect density and size observed by SIRM in both types of crystals. The total number of analyzed defects is also given in the last column. The observed reduction of the grown-in void concentration by a factor of about 50 in the 6 mΩcm crystal is in excellent agreement with the similar reduction of COP density observed for resistivities below 10 mΩcm (8). Heavy boron doping also has a strong effect on the position of the so called stacking fault-ring that indicates the crystal radius where the crystal changes from vacancy-rich to interstitial-rich (outside the ring) during crystal pulling (7,9). The results indicate a strong reduction of the thermal equilibrium concentration of vacancies, most probably related to the large compressive strain that is introduced in the Si lattice by the substitutional boron atoms. One should remark that the void size given in Table II for the BGCZ sample is the one obtained assuming that there is no absorption like for the lowly doped samples. A correction should be made for the absorption which leads to a decrease of the scattering intensity by a factor of about $e^{-2\alpha z}$, with z the probed depth and α the absorption coefficient which for the heavily doped sample is about 200 cm^{-1} at the 980 nm wavelength (10). Taking into account that the scattered intensity depends on the sixth power of the void size one obtains as real void size of about 47 nm, similar to that in the low doped material.

Typical optical micrographs of Secco etched samples are shown in Figure 2 for the GGCZ and BGCZ samples clearly illustrating the lower FPD density in the heavily boron doped material. Notice also the different appearance of the FPD's in the low resistivity material.

Table III gives an overview of the results after Secco etching. Due to the poorer surface quality after etching, SEPD's could not be observed in the heavily boron doped wafers. In general the recognition of SEPD's is much more difficult than of FPD's making the data for SEPD's also less reliable. A somewhat surprising result is the relatively high FPD concentration in the heavily boron doped samples which is similar to the defect density observed with SIRM. This might be an indication that most of the grown-in voids are multiple voids.

TABLE III. Measured FPD density (n_{FPD}), multiple void density (n_{mvoid}) and diameter (D_{mvoid}) and the total number of multiple voids (N_{tot}) observed by SIRM in the central area of wafers from two crystals doped with 10^{20} Ge atoms cm^{-3}.

Label	n_{FPD} (x10^5 cm^{-3})	n_{SEPD} (x10^5 cm^{-3})
GGCZ	4.38 ± 0.39	4.38 ± 0.57
Second experiment	4.21 ± 0.63	2.85 ± 0.94
GCZ	3.89 ± 0.88	5.0 ± 1.0
Second experiment	4.05 ± 0.38	3.07 ± 0.40
CZ	4.37 ± 0.42	3.676 ± 0.098
Second experiment	3.49 ± 0.20	7.88 ± 0.81
BGCZ	2.2 ± 1.0	Not observed
Second experiment	0.70 ± 0.60	Not observed
HB	1.80 ± 0.57	Not observed
Second experiment	1.62 ± 0.56	Not observed

As shown in Table IV, there is a clear effect of the crystal position on the grown in defect distributions. It is however not straightforward to pinpoint which individual process or material parameter is causing this change in defect distribution. As is obvious from Table IV also the pulling rate is indeed changing with crystal position and decreases from head to tail of the crystal. A lowering of the pulling rate v with constant thermal gradient G leads to a decreasing v over G ratio Γ and this is known to lead to a decrease of vacancy concentration. As also shown in Table IV, the change of pulling rate is however also accompanied by a change of interstitial oxygen and Ge concentration whereby both elements show an opposite trend: the oxygen concentration decreases towards the tail of the crystal while the Ge concentration increases. Both elements are known to play a role in vacancy thermal equilibrium concentration and thus probably also in void formation.

TABLE IV. Measured FPD density (n_{FPD}), multiple void diameter (D_{mvoid}) and density (n_{mvoid}) characterized by SIRM in the central area of wafers from head, middle and tail parts of crystal GGCZ.

Crystal Poisition	n_{FPD} (x10^5 cm^{-3})	n_{SEPD} (x10^5 cm^{-3})	n_{mvoid} (x10^5 cm^{-3})	D_{mvoid} (nm)	v (mm/min)	C_{OI} (x10^{17} cm^{-3})	C_{Ge} (x10^{20} cm^{-3})
Head	4.960 ± 0.037	3.58 ± 0.50	10 ± 16	104 ± 16	1.36	13.3	0.83 ± 0.01
Second experiment	5.09 ± 0.55	1.712 ± 0.081	-	-	-	-	-
Middle	4.298 ± 0.067	5.159 ± 0.019	5.7 ± 9.9	94 ± 15	1.27	10.6	1.1 ± 0.04
Second experiment	3.94 ± 0.32	2.73 ± 0.44	-	-	-	-	-
Tail	3.87 ± 0.14	4.42 ± 0.14	5.0 ± 8.9	108 ± 17	1.06	10.1	1.60 ± 0.02
Second experiment	3.61 ± 0.68	4.10 ± 0.83	-	-	-	-	-
Total crystal	4.29 ± 0.49	3.62 ± 0.94	7 ± 12	104 ± 16	1.23	11.3	1.22 ± 0.39

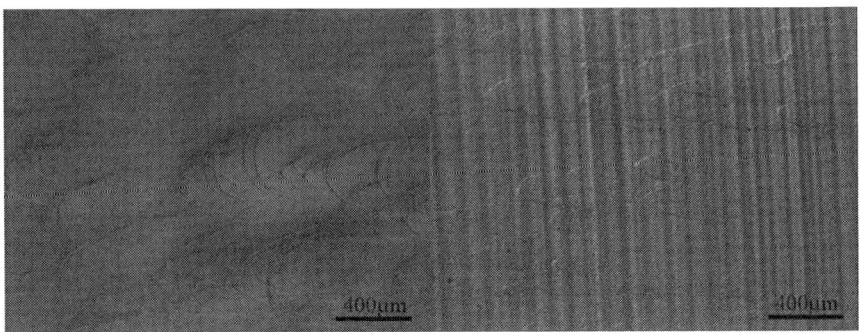

Figure 2. Optical micro-graphs of etched GGCZ (left) and BGCZ (right) samples. In the heavily boron doped sample the boron striations are clearly delineated.

Figure 3 is a graphical representation of the data in Table IV and shows the variation of multiple void (MVOID), FPD and SEPD densities for three different crystal positions in crystal GGCZ. Also the pulling rate v and the interstitial oxygen and Ge content C_{OI} and C_{Ge}, respectively, are shown. Both the multiple void and FPD concentrations decrease with decreasing pulling rate while the SEPD density increases. This is in agreement with the Voronkov theory which predicts that the vacancy concentration decreases with decreasing Γ while the self-interstitial concentration increases (11,12). The FPD concentration can be considered as a marker for the vacancy concentration as they correspond with voids that are formed by vacancy clustering while the SEPD defect are assumed to correspond with interstitial clusters and are thus a marker for the concentration of interstitials (13). On the other hand the Ge concentration increases towards the end of the crystal while the oxygen content decreases. Both effects could in principle also explain why the FPD and MVOID concentration decrease.

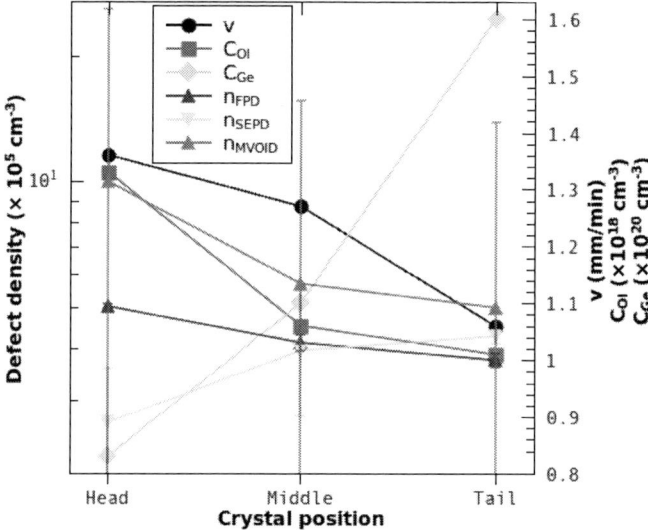

Figure 3. Variation of the SIRM multiple void, FDP and SEPD density as a function of the crystal position. Variation of the Ge and O concentration and the pulling speed with crystal position. Data are taken from Table IV.

Conclusions

Doping of Si with Ge concentrations in the range between 10^{16} and 10^{19} cm^{-3} is beneficial for the crystal and wafer quality as it improves the mechanical strength leading to lower wafer losses due to breaking, during wafer cutting and thermal processing. Furthermore, Ge doping suppresses thermal donor formation while at the same time it can strongly enhance oxygen precipitation and thus increase the internal gettering capacity of low oxygen content wafers which are both relevant advantages for device processing. A further advantage is that also the crystal pulling process itself does not need any modification for Ge concentrations below 10^{20} cm^{-3}.

Although there is most probably an impact of Ge doping on multiple void formation, doping with Ge or even changing the interstitial oxygen concentration are less effective to control single void formation than doping with nitrogen (14).

Further more, the reported impact of Ge doping on oxygen precipitation and on thermal donor formation is most probably not due to a reduction in the vacancy concentration immediately after solidification of the crystal but occurs at the lower temperatures when oxide nuclei are formed. At temperatures below the void nucleation temperature, which is estimated to be in the range between 1050 and 1100 °C, the free vacancy concentration that is playing a crucial role in oxide precipitate nucleation, might be changed by the presence of Ge atoms due to a.o. a shift in the VO and VO_2 balance by the interaction of Ge atoms with free vacancies and/or interstitial oxygen atoms.

Acknowledgments

J. Chen and J. Vanhellemont acknowledge the National Natural Science Foundation of China (NSFC, Grant nos. 50832006 and 0906001) and the Research Foundation-Flanders (FWO) for financial support.

References

1. C.A. Londos, A. Andrianakis, V. Emtsev, and H. Ohyama, *J. Appl. Phys.*, 105, 123508 (2009).
2. J. Vanhellemont, J. Chen, W. Xu, D. Yang, J.M. Rafi, H. Ohyama, and E. Simoen, *ECS Trans.*, 27, 1041 (2010).
3. J. Vanhellemont, J. Chen, J. Lauwaert, H. Vrielinck, W. Xu, D. Yang, J.M. Rafi, H. Ohyama, and E. Simoen, *J. Crystal Growth* (2010), doi:10.1016/j.jcrysgro.2010.11.024
4. J. Chen, D. Yang, H. Li, X. Ma, D. Tian, L. Li, and D. Que, *J. Cryst. Growth*, 306, 262 (2007).
5. J. Chen, Ph.D. Thesis, Zhejiang University, 2008 (in Chinese).
6. M. Arivanandhan, R. Gotoh, K. Fujiwara, and S. Uda, *J. Appl. Phys.*, 106, 013721 (2009).
7. J. Vanhellemont, X. Zhang, W. Xu, J. Chen, X. Ma, and D. Yang, *J. Appl. Phys.*, 108, 123501 (2010).
8. J. Vanhellemont, E. Dornberger, D. Gräf, J. Esfandyari, U. Lambert, R. Schmolke, W. von Ammon, and P. Wagner, in Proceedings of The Kazusa Akademia Park Forum on The Science and Technology of Silicon Materials,(Kazusa Akademia Park, Chiba, Japan, 1997), p. 173.
9. W. von Ammon, E. Dornberger, H. Oelkrug and H. Weidner, *J. Cryst. Growth*, 151, 273 (1995).
10. http://www.ioffe.rssi.ru/SVA/NSM/Semicond/Si/
11. V. V. Voronkov, *J. Cryst. Growth*, 59, 625 (1982).
12. J. Vanhellemont, P. Spiewak, K. Sueoka, and I. Romandic, *Phys. Stat. Sol. C*, 6, 1906 (2009).
13. T. Abe, *Materials Science and Engineering B*, 73, 16 (2000).
14. J. Vanhellemont, M. Suezawa, and I. Yonenaga, *J. Appl. Phys.*, 108, 016105 (2010).

ECS Transactions, 34 (1) 1159-1164 (2011)
10.1149/1.3567729 ©The Electrochemical Society

The Influence Of Silicon Orientation On Surface Blistering Behaviors For Molecular Hydrogen Ion Implantation

Y.C. Hsiao [a], J.H. Liang [a, b], C.M. Lin [c]

[a] Department of Engineering and System Science, National Tsing Hua University, Hsinchu 300, Taiwan, R.O.C.
[b] Institute of Nuclear Engineering and Science, National Tsing Hua University, Hsinchu 300, Taiwan, R.O.C.
[c] Department of Applied Science, National Hsinchu University of Education, Hsinchu 300, Taiwan, R.O.C.

This study attempted to investigate the silicon orientation effects on surface blistering behavior under the same implantation condition and thermal budget, Si<100>, Si<111>, and Si<110> were employed and implanted with molecular hydrogen ions (H_2^+) with a kinetic energy level 200 keV and molecular ion fluence of 2.5×10^{16} cm^{-2}. Following implantation, the variation occurred at specimen's surface under isothermal annealing in an atmosphere ambient were in-situ observed using optical microscopy (OM). Raman scattering spectroscopy (RSS) were adopted to detect radiation damage and secondary ion mass spectroscopy (SIMS) were used to measure the hydrogen trapping depth in the specimens. In our investigation, a post-annealing temperature fell into the regime of T_e-T_t for fabricating SOI material with difference silicon orientation is the most optimal under consideration.

Introduction

Recently, the Silicon-On-Insulator (SOI) material fabricated by <100> orientated silicon wafers were created for application of integrated circuit (IC) device industry and have been widely studied. However, in various fields such as micro-electro-mechanical system (MEMS) and nanowire MOSFETs, the use of the SOI material with different silicon orientation (e.g. Si<111> and Si<110>) compared to Si<100> are very important and have great potential for research [1, 2]. Among several methods of fabricating SOI material [3], the Smart-Cut^{TM} or (ion-cut) is the most beneficial technique and widely used in commercial fabrication due to its homogeneous thickness, good crystalline quality and smooth interface. The major procedures of ion cut technique divide in three stages: (1) hydrogen ion implantation, (2) wafer bonding (3) post-annealing with layer splitting. The critical point for layer splitting is implantation-induced defects microcracks coalesce during post-annealing treatment and form a continuous cleavage plane. Following by hydrogen ion implantation, post-annealing treatment driving the hydrogen atoms diffuse in the implantation-induced defects and got trapped causing surface blistering formation and growth [4, 5]. Also, silicon wafer orientation plays an important role and has an effect on ion cut process significantly [6]. Hence, an exhaustive investigation of wafer surface blistering behavior depend on silicon orientation is elemental for fabrication of the SOI material with different orientation. In this study, a

1159

fluence of 2.5×10^{16} cm^{-2}, 200 keV H$_2^+$ ion implanted into different silicon orientation (Si<100> Si<111> and Si<110>). After ion implantation, we used optical microscopy (OM), secondary ion mass spectrometer (SIMS) and raman scattering spectroscopy (RSS) to exam the specimens at various post-annealing temperature in order to investigate the influence of silicon orientation on surface blistering behavior relates the mechanism of Smart-Cut^{TM} technique.

Experiment

In this study, n-type <100>, <111>, and <110> Cz-grown silicon wafers with a resistivity of 2–6 Ω-cm were chosen to be implanted by molecular hydrogen (H$_2^+$) with kinetic energy level of 200 keV and molecular ion fluence of 2.5×10^{16} cm^{-2}. In order to prevent target materials from overheating, the incident ion beam current density was maintained lesser than 25 nA/cm^2. In addition, the specimens were tilted at an angle 7° with respect to the incident ion beam to minimize the ion channeling effect. Following implantation, the specimens were cut into small pieces of 1×1 cm^2 in size and subsequently by post annealing treatment at various temperatures (430~530°C). The specimens' surface blistering formation and growth were in-situ observed by optical microscopy (OM) with a magnification 500X (172×138 μm^2). Moreover, Raman scattering spectroscopy (RSS) and secondary ion mass spectrometry (SIMS) were, respectively, adopted herein to probe radiation damage and hydrogen trapping depth in the specimens.

Results and Discussion

Blistering formation and growth

The wafer surface blistering behavior occurred at isothermal post-annealing treatment were thermal activation controlled and revealed to nucleation and growth process. First, we used optical microscopy observing the specimen surface in-situ during anisothermal annealing treatment and determined the exfoliation temperature (T$_e$), which meant the appearance of the first crater detected. The T$_e$ for Si<100> is about 515°C compared to Si<111> 555°C and Si<110> 556°C. Then, for all the specimens, were in situ examined through isothermal annealing treatment at various temperatures lower the same temperature from T$_e$ (430°C~530°C). By these ways, we also could calculate activation energy of exfoliation for the three orientation silicon wafer, for Si<100> is 1.18 eV, Si<111> is 1.34 eV, and Si<110> is 1.55 eV. Figure 1, 2 illustrates the OM images and the exfoliation time at various temperatures of isothermal annealing treatment. The results show that the silicon wafer surface exfoliated much easier when post-annealing temperature above about 450°C for Si<100>, 510°C for Si<111>and Si<110>. We define these temperatures as transition temperature (T$_t$), is a temperature revealed to exfoliate hard transfer to exfoliate easily. In addition, the silicon wafer surface blistering behavior is combination of nucleation and growth process. The nucleation process is strongly dependent on the silicon areal number density, while the growth process is dominated by the silicon intra-planar spacing. The silicon areal number density for Si<100> Si<111> and Si<110> are $2/a_{Si}^2$, $4/\sqrt{3}\, a_{Si}^2$ and $2\sqrt{2}/a_{Si}^2$, respectively. However, the silicon intra-planar spacing for Si<100> Si<111> and Si<110> are $a_{Si}/4$, $a_{Si}/4\sqrt{3}$ and $a_{Si}/2\sqrt{2}$ (a_{Si}=5.43105 anstron). As a result, when the post-annealing temperature is below the T$_t$, say 430°C for Si<100>, 470°C for Si<111>of Si<100>, we predict that Si<100> can overcome the thermal activation barrier of exfoliation much easier and it has the smallest areal number density of the three oriented silicon wafer causing the nucleation time of blister is the shortest, so, the exfoliation rate from smallest to greatest is Si<100>, Si<111>, and Si<110>. But when the post-annealing temperature is above T_t, say 490°C for Si<100>, 530°C for Si<111> and Si<110>, all of the three can overcome the barrier, so it must consider both effects of areal number density and intra-planar spacing leading to the ascending order of exfoliation rate become Si<111>, Si<100>, and Si<110> Moreover, the detail information about crater such as average diameter (d_c), total amount (N_c), and the fraction of total craters area over the specimen area (f_c) are listed in the Table I.

Hydrogen-related defects

Figure 3 illustrates the RSS spectra in the Si-H stretch mode of as annealed specimens implanted with 200 keV and 2.5×10^{16} cm^{-2}. Silicon wafer orientation from left column to right column is <100>, <111>, and <110>, respectively. In those RSS spectra, there are several important wave bands of implantation-induced defects such as (1) 2160 ~2210 cm^{-1} related to the monovacancy defects VH$_{3,4}$ and (2) 2085 ~2130 cm^{-1} correlated to Si(100):H bonding configuration (i.e. extended H-terminated internal (100) surface, named platelet), (3) the mutivacancy defects V$_m$H$_n$ (broadband between 1880 to 2025 cm^{-1}) [7]. As can been see in figure 3, the intensity of the mutivacancy defects V$_m$H$_n$ for three oriented specimens are strongly decrease and has a tendency to form more stable phase like the monovacancy defects VH$_{3,4}$ and Si(100):H bonding configuration phase after post-annealing treatment. In addition, as stated in [8], the monovacancy defects VH$_3$ is the precursor of the Si(100):H bonding configuration phase. With further thermal budget applies, the hydrogen atoms arrangement of the Si(100):H configuration phase may change into H-H platelet, form highly pressure H$_2$-gas bubbles, then generate micocrack in damage layer, causing the formation of blisters and exfoliate finally. Moreover, the RSS spectra shows that the implantation-induced defects transformation against annealing temperature is quiet similar of the three oriented cases.

Hydrogen depth profile

Figure 4 illustrates the SIMS-measured hydrogen results of the three oriented silicon wafers. As it shows, for all three oriented specimens, the amount of hydrogen become lesser after post-annealing treatment which indicating that hydrogen atoms out diffusion from bulk to the specimen surface against the annealing temperature [9]. Furthermore, the hydrogen trapping depth shift which correlates closely to the depth of wafer cleavage when fabricating SOI materials is co-dominated by the driving force of channeling effect and damage effect. Channeling effect draws the hydrogen depth profile toward bulk and damage effect draws toward specimen surface in contrast. When the post-annealing temperature below T_t, the trapping depth shift is channeling effect dominated compared to damage effect for Si<111> and Si<110>. But, when the post-annealing temperature is above T_t, the trapping depth shift is opposite to the post-annealing temperature below T_t.

Conclusions

In conclusion, the behavior of blistering development with various is strongly depended on silicon orientation. According to our experiment results, we found that when the post-annealing temperature is below T_t, the surface exfoliation rate from smallest to greatest with Si<100>, Si<111>, and Si<110>, respectively. However, when the post-annealing is at the regime of T_e-T_t, the rank of surface exfoliation rate becomes Si<111>, Si<100>, and Si<110>.

In our investigation, a post-annealing temperature fell into the regime of T_a-T_f is the most optimal parameter for fabricating SOI materials by different silicon orientation.

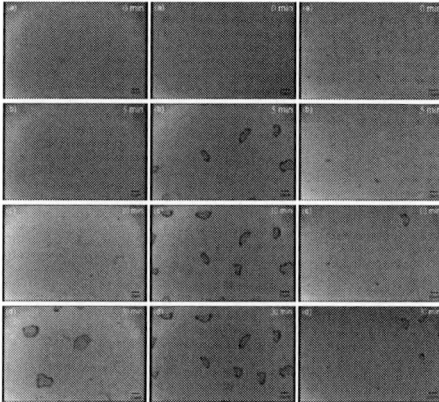

Figure 1. The OM images of the specimens implanted with 200 keV、2.5×10^{16} cm^{-2} molecular hydrogen ion under the same thermal budget. Silicon wafer orientation and the thermal budget from left column to right column is Si<100>, 490°C 30 min; Si<111>, 530°C 30 min; Si<110>, 530°C 30 min.

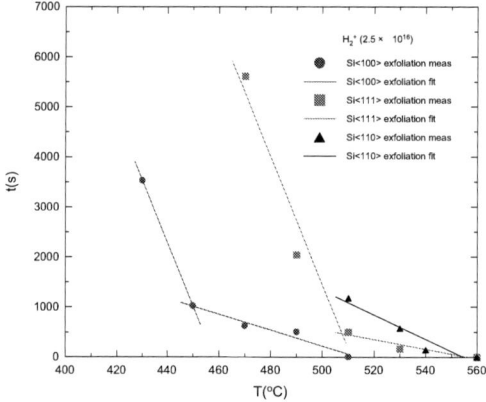

Figure 2. The silicon wafer surface exfoliation time at various post-annealing temperature.

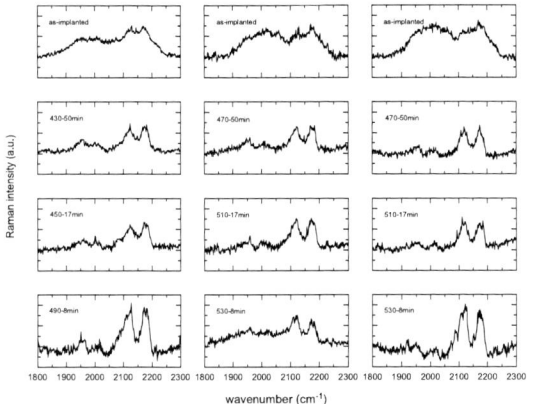

Figure 3. RSS spectra in the Si-H stretch mode of as-annealed specimens implanted with 200 keV、2.5×10^{16} cm^{-2} molecular hydrogen ion under the same thermal budget. Silicon wafer orientation from left column to right column is Si<100>, Si<111>, and Si<110>, respectively.

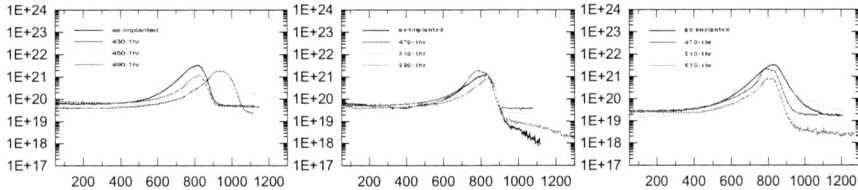

Figure 4. The hydrogen depth profile of the specimens implanted with 200 keV、2.5×10^{16} cm^{-2} molecular hydrogen ion under the same thermal budget. Silicon wafer orientation from left column to right column is Si<100>, Si<111>, and Si<110>, respectively.

TABLE I. Average diameter, total amount, the fraction of total exfoliation area over the specimen area of optically detected under the same thermal budget.

Si<100>				Si<111>				Si<110>			
$T(^{\circ}C)$	$N_c(1/cm^2)$	$d_c(\mu m)$	fc(%)	$T(^{\circ}C)$	$N_c(1/cm^2)$	$d_c(\mu m)$	$f_c(\%)$	$T(^{\circ}C)$	$N_c(1/cm^2)$	$d_c(\mu m)$	$f_c(\%)$
430	4213	9.79	0.32	470	0	12.63	0	470	0	0	0
450	33704	11.18	3.31	490	16852	12.80	2.1	490	0	0	0
470	42130	12.70	5.34	510	42130	12.71	5.47	510	12639	9.79	0.95
490	50556	12.77	6.48	530	50556	15.11	5.88	530	25278	9.61	1.83

Acknowledgments

The authors would like to appreciate Mr. J. Xia (Beijing Normal University, People's Republic of China) and Mr. C.H. Wang (National Tsing Hua University, Republic of China) for a great assistance with ion implantation and SIMS measurement. In addition, the authors also thank the financial support from National Science Council of the Republic of China so much.

References

1. C.w. Lin, H.A. Yang, W.C. Wang, W. Feng, J. Micromech. Microeng. 14 (2007) 1200.
2. J. Chen, "Hole Mobility Characteristics in Si Nanowire pMOSFETs on (110) Silicon-on-Insulator," *IEEE Electron Device Lett.*, vol. 31, no. 11, pp. 1181, Nov. 2010.
3. H.J. Woo, H.W. Choi, J.K. Kim, G.D. Kim, W.Hong, W.B. Choi and Y.H. Baec, Nucl Instr. and Meth. B, 241 (2005) 531-535.
4. L.J. Huang, Q.Y. Tong, Y.L. Chao, T.H. Lee, Appl. Phys. Lett. 74 (1999) 982.
5. M.K. Weldon, V.E. Marsico, Y.J. Chabal, A.Agarwal, D.J. Eaglesham, J. Sapjeta, W.L. Brown, D.C. Jacobson, Y. Caudano, S.B. Christman, E.E. Chaban, J. Vac. Sci. Technol. B15 (1997) 1065.
6. Y. Zeng, R. J. Welty, Z. F. Guan, K.V. Smith, P.M. Asbeck, E. T. Yu, S. S. Lau, C.H. Yun, A. B. Wengrow, N. W. Cheung, presented at the MRS spring meeting, symposium T (1999).
7. S. Personnic, F. Letertre, A. Tauzin, N. Cherkashin, A. Claverie, R. Fortunier, and H. Klocker, J. Appl. Phys. 103, 023508 (2008).
8. M.K. Weldon, M. Collot, Y.J. Chabal, V.C. Venezia, A. Agarwal, T.E. Haynes, D.J. Eaglesham, S.B. Christman, E.E. Chaban, Appl. Phys. Lett. 73 (1998) 3721.
9. C.H. Hu, J.H. Liang, C.M. Lin, "A study of the characteristics of surface blistering in hydrogen implanted (100) & (111) oriented silicon wafer," Proceedings of the 2007 Symposium on Nano Device Technology (SNDT 2007), Hsinchu, Taiwan, R.O.C., May 9-15, 2007.

Very high deposition rate of a-Si:H thin films by ECRCVD

H.F. Chiu[a], Y.S. Chang[a], J.Y. Wu[b], Y.S. Li[b], J.Y. Chang[c], C.C. Lee[c], I.C. Chen[d], and C.C. Su[e], Tomi T. Li[b]*

[a] Department of Materials Science and Engineering, National Tsing Hua University, Hsinchu 300, Taiwan, Republic of China
[b] Department of Mechanical Engineering, National Central University, Taoyuan 320, Taiwan, Republic of China
[c] Optical Science Center, National Central University, Taoyuan 320, Taiwan, Republic of China
[d] Institute of Material Science and Engineering, National Central University, Taoyuan 320, Taiwan, Republic of China
[e] Chung-Shan Institute of Science & Technology, Taoyuan 325, Taiwan, Republic of China

For the reduction of the manufacturing cost, a high deposition rate of amorphous silicon (a-Si:H) thin film in fabrication is very important. Thus high plasma density and low process temperature deposition technique is keen to develop. Another issue of a-Si:H thin film uses plasma enhanced CVD which causes plasma damage to the film. We use electron cyclotron resonance chemical vapor deposition (ECRCVD) to deposit a-Si:H layers with varying microwave power, magnetic field, and hydrogen dilution. The major advantages of ECRCVD are high deposition rate and remote plasma zone that can avoid surface damage. A high deposition rate more than 2 nm/sec was developed by ECRCVD. Fourier transform infrared spectroscopy (FTIR) is used for measuring the microstructure factor (R*) to interpret the effects of microwave power, magnetic field, and hydrogen dilution.

Introduction

Nowadays, plasma enhanced chemical vapor deposition (PECVD) has been widely used in fabrication of hydrogenated amorphous silicon (a-Si:H) thin film for silicon thin film solar cells. However, PECVD has the lower deposition rate in amorphous and microcrystalline silicon (μc-Si ~ 0.2 nm/s), generated a lot of ion bombardments on interface of the film and then needs to follow by various process treatments to remove these interfacial defects caused by ion bombardments (1). The proper process conditions were optimized to get the high microcrystalline silicon deposition rate (~2nm/s) by PECVD (2). Industries has developed technology for the high-rate deposition (3nm/s) of high-quality a-Si:H by using very high frequency (VHF) PECVD and achieved the 13.5% efficiency(3). Electron cyclotron resonance (ECR) plasma source (4,5) has great advantages such as high electron density (> 10^{12} cm^{-3}) (6), low temperature operation (< 200 °C), high deposition rate (> 1.5 nm/s for μc-Si), low gas pressure operation and low contaminations compared with other plasma sources. Compared to conventional PECVD, ECR offers lower operating pressures (typically in the mTorr range) and improved control of the deposition process. High plasma densities can be achieved giving the potential for high deposition rates and produce low cost and high efficiency silicon base solar cells(7).

Aside from the factors such as temperature, working pressure, and plasma density affecting the deposition rate, Summers et. al (8) found that change the ECR magnetic field could improve the deposition rate. In this paper, we present the results from the effects of different magnetic field, microwave power and hydrogen dilution on the high deposition rate of α-Si:H thin films.

Experimental

To achieve higher deposition rate, an ECRCVD shown in Figure 1 is used by changing power, magnetic field (main, outer, inner) and hydrogen dilution (H_2/SiH_4). The frequency of microwaves is 2.45 GHz and the power can be increased up to 1.2 kW. There are three coils, one is mail coil located at the upper place and the others are inner and outer coils under ECRCVD process chamber. Different magnetic configurations can be generated from different magnetic field combination within those three main, outer, and inner coils .All thin films were fabricated by ECRCVD under 10^{-6} mTorr. Two kinds of substrates including P-type wafer and glass were used, and three kinds of process gas Ar, H_2 and SiH_4 were delivered to deposit amorphous thin film. The deposition was processed under the following conditions as listed in TABLE I.

TABLE I. Parameters used for depositions.

Parameter	Growth condition
Base pressure	5×10^{-6} mTorr
Power	600 W ~ 1200 W
Substrate temperature	225 °C
H dilution ratio(H_2 /SiH_4)	0.2; 0.32
deposition time	300 s
Process pressure	5 mTorr
main magnetic field	32 A ~ 40 A
Ar	42 sccm

The film thickness was measured by DEKTAK which is a profilometer for measuring step heights or trench depths on a surface. The microstructure factor (R*, $Si-H_2/Si-H_2+Si-H$) was measured by FTIR. The R* presents the film's property which is calculated from the strength of Si-H ($2000cm^{-1}$) and $Si-H_2$ ($2100cm^{-1}$) peak.

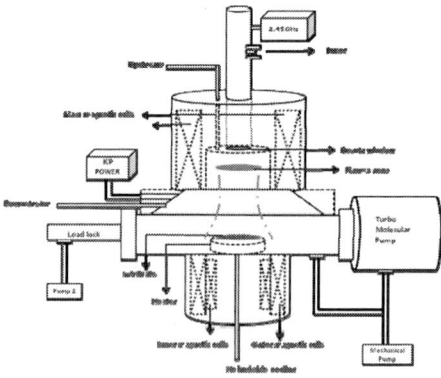

Figure 1. Schematic diagram of ECRCVD.

Microwave power effect

As shown in Figure 2, the deposition rate increases as the microwave power increases regardless of hydrogen dilution. A faster growth rate is obtained when the microwave power is higher. The similar results can be obtained in different magnetic configurations (Figure 2 and Figure 3). This result is the same as PECVD result (2). One of the reasons is that the plasma reaction particles will receive more energy as the microwave power increases, the kinetic energy of reaction particles increases, thus it accelerates the deposition rate. The other reason is that higher microwave power causes the SiH_4 dissociation increase under the higher plasma density. Therefore there will be more Si free radicals deposited on the substrate surface.

Hydrogen dilution effect

As showed in Figure 2, the deposition rate difference in range is from 0.05 to 0.1 nm/s when the ratio of H_2/SiH_4 decreases from 0.33 to 0.2 under the constant magnetic fields and microwave power. The main magnetic coil current is 32 A, outer magnetic coil current is 12 A and inner magnetic coil current is 22 A.Mullerova et. al (9) reported that increased crystallization of the film will cause the deposition rate lower and thus R* higher . However, increasing energy will also form more Si-H bonds resulting in R* of silicon amorphous film decreased. And decreasing R* is associated with reduction of undesirable microvoids (10, 11) as H_2/SiH_4 increasing from 0.2 to 0.33. As H_2/SiH_4 ratio is 0.33 and in 1000 W power or more, we will obtain the R * less than 0.1 which will provide better quality film. This film is good for devices fabrication as PECVD (12).

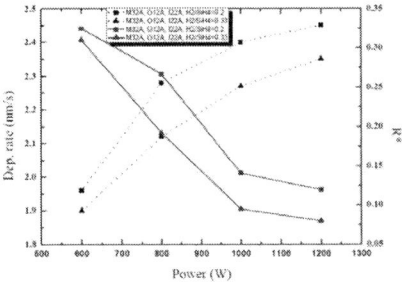

Figure 2. The relationship between power and deposition rate or R* under different H_2/SiH_4 ratio.

Magnetic field effect

In Figure 3, the different magnetic configurations have a significant impact on the deposition rate. Due to the special three magnetic field structures, the magnetic field configuration is different from other ECRCVD equipments. In the magnetic field configuration, magnetic field directions of the main magnetic coil and outer magnetic coil are all the same direction which is from up to downward and magnetic field direction of the inner magnetic coil is upward, thus the total magnetic field is downward. With the different magnetic field configuration, the ECR resonance (ECR Zone) position (875 G) is also different, as shown in Figure 4, where z axis is the distance between measured position and the quartz window. Location of ECR Zone will be closer to the wafer and farther away from the microwave source as the current of the main magnetic field increases. This result is opposite to general results in transitional ECRCVD (8), because ECR Zone location is close to the microwave source when the small main magnetic coil current is applied. We know that ECR Zone is plasma generating area, if close to the microwave source, microwave into the ECRCVD directly and its energy can be effectively absorbed response of the plasma reaction particles, which leads to higher dissociation and obtain more kinetic energy for plasma reaction particles, and effectively increasing the deposition rate of reaction particle. When the main magnetic coil current increases, it will make the ECR Zone moving down close to the wafer, and due to the microwave power entering ECRCVD chamber which has some distance with the ECR Zone area, that could easily lead to loss of microwave energy and decrease of deposition rate.We should also pay attention to the value of electric field in ECRCVD resonant cavity. As showed in Figure 3, as the main magnetic coil current is 40 A, the deposition rate is higher than it as main magnetic coil current of 35 A, but not as main magnetic coil current of 32 A. The main reason is that the position of ECR resonance (ECR Zone, 875 G) has different location with different currents of main magnetic coil. In the process of electromagnetic wave transmission, the electric field strength will vary with different locations. The power absorbed in a unit volume is related to electric field position, and thus absorbed power per differential volume is expressed as:

$$< P >_{abs} (\vec{r}) = \tfrac{1}{2} Re \left[\vec{E}(\vec{r}) \cdot (\bar{\bar{\sigma}}(\vec{r}) \cdot \vec{E}(\vec{r})^*) \right] \tag{1}$$

Where the discharge complex tensor conductivity $\bar{\bar{\sigma}}(\vec{r})$ and the electric field $\vec{E}(\vec{r})$ are functions of position $\vec{r}$ in the plasma and $< P >_{abs} (\vec{r})$ has units of power density. The energy of per unit volume plasma received from microwave energy is related to the value of electric field in different position. Microwave frequency is 2.45 GHz, and the wavelength λ is about 12.25 cm. When the main field current of 32 A, 35 A and 40 A, the location of the 875 G ECR Zone will be at about 1/4 λ, 3/4 λ and 1/8 λ, and the magnetic field distribution of the different positions in the different magnetic field shown in Figure 4. The energy magnitude of per plasma unit volume received microwave directly affects deposition rate. When the main magnetic coil current is 35 A, the position of the ECR Zone is located at area with the minimal electric field, and it will receive minimal deposition rate compared to the other different magnetic configurations. ECR Zone is located at area with the maximum value of electric field while main magnetic coil current is 32 A, thus obtaining the maximum deposition rate. A close to 2.4 nm/s deposition rate is obtained by changing the configuration of the main magnetic coil current at 32 A.

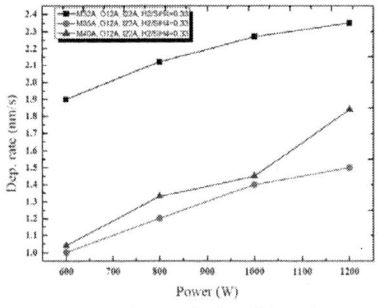

Figure 3. The relationship between power and deposition rate under different magnetic field configurations.

Figure 4. The magnetic field diagram when changing the main magnetic current.

The R* under different magnetic fields is showed in Figure 5.Main magnetic coil current is set at 35 A, more Si-H bonds are formed, causing lower R* which indicates a good amorphous silicon film is fabricated. From the result in Figure 5, R * less than 0.1 as microwave power up to 1200 W, it demonstrates that high deposition rate can be achieved , as well as high quality thin films can be obtained.

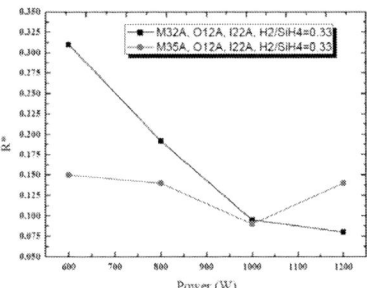

Figure 5. The relationship between power and R* for different magnetic field configurations.

On/Off Down magnetic coil effect

It is obviously that the deposition rate will decrease from 1.9 nm/s to 1.7 nm/s when the down magnetic coil current is set at off. Since the current of the down magnetic coil is in closure, it makes total downward magnetic field decrease. The dissociation rate of reactants will rise when the power is increasing, which will result in the deposition rate increase, and microwave power will dominate the deposition rate at this time because the impact of down magnetic field is decreasing.

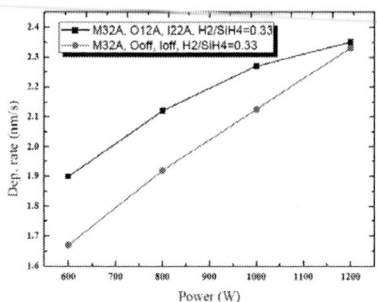

Figure 6. The relationship between the power and the deposition rate for closing the down magnetic field.

Conclusion

In this work, using ECRCVD can reach high deposition rate when we change the magnetic field configuration and microwave power. We found that it will affect the resonance position of ECRCVD plasma (ECR Zone 875G) when we change the main magnetic field. The large energy will transmit the plasma particles that results in plasma dissociation increasing. Microwave power will encounter to the plasma resonance region directly if ECR Zone closer to the microwave source when our resonant cavity position move upward to the top of ECRCVD resonant cavity chamber. Thereby, it can increase the amount of deposited particles and ECRCVD deposition rate, but we need to consider the relationship between the plasma resonance position and electric field.

In addition, we also found that microwave power can increase the deposition rate effectively. The reason is that particles gain more kinetic energy from the high microwave power injection and it makes the deposition rate increase. The magnetic field of ECRCVD is coupled by a main magnetic coil and two down magnetic coils, and they will affect total magnetic field direction. By setting up a configuration as the main magnetic field and outer field are in the same direction, which both are downward, and the inner magnetic field is upward. As the results of this configuration, the total magnetic field is downward and can lead the reacted ions have the same direction with the magnetic field. We can obtain high deposition rate when we raise both the main magnetic coil current and the down outer magnetic coil current, which increase downward magnetic field. This will drive reaction particles moving downward, thus a high deposition rate is formed. In this paper, we can obtain the highest deposition rate close to the 2.4 nm/s, and average growth rate is more than 1.5 nm/s. Some results of the high deposition rate with good quality film would be used for solar cell devices fabrication.

Acknowledgments

This work is supported by a grant from the fund of "National Science and Technology Nano Program –" (NSC-99-2120-M-008-003) under the National Science Council of Taiwan (R.O.C.).

References

1. W.G.J.H.M. van Sark, J. Bezemer,W.F., and van der Weg, *Surface and Coatings Technology*, **63 66,** 74-75 (1995).
2. Yan Ying Ong, Bang Tao Chen, Francis E.H. Tay, and Ciprian Iliescu, *Journal of Physics: Conference Series*, **34,** 812–817(2006).
3. Youji Nakano, Saneyuki Goya, Toshiya Watanabe, Nobuki Yamashita, and Yoshimichi Yonekura, *Thin Solid Films*, **506– 507**, 32 – 37(2006).
4. Ueda Y., and Kawai Y., *Appl. Phys Lett.*, **71**, 2100 (1997).
5. Koga M., Hishikawa Y., Tsuchiya H., and Kawai Y., *Thin Solid Films*, **506**, 499 (2006).
6. Yuri Glukhoy, Mahmud Rahman, Gotze Popov, Alexander Usenko, and Hans J. Walitzki, *Surface & Coatings Technology*, **196**, 172– 179 (2005).
7. G Ekanayake, S Summers, and H S Reehal, *Plasma CVD for Solar Cells 3rd World Conference on Photovoltaic Energy Conversion May 11-18, 2003 Osaka, Japan.*
8. S Summers, H S Reehal, and G H Shirkoohi, *J. Phys. D: Appl. Phys.*, **34**, 2782–2791 (2001).
9. J. Mullerova, S. Jurecka, and P. Sutta , *Solar Energy* , **80**, 667–674 (2006).
10. Wenhui Du, Xiesen Yang, Henry Povolny, Xianbo Liao, and Xunming Deng, *J. Phys. D: Appl. Phys.*, **38**, 838–842 (2005).
11. J. Mu ̈llerova, P. S ̌ utta, G. van Elzakker, M. Zeman, and M. Mikula, *Applied Surface Science* , **254**, 3690–3695 (2008).
12. J. Loffler, H-J. Muffler, C. Devilee, A. Gajovic, P. Dubcek, D. Gracin, and W.J. Soppe, *Presented at ICANS-21 Conference*, 5-9 September 2005, Lisbon, Portual.

Author Index

Adelmann, C.	473	Bryant, A.	81
Afanas'ev, V. V.	467	Bu, H.	37, 81
Ahmet, P.	87, 99, 483, 489		
Alian, A.	1017, 1065	Cai, J.	31, 37
Altamirano-Sanchez, E.	377	Cai, M.	997
Altimime, L.	509	Cai, X.	711
Ang, S. S.	893	Cai, Y.	9
Ang, Y.	119	Cao, C.	17, 155
Appelt, B. K.	857	Cao, Y.	743, 787
Arleo, P.	409	Carbonell, L.	515
Armini, S.	515	Caro, A.	515
Arnauts, S.	671	Caymax, M.	761, 1017, 1065
Arnold, J. C.	329	Chakrabarti, S.	125
Aw, T.	887	Chan, D.	479
		Chang, E. Y.	483
Bailly, F.	389	Chang, E.	1041
Baisie, E.	633	Chang, J. Y.	1097, 1165
Balda, J.	893	Chang, J.	399
Banerjee, G.	665	Chang, J.	1103
Bargallo Gonzalez, M.	725, 761	Chang, K.	285
Barnola, S.	389	Chang, K.	781
Basker, V.	81	Chang, S.	319, 335, 445, 405, 427
Baumann, R.	1059	Chang, Y. S.	1165
Beenakker, C.	913	Chang, Z.	237
Ben Jamaa, M.	1005	Chatterjee, S.	125
Berry, I.	433	Chattopadhyay, D.	125
Beyer, G.	515	Chen, C.	857
Beyne, E.	523	Chen, D.	43, 49, 75, 93, 361, 749
Bi, W.	819	Chen, F.	173
Bian, A.	743, 787	Chen, F.	609
Boeuf, F.	37	Chen, I. C.	1097, 1103, 1165
Borkulo, J. v.	873	Chen, J.	1151
Boullart, W.	377, 409	Chen, J. C.	479
Bouyssou, R.	389	Chen, J.	1135
Brammertz, G.	1017, 1041, 1065	Chen, J.	781
Brandt, D.	231	Chen, J.	149
Breach, C. D.	831	Chen, K.	919, 925
Brijs, B.	473	Chen, L.	937
Brun, P.	389	Chen, S.	583, 805

Chen, S. Y.	1145	Demuynck, S.	515
Chen, X.	985	Ding, X.	577
Chen, X.	173	Doh, J.	865
Chen, Y. P.	1097	Doris, B.	37, 81
Chen, Y.	1053	Drain, D.	243
Chen, Y.	479	Du, R.	985
Chen, Y.	719, 737	Du, W.	805
Chenc, B.	1053	Ducote, J.	389
Cheng, C.	479	Duyos Mateo, R.	627, 653
Cheng, K.	37		
Cheng, L.	395	Elshocht, S.	515
Cheng, L.	439	Endo, T.	257
Cheng, S.	1071	Eneman, G.	725, 761
Chevolleau, T.	389	Ercken, M.	377
Chew, Y. P.	887	Escorcia, O.	433
Chiu, H. F.	1165	Evans, T.	893
Chiu, T.	173	Everaert, J.	535
Cho, J.	81		
Choi, J.	879	Faltermeier, J.	81
Choi, S.	879	Fan, E.	285
Chris, C.	243	Fan, K.	529
Chu, T.	119	Fan, Q.	583, 811
Chu, Y.	1103	Fang, W.	503, 541, 571, 705
Chua, K.	985	Farrar, N.	231
Chung, J.	55	Frégonèse, S.	1005
Claeys, C.	725, 761	Franquet, A.	473
Clermidy, F.	1005	Fu, C.	811
Conard, T.	473	Fu, L.	445
Croes, K.	515	Fujitani, N.	257
Cui, H.	583, 805, 811	Fuller, N.	329
Cui, J.	103		
Cuypers, D.	671	Gaillardon, P.	1005
		Gang, D.	189
Dai, S.	61	Gao, H.	1117
Darlak, A.	329	Gao, R.	479
Darnon, M.	389	Ge, G.	683
David, T.	389	Ge, Q.	325
Daw Sun, J.	113	Geerpuram, D.	421
De Jaeger, B.	725	Geiger, D.	997
Dekoster, J.	455	Geissbühler, P.	433
Demand, M.	377, 409	Gessner, T.	1059
De Marchi, M.	1005	Gong, L.	1151
De Micheli, G.	1005	Gu, X.	1109

Gu, X.	627, 653	Hu, L. C.	1097
Gu, Y.	269, 303, 319	Hu, M.	335, 445
Guerrero, A.	249	Hu, X.	775
Guillermet, M.	389	Huang, B.	547, 551, 557, 563
Guo, R.	103, 719	Huang, C.	355
Guo, X.	277	Huang, M.	343
Guo, Y.	979	Huang, Q.	9
Guoxiang, L.	955	Huang, R.	9
Gupta Chatterjee, S.	125	Huang, S.	1135
		Huang, Y.	319, 427
Ha, J.	865	Huang, Y.	325
Han, S.	943	Hung, C. L.	919
Hane, M.	37	Huo, Z.	149
Hao, H.	103	Hurand, R.	389
Hao, J.	263		
Hao, J.	269, 303	Inada, A.	81
Haran, B.	37	Ishikawa, M.	329
Hardy, A.	473	Ishimaru, K.	37
Hashempour, Z.	67	Itoh, S.	825
Hattori, T.	99	Iwai, H.	87, 99, 483, 489
Hattori, T.	483, 489	Iwaniczko, E.	1129
Hattori, T.	371		
He, P.	775	Jagannathan, B.	31
He, Y.	719, 731, 737	Jagannathan, H.	81
Hegarty, M.	755	Jang, M.	865
Hendrianto, J.	349, 355	Jeong, Y.	55
Hendriks, R.	873	Ji, Y.	61
Herlambang, A. P.	979	Jia, N.	203
Heylen, N.	515	Jiang, D.	149
Heyns, M.	1065	Jiang, J.	775
Hikavyy, A. Y.	455, 761	Jiang, L.	609
Ho, B.	257	Jiang, T.	1109
Hoffmann, T.	455, 1017, 1065	Jing, J.	603, 711
Holliday, R.	831, 843	Johnson, C.	421
Hong, J.	285	Johnson, D.	421
Hong, S.	943	Joubert, O.	389
Hook, T.	37	Jourdan, N.	515
Horak, D.	329	Ju, G.	819
Horiguchi, N.	377	Ju, J.	719
Houssa, M.	467	Jurczak, M.	473
Hsiao, Y.	1159		
Hu, J.	173	Kakushima, K.	87, 99, 483, 489
Hu, K.	793	Kan, K.	1029

Kanakasabapathy, S.	81	Lee, C.	755
Kanaya, R.	223	Lee, H.	931
Kanda, T.	483	Lee, J. J.	1097
Kaneda, T.	99	Lee, J. H.	597
Kawakubo, M.	615	Lee, K.	335, 445
Kawanago, T.	99	Lee, K.	319, 405, 427
Kawanago, T.	489	Lee, S.	1135
Ke, D.	143	Lee, T.	831
Kettner, P.	1023	Lee, T.	479
Khakifirooz, A.	37	Leobandung, E.	37, 81
Khare, M.	37, 81	Leunissen, L.	647
Khare, P.	37	Li, C.	1145
Kho, E. S.	119	Li, D.	583, 811
Kim, B.	1023	Li, F.	335, 445
Kim, J.	879	Li, J.	361
Kim, K.	865	Li, M.	285
Kim, M.	943	Li, M.	93
Kim, S.	943	Li, M.	609
Kitayama, D.	99	Li, P.	677, 711
Kittl, J. A.	473	Li, T.	1103
Kleemeier, W.	37	Li, T. T.	1097, 1165
Kobayashi, D.	761	Li, X.	1145
Koo, S.	967	Li, X.	75
Kouda, M.	99	Li, X.	183
Koyanagi, T.	99	Li, X.	479
Kulkarni, P.	37, 81	Li, X.	1109
Kumar, A.	37	Li, X.	1047
Kuroki, S.	653	Li, Y. S.	1165
Kurstjens, R.	535	Li, Y.	1053
Kuss, J.	37	Li, Z.	633
Kweon, Y.	865, 879	Liang, J.	1159
		Liang, K.	439
Labelle, C.	329	Liang, L.	285
La Fontaine, B.	231	Liang, Q.	43, 49, 93
Lai, Y.	857	Liao, H. C.	919
Lai, Y.	577	Liao, X.	659
La Manna, A.	523	Lien, C.	1103
Lam, E. Y.	203	Liguo, Y.	705
Lamb III, J. E.	243	Lim, T.	985
Lay, Y.	937	Limaye, P.	523
Lee, C. C.	1165	Lin, B.	231
Lee, C.	865, 879	Lin, B.	215
Lee, C.	1103	Lin, B.	633

Lin, C.	1159	Liyun, Q.	901
Lin, C.	81	Lo, G.	55
Lin, D.	1065	Loewenhardt, P.	755
Lin, F.	819	Loo, R.	761
Lin, H.	1017	Loo, R.	455
Lin, J.	731	Loubet, N.	37
Lin, J.	755	Lowes, J. A.	249
Lin, L.	937	Lu, C.	711
Lin, P.	677, 711, 743, 775, 787	Lu, C.	919, 925
Lin, X.	155	Lu, C.	1041
Lin, X.	17	Lu, J.	737
Lin, X.	793	Lu, M.	907
Lin, Y. C.	483	Lu, X.	61
Lindain, J.	377	Lu, X.	991
Lindner, P.	1023	Luning, S.	37
Ling, H.	237	Luo, K.	603
Linville, J.	329	Luo, R.	557
Liou, J.	55	Luo, S.	433
Liu, A.	349	Luo, V.	399
Liu, B.	399	Luo, X.	991
Liu, B.	1053	Luo, Z.	43, 49, 107
Liu, C.	263, 269, 303	Luoh, T.	919, 925
Liu, D.	173	Lv, M.	355
Liu, J.	107	Lv, T.	285
Liu, J.	719		
Liu, J.	149	Ma, B.	439
Liu, J.	103	Ma, G.	731
Liu, J.	285	Ma, J.	775
Liu, J.	583, 811	Ma, X.	1151
Liu, J.	805, 991	Ma, Z.	137
Liu, L.	155	Ma, Z.	677, 711
Liu, L.	1123	Maitra, K.	81
Liu, M.	149	Mamatrishat, M.	99
Liu, Q.	37	Mantooth, A.	893
Liu, W.	55	Martin, I.	997
Liu, W.	439	Martinez, L.	421
Liu, X.	1123	Matthias, T.	1023
Liu, X.	155	Mehrotra, V.	639
Liu, X.	17	Mehta, S.	37
Liu, Y.	973	Men, L.	749
Liu, Z.	415	Meng, H.	277
Liu, Z.	901	Meng, X.	319, 427
Liu, Z.	167	Merckling, C.	1017

Merricks, D.	1035	Park, S.	879
Mertens, P.	671	Paruchuri, V.	81
Mignot, Y.	329	Patz, R.	329
Milenin, A.	409	Pei, H.	395
Miller, R.	81	Pender, J.	329
Min, L.	567	Peng, L.	399
Ming, X.	529	Peng, Y.	209
Monfray, S.	37	Philipossian, A.	621, 627, 659
Monsieur, F.	37	Plumhoff, J.	421
Morinaga, H.	591	Ponoth, S.	37
Moussa, A.	473	Possémé, N.	389
Nakashima, H.	1117	Qian, W.	173
Natori, K.	87, 99, 483, 489	Qiang, F.	355
Nemoto, T.	627, 653	Qiang, X.	793
Ng, H.	831	Qiao, C.	781
Nguyen, H.	1041	Qin, G.	137
Ning, G.	743	Qin, J.	415
Ning, J.	103	Qin, S.	9
Ning, T. H.	31		
Nishiyama, A.	99, 483, 489	Ren, J.	383
Niu, F.	439	Ren, Z.	75
Nyns, L.	1017, 1065	Ren, Z.	31
		Rice, A.	621, 659
O'Connor, I.	1005	Robison, R. R.	31
Oddou, J.	389	Roh, D.	433
Ohba, T.	1011	Rong, C.	949
Ohmi, T.	627, 653	Rosseel, E.	761
Ohmori, K.	87	Rowden, B.	893
Ohnishi, R.	257		
O'Neill, J.	81	Sakamoto, R.	257
Ong, P.	647	Sampson, R.	37
Opsomer, K.	473	Sampurno, Y.	621, 627, 659
Otto, T.	1059	Sato, S.	87
Owen, D. M.	737	Schaekers, M.	509, 535
		Schelen, J.	913
Page, M.	1129	Schirmer, K.	893
Pal, D.	967	Schulz, S.	1059
Pan, S. W.	1145	Seki, H.	825
Pang, H.	137	Seo, J.	137
Pang, K.	609	Shamiryan, D.	311
Paraschiv, V.	311	Shangguan, D.	997
Park, D.	31	Shao, C.	167

Shen, C.	901	Sugaya, T.	615
Shen, J.	167	Sugii, N.	99, 483, 489
Shen, M.	319, 427	Sullivan, D.	243
Shen, W.	683	Sun, J.	399
Shen, Z.	103	Sun, L.	137
Shetty, S.	737	Sun, W.	335, 445
Sheu, G.	979	Sunamura, H.	81
Shi, C.	355	Sung, J. C.	1029
Shi, J.	551	Sung, M.	1029
Shi, J.	563	Suzuki, T.	99
Shi, K.	399	Swerts, J.	473, 509, 515
Shi, X.	535		
Shi, X.	269, 303	Takahashi, H.	671
Shi, Y.	75	Takayanagi, M.	37
Shi, Z.	215	Takeda, T.	825
Shuai, Z.	143	Tamai, K.	591
Shum, D.	3	Tan, C.	683
Si, W.	415	Tan, C.	985
Sim, P.	967	Tan, Z.	949
Simoen, E.	725, 761	Tang, J.	769
Singh, N.	55	Tang, J.	913
Sioncke, S.	1017, 1065	Tang, J.	937
Skotnicki, T.	37	Tang, P.	9
Soh, Y.	985	Tang, S.	167
Soles, C. L.	931	Tang, X.	325
Song, H.	731	Tang, X.	811
Song, J.	355	Tang, Y.	9
Song, J.	189	Tang, Z.	719
Song, K.	25, 503, 571	Taofeng, Z.	383, 705
Song, L. J.	639	Teramoto, A.	653
Song, P.	113	Tia, S.	119
Song, X.	335	Tian, D.	1151
Song, Z.	1053	Tie, M.	183
Soussan, P.	523	Tielens, H.	473
Srivastava, A.	433	Tit, N.	161
Srivastava, R.	329	Tökei, Z.	509, 515
Standaert, T.	81	Tomita, Y.	653
Stesmans, A.	467	Tong, Y.	277
Struyf, H.	671	Tran, B.	1041
Su, B.	277	Trinh, H.	1041
Su, C. C.	1165	Tseng, A.	857
Sudargho, F.	659	Tsutsui, K.	99, 483, 489
Sugawa, S.	627, 653	Tu, H.	731

Van Bael, M. K.	473	Wang, X.	683
Van Elshocht, S.	473	Wang, Y.	209
Vancoille, E.	515	Wang, Y.	149
Vanhellemont, J.	1151	Wang, Y.	1123
Vanherle, W.	455	Wang, Y.	173
Vath III, C. J.	843	Wang, Z.	949
Vérove, C.	389	Warren, P.	799
Vinet, M.	37	Washburn, C.	249
Vogel, M.	1059	Wasisto, H.	979
Vos, I.	535	Watanabe, S.	615
Vos, R.	671	Wei, J. C.	343
		Wei, M.	61
Wada, M.	671	Wei, X.	195, 503, 541, 705
Waldfried, C.	433	Wei, X.	819
Waldron, N.	647	Weiwei, L.	691
Wan, X.	1053	Wensheng, Q.	143
Wang, A.	399	Westerman, R.	421
Wang, B.	103	Wimplinger, M.	1023
Wang, C.	603	Witters, L.	455
Wang, D.	1117	Witters, L.	647
Wang, F.	25, 195	Wu, B.	49
Wang, G.	725	Wu, B.	997
Wang, H.	107	Wu, G.	1053
Wang, J.	731	Wu, H.	49
Wang, J.	973	Wu, J. Y.	1165
Wang, J.	81	Wu, J.	719, 731, 737
Wang, K.	355	Wu, L.	699
Wang, L.	25, 195, 503, 571	Wu, Q.	269, 303
Wang, L.	277	Wu, W.	931
Wang, P.	1109		
Wang, P.	17, 155	Xi, Y.	985
Wang, Q.	1129	Xia, J.	439
Wang, Q.	195	Xiao, S.	277
Wang, Q.	149	Xiao, S.	173
Wang, S.	683	XIAO, W.	49
Wang, S.	383	Xie, C.	215
Wang, S.	793	Xie, D.	997
Wang, W.	1047	Xie, J.	1053
Wang, W.	1017, 1065	Xing, C.	677, 743, 787
Wang, W.	719	Xing, W.	189
Wang, W.	103	Xu, B. Y.	557
Wang, X.	405	Xu, B.	563
Wang, X.	445	Xu, C.	737

Xu, D.	61	Ye, S. X.	901
Xu, G.	749	Ye, T.	43, 49, 93
Xu, G.	495	Ye, Y.	365
Xu, J.	1053	Yeh, C.	81
Xu, J.	677	Yen, B.	755
Xu, M.	43	Yin, H.	43, 49, 107
Xu, N.	787	Yin, H.	361, 749
Xu, Q.	567	Yin, M.	415
Xu, Q.	361, 495, 749	Yin, Y.	329
Xu, S.	603	Yin, Z.	949
Xu, W.	1151	Ying, Z.	567
Xu, W.	731	Yiping, X.	955
Xu, Y.	303	Yoo, D.	865, 879
Xu, Y.	263, 269	Yoon, S.	597
Xu, Y.	1129	Young, T.	361
		Yu, H.	49
Yagishita, A.	37	Yu, S.	551
Yahyazadeh, R.	67	Yu, S.	563
Yamada, K.	87	Yu, T.	731, 737
Yamada, Y.	615	Yu, X.	1109
Yamaguchi, Y.	377	Yu, Z.	149
Yamamoto, K.	1117	Yu, Z.	209
Yamamoto, T.	37, 81	Yuan, H.	137
Yamashita, T.	81	Yuan, Z.	583, 805, 811
Yan, C.	699	Yusuff, H.	329
Yang, B.	25		
Yang, D.	1109, 1151	Zade, D.	483
Yang, H.	551	Zang, S.	17
Yang, H.	1117	Zanga, S.	155
Yang, H.	547, 557, 563	Zenbutsu, S.	825
Yang, L.	919, 925	Zeng, L.	551
Yang, L.	781	Zhang, B.	149
Yang, P.	119	Zhang, D.	17
Yang, S. F.	961	Zhang, D.	155
Yang, T.	919, 925	Zhang, H.	335, 445
Yang, T.	749	Zhang, H.	319, 405, 427
Yang, X.	149	Zhang, H.	893
Yang, Y.	365	Zhang, J.	285
Yao, E.	285	Zhang, J.	711
Yaoming, S.	955	Zhang, J.	1047
Yaoying, D.	113	Zhang, J.	395
Ye, B.	731	Zhang, J.	1047
Ye, L.	583, 805, 811	Zhang, J.	209

Zhang, K.	25, 195, 383, 503, 541, 571, 705, 793	Zhuang, Y.	621, 627, 659
Zhang, K.	137	Zimmer, T.	1005
Zhang, L.	991		
Zhang, L.	103		
Zhang, L.	365		
Zhang, M.	149		
Zhang, P.	365		
Zhang, Q.	1077, 1087		
Zhang, S.	1077		
Zhang, S.	61		
Zhang, W.	1071		
Zhang, W.	683		
Zhang, W.	523		
Zhang, X.	633		
Zhang, X.	1151		
Zhang, Y.	503, 541		
Zhang, Y.	737		
Zhao, C.	361, 749		
Zhao, G.	479, 769		
Zhao, H.	743		
Zhao, J.	541		
Zhao, L.	949		
Zhao, S.	943		
Zhao, Z.	769		
Zhensheng, Z.	955		
Zhijie, H.	349		
Zhijun, L.	955		
Zhong, H.	93		
Zhong, M.	609, 1053		
Zhou, A.	103		
Zhou, B.	1145		
Zhou, H.	137		
Zhou, H.	787		
Zhou, J.	893		
Zhou, J.	335, 445		
Zhou, J.	103		
Zhou, K.	329		
Zhou, W.	137		
Zhou, Y.	329		
Zhu, H.	43, 49, 93, 107		
Zhu, N.	1053		
Zhu, Z.	107, 243		